Amorphous and Polycrystalline Thin-Film Silicon Science and Technology–2011

MATERIALS RESEARCH SOCIETY
SYMPOSIUM PROCEEDINGS VOLUME 1321

Amorphous and Polycrystalline Thin-Film Silicon Science and Technology–2011

Symposium held April 25–29, 2011, San Francisco, California, U.S.A.

EDITORS

Baojie Yan
United Solar Ovonic LLC
Troy, Michigan, U.S.A.

Qi Wang
National Renewable Energy Laboratory
Golden, Colorado, U.S.A.

Helena Gleskova
University of Strathclyde
Glasgow, United Kingdom

Chuang Chuang Tsai
National Chiao Tung University
Hsinchu, Taiwan

Seiichiro Higashi
Hiroshima University
Higashi-Hiroshima, Japan

Materials Research Society
Warrendale, Pennsylvania

CAMBRIDGE
UNIVERSITY PRESS

University Printing House, Cambridge CB2 8BS, United Kingdom

One Liberty Plaza, 20th Floor, New York, NY 10006, USA

477 Williamstown Road, Port Melbourne, VIC 3207, Australia

314-321, 3rd Floor, Plot 3, Splendor Forum, Jasola District Centre, New Delhi - 110025, India

103 Penang Road, #05-06/07, Visioncrest Commercial, Singapore 238467

Cambridge University Press is part of the University of Cambridge.

It furthers the University's mission by disseminating knowledge in the pursuit of education, learning and research at the highest international levels of excellence.

www.cambridge.org
Information on this title: www.cambridge.org/9781605112985

Materials Research Society
506 Keystone Drive, Warrendale, PA 15086, USA
http://www.mrs.org

First published 2012

CODEN: MRSPDH

A catalogue record for this publication is available from the British Library

ISBN 978-1-605-11298-5 Hardback

CONTENTS

*Invited Paper

*Invited Paper

POLYCRYSTALLINE FILMS

THIN FILM SILICON ALLOYS

SIMULATION AND CHARACTERIZATION

*Invited Paper

NANOSTRUCTURES

*Invited Paper

GROWTH MECHANISM

SENSORS AND NOVEL DEVICES

*Invited Paper

*Invited Paper

PREFACE

This volume includes sixty-eight papers presented in the 2011 MRS Spring Meeting Symposium A, “Amorphous and Polycrystalline Thin Film Silicon Science and Technology – 2011”, which took place April 25-29, in San Francisco, California. The symposium covers the science and technology of thin-film silicon based materials and devices. The symposium traditionally started off on April 25 with an extremely well-attended full-day tutorial aimed at young researchers and people new to the field. The tutorial was lectured by Profs. Andrew Flewitt and Arokia Nathan. During the four days of fourteen oral sessions and two evenings of poster presentations, seventeen invited talks reviewed the recent progress and addressed the scientific and technical issues in the field. The oral and poster presentations reported new results in various areas, covering fundamental studies and technology advances.

Among various applications, solar cells for photovoltaic solar energy and thin-film transistors for flat-panel display have been the two major driving forces for research and development of thin-film silicon materials and devices. In the last few years, the thin-film silicon community has mainly focused on thin-film silicon solar cells to address the issues of efficiency, manufacturing capability and manufacturing cost. This year Symposium A held focused sessions on the topic of solar cell efficiency. Microcrystalline (μc-Si:H) or nanocrystalline silicon (nc-Si:H) offers the potential for improving the cell efficiencies. Dr. Friedhelm Finger (Forschungszentrum Jülich, Germany) reviewed the recent efficiency improvement in μc-Si:H solar cells with an emphasis on μc-SiC:H and μc-SiO:H doped layers and effective light management. Dr. Finger expected that over 14% efficiency with a-Si:H/μc-Si:H tandem solar cells will be attained soon. Prof. Miro Zeman (Delft University of Technology, The Netherlands) showed that 23% efficiency is achievable using a-SiC:H/a-SiGe:H/nc-Si:H triple-junction structure with advanced light trapping to enhance the photon harvesting of the sun light. Along this line, many new light trapping and light management approaches have been investigated, including plasmonic light scattering using metal and dielectric nano-particles and photonic structures. Theoretical and simulation studies show that the classical limit of $4n^2$ can be exceeded using advanced light trapping techniques. Dr. Takuya Matsui (AIST, Japan) presented their recent progress in developing μc-SiGe:H materials as low bandgap materials to absorb long wavelength light, which provides a new material for high efficiency solar cells. Significant progress in advancing the nc-Si:H technology for mass production has been made. Dr. Arindam Banerjee of United Solar Ovonic LLC (Michigan, USA) reported achieving initial 12% and stable 11.2% encapsulated module (400 cm^2) efficiencies with an a-Si:H/nc-Si:H/nc-Si:H triple-junction structure. The high module efficiencies are new world records for thin film silicon solar modules measured by the National Renewable Energy Laboratory. In addition, Symposium A had three sessions on thin film transistors, sensors and other novel devices reporting advances in these areas. To improve the device quality significant fundamental studies have been presented, especially advanced microscopic characterization (Dr. A. Fejfar, Academy of Sciences, Czech Republic) and simulations (Prof. D. Drabold, Ohio University, USA). The presentations covered the thin-film silicon materials ranging from amorphous to nano- and micro-structured materials, and polycrystalline thin films. The unique optical properties of black silicon attracted significant attention for its potential application in solar cells as an effective light trapping material.

We had a very successful and enjoyable symposium. The number of presentations and attendees reflect the great need for development of thin-film silicon materials and devices. Unique and advanced results ensured the high quality of the symposium. As the organizers of Symposium A, we greatly acknowledge the invaluable contributions of the authors of oral and poster presentations, especially those who made written contributions to this volume

The symposium organizers thank all the people evolved in the Symposium before, during, and after the conference. The organizers greatly appreciate the program committee members of V. Chu (INESC, Portugal), A. Fejfar (Academy of Sciences of the Czech Republic, Czech Republic), F. Finger (Forschungszentrum Jülich, Germany), A. Flewitt (University of Cambridge, United Kingdom), T. Matsui (AIST, Japan), E.A. Schiff (Syracuse University, USA), P. Stradins (NREL, USA), J. Robertson (University of Cambridge, United Kingdom), and M. Zeman (Delft University of Technology, The Netherlands). They kindly reviewed all of the abstracts, which helped the organizers to prepare an interesting program. Special appreciation goes to all of the referees for their careful review of papers in the proceedings and valuable feedback given to the authors. We sincerely thank Mary Ann Woolf, who supervised and managed the abstract and manuscript reviewing process. Her experience and hard work allowed for smooth and timely production of this volume. The MRS staff provided friendly and professional support throughout the organization of the Symposium and Proceedings

On behalf of all the participants, we thank the generous financial support of our corporate sponsors: ITRI, NREL, ULVAC Inc., and United Solar Ovonic LLC.

Baojie Yan
Qi Wang
Helena Gleskova
Chuang-Chuang Tsai
Seiichiro Higashi

September 2011

MATERIALS RESEARCH SOCIETY SYMPOSIUM PROCEEDINGS

Volume 1321 — Amorphous and Polycrystalline Thin-Film Silicon Science and Technology—2011, B. Yan, Q. Wang, H. Gleskova, C.C. Tsai, S. Higashi, 2011, ISBN 978-1-60511-298-5
Volume 1322 — Third-Generation and Emerging Solar-Cell Technologies, Y. Lu, J.M. Merrill, M.T. Lusk, S. Bailey, A. Franceschetti, 2011, ISBN 978-1-60511-299-2
Volume 1323 — Advanced Materials Processing for Scalable Solar-Cell Manufacturing, H. Ji, B. Ren, L. Tsakalakos, 2011, ISBN 978-1-60511-300-5
Volume 1324 — Compound Semiconductors for Energy Applications and Environmental Sustainability—2011, F. Shahedipour-Sandvik, L.D. Bell, K. Jones, B. Simpkins, D. Schaadt, M. Contreras, 2011, ISBN 978-1-60511-301-2
Volume 1325 — Energy Harvesting—Recent Advances in Materials, Devices and Applications, R. Venkatasubramanian, H. Liang, H. Radousky, J. Poon, 2011, ISBN 978-1-60511-302-9
Volume 1326E — Renewable Fuels and Nanotechnology, H. Idriss, 2011, ISBN 978-1-60511-303-6
Volume 1327 — Complex Oxide Materials for Emerging Energy Technologies, J.D. Perkins, A. Ohtomo, H.N. Lee, G. Herranz, 2011, ISBN 978-1-60511-304-3
Volume 1328E — Electrochromic Materials and Devices, M. Bendikov, D. Gillaspie, T. Richardson, 2011, ISBN 978-1-60511-305-0
Volume 1329E — Nanoscale Heat Transfer—Thermoelectrics, Thermophotovoltaics and Emerging Thermal Devices, K. Nielsch, S.F. Fischer, B.J.H. Stadler, T. Kamins, 2011, ISBN 978-1-60511-306-7
Volume 1330E — Protons in Solids, V. Peterson, 2011, ISBN 978-1-60511-307-4
Volume 1331E — Frontiers of Solid-State Ionics, K. Kita, 2011, ISBN 978-1-60511-308-1
Volume 1332E — Interfacial Phenomena and *In-Situ* Techniques for Electrochemical Energy Storage and Conversion, H. Li, 2011, ISBN 978-1-60511-309-8
Volume 1333E — Nanostructured Materials for Energy Storage, J. Lemmon, 2011, ISBN 978-1-60511-310-4
Volume 1334E — Recent Developments in Materials for Hydrogen Storage and Carbon-Capture Technologies, R. Zidan, 2011, ISBN 978-1-60511-311-1
Volume 1335 — Materials, Processes, and Reliability for Advanced Interconnects for Micro- and Nanoelectronics—2011, M. Baklanov, G. Dubois, C. Dussarrat, T. Kokubo, S. Ogawa, 2011, ISBN 978-1-60511-312-8
Volume 1336E — Interface Engineering for Post-CMOS Emerging Channel Materials, Y. Kamata, 2011, ISBN 978-1-60511-313-5
Volume 1337 — New Functional Materials and Emerging Device Architectures for Nonvolatile Memories, D. Wouters, O. Auciello, P. Dimitrakis, Y. Fujisaki, E. Tokumitsu, 2011, ISBN 978-1-60511-314-2
Volume 1338E — Phase-Change Materials for Memory and Reconfigurable Electronics Applications, B.-K. Cheong, P. Fons, B.J. Kooi, B.-S. Lee, R. Zhao, 2011, ISBN 978-1-60511-315-9
Volume 1339E — Plasma-Assisted Materials Processing and Synthesis, J.L. Endrino, A. Anders, J. Andersson, D. Horwat, M. Vinnichenko, 2011, ISBN 978-1-60511-316-6
Volume 1340E — High-Speed and Large-Area Printing of Micro/Nanostructures and Devices, T. Sekitani, 2011, ISBN 978-1-60511-317-3
Volume 1341 — Nuclear Radiation Detection Materials—2011, A. Burger, M. Fiederle, L. Franks, K. Lynn, D.L. Perry, K. Yasuda, 2011, ISBN 978-1-60511-318-0
Volume 1342 — Rare-Earth Doping of Advanced Materials for Photonic Applications—2011, V. Dierolf, Y. Fujiwara, T. Gregorkiewicz, W.M. Jadwisienczak, 2011, ISBN 978-1-60511-319-7
Volume 1343E — Recent Progress in Metamaterials and Plasmonics, G.J. Brown, J. Pendry, D. Smith, Y. Lu, N.X. Fang, 2011, ISBN 978-1-60511-320-3
Volume 1344 — Functional Two-Dimensional Layered Materials—From Graphene to Topological Insulators, A.A. Balandin, A. Geim, J. Huang, D. Li, 2011, ISBN 978-1-60511-321-0
Volume 1345E — Nanoscale Electromechanics of Inorganic, Macromolecular and Biological Systems, J. Li, S.V. Kalinin, M.-F. Yu, P.S. Weiss, 2011, ISBN 978-1-60511-322-7
Volume 1346 — Micro- and Nanofluidic Systems for Materials Synthesis, Device Assembly and Bioanalysis—2011, R. Fan, J. Fu, J. Qin, A. Radenovic, 2011, ISBN 978-1-60511-323-4
Volume 1347E — Nanoscale Heat Transport—From Fundamentals to Devices, A. McGaughey, M. Su, S. Putnam, J. Shiomi, 2011, ISBN 978-1-60511-324-1

MATERIALS RESEARCH SOCIETY SYMPOSIUM PROCEEDINGS

Volume 1348E — Hybrid Interfaces and Devices, D.S. Ginley, N.R. Armstrong, G. Frey, R.T. Collins, 2011, ISBN 978-1-60511-325-8
Volume 1349E — Quantitative Characterization of Nanostructured Materials, A. Kirkland, 2011, ISBN 978-1-60511-326-5
Volume 1350E — Semiconductor Nanowires—From Fundamentals to Applications, V. Schmidt, L.J. Lauhon, T. Fukui, G.T. Wang, M. Björk, 2011, ISBN 978-1-60511-327-2
Volume 1351 — Surfaces and Nanomaterials for Catalysis through *In-Situ* or *Ex-Situ* Studies, F. Tao, M. Salmeron, J.A. Rodriguez, J. Hu, 2011, ISBN 978-1-60511-328-9
Volume 1352 — Titanium Dioxide Nanomaterials, X. Chen, M. Graetzel, C. Li, P.D. Cozzoli, 2011, ISBN 978-1-60511-329-6
Volume 1353E — The Business of Nanotechnology III, L. Merhari, M. Biberger, D. Cruikshank, M. Theelen, 2011, ISBN 978-1-60511-330-2
Volume 1354 — Ion Beams—New Applications from Mesoscale to Nanoscale, J. Baglin, D. Ila, G. Marletta, A. Öztarhan, 2011, ISBN 978-1-60511-331-9
Volume 1355E — Biological Hybrid Materials for Life Sciences, L. Stanciu, S. Andreescu, T. Noguer, B. Liu, 2011, ISBN 978-1-60511-332-6
Volume 1356E — Microbial Life on Surfaces—Biofilm-Material Interactions, N. Abu-Lail, W. Goodson, B.H. Lower, M. Fornalik, R. Lins, 2011, ISBN 978-1-60511-333-3
Volume 1357E — Biomimetic Engineering of Micro- and Nanoparticles, D. Discher, 2011, ISBN 978-1-60511-334-0
Volume 1358E — Organic Bioelectronics and Photonics for Sensing and Regulation, L. Torsi, 2011, ISBN 978-1-60511-335-7
Volume 1359 — Electronic Organic and Inorganic Hybrid Nanomaterials—Synthesis, Device Physics and Their Applications, Z.-L. Zhou, C. Sanchez, M. Popall, J. Pei, 2011, ISBN 978-1-60511-336-4
Volume 1360 — Synthesis and Processing of Organic and Polymeric Materials for Semiconductor Applications, A.B. Padmaperuma, J. Li, C.-C. Wu, J.-B. Xu, N.S. Radu, 2011, ISBN 978-1-60511-337-1
Volume 1361E — Engineering Polymers for Stem-Cell-Fate Regulation and Regenerative Medicine, S. Heilshorn, J.C. Liu, S. Lyu, W. Shen, 2011, ISBN 978-1-60511-338-8
Volume 1362 — Carbon Functional Interfaces, J.A. Garrido, K. Haenen, D. Ho, K.P. Loh, 2011, ISBN 978-1-60511-339-5
Volume 1363 — Fundamental Science of Defects and Microstructure in Advanced Materials for Energy, B.P. Uberuaga, A. El-Azab, G.M. Stocks, 2011, ISBN 978-1-60511-340-1
Volume 1364E — Forum on Materials Education and Evaluation—K-12, Undergraduate, Graduate and Informal, B.M. Olds, D. Steinberg, A. Risbud, 2011, ISBN 978-1-60511-341-8
Volume 1365 — Laser-Material Interactions at Micro/Nanoscales, Y. Lu, C.B. Arnold, C.P. Grigoropoulos, M. Stuke, S.M. Yalisove, 2011, ISBN 978-1-60511-342-5
Volume 1366E — Crystalline Nanoporous Framework Materials—Applications and Technological Feasibility, M.D. Allendorf, K. McCoy, A.A. Talin, S. Kaskel, 2011, ISBN 978-1-60511-343-2
Volume 1367E — Future Directions in High-Temperature Superconductivity—New Materials and Applications, C. Cantoni, A. Ballarino, K. Matsumoto, V. Solovyov, H. Wang, 2011, ISBN 978-1-60511-344-9
Volume 1368 — Multiferroic, Ferroelectric, and Functional Materials, Interfaces and Heterostructures, P. Paruch, E. Tsymbal, M. Rzchowski, T. Tybell, 2011, ISBN 978-1-60511-345-6
Volume 1369E — Computational Studies of Phase Stability and Microstructure Evolution, Z.-K. Liu, 2011, ISBN 978-1-60511-346-3
Volume 1370 — Computational Semiconductor Materials Science, L. Liu, S.-H. Wei, A. Rubio, H. Guo, 2011, ISBN 978-1-60511-347-0

Prior Materials Research Society Symposium Proceedings available by contacting Materials Research Society

Solar Cells

Mater. Res. Soc. Symp. Proc. Vol. 1321 © 2011 Materials Research Society
DOI: 10.1557/opl.2011.1092

High Efficiency, Large Area, Nanocrystalline Silicon Based, Triple-Junction Solar Cells

A. Banerjee, T. Su, D. Beglau, G. Pietka, F. Liu, B. Yan, J. Yang, and S. Guha
United Solar Ovonic LLC, 1100 West Maple Road, Troy, MI, 48084, U.S.A.

ABSTRACT

We have fabricated large-area, thin-film multijunction solar cells based on hydrogenated amorphous silicon (a-Si:H) and nanocrystalline silicon (nc-Si:H) made in a large area batch reactor. The device structure consisted of an a-Si:H/nc-Si:H/nc-Si:H stack on Ag/ZnO back reflector coated stainless steel substrate, deposited using our proprietary High Frequency (HF) glow discharge technique. For the nc-Si:H films, we investigated two deposition rate regimes: (i) low rate <1 nm/s and (ii) high rate >1 nm/s. We optimized the deposition parameters, such as pressure, gas flow, dilution, and power. We did SIMS analysis on the optimized films, and found the impurity concentrations were one order of magnitude lower than the films made with the conventional RF process. In particular, the oxygen concentration is reduced to $\sim 10^{18}$ cm^{-3}. This value is among the lowest oxygen concentration reported in literature. The low impurity content is attributed to proprietary cathode hardware and the optimized deposition process. During the initial optimization and investigative phase, we fabricated small-area (0.25 cm^2 and 1.1 cm^2) cells. The information obtained from the initial phase was used to fabricate large-area (aperture area 400 cm^2) cells, and encapsulated the cells using the same flexible encapsulants that are used in our commercial product. We have light soaked the low-rate and high-rate encapsulated modules. The highest initial efficiency of the low-rate modules is 12.0% as confirmed by NREL. The highest corresponding stable efficiency attained for the low-rate samples cells is 11.35%. For the high-rate small-area (1.1 cm^2) cells, the highest initial active-area efficiency and corresponding stable efficiency attained are 13.97% and 12.9%, respectively. We present the details of the research conducted to develop the low- and high-rate cells and modules.

INTRODUCTION

nc-Si:H material is a promising candidate to replace a-SiGe:H in multijunction thin film silicon solar cells [1-3]. In view of its indirect bandgap, the nc-Si:H layer must be much thicker than its amorphous counterparts to effectively absorb the incident radiation. Typical thicknesses for a nc-Si:H based multijunction cell is 2-5 μm, compared with <0.5 μm for a corresponding a-Si:H/a-SiGe:H/a-SiGe:H triple-junction structure. While for commercial viability, the nc-Si:H layer must be deposited at a high rate, one must investigate the highest efficiency attainable with and without manufacturing constraint of deposition rate. In this paper, we report on this two-pronged strategy of fabricating large-area, high-efficiency a-Si:H/nc-Si:H/nc-Si:H solar cells at low and high rates.

To fabricate high-efficiency, large-area nc-Si:H based multijunction solar cells, it is necessary to first fabricate high quality nc-Si:H based component cells that exhibit excellent spatial uniformity of device thickness and device performance over a large area [4]. In particular, the bottom cell must meet stringent requirements in terms of short-circuit-current density (J_{sc}), open-circuit voltage (V_{oc}), and fill factor (FF). It has been suggested that oxygen incorporated in the nc-Si:H absorbing layer is the major cause of loss in the long wavelength response: quantum efficiency (QE) at long wavelength and fill factor under red light [5,6]. We have developed a proprietary HF glow discharge technology to deposit the devices with good large area spatial uniformity and low impurity content. We have attained similar levels of oxygen concentration in nc-Si:H layers deposited at both low and high rates.

EXPERIMENT

Large-area nc-Si:H single-junction and a-Si:H/nc-Si:H/nc-Si:H triple-junction solar cells were deposited using the HF process in a large-area batch reactor. The substrate consisted of a ~38 cm x 35 cm piece of stainless steel coated with a back reflector of Ag/ZnO. For the nc-Si:H films, we investigated two deposition rate regimes: (i) low rate <1 nm/s and (ii) high rate >1 nm/s. We optimized the deposition parameters, such as pressure, gas flow, dilution, and power. During the initial optimization phase, the large-area coated substrate was cut into smaller pieces and ITO was deposited through an evaporation mask to delineate the device area. Two different device areas were used: 0.25 cm^2 and 1.1 cm^2. For the 0.25 cm^2 devices, grid fingers were vacuum deposited for current collection. For the 1.1 cm^2 devices, grid wires and bus bars were applied for current collection. Device thickness, I-V characteristics, QE response, and their spatial uniformity were measured. Representative samples were sent out for impurity analysis and depth profiling using the SIMS technique. Finally, large-area (aperture area of 400 cm^2) triple- junction cells were deposited. Grid wires and bus bars were applied for current collection. The large-area cells were encapsulated using our proprietary flexible encapsulants. The 400 cm^2 module is shown in Fig. 1.

Figure 1. A 400 cm^2 encapsulated module.

I-V characteristics of the large-area encapsulated modules were measured using a Spire solar simulator. The modules were light soaked under one-sun intensity at 50 °C and open-circuit conditions for 1000 hours. The modules in the initial and light stabilized states were sent to NREL for efficiency confirmation.

DISCUSSION

Impurity levels in absorbing layers

Reducing incorporation of impurities in the nc-Si:H absorbing layers is crucial to achieving high efficiency. It is well known that high levels of impurities, especially oxygen, in the absorbing layer can produce additional defect states that cause inferior long wavelength response. By continuing optimization of hardware, especially the HF cathode design, and deposition process, we have reduced the impurities in the absorbing layers to much lower levels. More importantly, we achieved very low impurity levels with both low and high deposition rates. Table I shows the impurity concentrations in the absorbing layer in the nc-Si:H bottom cells made at both low (<1 nm/s) and high (>1 nm/s) deposition rates.

Table I. Concentration of major impurities in the absorbing layer in nc-Si:H bottom cell made at low (<1 nm/s) and high (>1 nm/s) deposition rates. Shaded rows represent cells at low rate.

Run # 3D	Impurity concentration (atom/cm^3)				
	oxygen	**carbon**	**nitrogen**	**boron**	**fluorine**
3D10451	2×10^{18}	6×10^{16}	5×10^{15}	$<1\times10^{16}$	2×10^{15}
3D10455	1.2×10^{18}	6×10^{16}	5×10^{15}	$<1\times10^{16}$	2×10^{15}
2B18283	1.5×10^{18}	4×10^{17}	1×10^{17}	$<1\times10^{16}$	2×10^{15}
2B18687	1.5×10^{18}	4×10^{17}	1×10^{17}	$<1\times10^{16}$	2×10^{15}

In Table I, the shaded rows represent cells made at low rate. The oxygen concentration is similar in cells deposited at both low and high rates. The low rate deposition produces very low concentrations of nitrogen and carbon. These are among the lowest impurity levels reported in nc-Si:H based cells.

Low rate (<1 nm/s) cell: initial performance

Table II lists the initial efficiencies of the large-area a-Si:H/nc-Si:H/nc-Si:H triple-junction modules with an aperture area of 400 cm^2 made at low deposition rate (<1 nm/s). These modules were measured at United Solar (USO) and at NREL under respective Spire solar simulators. The initial aperture-area efficiency ranges between 11.1-12.0% as measured by NREL. The agreement in efficiency between the two laboratories is within 4%. Modules 10554 and 10555 exhibit an efficiency of 12.0%. Figure 2 shows the NREL initial I-V characteristics of the two modules. This is the highest value as confirmed by NREL for a module of this size. The results also demonstrate the capability of the nc-Si:H technology to make high efficiency solar cells.

Table II. Initial efficiency of large-area (aperture area ~400 cm^2) modules made at low deposition rate (<1 nm/s).

Module # 3D	Measurement	Area (cm^2)	Temp. (°C)	V_{oc} (V)	I_{sc} (A)	FF (%)	Efficiency (%)
10527	USO	400	24.2	1.87	3.43	71.7	11.50
	NREL	399.8	25.2	1.882	3.525	67.1	11.10
10529	USO	400	25.4	1.912	3.36	72.5	11.67
	NREL	399.8	25.1	1.925	3.5	68.9	11.60
10554	USO	400	24.5	1.936	3.41	73.2	12.09
	NREL	399.8	25.1	1.948	3.57	69.2	12.00
10555	USO	400	25.3	1.906	3.53	71.3	12.02
	NREL	399.8	25.1	1.92	3.676	68.0	12.00

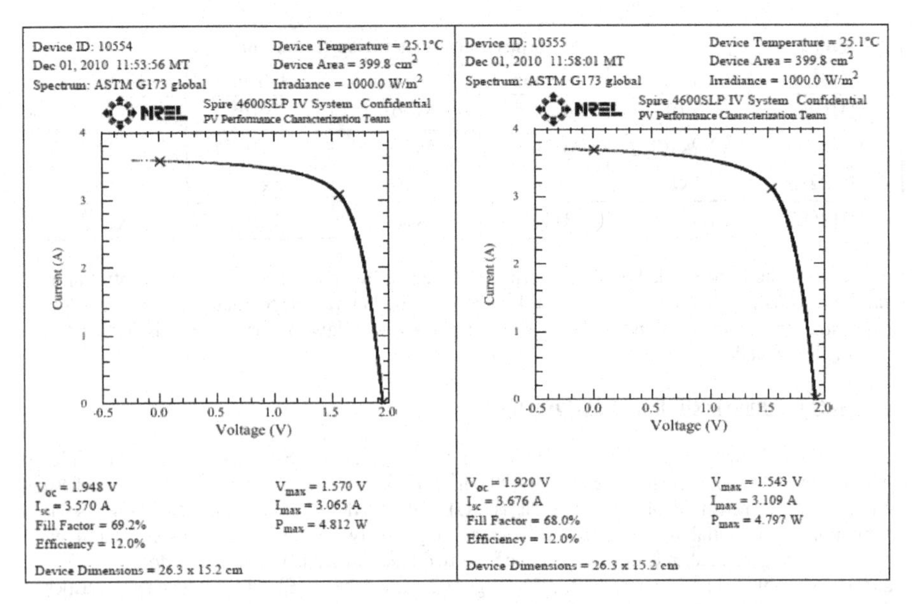

Figure 2. Initial J-V characteristics measured at NREL of the two highest efficiency (12.0%) a-Si:H/nc-Si:H/nc-Si:H encapsulated modules 10554 and 10555 shown in Table II.

High rate (>1 nm/s) cell: initial performance

We have fabricated small-area (1.1 cm^2) a-Si:H/nc-Si:H/nc-Si:H triple-junction cells deposited at high rate (>1 nm/s). Table III shows the initial active-area J-V characteristics of the unencapsulated cells and the QE values for the component cells and the total device. The efficiency (last column) is the active-area efficiency as calculated from the V_{oc}, FF, and limiting-current QE values in Table III. The efficiency is in the range of 13.67-13.97%.

Table III. Initial active-area efficiency of 1.1 cm^2 unencapsulated cells deposited at >1 nm/s as measured at USO.

Cell ID	V_{oc} (V)	FF	QE (mA/cm^2)				Efficiency (%)
			Top	Middle	Bottom	Total	
18730D3-5	1.99	0.762	9.42	9.7	9.13	28.3	13.81
18730D3-6	2.03	0.764	9.55	9.4	8.93	27.9	13.80
18730D3-4	1.97	0.764	9.49	9.73	9.25	28.5	13.90
18708B2-2	2.01	0.763	9.54	9.16	9.11	27.8	13.97
18708B3-4	2.01	0.760	9.53	9.14	9.03	27.7	13.80
18708B3-5	2.02	0.766	9.79	8.98	8.83	27.6	13.67

Stability studies

The a-Si:H/nc-Si:H/nc-Si:H triple-junction modules (400 cm^2) for the low-rate case (Table II) and triple-junction unencapsulated cells (1.1 cm^2) high-rate case (Table III) have been light soaked to attain stable values. The stable efficiency of the low-rate modules in Table II is in the range of 10.8-11.3%. The best cell, as measured at USO, showed stable characteristics of V_{oc} = 1.90 V, J_{sc} = 8.52 mA/cm^2, FF = 0.699, and efficiency = 11.35%.

The stable active-area efficiency of the high-rate cells in Table III is in the range of 12.4-12.9%. The best cell, as measured at USO, showed stable characteristics of V_{oc} = 1.980 V, J_{sc} = 9.04 mA/cm^2, FF = 0.719, and efficiency = 12.86 %. Both types of modules/cells have been sent to NREL for confirmation of efficiency, and will be reported when available.

CONCLUSIONS

We have fabricated a-Si:H/nc-Si:H/nc-Si:H triple junction solar cells and modules at low rate (<1 nm/s) and high rate (>1 nm/s) using our HF process. The highest initial, aperture-area efficiency attained for 400 cm^2 encapsulated modules is 12.0%, as confirmed by NREL. The corresponding stable efficiency, as measured at USO, is ~11.3%. For high rate, small-area (1.1 cm^2) unencapsulated cells, we have attained initial active-area efficiency of 13.97% and stable efficiency of 12.86%.

ACKNOWLEDGMENTS

The authors thank X. Xu and J. Zhang for important contributions. The authors also thank G. DeMaggio for useful discussions and data analysis, D. Wolf, N. Jackett, L. Sivec, C. Worrel, J. Piner, Y. Zhou, T. Palmer, J. Owens, R. Caraway, and B. Hartman for sample preparation and measurements. The work was supported by US DOE under the Solar America Initiative Program Contract No. DE-FC36-07GO17053.

REFERENCES

1. J. Meier, R. Flückiger, H. Keppner, and A. Shah, Appl. Phys. Lett. **65**, 860 (1994).
2. B. Yan, G. Yue, and S. Guha, Mat. Res. Soc. Symp. Proc. Vol. **989**, 335 (2007).
3. O. Vetterl, F. Finger, R. Carius, P. Hapke, L. Houben, O. Kluth, A. Lambertz, A. Mück, B. Rech and H. Wagner, Sol. Ener Mater abd Sol Cells, **62**, 97 (2000).
4. X. Xu, Y. Li, S. Ehlert, T. Su, D. Beglau, D. Bobela, G. Yue, B. Yan, J. Zhang, A. Banerjee, J. Yang, and S. Guha, Mat. Res. Soc. Symp. Proc. Vol. **1153** (2009).
5. M. Kondo, T. Matsui, Y. Nasuno, C. Niikura, T. Fujibayashi, A. Sato, A. Matsuda, and H. Fujiwara, Proc. 31st IEEE PVSC (IEEE, New York, 2005), p. 1377 (and references therein).
6. P. Torres, J. Meier, R. Fluckiger, U. Kroll, J. A. A. Selvan, H. Keppner, A. Shah, S. D. Littlewood, I. E. Kelly, and P. Giannoules, Appl. Phys. Lett. **69**, 1373 (1996).

Mater. Res. Soc. Symp. Proc. Vol. 1321 © 2011 Materials Research Society
DOI: 10.1557/opl.2011.803

THIN FILM SILICON SOLAR CELLS UNDER MODERATE CONCENTRATION

L.M. van Dam, W.G.J.H.M. van Sark, R.E.I. Schropp
Utrecht University, Faculty of Science, Debye Institute for Nanomaterials Science,
Nanophotonics - Physics of Devices, P.O. Box 80.000, 3508 TA Utrecht, The Netherlands,
Tel: 030 253 3170, e-mail: r.e.i.schropp@uu.nl

ABSTRACT

There are only very few reports on the effects of concentration in thin film silicon-based solar cells. Due to the presence of midgap states, a fast decline in fill factor was observed in earlier work. However, with the advent of more stable and lower defect density protocrystalline silicon materials as well as high quality micro-/nanocrystalline silicon materials, it is worth revisiting the performance of cells with these absorber layers under moderately concentrated sunlight. We determined the behavior of the external *J-V* parameters of pre-stabilized substrate-type (n-i-p) amorphous and microcrystalline solar cells under moderate concentrations, between 1 sun and 21 suns, while maintaining the cell temperature at 25°C. It was found that the cell efficiency of both the amorphous and the microcrystalline cells increased with moderate concentration, showing an optimum at approximately 2 suns. Furthermore, the enhancement in efficiency for the microcrystalline cells was larger than for the amorphous cells. We show that the V_{oc}'s up to 0.63 V can be reached in microcrystalline cells while *FF*'s only decrease by 9%. The effects have also been computed using the device simulator ASA, showing qualitative agreement.

INTRODUCTION

Recently, in the quest for higher efficiencies for thin film solar cells, much emphasis is placed on light trapping or absorption enhancement techniques, such as the use of plasmonic or diffractive back contacts [1] and luminescent concentrators [2], to concentrate the incident light in cells with a thinner absorber layer or with a smaller area. Even in less advanced schemes, merely optical concentration of light can yield higher efficiencies. There are only very few reports on the effects of concentration in thin film silicon-based solar cells [3]. Due to the presence of midgap states, a fast decline in fill factor was observed in earlier work. However, with the advent of more stable and lower defect density protocrystalline silicon materials as well as high quality micro-/nanocrystalline silicon materials, as well as the increasing concentration ratios obtained by novel light management techniques, it is worth revisiting the performance of cells with these absorber layers under moderately concentrated sunlight. Furthermore, local generation rates in thin film silicon solar cells with a plasmonic back reflector are much enhanced and the study of the effect of external optical concentration can help to optimize their design.

We determined the behavior of the external *J-V* parameters of pre-stabilized substrate-type (n-i-p) amorphous and microcrystalline solar cells under moderate concentrations, between 1 sun and 21 suns, while maintaining the cell temperature at 25°C. Maintaining the cell temperature is an important requirement when measuring under concentrated illumination, because rapidly increasing temperatures can influence the *J-V* measurement and too high temperatures can damage the cell [4]. We apply extensive active cooling by means of Peltier devices and selected solar cells deposited in the substrate (n-i-p) configuration on stainless steel, because they can be cooled very accurately at the rear side of the substrate, while maintaining easy access to the electrical contacts. It should be noted that these cells already comprise some sort of local light concentration due to the use of textured surfaces.

In addition, we computed *J-V* characteristics using the one-dimensional model for semiconductor device simulation ASA, which is a product of the Delft University of Technology [5]. Simulations were performed with cell parameters approximating the measured cells and illuminations between 1 sun and 21 suns.

METHODS AND MATERIALS

Two types of thin-film photovoltaic cells were investigated: a single junction hydrogenated amorphous silicon cell (a-Si:H) and a single junction hydrogenated microcrystalline silicon cell (μc-Si:H), both in substrate (n-i-p) configuration. The intrinsic layer in the a-Si:H cell had protocrystalline nature. The estimated thicknesses of the n, i and p layer of the amorphous silicon cells were 80 nm, 500 nm, and 20 nm, respectively, and were deposited at 195°C, 250°C, and 160°C, respectively. The intrinsic layer was deposited by Hot-Wire Chemical Vapor Deposition. The estimated thicknesses of the n, i and p layer of the microcrystalline silicon cells were 27 nm, 2000 nm and 20 nm, respectively. The i-layer was deposited by Hot-Wire Chemical Vapor Deposition at a temperature of 250°C, the other layers at 195°C. The microcrystalline cells were applied on a stainless steel substrate and have an Ag and ZnO back reflecting contact. The square cells have either a V-shaped top contact or a grid-type contact. In both cases, the total area is 16 mm² and the active area is 13 mm².

AM1.5 light was produced by a WACOM dual source solar simulator, which has a uniformity of the irradiated area within ±3%, an illumination stability of < 1% reproducibility and a spectral match of ±3.0% for 350-500 nm, ±2.0% for 500 – 800 nm and 4.5% for 800 – 1100 nm [6]. Increasing the light intensity was done by means of a polymer hybrid Fresnel-prismatic lens, a lens with a surface that partly consists of square planes and partly of curved planes to refract the light. The dimensions of the lens are 160 mm by 160 mm. This lens (initially developed for use in a concentrating photovoltaic module [7]) was kindly supplied by the Italian National Agency for New Technologies, Energy and sustainable economic development (ENEA) and has a transparency of 82-83% according to their specifications [8]. This lens was mounted on a labjack between the solar simulator and the solar cells. The illumination intensity (concentration ratio) of the incident light on the cell was varied by varying the distance (x) between the lens and the solar cell, where $x <$ the focal distance of the lens (f) (Figure 1).

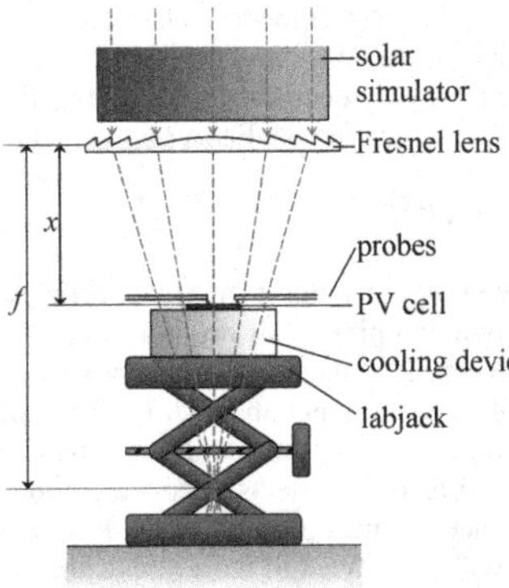

Figure 1. Schematic drawing of the measurement setup, showing the light path from the solar simulator, passing through the Fresnel lens where it is converged and reaches the PV cell. The labjack is used to modify the distance from the lens to the cell (x). Distance f is the focal distance of the lens.

Calibration of the light intensity was done with the use of a Coherent Inc. LaserCheck, a hand-held laser power density meter. This device was placed at the location of the solar cells and the light intensity was measured with the lens at a range of distances. Comparing the measured light intensities with the measured light intensity at 1 sun provided the intensity in relation to the distance to the lens. To protect the power density meter, a neutral density filter of 10% transparency was used. The results were fitted to a simple power function with an R^2 of 0.9997 to provide continuous data.

Cooling of the solar cells by merely forcing a stream of ambient air of approximately 17°C along the solar cell proved to be insufficient. The substrate configuration of the solar cells provided the possibility to actively cool the cells at the stainless steel back side. In addition, the stainless steel substrate of the cells helps to disperse the heat and to stabilize the temperature of the cell. The cells were attached to an aluminum slab with the use of a heat sink compound to ease the transfer of heat from the stainless steel substrate to the aluminum slab. On the other side of the slab, two PolarTEC Peltier devices were attached to actively extract a maximum of 83.4 W of heat each from the block. Two CoolerMaster Hyper 101 convector cooling units, each capable of removing a maximum of 100 W heat, disposed of the heat from the Peltier devices. A Newport Electronics thermal control unit measured the cell temperature by means of a thermocouple wire on the stainless steel substrate and activated the Peltier devices. This way, the temperature could be kept within the range of 24.5 – 25.0°C at all times. Illuminated J-V characteristics were measured by a Keithley 238 source-measure unit using a four point probe configuration at a voltage between -0.3 V and 1.0 V in steps of 100 mV.

The input parameters for the simulations were defined by a standardized amorphous photovoltaic cell in a superstrate configuration from Delft University of Technology, which were adapted to have an approximate match to the specific amorphous solar cell that was used for the measurements. This was done in the settings file by removing the glass and the transparent conductive oxide (TCO) layer (because we were using substrate type cells) and adjusting the thickness of the p, i and n layers. The light intensity was varied by multiplying the input spectrum, AM1.5, 100 mW/cm², by 1 to 21.

RESULTS AND DISCUSSION

Measurements

The measured illuminated *J-V* characteristics of both the amorphous and the microcrystalline silicon cells are shown in figures 2a and 2b, respectively, both normalized to the J_{sc} of their individual reference measurement under standard AM1.5, 100 mW/cm^2 conditions. The arrows in the graphs indicate the trends when increasing the illumination intensity. Both the amorphous and the microcrystalline silicon cells show an increase in V_{oc}, but the amorphous silicon cell shows in addition a strong change in the curvature, affecting the fill factor. The decrease in fill factor for the microcrystalline silicon cells is

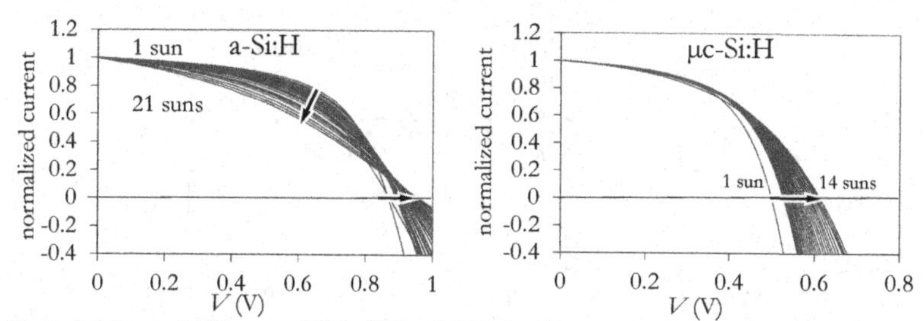

Figure 2. Measured *J-V* curves of the a-Si:H cell G6 (a) and the μc-Si:H cell I6 (b) at light intensities ranging from, respectively, 1 to 21 suns and 1 to 14 suns. The measured current densities are normalized to the J_{sc} of their respective reference measurement under standard AM1.5, 100 mW/cm^2 conditions.

significantly smaller than that for the amorphous silicon cells. The fill factor of the amorphous silicon cell drops from 0.56 (at 1 sun) to 0.41 (at 14 suns), a drop of 27%, whereas the fill factor of the microcrystalline cell only drops from 0.55 (at 1 sun) to 0.46 (at 14 suns), a drop of only 16%. The relative gain in V_{oc} for the μc-Si:H solar cell is considerably higher than that for the a-Si:H cell. Studied over the intensity range of 1 to 14 suns, the gain is 21%, while that for a-Si:H cells is only 9.0%. This is also depicted in figures 3a and 3b, where both the V_{oc} and the fill factor are plotted versus the light intensity.

The measurements show that the V_{oc} of the a-Si:H cells increases logarithmically from 0.87 V at 1 sun to a maximum close to 1.0 V at the highest intensity studied. The μc-Si:H cells show a similar behavior, only the increase in V_{oc} is more remarkable, i.e., 22% from 0.50 V at 1 sun to 0.61 V at 12 suns. When the AM1.5 efficiencies, calculated from the J-V characteristics and the calibrations, are plotted, both the amorphous and the microcrystalline silicon cells show an increase in efficiency, which is optimal at a light intensity of around 2 suns (Figure 4). The efficiency of the a-Si:H cells appears to have an optimum at a relative illumination of 1.9 times AM1.5, peaking at 7.3%, where the standard one sun efficiency is measured at 5.5%. At higher light intensities, the efficiency decreases.

For the μc-Si:H cells, the peak efficiency is measured at 9.4% at a relative illumination of 2.1 times AM1.5, where the standard one sun efficiency is measured to be 7.1%.

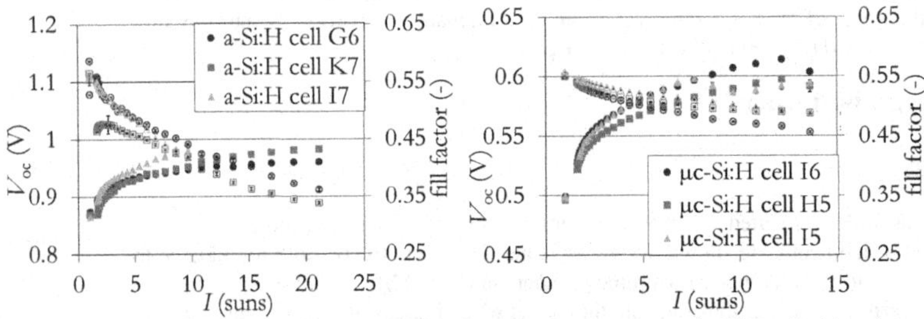

Figure 3. V_{oc} and fill factor plotted versus the light intensity for the measured a-Si:H cells (a) and μc-Si:H cells (b). Filled markers show the V_{oc}, open markers show the fill factor.

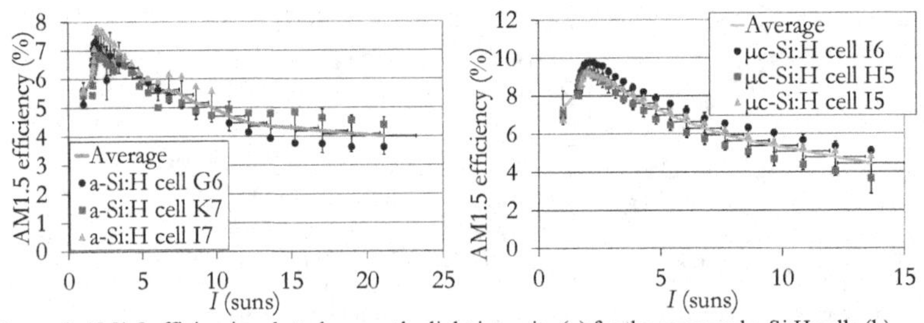

Figure 4. AM1.5 efficiencies plotted versus the light intensity (a) for the measured a-Si:H cells (b) and μc-Si:H cells.

Simulations

The ASA input parameters for a standardized a-Si:H cell refer to a cell in the initial, as deposited state. The defect density is computed from the Defect Pool Model [9]. This is at variance with the cells in the experiments, which were in the light-soaked, stabilized state. This explains why the *FF*'s in the simulation are higher. At the time of this publication we did not have available input parameters for light-soaked and stabilized cells. Nevertheless, the computer simulation clearly shows trends in *FF* and V_{oc} that are similar to our experiments. The optimum in the efficiency-light intensity plot occurs at the same concentration of 2 suns, although the optimum in the efficiency is less pronounced.

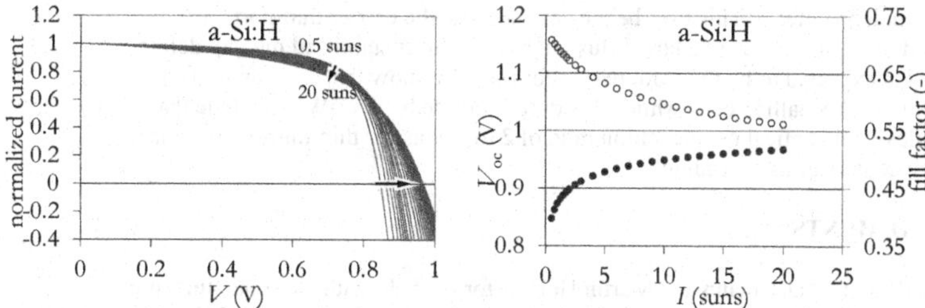

Figure 5. (a) Simulated *J-V* curves of the a-Si:H cell at light intensities ranging from 0.5 to 20 suns. The calculated current densities are normalized to the J_{sc} of the reference simulation under standard AM1.5, 100 mW/cm² conditions; (b) V_{oc} and fill factor plotted versus the light intensity for the simulated a-Si:H cell. Filled markers show the V_{oc}, open markers show the fill factor.

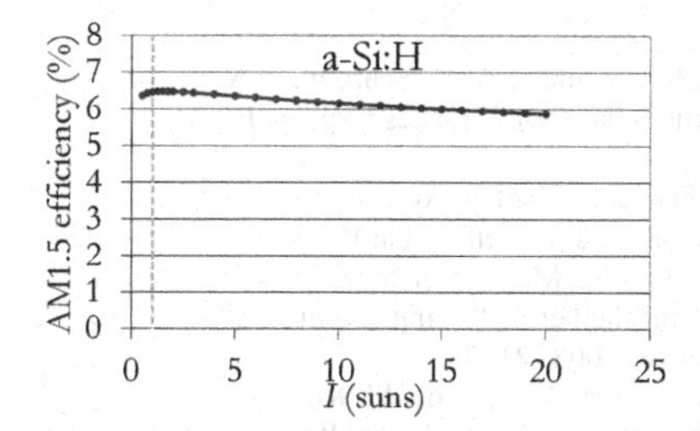

Figure 6. AM1.5 efficiencies plotted as a function of light intensity for the simulated a-Si:H cell. The dashed line marks the 1 sun intensity level.

Discussion

Possible causes for the behavior of the efficiency might be found in the defect density and the presence of midgap states. Also, the investigated cells were not designed for use under concentrated light, so grid contacts were not optimized to handle large currents. The hybrid lens that was used in the measurement setup was designed and optimized for use at the focal distance, so use at other distances might have resulted in some inhomogeneity of the light and possibly small measurement errors (as indicated in figure 4).

The gains in efficiency, especially in the microcrystalline silicon cells offer opportunities for use in concentrating solar systems, such as a luminescent concentrator. The increase in V_{oc} in the thin film cells, and in the microcrystalline silicon cell in particular, looks promising in particular for tandem and triple cells, where an even higher increase in V_{oc} can be expected. Further work is in progress on computer simulation of the microcrystalline silicon cell and multijunction cells, as well as experimental verification.

CONCLUSIONS

It was found that the cell efficiency of amorphous silicon and microcrystalline silicon solar cells increases with increasing light concentration, showing an optimum at approximately 2 suns. The effects have also been computed using the device simulator ASA, showing qualitative agreement. Furthermore, the enhancement in efficiency for the microcrystalline cells is larger than for the amorphous cells. The μc-Si:H cells appear to be very well suited for use in concentrators, as the fill factor is much less affected by the recombination in midgap defects. This may be the case because the carrier transport mechanism is predominantly determined by diffusion, but the lower stabilized midgap defect density in μc-Si:H is expected to be the main reason for this. We show that V_{oc}'s up to 0.63 V can be reached in microcrystalline cells while *FF*'s only decrease by 9%. We conclude that it is meaningful to design an optical concentration ratio of 2 to 5 suns for thin film silicon solar cells and optimize contact grids accordingly.

ACKNOWLEDGEMENTS

We thank Roberto Ackermann and Martin Huijzer for the help with the solar simulator set up, Karine van der Werf for the production of the solar cells, Serge Solntsev and Miro Zeman (TU Delft) for help with the ASA simulations and Francesco Roca (ENEA) for the kind supply of the Fresnel lens.

REFERENCES

1. V.E. Ferry, M.A. Verschuuren, H.B.T. Li, E. Verhagen, R.J. Walters, R.E.I. Schropp, H.A. Atwater, A. Polman, "Light Trapping in Ultrathin Plasmonic Solar Cells", Optics Express **18** (2010) A237.
2. W.G.J.H.M. van Sark, K.W.J. Barnham, L.H. Slooff, Amanda J. Chatten, Andreas Büchtemann, Andreas Meyer, S.J. McCormack, R. Koole, Daniel.J. Farrell, Rahul Bose, Evert E. Bende, Antonius R. Burgers, Tristram Budel, Jana Quilitz, Manus Kennedy, Toby Meyer, Celso de Mello Donegá, A. Meijerink, D.A.M. Vanmaekelbergh, "Luminescent solar concentrators – A review of recent results", Optics Express **16** (2008) 21773.
3. S. Kasashima, R. Uzawa, I.A. Yunaz, Y. Kakihara, S. Miyajima, A. Yamada, M. Konagai, "Amorphous silicon solar cells with novel mesh substrates for concentrator photovoltaic application", 35th IEEE Photovoltaic Specialists Conference, 2009, pp. 47-48.
4. M.A. Green, "General Temperature Dependence of Solar Cell Performance and Implications for Device Modelling", Progress in Photovoltaics **11** (2003) 333.
5. G. Munyeme, G.K. Chinyama, M. Zeman, R.E.I. Schropp, and W.F. van der Weg, "Modelling the light induced metastable effects in amorphous silicon", Physica Status Solidi C 5 (2008) 606.
6. Voss Electronic GmbH, "Manual, Solar Simulator, Model: WXS-140-SUPER"
7. ENEA, "Integrated Structural Element for Concentrating Photovoltaic Module", Patent WO 2006/070425.
8. F. Roca, "Le attività ENEA sulla concentrazione e per la tecnologia del cSi", website ENEA (2007), presentation handout.
9. M. J. Powell and S. C. Deane, "Defect-pool model and the hydrogen density of states in hydrogenated amorphous silicon", Phys. Rev. B 53 (1996)10121.

Mater. Res. Soc. Symp. Proc. Vol. 1321 © 2011 Materials Research Society
DOI: 10.1557/opl.2011.928

Effect of Bandgap Grading on the Performance of a-$Si_{1-x}Ge_x$:H Single-Junction Thin-Film Solar Cells

H. J. Hsu, C. M. Wang, C. H. Hsu and C. C. Tsai
Department of Photonics, National Chiao Tung University, Hsinchu, Taiwan

ABSTRACT

In this work, the effect of bandgap grading of hydrogenated amorphous silicon germanium (a-$Si_{1-x}Ge_x$:H) absorber near the p/i and the i/n interfaces was investigated. The a-$Si_{1-x}Ge_x$:H single-junction solar cells were improved by applying both p/i grading and i/n grading. Our results showed that both the p/i and the i/n grading can increase the open-circuit voltage (V_{OC}) as compared to the cell without grading. The i/n grading can further improve the FF. Presumably the potential gradient created by the i/n grading can facilitate the hole transport thus it can improve the FF. However, the J_{SC} decreased as the i/n grading width increased. The reduction of J_{SC} was due to the loss in the red response, which can be attributed to the replacement of lower bandgap material by the larger ones. Combining the effects of V_{OC}, J_{SC} and FF, a suitable thickness of the p/i and the i/n grading was 20 nm and 45 nm, respectively. Finally, the grading structures accompanied with further optimization of doped layers were integrated to achieve a cell efficiency of 8.59 %.

INTRODUCTION

Silicon-based thin-film solar cell has the potential to be cost-effective for large-scale manufacturing among various types of solar cells. However, the thin-film silicon solar cells are generally low in efficiency compared to the crystalline silicon solar cells. Moreover, the light-induced degradation, also known as the Staebler-Wronski effect (SWE) [1] has influence on the amorphous material which limits the thickness of the absorber layer (i-layer). One effective approach to improve cell efficiency meanwhile minimize SWE is the adoption of the multi-junction structure. Based on the concept of spectrum splitting, each sub-cells with a distinct bandgap are connected in series to extend the absorption range. In the development of low-gap materials, a-$Si_{1-x}Ge_x$:H has received much attention due to its tunable bandgap and high absorption coefficient. The incorporation of germanium into the a-Si:H network narrows the bandgap and enhances the optical absorption. However, the microscopic structure becomes defective due to the low diffusivity of Ge related precursors on the growing surface. As a result, the carrier transport generally degrades as Ge content increases. To improve the material quality and hence the cell performance, the hydrogen dilution was used [2-3].

The a-$Si_{1-x}Ge_x$:H thin films has been characterized and optimized in our previous work [4] using the hydrogen dilution. In practical applications, one of the major challenges in a-$Si_{1-x}Ge_x$:H single-junction solar cell is the bandgap discontinuity between the a-$Si_{1-x}Ge_x$:H i-layer and the a-Si:H doped layer. Such discontinuity at both the p/i and the i/n interfaces creates a potential barrier against the carrier transport. Moreover, the interfaces may contain considerable defects due to the lattice mismatch. These interface states may trap charge carriers, which acts as an opposite force to the build-in electric field. As a result, the build-in electric field may be distorted and the carrier transport may be further deteriorated. According to the mentioned above, Yang et al. have proposed that the drawbacks related to the bandgap discontinuity can be alleviated by the bandgap grading in the intrinsic layer near the interfaces [5]. However, the detailed grading structure such as grading width was not included in the paper. Several groups have investigated the appropriate structure of the bandgap grading by computer simulation [6-8]. In this work, we carried out experiments and focused on the effect of the bandgap grading width on the performance of single-junction a-$Si_{1-x}Ge_x$:H solar cells.

EXPERIMENT

The deposition was carried out in a single chamber system composed of a load-lock, a transfer and a deposition chamber. The silicon-based thin films were deposited in a 27.12 MHz radio-frequency plasma-enhanced chemical vapor deposition (PECVD) system with a gas mixture of SiH_4, GeH_4, B_2H_6, PH_3 and H_2. The a-$Si_{1-x}Ge_x$:H single junction solar cells were deposited on the SnO_2:F substrates in a superstrate configuration. The bandgap profiling was achieved by continuously varying the germane-to-silane ratio during the deposition process. After the deposition, the cells were transferred to deposit the TCO/Ag back contacts.

The deposited thin-films were measured by UV/VIS spectroscopy to obtain their optical properties. The transmittance spectra were then analyzed by Tauc's method to obtain the optical bandgap. The germanium content was measured by the X-ray photoelectron spectroscopy (XPS). The J-V characteristics and the external quantum efficiency (EQE) of solar cells were measured by an AM1.5G illuminated I-V measurement system and a mono-collimated light source, respectively.

RESULTS AND DISCUSSION

Figure 1 (a) shows the effect of H_2 dilution on the film Ge content and the bandgap as a function of the germane (GeH_4) gas flow ratio (R). The germane gas flow ratio was defined as a ratio of GeH_4 gas flow rate to SiH_4 and GeH_4 flow rate. As the germane gas flow ratio increased from 8.3% to 16.7% at a fixed H_2 dilution ratio, the Ge content increased and the bandgap decreased accordingly. The increase in the GeH_4 source gas will increase the portion of GeH_n precursors in the gas phase and hence increase the film Ge content. On the other hand, Fig. 1(b) demonstrates the effect of H_2 dilution ratio on the enhancement factor as a function of GeH_4 gas flow ratio. The enhancement factor was defined as the ratio of the film Ge content to the germane gas flow ratio. Even without the presence of H_2, the enhancement factor is larger than 1 for different germane gas flow ratio. This suggests that the Ge was more effective to incorporate into the film than Si. However, at a fixed H_2 dilution ratio, the enhancement factor decreased as the GeH_4 gas flow ratio increased. The incorporation of Ge became less efficient at higher GeH_4 gas flow ratio.

Moreover, as the H_2 dilution ratio increased from 0 to 6, the film Ge content increased and the bandgap decreased, as shown in Fig. 1(a). The mechanism of H_2 on film Ge content is still not well understood. One possible reason is that H_2 may increase the Ge incorporation. The other one is that H_2 may reduce the etching rate of the GeH_n precursors at the growing surface.

The same trend can also be observed in the Fig. 1(b). At a fixed GeH_4 gas flow ratio, the increase in the H_2 dilution ratio leads to the increase in the film Ge content and thus increasing the enhancement factor. The result indicates that H_2 is more effective in incorporation of Ge than Si. Therefore, H_2 indeed played a crucial role in either gas phase or surface reaction. More efforts should be delivered to clarify the role of H_2 in the plasma and the effect of H_2 on the film Ge content. Based on these results, the graded bandgap can be achieved by varying GeH_4 gas flow ratio and H_2 dilution ratio.

Our results show that the graded bandgap at both the p/i and the i/n interfaces enhanced the cell efficiency as compared to cells with constant bandgap of 1.55 eV. According to the results, the increase in the efficiency is due to the enhancement in V_{OC} and FF, which may arise from a better short-wavelength absorption and carrier transport, respectively. In order to assess the effect of bandgap grading in the absorber near both the p-layer and the n-layer on the cell performance, the grading width was systematically investigated.

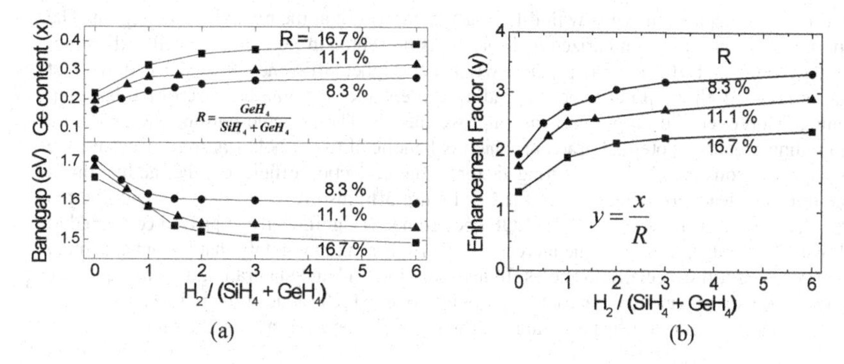

Figure 1. (a) The effect of H_2 dilution on film Ge content and bandgap as a function of germane gas flow ratio (R); (b) The relation between the H_2 dilution ratio and the enhancement factor (y).

Figure 2(a) and 2(b) illustrates an example of the bandgap structure of a-$Si_{1-x}Ge_x$:H single-junction solar cell without and with bandgap grading in the p/i and the i/n interfaces, respectively. The lowest optical bandgap of the a-$Si_{1-x}Ge_x$:H i-layer was 1.55 eV, while the bandgap of the a-Si:H doped layers have a bandgap of around 1.75 eV. To investigate the appropriate grading structure, the p/i grading width was varied from 0 to 30 nm and the i/n grading width was varied from 0 to 120 nm. The total thickness of the a-$Si_{1-x}Ge_x$:H was in a range from 200 nm to 230 nm.

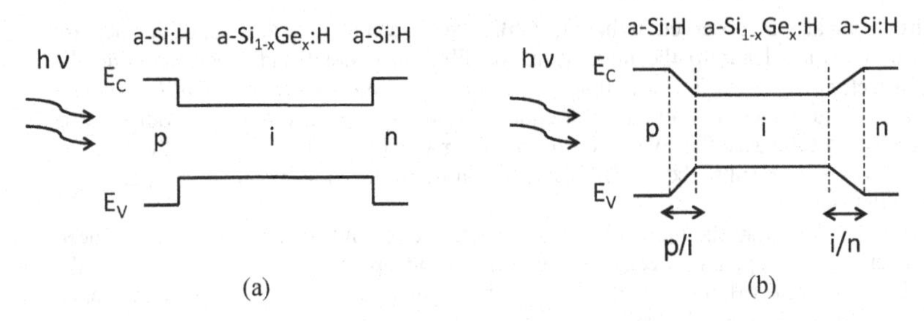

Figure 2. The band structure of the a-$Si_{1-x}Ge_x$:H cell (a) without bandgap grading, or (b) with bandgap grading at both the p/i and the i/n interfaces.

Figure 3 compares the cells with different grading width at the p/i and the i/n region. The corresponding results were summarized in Table 1. As presented in Fig. 3(a), the cell with 20 nm p/i grading width and 105 nm i/n grading width has higher efficiency than the cell without bandgap grading. Both the p/i and the i/n grading can enhance the V_{OC} because of the bandgap broadening. Moreover, the i/n grading can improve the FF. The possible reason may be due to the i/n grading creates a potential gradient which is beneficial to the hole transport. Because the hole transport is considered as the limiting factor in the collection efficiency, the facilitation of the hole transport leads to an increase in the FF. Figure 3(b) also revealed that although the red response decreased, the peak value of the EQE spectrum was significantly improved compared to the cell without bandgap grading. The increase in the EQE peak may imply that the enhancement in the carrier transport especially for holes. In addition, the i/n bandgap grading was achieved by replacing the narrower bandgap material by the wider one, which reduced the red response. As a result, the i/n grading can significantly improve the FF while the J_{SC} is almost the same.

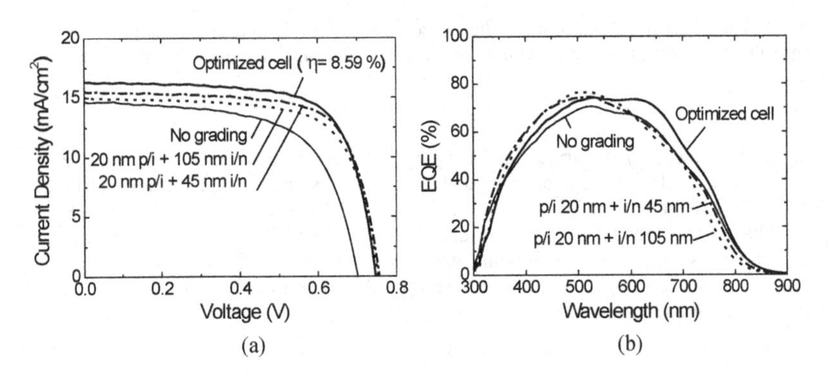

Figure 3. (a) The J-V characteristics and (b) quantum efficiency of the a-$Si_{1-x}Ge_x$:H single-junction solar cells with different grading structures.

Furthermore, as the i/n grading width reduced to 45 nm as compared to the cell with grading width of 105 nm, the J_{SC} and the FF can be further improved. The thinner the i/n grading width is equivalent to the increase in the thickness of the lowest bandgap material, which can increase the absorption of red photons. As confirmed by the EQE measurement, the increase in the J_{SC} is mainly due to the improved red response which can be seen in Fig. 3(b). However, our previous result revealed that as the i/n grading width reduced to 0 nm, the FF and the V_{OC} dropped while the J_{SC} remained almost the same. The decrease in the V_{OC} came from the bandgap narrowing as expected. To our surprise, the J_{SC} saturated as more low-gap materials presented in the i/n region. The results may imply that the hole transport was degraded due to the lack of potential gradient and deteriorated film quality in the i/n region. The drop in the FF can also be a support of such explanation. As a result, the experimental results showed a suitable thickness of the i/n grading of 45nm in this study.

Finally, the cell was further optimized by slightly varying the absorber layer thickness and narrowing the optical bandgap of the i/n grading region. As can be seen in the Fig. 3(a), the J_{SC} was significantly increased without reducing the FF. Figure 3(b) also shows that the red response was significantly improved due to the bandgap narrowing of the i/n grading width. The a-$Si_{1-x}Ge_x$:H single-junction solar cell having an efficiency of 8.59% was achieved, with V_{OC} = 0.75 V, J_{SC} = 16.31 mA/cm^2 and FF = 70.38 %.

Table 1. J-V characteristics of a-$Si_{1-x}Ge_x$:H single-junction solar cells with different grading structures.

	V_{OC} (V)	J_{SC} (mA/cm^2)	FF (%)	η (%)
No grading	0.68	15.07	61.16	6.31
20 nm p/i + 105 nm i/n	0.75	15.08	67.94	7.68
20 nm p/i + 45 nm i/n	0.75	15.52	70.51	8.28
Optimized cell	0.75	16.31	70.38	8.59

CONCLUSIONS

We have found that both the p/i and the i/n grading were effective in increasing the cell efficiency. The mechanisms of the enhancement at the p/i and the i/n interfaces were different. Both the p/i and the i/n grading can enhance the V_{OC} due to the bandgap broadening, but they would also reduce the J_{SC} because of the suppressed light absorption. The i/n grading can improve the FF while the p/i grading may reduce it. The i/n grading seemed to facilitate the hole transport whereas the p/i grading seemed to hinder it. Combining the improvement in V_{OC}, J_{SC} and FF, the suitable thicknesses of the p/i and the i/n grading width were 20 nm and 45 nm in this study. Further optimization of the doped layer in conjunction with the p/i and the i/n gradings achieved an a-$Si_{1-x}Ge_x$:H single-junction cell with η = 8.59%, V_{OC} = 0.75 V, J_{SC} = 16.31 mA/cm^2 and FF = 70.38 %.

ACKNOWLEDGMENTS

This work was sponsored by the Center for Green Energy Technology at the National Chiao Tung University and the National Science and Technology Program-Energy (100-3113-E-009-007-CC2).

REFERENCES

1. D. L. Staebler and C. R. Wronski, Appl. Phys. Lett. **31**, 292 (1977).
2. A. R. Middya, S. Ray, S. J. Jones, and D. L. Williamson, J. Appl. Phys. **78**, 4966 (1995).
3. M. Shima, A. Terakawa, M. Isomura, M. Tanaka, S. Kiyama, and S. Tsuda, Appl. Phys. Let. **71**, 84 (1997).
4. C. M. Wang, Y. T. Huang, K. H. Yen, H. J. Hsu, C. H. Hsu, H. W. Zan, and C. C. Tsai, Mat. Res. Soc. Symp. Proc. **1245**, 85 (2010)
5. J. Yang, A. Banerjee, and S. Guha, Appl. Phys. Lett. **70**, 2975 (1997).
6. R. J. Zambrano, F. A. Rubinelli, J. K. Rath, and R. E. I. Schropp, J. Non-Cryst. Sol. **299**, 1131 (2002).
7. R. J. Zambrano, F. A. Rubinelli, W. M. Arnoldbik, J. K. Rath, and R. E. I. Schropp, Sol. Energy Mater. Sol. Cells **81**, 73 (2004).
8. J. Zimmer, H. Stiebig, and H. Wagner, J. Appl. Phys. **84**, 611 (1998).

Mater. Res. Soc. Symp. Proc. Vol. 1321 © 2011 Materials Research Society
DOI: 10.1557/opl.2011.1149

High-Efficiency Microcrystalline Silicon and Microcrystalline Silicon-Germanium Alloy Solar Cells

Takuya Matsui and Michio Kondo
Research Center for Photovoltaic Technologies, National Institute of Advanced Industrial Science and Technology (AIST), Tsukuba, Ibaraki, 305-8568, Japan

ABSTRACT

This paper presents our material studies on hydrogenated microcrystalline silicon (μc-Si:H) and microcrystalline silicon-germanium alloy (μc-$Si_{1-x}Ge_x$:H) thin films for the development of high efficiency *p-i-n* junction solar cells. In μc-Si:H solar cells, we have evaluated the structural properties of the intrinsic μc-Si:H layers grown by plasma-enhanced chemical vapor deposition at high deposition rates (>2 nm/s). Several design criteria for the device grade μc-Si:H are proposed in terms of crystallographic orientation, grain size and grain boundary passivation. Meanwhile, in μc-$Si_{1-x}Ge_x$:H solar cells, we have succeeded in boosting the infrared response of solar cell upon Ge incorporation up to x~0.2. Nevertheless, a degradation of solar cell parameters is observed for large Ge contents (x>0.2) and thick *i*-layers (> 1 μm), which is attributed to the influence of the Ge dangling bonds that act as acceptorlike states in undoped μc-$Si_{1-x}Ge_x$:H. To improve the device performance, we introduce an oxygen doping technique to compensate the native defect acceptors in μc-$Si_{1-x}Ge_x$:H *p-i-n* solar cells.

INTRODUCTION

Hydrogenated microcrystalline silicon (μc-Si:H) grown by plasma-enhanced chemical vapor deposition (PECVD) is extensively employed as a stable bottom cell absorber in hydrogenated amorphous silicon (a-Si:H)/μc-Si:H double junction tandem solar cells [1-5]. As μc-Si:H is a material exhibiting indirect optical transition, a relatively thick absorber layer (>2 μm) is necessary for efficient absorption in the infrared wavelengths, which in turn requires the high-rate deposition for industrial production. To achieve both high deposition rate and high efficiency, we have developed a deposition process based on a combination of high-pressure depletion (HPD) [6] and very-high-frequency (VHF) [1,7] glow discharge techniques. This process allows growing μc-Si:H at high rates (> 2 nm/s) while preserving excellent film qualities in terms of compact microstructure and less post-deposition oxidation [8]. As a result, efficiencies of 8-9% have been demonstrated for the μc-Si:H single junction *p-i-n* solar cells at deposition rates between 2 and 3 nm/s.

Despite the successful material combination of a-Si:H and μc-Si:H, the stabilized efficiency of the a-Si:H/μc-Si:H tandem solar cells is still limited as low as ~12% [9]. The one of the major limitations of efficiency is the weak light absorption in the infrared wavelengths in μc-Si:H bottom cell. To extend the spectral sensitivities of solar cells into longer wavelengths, hydrogenated microcrystalline silicon-germanium alloys (μc-$Si_{1-x}Ge_x$:H) have been proposed [10-13] as a low-band-gap absorber in multijunction structures such as a-Si:H/μc-$Si_{1-x}Ge_x$:H [14] and a-Si:H/μc-Si:H/μc-$Si_{1-x}Ge_x$:H [15]. In single junction devices, we have demonstrated efficient (~7-8%) μc-$Si_{1-x}Ge_x$:H (x~0.1-0.17) *p-i-n* solar cells with markedly higher short-circuit current densities than for μc-Si:H (x=0) solar cells due to an enhanced infrared absorption. Nevertheless, the photocarrier collection is degraded significantly when increasing either Ge content (x>0.2) or cell thickness (t_i> 1 μm). We attributed the inferior performance of such solar cells to the influence of the Ge-related acceptorlike states that strongly distort the built-in electric field in the *p-i-n* solar cells [16]. Recently, we have developed a counter doping technique to compensate the acceptorlike states for further improvement of the μc-$Si_{1-x}Ge_x$:H solar cells.

In this paper, we review our research and progresses in μc-Si:H and μc-$Si_{1-x}Ge_x$:H thin film solar cells. We will show that the solar cell properties of these materials are differently affected by the impurity incorporation such as oxygen.

EXPERIMENT

Solar cells fabricated in this study consist of glass/transparent conductive oxide (TCO)/p – i – n /ZnO/Ag. As front TCO layer, we employed the chemically-etched ZnO:Ga films with an antireflective TiO_2-ZnO coating. For μc-Si:H solar cells, all μc-Si:H *p*-, *i*-, and *n*-layers were subsequently deposited by the capacitively-coupled 70-100 MHz plasma-enhanced chemical vapor deposition (PECVD) in an ultrahigh vacuum (UHV) multichamber system. The μc-Si:H *i*-layers were deposited using a gas mixture of SiH_4 and H_2 ($[H_2]/[SiH_4]$~50) under different gaseous pressures in the range from 4 to 9 Torr. The detail preparation conditions for high-rate μc-Si:H deposition have been described elsewhere [8]. For μc-$Si_{1-x}Ge_x$:H *p-i-n* solar cells, Ge was incorporated into only absorber *i*-layers in the composition range from 10 to 35% by adding GeH_4 gas during the *i*-layer deposition. The μc-$Si_{1-x}Ge_x$:H *i*-layers were also grown in the UHV PECVD reactor under moderate low power (~20 mW/cm^2) and low pressure (1.5 Torr) conditions.

The solar cells were characterized by current density-voltage (*J-V*) and spectral response measurements under standard air mass 1.5 (100 mW/cm^2) and white-biased monochromatic light illuminations, respectively. Material investigations were carried out by using Raman scattering spectroscopy (λ=633 nm), Fourier transform infrared spectroscopy (FT-IR), secondary-ion mass spectroscopy (SIMS), transmission electron microscopy (TEM), x-ray diffraction (XRD), electron spin resonance (ESR) and Hall-effect measurements.

RESULTS AND DISCUSSION

μc-Si:H solar cells prepared at high deposition rates

Table I summarizes the illuminated *J-V* parameters of the μc-Si:H *p-i-n* solar cells prepared at various *i*-layer deposition rates. Sample 1 is a standard cell prepared at a low deposition rate (0.8 nm/s), and sample 2 is a cell prepared at an increased deposition rate of 2.0 nm/s by increasing discharge power (~100 mW/cm^2) and SiH_4 flow rate, under a constant gaseous pressure of 4 Torr. For high-deposition-rate (>2.0 nm/s) solar cells, we have also performed μc-Si:H deposition at higher pressures of 7 Torr (sample 3) and 8 Torr (sample 4), while the other deposition parameters such as discharge power and electrode gap were simultaneously adjusted to keep the SiH_4-depletion condition and yield the similar film crystallinity (Raman crystallinity: $I_{520cm^{-1}}/I_{480cm^{-1}} \sim 5\text{-}7$). It should be mentioned that the optimum range of deposition parameters becomes narrower under such high-pressure conditions. In particular, the uniformity of the electrode gap is found to be very critical in obtaining the uniform distribution of the material properties, even though the material is prepared in a small-area laboratory deposition system (electrode size: 6 inch diameter). In Table I, all solar cell parameters, i.e., short-circuit current density (J_{sc}), open-circuit voltage (V_{oc}), fill factor (FF) and conversion efficiency (η) are found to show strong dependence on deposition rate for samples prepared at low pressure (4 Torr). In particular, J_{sc} decreases drastically when deposition rate is increased from 0.8 to 2.0 nm/s. From the spectral response measurement, the poor J_{sc} of sample 2 is due to the low quantum efficiencies (QEs) in the long wavelength region (>600 nm), as will be shown later. As for sample 3, however, a remarkable improvement in the infrared response is observed when applying the HPD condition (7 Torr), leading to high J_{sc} as well as high efficiency comparable to those obtained at low deposition rates. After the optimization of electrode gap and SiH_4 flow rate, we obtained η~8% at a pressure of 8 Torr at an increased deposition rate of 3.0 nm/s (sample 4). Further progress in solar cell efficiency has been achieved by the improvement of TCO/*p* and *p*/*i* interfaces. As a result, we obtained a 9.1% efficient solar cell at a deposition rate

Table I. Illuminated *J-V* parameters of the μc-Si:H *p-i-n* solar cells (0.25 cm^2) for *i*-layers deposited at different deposition rates (R_d) under low and high-pressure conditions.

Sample	R_d (nm/s)	Pressure (Torr)	J_{sc} (mA/cm^2)	V_{oc} (V)	*FF*	η (%)
1	0.8	4.0	22.2	0.527	0.71	8.3
2	2.0	4.0	15.5	0.467	0.62	4.5
3	2.1	7.0	22.3	0.530	0.69	8.2
4	3.0	8.0	22.1	0.530	0.67	7.9
5	2.3	9.0	23.7	0.528	0.73	9.1

of 2.3 nm/s (sample 5).

It has been revealed that the HPD condition provides not only high efficiencies but also an excellent durability against the post-deposition oxidation in the μc-Si:H absorber layer. Figure 1 shows the QE spectra of the high-deposition-rate (~2 nm/s) μc-Si:H solar cells prepared at pressures of 4 and 9 Torr as a function of the duration of air exposure. In this experiment, samples were stored in the dark atmosphere at room temperature. In Fig. 1, the solar cell prepared at high pressure (9 Torr) remains completely stable upon prolonged air exposure. In contrast, we observed a significant cell degradation for the μc-Si:H solar cell prepared at low pressure (4 Torr). The pronounced QE reduction is seen in the long wavelength region ($\lambda > 600$ nm), which results in a reduction of J_{sc} by more than 50 % with respect to the initial value. The QE spectra exhibit a characteristic bend at λ~600-700 nm, which are similar to those reported earlier in the μc-Si:H solar cells contaminated with oxygen [17]. Although the degraded J_{sc} can be slightly recovered upon annealing (~200°C), the changes in solar cell parameters are not reversible unlike the light-induced degradation of a-Si:H solar cells. These results reveal that the atmospheric impurity incorporation takes place during the air exposure, which gives rise to an irreversible degradation for the μc-Si:H solar cells prepared at high rates under low-pressure conditions.

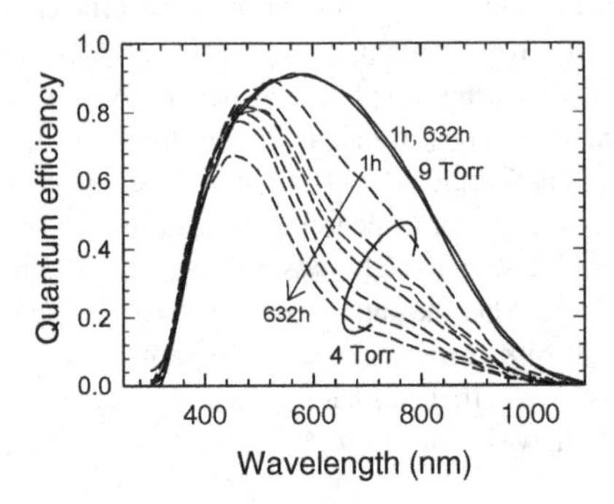

Figure 1. Influence of air exposure on QE spectra of the μc-Si:H solar cells prepared at different deposition pressures of 4 Torr (dashed lines) and 9 Torr (solid lines).

Figure 2 shows the IR absorption spectra of the μc-Si:H films grown at different deposition pressures of (a),(b) 4 Torr and (c), (d) 9 Torr. The FT-IR measurements were carried out after exposing samples to room air for various durations [18]. In Fig. 2(b), the low-grade μc-Si:H film grown at 4 Torr exhibits small narrow peaks at 2083 cm^{-1} and 2101 cm^{-1} which are associated with the monohydride (SiH) and dihydride (SiH_2) stretching modes at crystalline silicon surface, respectively. The appearance of such narrow peaks indicates that there exist micro-pores or

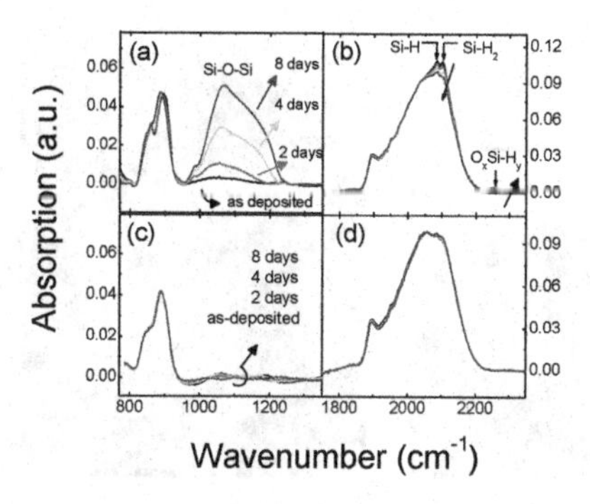

Figure 2. IR absorption spectra of the μc-Si:H films grown at different deposition pressures of (a),(b) 4 Torr and (c), (d) 9 Torr.

voids at grain boundaries and their surfaces are passivated by hydrogen. These narrow peaks disappear in several days while the $O_xSi\text{-}H_y$ mode (2250 cm^{-1}) and Si-O-Si stretching mode (1000-1300 cm^{-1}) appear instead. These results indicate that atmospheric impurities diffuse via the porous grain boundaries after the air break and the hydrogenated crystalline grain boundaries are replaced with the oxide surface. Vepřek *et al* have reported a similar post-deposition oxidation behavior in μc-Si:H films grown by chemical transport technique [19]. In contrast, for the device-grade μc-Si:H film grown at high pressure shown in Figs. 2 (c) and (d) , the narrow $Si\text{-}H_n$ stretching modes are absent and no post-deposition oxidation is observed over the prolonged air exposure. Based on these results, we conclude that the device-grade μc-Si:H films reflect compact grain boundaries that are most likely passivated by amorphous tissues rather than by hydrogen atoms.

TEM and XRD studies provide additional support to the above conclusions. Figure 3 shows the bright-field plane-view TEM images and the corresponding XRD patterns of the same μc-Si:H samples evaluated in the FT-IR measurement. As shown in Fig. 3(a), the low-grade μc-Si:H film grown at low pressure (4 Torr) is composed of the fine crystalline grains with typical grain diameters as small as 30 nm. The grain boundaries are clearly identified by the presence of the narrow cracks between isolated grains. In contrast, as shown in Fig. 3(b), the device-grade μc-Si:H film grown at high pressure (9 Torr) exhibits a denser grain arrangement with larger grain size distributions (30-50 nm). Such grain growth appears to form large conglomerates (~0.5-1 μm). These TEM observations are consistent with the XRD results. In XRD patterns, a structural change occurs from random to (220) preferential orientation and the grain size estimated from the line width of the (220) diffraction peak increases from 14 to 25 nm as the deposition pressure increases. A similar microstructural dependence of solar cell performance has been reported for low-deposition-rate (<0.5 nm/s) samples where the (220) preferential

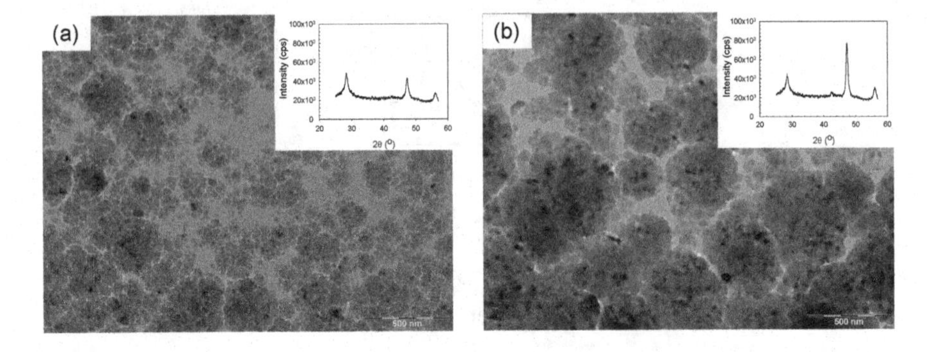

Figure 3. Bright-field plane-view TEM images of the μc-Si:H films prepared at deposition pressures of (a) 4 Torr and (b) 9 Torr. The corresponding XRD patterns are also shown.

orientation is a necessary condition in attaining high-efficiency μc-Si:H solar cells [20].

μc-$Si_{1-x}Ge_x$:H solar cells

Figure 4 shows the illuminated *J-V* parameters of the μc-$Si_{1-x}Ge_x$:H *p-i-n* solar cells as a function of Ge content in the *i*-layer. The composition of the *i*-layer was changed between x=0 and 0.27 by varying GeH_4 gas flow rate ([GeH_4]) from 0 to 0.42 sccm at a fixed SiH_4 flow rate ([SiH_4]) of 5.4 sccm. In this composition range, the deposition rate of the *i*-layer increased from

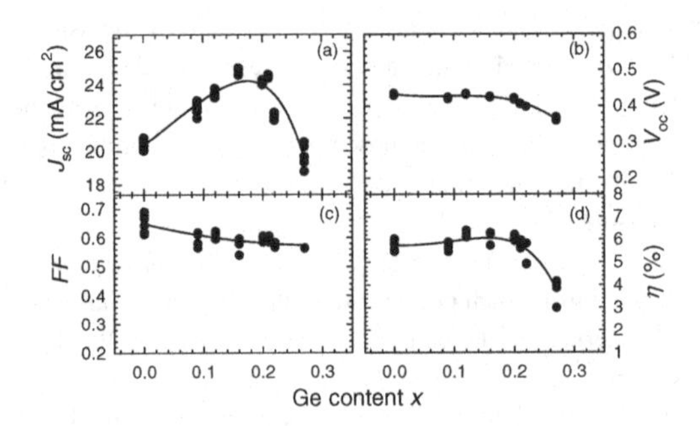

Figure 4. *J-V* parameters of the 1-μm-thick μc-$Si_{1-x}Ge_x$:H *p-i-n* solar cells (1 cm^2) under 1-sun illumination condition (AM1.5, 100 mW/cm^2) as a function of Ge content in the *i*-layer.

0.26 to 0.37 nm/s with increasing GeH_4 flow rate while the thickness was kept constant at 1.0 μm. It should be addressed that Ge contents in the films are larger by a factor of 4 than the gas-phase GeH_4 concentrations, $[GeH_4]/([SiH_4]+[GeH_4])$, indicating the efficient gas utilization of GeH_4 than for SiH_4. In Fig. 4(a), J_{sc} increases monotonically with increasing Ge content in the composition range between x=0 and ~0.2. The maximum J_{sc} of ~25 mA/cm^2 is obtained for x=0.15-0.2, which is larger by about 5 mA/cm^2 than J_{sc} of the μc-Si:H (x=0) solar cells. As will be shown later, the increase in J_{sc} comes from the enhancement in the QEs in the infrared wavelengths. With further increasing Ge content (x>0.2), however, J_{sc} rapidly drops to lower values. In Figs. 4(b) and (c), on the other hand, the solar cells show a gradual decrease in V_{oc} and FF with increasing Ge content up to x~0.2. In particular, the decrease in V_{oc} for x>0.2 is much greater than the band gap narrowing observed in unstrained bulk crystalline $Si_{1-x}Ge_x$ alloys [21], indicating that the solar cell performance is dominated by the charge carrier recombination in the regime of large Ge contents. As a result, as shown in Fig. 4(d), efficiency is almost independent of alloy composition between x=0 and ~0.2 whereas it falls drastically with further Ge incorporation. So far, we have achieved an efficiency of 6.33% (J_{sc}=24.1 mA/cm^2, V_{oc}=0.427, FF=0.616) at x=0.2. The light soaking test revealed no light-induced degradation for the μc-$Si_{0.8}Ge_{0.2}$:H solar cells. Further efficiency improvement has been achieved at smaller Ge contents (x~0.1) with improved front ZnO:Ga layer [15] and back reflector. While reducing the Ge content from x=0.2 to 0.1, the relatively large J_{sc} (25.5 mA/cm^2) is maintained and an efficiency of 8.2% is obtained using a 1.1-μm-thick μc-$Si_{0.9}Ge_{0.1}$:H i-layer.

Figure 5 shows the QE spectra of the 1-μm-thick μc-$Si_{1-x}Ge_x$:H p-i-n solar cells with different alloy compositions (x=0.12, 0.17, 0.20) in the i-layer. The typical QE spectra of the 1- and 2-μm-thick μc-Si:H solar cells are also shown for comparison. In Fig. 5, the spectral sensitivity of solar cell extends systematically towards longer wavelengths as the Ge content increases in the i-layer in the range of x from 0 to 0.2. In particular, the μc-$Si_{1-x}Ge_x$:H solar cells

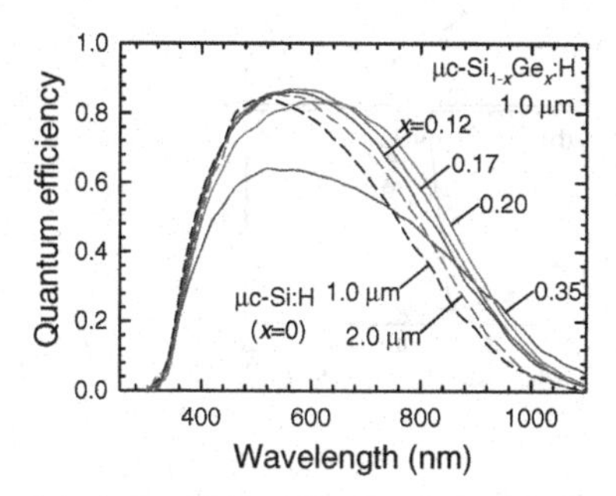

Figure 5. QE spectra of the 1-μm-thick μc-$Si_{1-x}Ge_x$:H single junction solar cells for various Ge contents (x=0.12, 0.17, 0.20) in the i-layer (solid lines). QE spectra of the μc-Si:H solar cells comprising 1- and 2-μm-thick i-layers are also shown for comparison (dashed lines).

(0.12≤x≤0.2) exhibit infrared sensitivities even higher than those of the double-thickness μc-Si:H solar cell. For example, the QE of the μc-$Si_{0.8}Ge_{0.2}$:H solar cell at λ=900 nm is higher by a factor of 1.4 than that of the double-thickness μc-Si:H solar cell, although the absorption coefficient at this wavelength increases by only a factor of 1.2 when incorporating 20 at.% of Ge. Our internal quantum efficiency (IQE) analysis suggests that the absorption enhancement in the *i*-layer leads to the less parasitic absorption loss in the front and rear electrodes because the relative absorption in the *i*-layer becomes larger than the absorption in electrodes during the multiple internal reflection of lights in the textured solar cells. In Fig. 5, on the other hand, a decrease in QEs in the short wavelength region (λ<650 nm) is observed when increasing Ge content from x=0.17 to 0.2. The Ge incorporation larger than x=0.2 results in a further drop in the QE peak, which is responsible for the decrease in J_{sc} for x>0.2 shown in Fig. 4(a). The ESR measurements on μc-$Si_{1-x}Ge_x$:H [22] revealed that degraded solar cell performance for x>0.2 can be partly attributed to the increased neutral Ge-dangling-bond defects that act as recombination centers of photogenerated carriers. In addition, we found the strong injection-level-dependent *p-i* interface recombination in the μc-$Si_{1-x}Ge_x$:H *p-i-n* solar cells when x>0.35 [16]. Spectral response measurements under various illumination conditions revealed that the built-in field in the *i*-layer is highly distorted by the negative space charge generated near the *p-i* interface.

In our previous study, it has been suggested that undoped μc-$Si_{1-x}Ge_x$:H contains numerous concentration of acceptorlike states [16]. To gain insights into *p*-type nature of Ge, we investigate the electronic properties of pure Ge:H films deposited from the GeH_4-H_2 glow discharge. Figure 6 shows (a) ESR spin density and (b) carrier concentration of the μc-Ge:H films as a function of the Raman crystallinity factor I_R defined as an intensity ratio $I_{300cm^{-1}}/I_{270cm^{-1}}$.

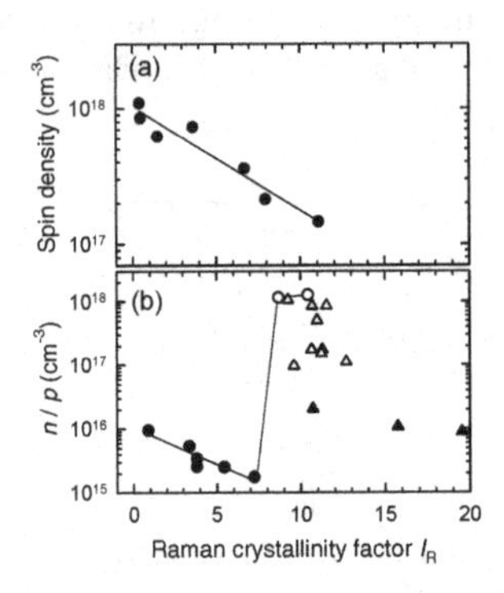

Figure 6. (a) ESR spin density and (b) electron (closed)/hole (open) concentration of the Ge:H films plotted versus the Raman crystallinity factor I_R. The circles and triangles represent the Ge:H films prepared by using GeH_4-H_2 and GeH_4-H_2-CO_2 gas mixtures, respectively.

In Fig. 6(a), it is found that the higher crystalline fraction of Ge results in smaller ESR spin densities. With further increasing crystalline fraction (I_R>11), the ESR signal intensity falls below the detection limit. As shown in Fig. 6(b), on the other hand, μc-Ge:H films exhibit *n*-type conduction in the less crystalline region while the electron concentration decreases gradually with I_R and the majority carrier changes from electrons to holes at I_R~8. At this electrical transition, the hole concentration increases by three orders of magnitude. These results indicate that (i) an ESR signal stems only from the neutral Ge dangling bonds in amorphous phase, and (ii) an increase of crystalline fraction shifts the Fermi level toward the valence band. Based on above, it is plausible that the Ge dangling bonds at crystalline grain boundaries act as negatively charged acceptors and thus generate free holes in the crystalline grains. Quite interestingly, we find that an introduction of CO_2 gas during the μc-Ge:H deposition drastically reduces the hole concentration. The triangles in Fig. 6(b) are the data obtained by doping of CO_2 gas with different CO_2/GeH_4 ratios between 0.18 and 3.9. As the CO_2/GeH_4 ratio increases, the hole concentration decreases by two orders of magnitude and the carrier sign converts from *p*- to *n*-type when CO_2/GeH_4>1.5. We attribute this effect to the incorporation of oxygen rather than carbon atoms because the carbon-contained μc-Ge:H is also reported to show strong *p*-type conduction [23].

Figure 7 shows the QE spectra of the μc-$Si_{1-x}Ge_x$:H *p-i-n* solar cells [(a): x=0.3, t_i=1.0 μm and (b): x=0.1, t_i=3.4 μm)] for different gas-phase CO_2 concentration, C_{CO2}=[CO_2]/([SiH_4]+[GeH_4]), during the deposition of the *i*-layer. In Fig. 7(a), the QE spectrum of the μc-$Si_{0.7}Ge_{0.3}$:H solar cell is markedly improved over the entire spectral region by introducing C_{CO2}=2.4%. With further increasing C_{CO2}, however, the QEs for λ>600 nm decrease down to below the original values. As shown in Fig. 7(b), a similar effect is seen for solar cells with a smaller Ge content (x=0.1) but a thicker *i*-layer thickness (t_i=3.4 μm). By introducing C_{CO2}=0.3 %, the enhancement of QEs particularly in the short wavelength region (λ<700 nm) is attained. A little reduction is visible in the infrared response, indicating that the optimum C_{CO2}

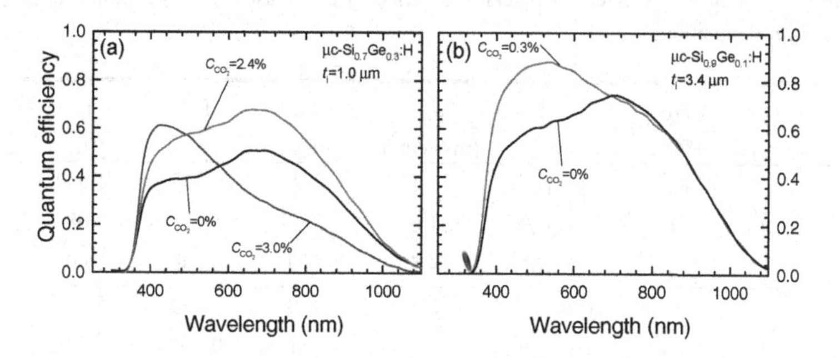

Figure 7. QE spectra of the μc-$Si_{1-x}Ge_x$:H *p-i-n* solar cells [(a) x=0.3, t_i=1.0 μm and (b) x=0.1, t_i=3.4 μm] for different CO_2 concentrations during the *i*-layer deposition.

lies below 0.3% in this solar cell. In Table II, the J-V parameters of these solar cells are summarized together with the oxygen concentrations measured by SIMS. For μc-$Si_{0.7}Ge_{0.3}$:H solar cells, the oxygen concentration that gives the highest J_{sc} is found in the order of ~10^{20} cm^{-3}. In this solar cell, J_{sc} increases by 5 mA/cm^2 with respect to the cell without intentional oxygen incorporation. In addition, we observed the slight improvement in V_{oc} and FF. The oxygen incorporation is also effective for μc-$Si_{0.9}Ge_{0.1}$:H solar cells, while the optimum oxygen concentration is by an order of magnitude smaller than that for μc-$Si_{0.7}Ge_{0.3}$:H solar cells.

From the ESR and Hall-effect experiments on pure μc-Ge:H films, it is suggested that oxygen incorporation provides compensation of the Ge-dangling-bond acceptors. When the oxygen is incorporated into μc-$Si_{1-x}Ge_x$:H alloys, oxygen is expected to react with Si preferentially because of the preference of silicon oxides formation against the germanium oxides [24]. As oxygen acts as a donor in silicon, we conclude that the improved device performance by oxygen incorporation is mainly occurred by the compensation of the Ge-dangling-bond acceptors by oxygen donors that mostly bound to Si atoms. In QE spectra shown in Fig. 7(a), the solar cells with excess oxygen concentrations exhibit significant recombination loss in the long wavelength region, as often observed in μc-Si:H solar cells contaminated with oxygen. Kilper *et al.* reported a critical oxygen concentration of ~10^{19} cm^{-3} above which the infrared response of μc-Si:H solar cell is degraded drastically [25]. Furthermore, no beneficial effect has been found in incorporating oxygen into μc-Si:H solar cells even if the oxygen concentration is below the critical level. Thus, the improvement of device performance by oxygen incorporation is uniquely found for μc-$Si_{1-x}Ge_x$:H solar cells unlike what was observed in other silicon-based thin film solar cells.

Table II. J-V parameters of the *p-i-n* junction μc-$Si_{1-x}Ge_x$:H solar cells (1 cm^2) for different Ge contents measured under 1-sun illumination condition (AM1.5, 100 mW/cm^2). The oxygen concentration [O] in the μc-$Si_{1-x}Ge_x$:H *i*-layers was varied by gas phase CO_2 concentration, C_{CO2}, during the *i*-layer deposition.

x	t_i (μm)	C_{CO2} (%)	[O] (cm^{-3})	J_{sc} (mA/cm^2)	V_{oc} (mV)	FF	η (%)
0.3	1.0	0	1.0×10^{18}	16.1	315	0.377	1.91
		2.4	1.0×10^{20}	21.6	330	0.432	3.08
		3.0	1.2×10^{20}	14.5	319	0.346	1.60
0.1	3.4	0	3.0×10^{17}	22.7	449	0.575	5.86
		0.15	5.5×10^{18}	27.5	451	0.504	6.24
		0.3	1.1×10^{19}	26.5	454	0.528	6.35

CONCLUSIONS

We have studied the material properties of μc-Si:H and μc-$Si_{1-x}Ge_x$:H films grown by UHV PECVD and relate them to their device performance of *p-i-n* solar cells. For μc-Si:H, it is found that the deposition pressure is a key parameter that determine the microstructural properties of the μc-Si:H prepared under high-rate deposition condition. The high efficiency μc-Si:H solar cells are found to obtain when the material exhibits compact grain boundaries passivated by the surrounding amorphous tissues. For μc-$Si_{1-x}Ge_x$:H, we have demonstrated that μc-$Si_{1-x}Ge_x$:H single junction device (x~0.2) provides enhanced infrared response at the reduced absorber thickness, offering the potential advantage over μc-Si:H for bottom cell application. Unlike the μc-Si:H solar cells, the oxygen incorporation in the μc-$Si_{1-x}Ge_x$:H solar cells is found to be effective in improving the solar cell parameters. The improved device performance is ascribed to the enhanced built-in field in the *i*-layer as a consequence of the compensation of the Ge-dangling-bond negative space charge by the positively-charged oxygen donors.

ACKNOWLEDGMENTS

The authors are grateful to A.H.M. Smets for the research collaboration in the microstructure analysis of μc-Si:H. They also thank H. Jia for TCO development, C.W. Chang for ESR measurements, T. Takada, T. Asakusa and K. Mizuno for help in sample preparation and characterization. They acknowledge A. Matsuda, H. Sai, H. Fujiwara, A. Masuda, Y. Takeuchi and M. Isomura for research support and useful discussion. This work was supported by the New Energy and Industrial Technology Development Organization (NEDO), Japan.

REFERENCES

1. J. Meier, S. Dubail, R. Flückiger, H. Keppner and A. Shah, Appl. Phys. Lett. **65**, 860 (1994).
2. K. Yamamoto, IEEE Trans. Electron Devices **46**, 2041 (1999).
3. K. Saito, M. Sano, A. Sakai, R. Hayashi and K. Ogawa, *Tech. Dig. 20th Int. PVSEC, Jeju, Korea* (2001) p. 429.
4. Y. Nasuno, M. Kondo and A. Matsuda, Jpn. J. Appl. Phys. **41**, 5912 (2002).
5. O. Vetterl, R. Carius, L. Houben, C. Scholten, M. Luysberg, A. Lambertz, F. Finger and H. Wagner, Mater. Res. Soc. Symp. Proc. **609**, A15.2.1 (2000).
6. M. Kondo, M. Fukawa, L. Guo and A. Matsuda, J. Non-Cryst. Solids **266-269**, 84 (2000).
7. F. Finger, U. Kroll, V. Viret, A. Shah, W. Beyer, X. -M. Tang, J. Weber, A. Howling and Ch. Hollenstein, J. Appl. Phys. **71**, 5665 (1992).
8. T. Matsui, M. Kondo and A. Matsuda, Jpn. J. Appl. Phys. **42**, L901 (2002).
9. J. Bailat, L. Fesquet, J-B. Orhan, Y. Djeridane, B. Wolf, P. Madliger, J. Steinhauser, S. Benagli, D. Borrello, L. Castens, G. Monteduro, M. Marmelo, B. Dehbozorghi, E. Vallat-Sauvain, X. Multone, D. Romang, J-F. Boucher, J. Meier, U. Kroll,M. Despeisse, G. Bugnon,

C. Ballif, S. Marjanovic, G. Kohnke, N. Borrelli, K. Koch, J. Liu, R. Modavis, D. Thelen, S. Vallon, A. Zakharian and D. Weidman, *Proc. of 5th WCPEC, Valencia, Spain* (2010) p. 2720.

10. G. Ganguly, T. Ikeda, T. Nishimiya, K. Saitoh, M. Kondo and A. Matsuda, Appl. Phys. Lett. **69**, 4224 (1996).
11 R. Carius, J. Fölsch, D. Lundszien, L. Houben and F. Finger, Mater. Res. Soc. Symp. Proc. **507**, 813 (1998).
12. M. Isomura, K. Nakahata, M. Shima, S. Taira, K. Wakisaka, M. Tanaka and S. Kiyama, Sol. Energy Mater. & Sol. Cells **74**, 519 (2002).
13. T. Matsui, T. Takada, C. W. Chang, M. Isomura, H. Fujiwara and M. Kondo, Appl. Phys. Express **1**, 031501 (2008).
14. T. Matsui, H. Jia and M. Kondo, Prog. Photovolt: Res. Appl. **18**, 48 (2010).
15. T. Matsui, H. Jia, M. Kondo, K. Mizuno, S. Tsuruga, S. Sakai and Y. Takeuchi, *Proc. of 35th IEEE-PVSC, Hawaii,* (2010) p. 000311.
16. T. Matsui, C.W. Chang, M. Kondo, K. Ogata and M. Isomura, Appl. Phys. Lett. **91**, 102111 (2007).
17. P. Torres, J. Meier, R. Flückiger, U. Kroll, J. A. Anna Selvan, H. Keppner, A. Shah, S. D. Littlewood, I. E. Kelly and P. Giannoulès, Appl. Phys. Lett. **69**, 1373 (1996).
18. A.H.M. Smets, T. Matsui and M. Kondo, Appl. Phys. Lett. **92**, 033506 (2008).
19. S. Vepřek, Z. Iqbal, R. O. Kühne, P. Capezzuto, F-A. Sarott and J. K. Gimzewski, J. Phys. C: Solid State Phys. **16**, 6241 (1983).
20. T. Matsui, M. Tsukiji, H. Saika, T. Toyama and H. Okamoto, Jpn. J. Appl. Phys. **41**, 20 (2002).
21. R. Braunstein, A. R. Moore, and F. Herman, Phys. Rev. **109**, 695 (1958).
22. C.W. Chang, T. Matsui and M. Kondo, J. Non-Cryst. Solids **354**, 2365 (2008).
23. Y. Yashiki, S. Miyajima, A. Yamada and M. Konagai, Jpn J. Appl. Phys. **46**, 2865 (2007).
24. S. Margalit, A. Bar-Lev, A. B. Kuper, H. Aharoni and A. Neugroschel, J. Cryst. Growth **17**, 288 (1972).
25. T. Kilper, W. Beyer, G. Bräuer, T. Bronger, R. Carius, M. N. van den Donker, D. Hrunski, A. Lambertz, T. Merdzhanova, A. Mück, B. Rech, W. Reetz, R. Schmitz, U. Zastrow and A. Gordijn, J. Appl. Phys. **105**, 074509 (2009).

Mater. Res. Soc. Symp. Proc. Vol. 1321 © 2011 Materials Research Society
DOI: 10.1557/opl.2011.805

V_{oc} Saturation Effect in High-temperature Hydrogenated Polycrystalline Silicon Thin-film Solar Cells

Hidayat Hidayat[1, 2,*], Per I. Widenborg[2, 1], and Armin G. Aberle[2, 1]
[1] Electrical and Computer Engineering, National University of Singapore, Singapore
[2] Solar Energy Research Institute of Singapore, National University of Singapore, Singapore
*Electronic mail: hidayat@nus.edu.sg

ABSTRACT

Hydrogenation of polycrystalline silicon thin-film solar cells is performed to improve the one-sun open-circuit voltage (V_{oc}) of the device. V_{oc} is found to increase linearly with increasing hydrogenation temperature and then saturates. For planar and textured samples, the V_{oc} saturates at about 340 °C and 307 °C respectively. The low hydrogenation temperature helps to lower thermal budget during industrial process. Arrhenius plot of V_{oc} prior to the saturation shows that the textured samples have lower activation energies than the planar sample. The activation energies of samples 188 (planar), 788 (textured) and 888 (textured) are 1.31 eV, 0.86 eV and 0.92 eV, respectively. The lower activation energies of the textured samples could be due to the shorter diffusion thickness and the increased surface area that is exposed to the hydrogen plasma.

INTRODUCTION

Polycrystalline silicon (poly-Si) thin-film solar cells have the potential of achieving a conversion efficiency of more than 13% using a simple solar cell structure. The highest efficiency so far is 10.5%, achieved by CSG Solar [1]. This technology has the potential to reach low fabrication costs due to several advantages, such as the use of relatively inexpensive large-area glass substrates and monolithic series interconnection of the solar cells to form a solar module. In addition, the abundance and non-toxicity of silicon make it a promising candidate for large-volume production.

However, compared to silicon wafer solar cells, the poly-Si thin-film material quality suffers from a high defect density, such as dangling bonds at the grain boundaries. One way to improve the quality of poly-Si thin-film material is to passivate these dangling bonds by diffusing hydrogen into the film, a method that has been found to improve the device performance of transistors and solar cells. This process is termed as hydrogenation. Deuterium, an isotope of hydrogen is sometimes used as well, because it is easily detected in secondary-ion mass spectroscopy (SIMS) measurements. The terms deuteration and hydrogenation are sometimes interchangeably used, as there is no significant difference found in terms of diffusion, defect passivation and defect generation [2].

Two main factors that affect the hydrogenation process are the hydrogenation temperature and hydrogenation time. It was found in the experiment done by Nickel et al. [3] that with increasing temperature, the deuterium diffuses deeper into the bulk of the film and the surface concentration of deuterium drops from $1x10^{21}$ cm^{-3} at 250 °C to $2.6x10^{19}$ cm^{-3} at 450 °C. Decreasing the hydrogenation time from 30 minutes to 10 minutes also reduces the peak deuterium concentration, by about two orders of magnitude at a temperature of 350 °C. Extensive studies were done to reveal the behavior of the diffusion mechanism and concentration profiles using the SIMS characterization method. However, little has been studied on the impact of hydrogenation on the device performance, except for some works done by Terry et al. [4] and Gorka et al. [5] on different types of solar cells and plasma processes.

In this paper, the impact of the hydrogenation temperature on the V_{oc} of the devices is investigated. Hydrogen atoms and ions are generated via a microwave-powered plasma in a vacuum chamber, using a mixture of argon and hydrogen gases. Characterization of the samples uses the so-called Suns-V_{oc} method [6], which gives the one-sun V_{oc} and the pseudo fill factor (*pFF*) of the device.

EXPERIMENT

In this work, about 2 μm thick a-Si:H precursor diodes were deposited by PECVD onto 3.3 mm thick planar and textured borosilicate glass substrates, followed by solid phase crystallization (SPC) to form a polycrystalline diode. The glass texture was realized with the aluminum induced texturing (AIT) method [7]. The final sample structure is: glass/70 nm SiN/100 nm n^+ layer/ 2 μm p^- layer/ 100 nm p^+ layer. The sample is then heated to above 900 °C for a short period of time (rapid thermal annealing, RTA) to activate the dopants and anneal crystal defects. Then, the sample is hydrogenated to passivate a large fraction of the remaining defects. We are using an AK800 system from Roth and Rau, Germany, for hydrogenation. The hydrogen plasma generator has two linear microwave plasma sources. Typically, the V_{oc} of our devices is about 200 mV prior to hydrogenation and above 400 mV after a 30-minute hydrogenation process above 400 °C.

Prior to hydrogenation step, the sample is dipped into a 5% hydrofluoric acid solution to remove the thin layer of silicon oxide. The sample is then heated to the set temperature, followed by 5 minutes for stabilization of the glass temperature. Then, a mixture of argon and hydrogen gases is introduced into the chamber with flow rates of 30 and 90 sccm, respectively, creating a pressure of 4.0×10^{-2} mbar. The plasma sources are then turned on. The plasma is pulsed at 8/8 ms on/off time. The heating is stopped after 30 minutes of hydrogenation at plateau temperature, also known as hydrogenation temperature (T_H), to allow the sample to cool down before it is taken out of the chamber. The plasma generation continues during this cool-down period. The overall temperature profile of the hydrogenation process is shown in Figure 1.

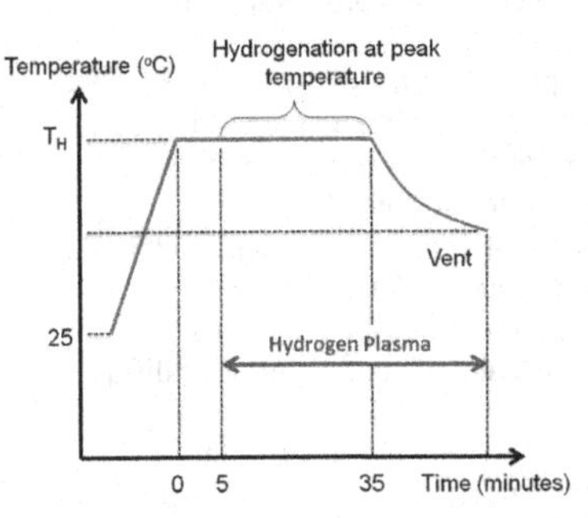

Figure 1. Temperature profile of the hydrogenation process. Also shown is the period during which the plasma is on.

The samples are characterized with the Suns-V_{oc} method, using a light pulse from a flash light. The decaying flash light intensity is measured with a calibrated reference solar cell, giving the light intensity (in suns) that falls onto the sample [6]. Five points are measured on each sample. The average V_{oc} and *pFF* values are then plotted as a function of the hydrogenation temperature. After the measurement, the sample is baked at 610 °C to drive out the hydrogen and thus making it re-usable for another hydrogenation experiment. After the 610 °C hydrogen drive-out step, the V_{oc} and the *pFF* of all investigated samples were roughly the same.

RESULTS and DISCUSSION

Three different samples were used for this study: One planar sample and two textured samples. Figure 2 shows the results obtained from these samples. It can be seen that the V_{oc} initially increases with increasing hydrogenation temperature and then saturates. The saturation voltage for each sample is obtained by averaging the V_{oc} data points that are within 5% of the maximum V_{oc} obtained. The data points prior to this V_{oc} saturation are fitted with a linear fit. The intersection between average saturated V_{oc} and the linear fit line gives the saturation temperature. For planar and textured samples, the V_{oc} saturates at about 340 ± 11 °C and 307 ± 9 °C, respectively. The V_{oc} of both planar and textured samples saturates at about 450 mV. This indicates that both sample types have similar film quality. The saturated V_{oc} is termed as V_{sat} and the temperature where it starts to saturate is termed as T_{sat}. The *pFF* profiles behave similarly as the V_{oc} profiles. The results are summarized in Table I. For sample 788, the V_{sat}, the linear fit line prior to V_{sat}, and T_{sat} are also shown in Figure 2.

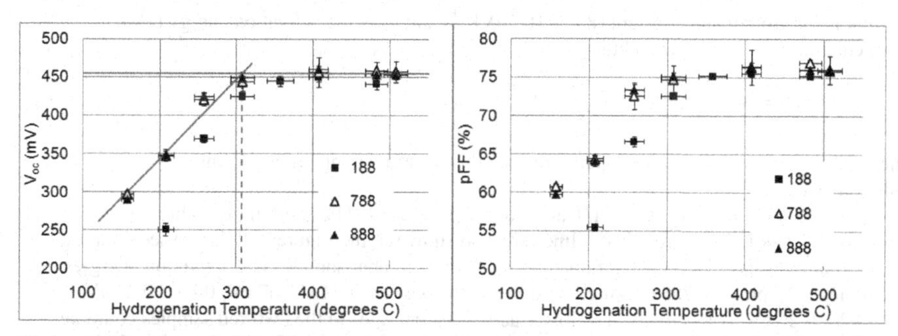

Figure 2. The V_{oc} and *pFF* of the samples versus the substrate temperature during hydrogenation (square symbols = planar sample, triangles = textured samples). The V_{sat}, linear fit line prior to V_{sat}, and T_{sat} of sample 788 are also shown.

Table I. Summary of the results obtained on the three investigated samples.

Sample name	Type	Area (cm^2)	Hydrogenation time (mins)	Activation Energy (eV)	V_{sat} (mV)	T_{sat} (°C)
188	Planar	100	30	1.31 ± 0.67	445	340 ± 11
788	Textured	100	30	0.86 ± 0.25	454	307 ± 9
888	Textured	100	30	0.92 ± 0.26	454	301 ± 9

The hydrogen concentration introduced by a plasma process can typically exceed the concentration of dangling bonds by two orders of magnitude. In addition, passivation efficiency also increases with increasing temperature [8]. However, in Nickel's experiment the thickness of the film was 0.55 μm, while in the present work it is about 2 μm. It could well be that the p-n junction of our cells, sitting nearer to the glass side, may not be well passivated. Increasing the hydrogenation temperature further does not passivate these defects well. This could explain why increasing the hydrogenation temperature does not increase the V_{oc} further. However, this does not imply that the saturated V_{oc} is the limit for the achievable V_{oc}. Increasing the hydrogenation time as well as the concentration of hydrogen atoms in the plasma could also increase the passivation depth, which may result in an improved V_{oc}. It is also known that hydrogenation process introduces defects such as hydrogen platelets and silicon dangling bonds [9,10].

In a simple model of a p-n junction solar cell, the V_{oc} is proportional to the logarithm of the minority carrier lifetime in the cell's absorber layer. Since our absorber layer is lightly doped with boron, the corresponding lifetime is the electron lifetime. V_{oc} can then be approximated as

$$V_{oc}/V_t \propto \ln(\tau_e) \quad , \qquad (1)$$

where V_t is the thermal voltage (25.7 mV at 300 K) and τ_e is the electron lifetime.

The V_{oc} data points in Figure 2 prior to saturation can be fitted empirically with an Arrhenius plot, as described below.

$$lnY = lnY_o - \frac{E_a}{kT} \quad , \qquad (2)$$

where k is the Boltzmann constant (8.62×10^{-5} eVK^{-1}) and E_a is the activation energy (eV). From equations (1) and (2) we obtain

$$\frac{V_{oc}}{V_t} = A - \frac{E_a}{kT_H} \quad , \qquad (3)$$

where T_H is the hydrogenation temperature (i.e., the substrate temperature during hydrogenation).

The results of V_{oc}/V_t versus 1/T are plotted in Figure 3. The resulting E_a values are included in Table I. The Arrhenius fit lines are also shown in the Figure 3. It can be seen that the textured samples have lower activation energies than the planar sample. The activation energies of samples 188 (planar), 788 (textured) and 888 (textured) are 1.31 ± 0.67 eV, 0.86 ± 0.25 eV and 0.92 ± 0.26 eV, respectively. The lower activation energies in the textured samples could be explained by the increase in the surface area exposed to the hydrogen plasma. From the atomic force microscope (AFM) scan result of area 20 x 20 μm^2, the surface area increases by about 10%. The increase in surface area gives the hydrogen atoms more entry points to diffuse into the film. Another possible reason could be due to the decrease in diffusion path of atomic hydrogen to reach the junction [11]. Figure 4 helps to illustrate the abovementioned factors.

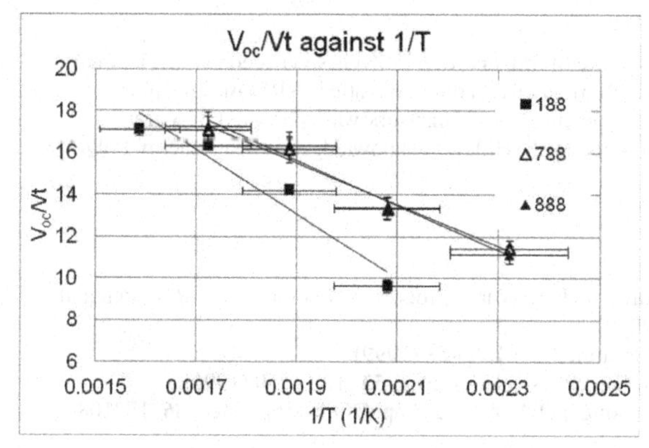

Figure 3. V_{oc}/V_t versus the inverse of the hydrogenation temperature (square symbols = planar sample, triangles = textured samples). The Arrhenius fit lines are also shown.

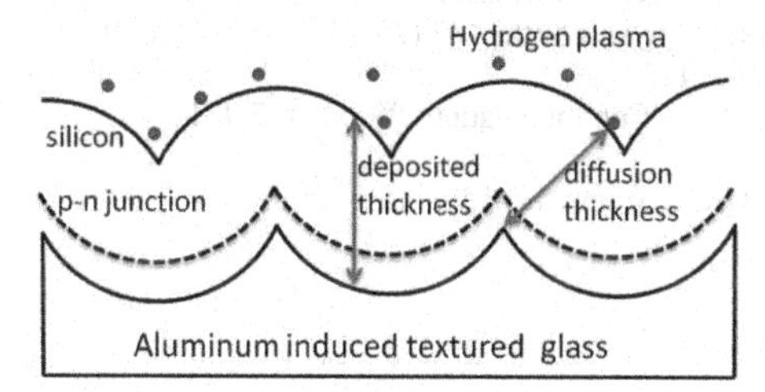

Figure 4. Illustration of deposited films on textured glass substrate. The shorter diffusion thickness compared to the deposited thickness could contribute to the better passivation of the p-n junction region and the emitter layer.

CONCLUSIONS

Hydrogenation temperatures of about 400 °C and a hydrogenation time of at least 30 minutes are sufficient to saturate the V_{oc}. These hydrogenation conditions are favorable for industrial processes that require high throughput and a low thermal budget. This temperature is also much lower than the strain point of the Borofloat glass used in this study (about 520 °C). Hydrogenation temperatures above the strain point of the glass could possibly degrade the material quality of the poly-Si film.

The V_{oc} data points prior to the saturation are fitted empirically with an Arrhenius plot. The results show that the textured samples have a lower activation energy than the planar sample. The activation energies of samples 188 (planar), 788 (textured) and 888 (textured) are 1.31 ± 0.67 eV, 0.86 ± 0.25 eV and 0.92 ± 0.26 eV, respectively. The lower activation energies of textured samples could be due to the shorter diffusion thickness and the increase in surface area exposed to the hydrogen plasma.

ACKNOWLEDGMENTS

The Solar Energy Research Institute of Singapore (SERIS) is sponsored by the National University of Singapore (NUS) and the National Research Foundation (NRF) of Singapore through the Singapore Economic Development Board. This work was sponsored by a grant (Clean Energy Research Program) from the NRF. Hidayat acknowledges a Clean Energy PhD scholarship from the NRF.

REFERENCES

1. M. Green, K. Emery, Y. Hishikawa and W. Warta, Progress in Photovoltaics: Research and Applications **18** (2), 144-150 (2010).
2. N. Nickel, Semiconductors and Semimetals **61**, 86-88 (1999).
3. N. Nickel, W. Jackson and J. Walker, Physical Review B **53** (12), 7750 (1996).
4. M. Terry, A. Straub, D. Inns, D. Song and A. Aberle, Applied Physics Letters **86**, 172108 (2005).
5. B. Gorka, B. Rau, K. Lee, P. Dogan, F. Fenske, E. Conrad, S. Gall and B. Rech, in 22nd European Photovoltaic Solar Energy Conference (Milan, Italy).
6. R. Sinton and A. Cuevas, Applied Physics Letters **69**, 2510 (1996).
7. P. Widenborg and A. Aberle, Advances in OptoElectronics **7** (2007).
8. N. Nickel, N. Johnson and W. Jackson, Applied Physics Letters **62** (25), 3285-3287 (2009).
9. N. Nickel, Semiconductors and Semimetals **61**, 123 (1999).
10. N. Nickel, G. Anderson and J. Walker, Solid State Communications **99** (6), 427-431 (1996).
11. M. Keevers, T. Young, U. Schubert and M. Green, (unpublished).

Mater. Res. Soc. Symp. Proc. Vol. 1321 © 2011 Materials Research Society
DOI: 10.1557/opl.2011.931

Improvement of Single-Junction a-Si:H Thin-Film Solar Cells Toward 10% Efficiency

P. H. Cheng, S. W. Liang, Y. P. Lin, H. J. Hsu, C. H. Hsu and C. C. Tsai
National Chiao Tung University, Hsinchu, Taiwan

ABSTRACT

The hydrogenated amorphous silicon (a-Si:H) single-junction thin-film solar cells were fabricated on SnO_2:F-coated glasses by plasma-enhanced chemical vapor deposition (PECVD) system. The boron-doped amorphous silicon carbide (a-SiC:H) was served as the window layer (p-layer) and the undoped a-SiC:H was used as a buffer layer (b-layer). The optimization of the p/b/i/n thin-films in a-Si:H solar cells have been carried out and discussed. Considering the effects of light absorption, electron-hole extraction and light-induced degradation, the thicknesses of p, b, n and i layers have been optimized. The optimal a-Si:H thin-film solar cell having an efficiency of 9.46% was achieved, with V_{OC}=906 mV, J_{SC}=14.42 mA/cm^2 and FF=72.36%.

INTRODUCTION

Hydrogenated amorphous silicon (a-Si:H) is one of the promising materials for thin-film solar cell applications [1]. The large bandgap and the high absorption coefficient make it suited for single- and multi- junction solar cells. The undoped layer (i-layer) in the a-Si:H thin-film solar cell is the key element to generate photo-current. However, the a-Si:H suffers certain degree of light-induced degradation (also known as Staebler-Wronski effect, SWE [2,3]) which leads to a reduction in efficiency. The degree of degradation can be reduced by the consideration of the absorber thickness and the film quality (e.g. microstructure parameter or conductivity).The dependence of the absorber thickness on efficiency is a trade-off between optical absorption and photo-generated carrier collection.

Moreover, the doped layer (n-layer, p-layer) in the a-Si:H thin-film solar cell should consider not only the electrical properties but also the optical influences. The boron-doped amorphous silicon carbide (a-SiC:H) has been served as the window layer (p-layer) for its wider bandgap and reasonable conductivity [4]. The thickness of the window layer should be minimized to reduce optical losses while keeping sufficient built-in field for carrier extraction [5, 6]. The band offset between p-layer (~2.0 eV) and i-layer (~1.75 eV), however, induces defects at the p/i interface [7]. This defective heterojunction acts as a recombination center which limits the performance of the cells due to recombination losses [8]. A high quality buffer layer has therefore been inserted to alleviate the bandgap difference as well as the defects between p/i interfaces.

In this study, experiments were carried out to improve the undoped a-Si:H and the doped a-SiC:H. Effect of the thicknesses of the p/i/n layers were also investigated to optimized the cell performance. Efforts were integrated to achieve a high efficiency solar cell.

EXPERIMENT

In this work, a-Si:H solar cells were fabricated by the radio frequency (27.12 MHz) plasma-enhanced chemical vapor deposition (PECVD) system. It is a commercial load-lock system with a maximal substrate size of 20 × 20 cm^2, and an equipped robotic arm for the transport of substrate from the load-lock chamber to the deposition chamber. All 3 layers (p-i-n) were prepared in a single deposition chamber, while the cells were patterned as 5 × 5 mm^2 for I-V measurement. A 3.9 mm-thick SnO_2:F-coated glass was used as a substrate. A superstrate configuration and a back reflector were used. The thin-films were prepared by using diborane (B_2H_6), phosphine (PH_3), silane (SiH_4), methane (CH_4) and hydrogen (H_2). The deposition conditions of a-Si:H solar cells were listed in Table 1.

Electrical and optical measurements were carried out to investigate the dark and photo conductivity (σ_{dark}, σ_{photo}) and the optical bandgap (E_g) of each layer. The microstructure parameter (R) was defined as the ratio of the integrated intensity of SiH_2 mode at 2100 cm^{-1} to that of the SiH_2 and SiH modes at 2100 cm^{-1} and 2000 cm^{-1}, as obtained from Fourier transform infrared spectroscopy (FTIR). The a-Si:H solar cells with were characterized by an I-V measurement system under AM1.5G illumination.

Table 1. The deposition conditions of the a-Si:H thin-film solar cells in PECVD

Substrate temperature (°C)	180-200
Electrode to substrate (E/S) distance (mm)	14-25
Pressure (Pa)	40-133
RF power (W)	20-30
SiH_4 flow rate (sccm)	40-50
H_2 flow rate (sccm)	40-100

RESULTS AND DISCUSSION

Figure 1 shows the σ_{dark}, σ_{photo} and R as a function of the electrode to substrate (E/S) distance of the undoped a-Si:H. As the E/S spacing increased from 14 mm to 25 mm, the peak located at 2100 cm^{-1} increased, leading to an increase in R from 0.04 to 0.29. The SiH_2 bonding configuration often indicates a weakly bonding structure associated with the light-induced degradation [9]. Silane molecules are ionized by non-elastic collision of high energy electrons before arriving at the surface of substrate. The larger E/S spacing increases the probability of the ionization, and more gas phase reactions leads to a less dense and defective structure [10]. The higher R corresponds to a higher SiH_2 content, and a more defective film. It seems that the quality of the thin-film could be further improved with even smaller E/S spacing, had it not been limited by the deposition system. Despite the dependence of the E/S spacing on σ_{dark}, and σ_{photo} showed no significant change, a better photo-to-dark conductivity ratio of 1.91×10^5 appeared as the E/S spacing was 14 mm.

The performance of a-Si:H solar cell as a function of i-layer thickness is illustrated in Fig. 2. There was no significant difference in V_{OC} with the increasing thickness of the i-layer. On the other hand, the J_{SC} increased from 12.9 mA/cm^2 to 15.17 mA/cm^2 as the thickness of the i-layer increased from 250 nm to 600 nm. The improvement in J_{SC} should be due to the further absorption by a thicker layer. The FF, however, degraded with the increase of the thickness due to the reduced electric field which led to the weak collection of photocurrent. The efficiency of the thicker absorbers could possibly be further improved, since the doped layers were optimized for 300 nm-thick device so far. Considering the light-induced degradation, an i-layer thickness of 300 nm was selected for further optimization.

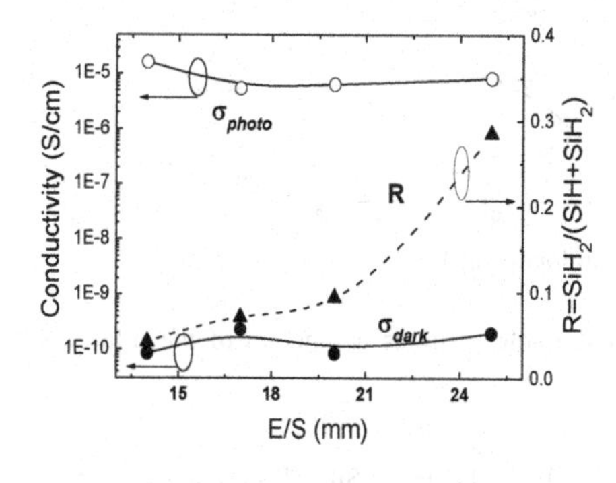

Figure 1. The dependence of the electrode-to-substrate (E/S) spacing on σ_{dark}, σ_{photo} and microstructure parameter (R).

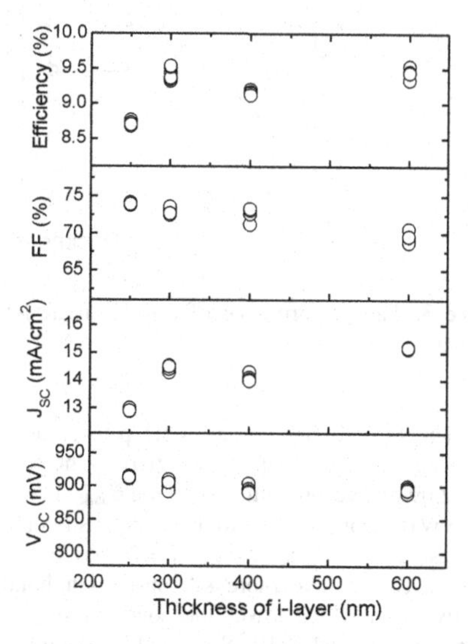

Figure 2. The performance of a-Si:H single-junction solar cells as a function of i-layer thickness.

The n-type a-Si:H located in the back side of the device has less importance as compared to the window layer. Nevertheless it still requires optimization to have better doping level and conductivity to reduce the thickness, as well as to reduce the optical absorption. For the deposition of n-type a-Si:H, the PH_3-to-SiH_4 flow rate ratio was changed from 2×10^{-3} to 5×10^{-3}. It was found that the σ_{dark} and E_g had no significant change in such a range of flow rate ratio. The E_g and σ_{dark} were around 1.75 eV and 10^{-2} S/cm, respectively.

Furthermore, the effect of the n-layer thickness on cell performance was investigated, as can be seen in Fig. 3. There is no significant difference observed as the n-layer thickness increased from 20nm to 30nm. Nevertheless, decreases in V_{OC}, FF and efficiency are shown as the n-layer thickness decreases to 15nm. Such a reduction should be due to the decreased build-in field caused by the insufficient n-layer thickness. An optimized n-type a-Si:H with conductivity of 10^{-2} S/cm and thickness of 20 nm was suggested for the a-Si:H cell having 300 nm absorber.

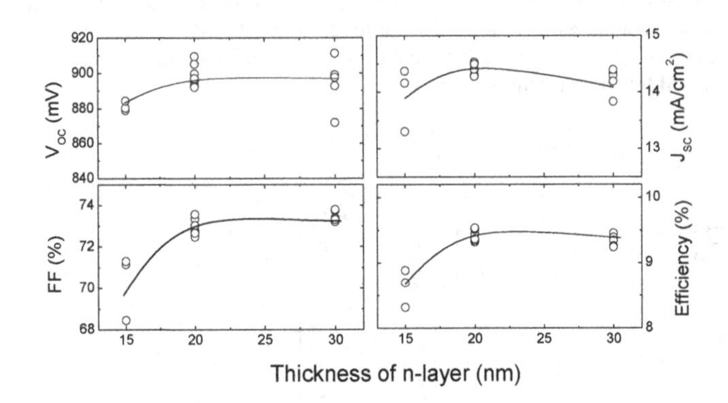

Figure 3. The performance of a-Si:H single-junction solar cells as a function of the n-layer thickness.

The E_g and σ_{dark} of a-SiC:H p-layer as a function of CH_4-to-SiH_4 flow rate ratio are illustrated in Fig. 4. As the flow rate ratio increased from 1.27 to 3.27, the E_g increased from 1.97 eV to 2.09 eV, accompanied with the σ_{dark} decreased from 2.9×10^{-5} S/cm to 1.9×10^{-7} S/cm. More carbon was incorporated into amorphous silicon film with increasing concentration of CH_4 in the gas phase. However, the conductivity decreased with increasing carbon content in the amorphous silicon network. The more silicon-carbon bonding corresponds to a higher E_g, but a more defective film which lower the conductivity. As the CH_4-to-SiH_4 flow rate ratio was 1.27, a better σ_{dark} around 2.9×10^{-5} S/cm and E_g around 1.97 eV were achieved for the a-SiC:H film.

The bandgap discontinuity between the two different energy bands induces defects at the p/i interface. To improve the carrier transport barrier and reduce the recombination losses, an a-SiC:H buffer layer with bandgaps between 2.00 eV and 1.75 eV was inserted at the p/i interface. By optimizing the thicknesses of the buffer layer with compromised bandgap and conductivity, a relative increase of 4.5% in single-junction cell efficiency was attained [11]. The improvement was mainly due to the increase of J_{SC} and FF, while V_{OC} has no change in the study.

The performance of the a-Si:H single-junction solar cells as a function of the p-layer thickness is shown in Fig. 5. There is no significant difference observed from the FF. The V_{OC} saturated when the thickness of p-layer was larger than 8nm. The J_{SC} increased as the thickness increased from 5 nm to 8 nm, but decreased as the thickness increased from 8 nm to 12 nm. The reduction in J_{SC} as the thickness less than 8 nm should be resulted from the degraded built-in field and a less uniform surface coverage. On the contrary, a thicker p-layer would have better V_{OC} and FF, but may reduce the absorption in i-layer which resulted in a reduced J_{SC}. The results shown in Fig. 5 reveal an optimized p-layer thickness of around 8nm. The optimal a-Si:H thin-film solar cell having an efficiency of 9.46% was achieved, with V_{OC}=906 mV, J_{SC}=14.42 mA/cm^2 and FF=72.36%.

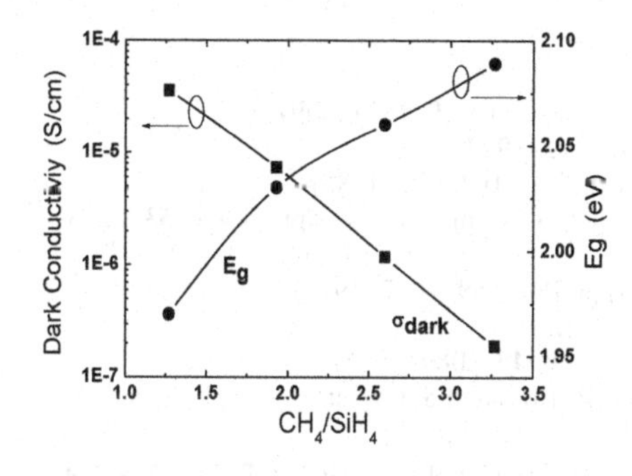

Figure 4. The optical bandgap and dark conductivity of the a-SiC:H p-layer as a function of the CH_4-to-SiH_4 flow rate ratio.

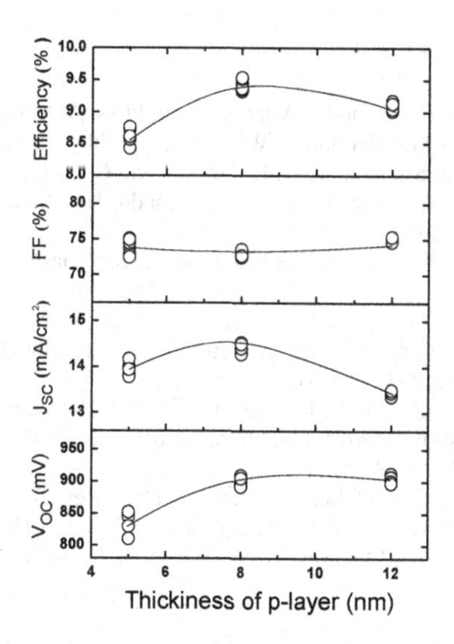

Figure 5. The performance of a-Si:H single-junction solar cells as a function of the p-layer thickness

CONCLUSIONS

The a-Si:H single-junction thin-film solar cells were fabricated on SnO_2:F-coated glasses by PECVD system. The boron-doped a-SiC:H was served as the window layer and the undoped a-SiC:H was used as the buffer layer. For the a-SiC:H p-layer, a σ_{dark} of 2.9×10^{-5} S/cm and E_g of 1.97 eV were achieved as the CH_4-to-SiH_4 flow rate ratio was 1.27. In the a-Si:H i-layer, a σ_{photo} of 1.6×10^{-5} S/cm and R of 0.04 were achieved by changing the E/S spacing to 14mm. The optimized thin-films were integrated and the thickness of each layer was optimized. The optimal a-Si:H thin-film solar cell having an efficiency of 9.46% was achieved, with V_{OC}=906 mV, J_{SC}=14.42 mA/cm^2 and FF=72.36%.

ACKNOWLEDGMENTS

This work was sponsored by the Center for Green Energy Technology at the National Chiao Tung University and National Science and Technology Program-Energy (100-3113-E-009-007-CC2).

REFERENCES

1. B. Rech and H. Wagner, Appl. Phys. A: Mater. Sci. Process. **69**, 155 (1999).
2. D. Staebler and C. Wronski, Appl. Phys. Lett. **31**, 292 (1977).
3. M. Stutzmann, W. B. Jackson and C. C. Tsai, Phys. Rev. B **32**, 23 (1985).
4. Y. Tawada, K. Tsuge, M. Kondo, H. Okamoto and Y. Hamakawa, J. Appl. Phys. **53**, 5273 (1982).
5. J. Arch, F. Rubinelli, J. Hou and S. Fonash, J. Appl. Phys. **69**, 7057 (1991).
6. U. Dutta and P. Chatterjee, J. Appl. Phys. **96**, 2261 (2004).
7. R. Arya, A. Catalano and R. Oswald, Appl. Phys. Lett. **49**, 1089 (1986).
8. S. Guha, J. Yang, A. Pawlikiewicz, T. Glatfelter, R. Ross and S. Ovshinsky, Appl. Phys. Lett. **54**, 2330 (1989).
9. C. C. Tsai, C. H. Hsu, H. J. Hsu, P. H. Cheng, C. M. Wang and Y. T. Huang, paper presented at the Third International Workshop on Thin-Film Silicon Solar Cells, Nagasaki, Japan, 11-14 October 2010.
10. G. Ganguly and A. Matsuda, Phys. Rev. B **47**, 3661 (1993).
11. C. H. Hsu, P. H. Cheng, Y. P. Lin, H. J. Hsu and C. C. Tsai, Mat. Res. Soc. Proc. of 2011 Spring Meeting, Symposium A (to be published).

Mater. Res. Soc. Symp. Proc. Vol. 1321 © 2011 Materials Research Society
DOI: 10.1557/opl.2011.1093

Semiconducting Polymer and Hydrogenated Amorphous Silicon Heterojunction Solar Cells

A. R. Middya and Eric A. Schiff

Department of Physics, Syracuse University, Syracuse, NY 13244 - 1130

ABSTRACT

In this work, we report on investigation of p-type semiconducting polymer, {poly(3,4 polyethylenedioxythiophene)-poly(styrenesulfonate)} (PEDOT:PSS) as the p-layer in NIP and PIN hydrogenated amorphous silicon (a-Si:H) solar cells. The rectification ratio of solution-casted diode is ~ 10, it increases to $3x10^4$ when PEDOT:PSS is deposited by Spin Coating technique. We observed additional photovoltaic effect when light is illuminated through polymer side. So far, best solar cells characteristics observed for PEDOT:PSS/a-Si:H hybrid solar cells are $V_{oc} \approx 720$ mV and $J_{sc} \approx 1 - 2$ mA/cm^2.

INTRODUCTION

Amorphous silicon (a-Si:H) thin films is one of the promising candidates for the application of low-cost solar cells. However, the highest efficiency of amorphous silicon (a-Si) based solar cells is at present limited to 12% [1]. An increase in open-circuit voltage (V_{oc}) is a promising route to increase the power output of a-Si:H solar panel. In principle, increase in V_{oc} should be achievable through increase in built-in potential to 1.3 eV - 1.4 eV from it's present value of 1.2 eV [2]. In this work, we investigated whether semiconducting polymer can perform as a better p-layer in NIP or PIN a-Si:H solar cells. We attempted *molecularly doped* p-type semiconducting polymer, {poly(3,4-polyethylenedioxythiophene)-poly(styrenesulfonate)} (PEDOT:PSS) [3,4] as the p-layer in NIP and PIN a-Si:H solar cells. The PEDOT:PSS has simultaneously high conductivity (up to 30 S/cm) and high transparency (T), T > 90% for 6 μm Spin Coated film. The combination of these properties are better than the inorganic p-layer (a-Si:H, μc-Si or a-SiC:H) used in state-of-art PIN or NIP a-Si:H solar cells. We would like to investigate additional effect whether p-type semiconducting polymer, PEDOT:PSS can increase the conduction band offset at P/I interface besides increase in built-in potential. An increase in either, or both of these device parameters is expected to lead to improvement in V_{oc}. So far, few attempts were made to investigate the feasibility of semiconducting polymer as the p-layer in a-Si:H or c-Si solar cells [5,6]. The p-type semiconducting polymers have also technological advantages over inorganic p-layer, for example, it is easily processable over large area by low-cost printing technique instead of vacuum based technology. In this work, we investigated semiconducting polymer, PEDOT:PSS as the p-layer which is not p-type silicon thin-film, or p-type silicon based alloy, we did experiment to see how PEDOT:PSS works in NIP or PIN a-Si:H solar cells.

EXPERIMENTAL DETAILS

The state-of-the-art amorphous silicon (a-Si:H) absorber layer and n-type a-Si:H layer have been deposited on top of stainless steel substrate as well as transparent conducting oxide (TCO) coated glass substrate by radio frequency plasma-enhanced chemical vapor deposition (RF PECVD) technique. The thickness of the intrinsic absorber layer and n-type layer are 300 nm and 50 nm respectively. Fig. 1 shows the schematic diagram of the PEDOT:PSS/a-Si:H heterojunction solar cells device structure. The NIP solar cells device has been completed by solution-casted technique as well as Spin Coating technique by putting PEDOT:PSS on top of a-Si:H i-layer. Fig. 1 shows solar cells structure where PEDOT:PSS layer was fabricated on top of a-Si:H by solution-casted technique. The PEDOT:PSS, the commercial name BAYTRON P was supplied by the Bayer AG . We removed the native oxide (SiO_2) layer on top of a-Si:H by etching with hydrofluoric (HF) acid before the polymer layer was solution-casted on top of a-Si:H. The thickness of PEDOT:PSS could not be controlled because of the solution-casting technique, the thickness of PEDOT:PSS is ~ 10 μm. Thin layer of p-type PEDOT:PSS was also deposited on top of intrinsic a-Si:H layer by Spin Coating technique. The thickness of the Spin Coated layer is ~ 500 nm - 1 μm. The entire PEDOT:PSS/a-Si:H/n-type a-Si:H structure was cured by annealing for 1 hour at atmospheric pressure at 100°C. The current-voltage (*J-V*) characteristic of the diode has been measured under 100 mW/cm^2 illumination at 30°C. The dark conductivity of the polymer has been measured by co-planner geometry using silver (Ag) electrode. The optical transmission of the PEDOT:PSS films have been measured by UV-VIS spectroscopy (Shimadzu, UV-160).

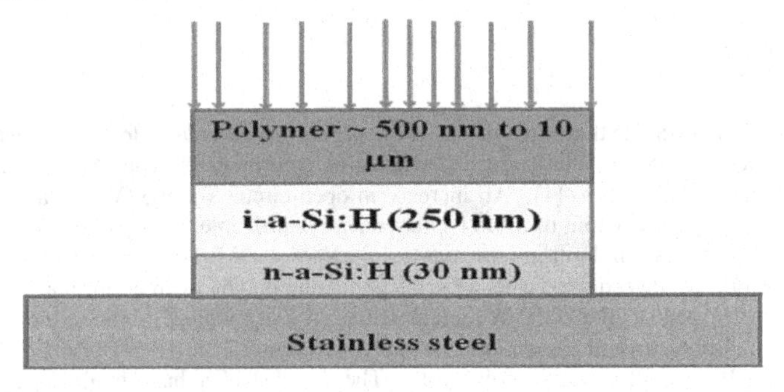

Fig. 1: Schematic diagram of NIP PEDOT:PSS/a-Si:H heterojunction solar cells.

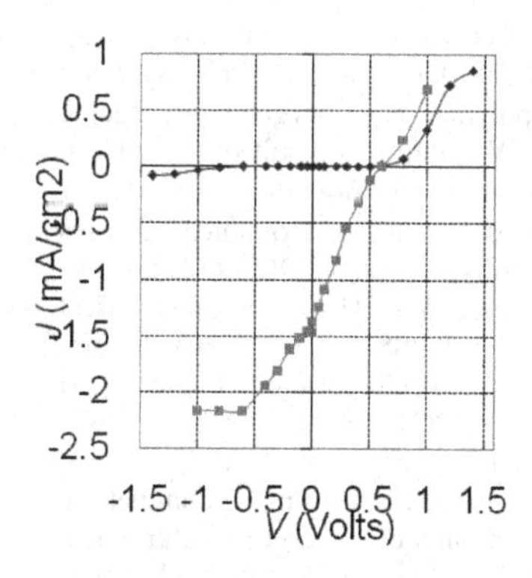

Fig. 2: Dark and light characteristics of NIP PEDOT:PSS/a-Si:H solar cells.

DISCUSSION

Fig. 2 shows dark and light J-V characteristics of PIN PEDOT:PSS/a Si:H diode. Fig. 2 shows formation of rectifying junction between p-type PEDOT:PSS and a-Si:H i-layer with reasonable rectification ratio 10 at $V_B \approx \pm 1$ Volts.

The leakage of the diode is minimum until $V_B \approx -1.5$ Volts. When these diodes are illuminated through p-layer, they showed solar cells characteristic with open-circuit voltage (V_{oc}) ~ 600 mV and short-circuit current (J_{sc}) ~ 1.5 mA/cm^2. Unlike the case of bulk or thin-film solar cells with all inorganic layers, light J-V of PEDOT:PSS/a-Si:H solar cells crosses dark J-V. This is unusual characteristics, indicating complex dark and phototransport properties across the interface between 1-D semiconducting polymer, PEDOT:PSS and tetrahydrally bonded a-Si:H semiconductor. The relatively low value of J_{sc} (1 -2 mA/cm^2) is most likely due to absorption of light in thick PEDOT:PSS layer. The thickness of PEDOT:PSS layer is 10 μm. A rough estimate of the transmittance of PEDOT:PSS layer at 633 nm (He-Ne laser) may be obtained from the ratio of the reverse-bias photocurrent to the incident laser power; we find $T \approx 0.1\%$. It means light absorbed in a-Si:H is 0.1 mW/cm^2. We obtained V_{oc} ~ 600 mV with 0.1 mW/cm^2 illumination, there must be other effect contributing to the V_{oc} that is different from photovoltaic effect, it might be originating at the interface of p-type PEDOT:PSS and a-Si:H i-layer.

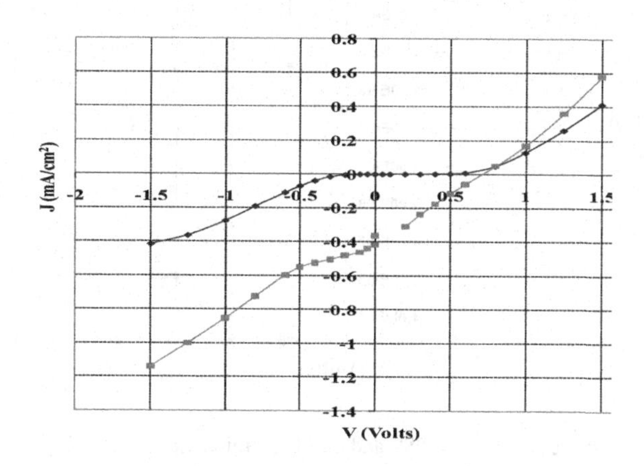

Fig. 3: Dark and light I-V characteristics of PIN PEDOT:PSS/a-Si:H heterojunction solar cells.

Now question arises where from additional photovoltaic at the P/I interface effect is coming when we made heterojunction solar cells combining p-type PEDOT:PSS with a-Si:H i-layer and n^+-layer (a-Si:H). There is an effect at the interface of polymer and inorganic or organic semiconductor or insulator, if polymer is illuminated with UV light, as a result of this effect, photo-induced charge transfer takes at the interface of polymer and organic and inorganic semiconductor [7]. Similar effect is taking place in our case, as a results, photoinduced charges are transferring from PEDOT:PSS to a-Si:H layer, consequently, p-type PEDOT:PSS layer is depleted with negative charge and a-Si:H layer gets more electrons. Thus, additional effect similar to photovoltaic effect takes place at the interface of PEDOT:PSS and a-Si:H which leads to much higher V_{oc} compared to light absorbed in the i-layer. We would get much higher V_{oc} if we get p-type PEDOT:PSS with higher conductivity, present PEDOT:PSS we got from Bayer AG, which is may be slightly photoconductive, we found that evidence at Au/PEDOT:PSS, metal-semiconductor junction, it shows diode and solar cells characteristics under illumination, indicating low doping level in PEDOT:PSS. We verified photoinduced charge transfer effect with UV illumination by sending light through glass/TCO in PIN a-Si:H solar cells with p-type PEDOT:PSS as the p-layer. Fig. 3 shows dark and light J-V characteristics of PIN solar cells. Fig. 3 shows rectification ratio ~ 10 at $V_B \approx \pm 1$ Volt. Light J-V characteristics of this PIN PEDOT:PSS/a-Si:H heterojunction solar cells shows V_{oc} ~ 720 mV and J_{sc} ~ 0.5 mA/cm^2. We have not deposited TCO on top of PEDOT:PSS, when we will use TCO, J_{sc} will be much higher than present value. Low J_{sc} (~ 0.5 mA/cm^2) can not be attributed to low transmittance since diode was illuminated through glass/TCO layer, i.e. fully illuminated, J_{sc} did not increase proportional to light intensity. This results confirms that the origin of low J_{sc} in PEDOT:PSS/a-Si:H heterojunction solar cells is different than the light intensity absorbed at the i-layer. Fig. 4 shows dark I-V of PEDOT:PSS/a-Si:H hetrojunction solar cells on stainless steel substrate

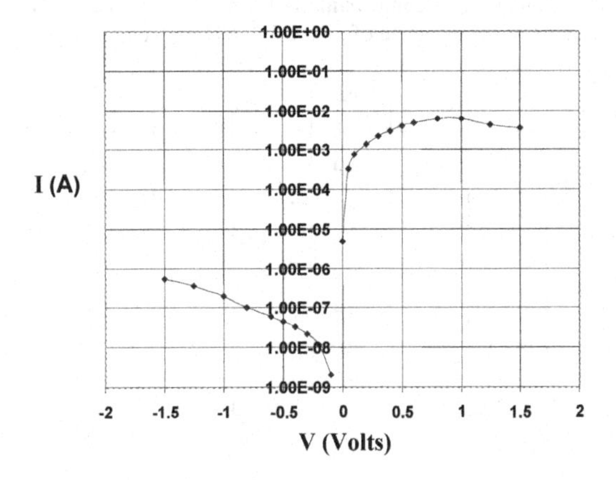

Fig. 4: Dark I-V characteristics of PEDOT:PSS and a-Si:H hetrojunction solar cells.

where PEDOT:PSS is Spin Coated on top of a-Si:H. The rectification ratio of Spin Coated diode is $3x10^4$, this value are much larger than that of diode fabricated by solution-casting technique. Surprisingly enough, the conductivity (σ_d) derived from planner configuration is ~ $4x10^{-4}$ S/cm for solution-casted PEDOT:PSS thin-film and σ_d ~ $6.7x10^{-5}$ S/cm for Spin Coated PEDOT:PSS respectively. Generally, when conductivity of p-layer is higher, it always gives better junction and better rectification ratio for diode. However, we observed lower rectification ratio for diode with p-type PEDOT:PSS having higher conductivity. It is very different results than our knowledge from diode fabricated with all inorganic layers. Here lies the importance of new physics at the interface of organic-inorganic semiconductor. The Spin Coated PEDOT:PSS/a-Si:H solar cells (Fig. 4) when illuminated through polymer layer, we observed V_{oc} ~ 540 mV and J_{sc} ~ 1 $\mu A/cm^2$. Lower V_{oc} in case of Spin Coated solar cells is consistent with lower conductivity of PEDOT:PSS when it is deposited by Spin Coated technique. Fig. 5 shows variation of open-circuit voltage with annealing temperature after PEDOT:PSS is put down on top of a-Si:H i-layer. It has been observed that V_{oc} increases with annealing temperature upto $100^{\circ}C$ and then it decreases continuously. Therefore, the optimum annealing temperature is $100^{\circ}C$ for the fabrication of PEDOT:PSS/a-Si:H hybrid solar cells. Fig. 6 shows the variation of light I-V characteristics of PEDOT:PSS/a-Si:H heterojunction solar cells with light intensity. For these diodes, native oxide layer was not etched on top of a-Si:H layer. All the light I-V measurements were done under white light. We observed surprising results. All the light I-V curves are separated in the forward bias and they appear according to light intensity. In case of state-of-art a-Si:H solar cells, all the curves merge with dark I-V because depletion region collapses under forward bias for $V_B > V_{oc}$. We found photocurrent increases with light intensity in the forward bias for polymer/a-Si:H solar cells. When the depletion region collapses, no photocurrent should be collected.

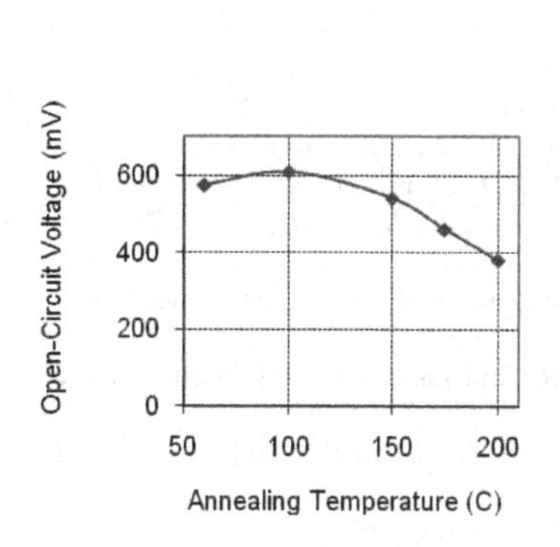

Fig. 5: Variation of V_{oc} with annealing temperature.

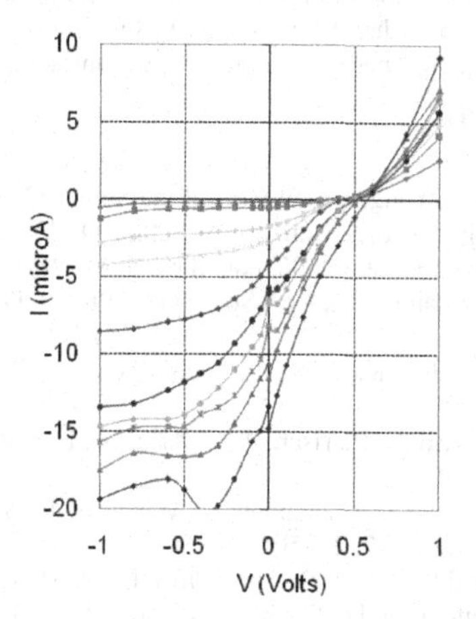

Fig. 6: Variation of light I-V with intensity.

This indicates additional effect at the interface, i.e. photoinduced charge transfer takes place from polymer to SiO_2 coated a-Si:H layer. These photoinduced charges can be added to photogenarated charges in the i-layer. As a results of this photoinduced charge transfer, photocurrent increases with light intensity under forward bias. In order to quantify this effect we should consider the data presented in Fig. 2. If we compare light absorbed in a-Si:H i-layer (Fig. 2) ~ 0.1 mW/cm^2 and open-circuit voltage produced ~ 600 mV, it is clear that photoinduced charge transfer contributing to open-circuit voltage (V_{oc}). This is an important observation in the field of science and technology of photovoltaic. It is expected that researcher will take advantage of photoinduced charge transfer effect in polymer to increase the efficiency of solar cells.

CONCLUSIONS

Semiconducting polymer, PEDOT:PSS works fairly well as the p-layer of a-Si:H solar cells. We observed rectification ratio is $3x10^4$ for Spin Coated PEDOT:PSS/a-Si:H diode. By comparing NIP and PIN structure, we clarified an additional effect is present at P/I interface besides photovoltaic effect, contributing to open-circuit voltage (V_{oc}). We observed, for the first time, light J-V crosses dark J-V in PEDOT:PSS/a-Si:H heterojunction solar cells. So far, best solar cells parameters obtained for PEDOT:PSS/a-Si:H hybrid solar cells are $V_{oc} \approx 720$ mV and $J_{sc} \approx 1 - 2$ mA/cm^2.

ACKNOWLEDGEMENT

We thank Dr. Qi Wang of National Renewable Energy Laboratory (NREL) for providing amorphous silicon structures used in this work. We thank Dr. Jill Simpson and Dr. Kelly Douglas of Bayer Corporation for providing supplies of PEDOT-PSS and for discussions of this material. This works has been supported by Thin-Film Photovoltaics Partnership Program of National Renewable Energy Laboratory (Subcontract XAK- 8-17619-23).

REFERENCES

1. S. Guha and J. Yang, J. Non-Crystalline Solids **352,** 1917 (2006).
2. L. Jiang, J. H. Lyou, S. Rane, E. A. Schiff, Q. Wang, and Q. Yuan, in Amorphous and Heterogeneous Films - 2000, edited by H. M. Branz, R. W. Collins, S. Guha, H. Okamoto, and M. Stutzmann (Mat. Res. Soc. Symp. Proc. **609**, Pittsburgh, PA, 2001), pp. A18.3.1 - A18.3.12.
3. T. M. Brown, J. S. Kim, R. H. Friend, F. Caciali, R. Daik and W. J. Feast, Appl. Phys. Letts. **75**, 1679 (1999).
4. M. Gransstroem, K. Patrisch, A. C. Arias, A. Lux, M. R. Andersson and R. H. Friend, Nature **395**, 257 (1008).
5. E. L. Williams, G. E. Jabbour, Q. Wang, S. E. Shaheen, D. S. Gingley and E. A. Schiff, Appl. Phys. Letts. **87**, 223504 (2005).
6. W. Wang and E. A. Schiff, Appl. Phys. Letts. **91**, 133504 (2007).
7. A. D. Pasquiera, D. D. T. Mastrogiovanni, L. A. Klein, T. Wang, E. Garfunkel, Appl. Phys. Letts. **91**, 183501 (2007).

Mater. Res. Soc. Symp. Proc. Vol. 1321 © 2011 Materials Research Society
DOI: 10.1557/opl.2011.806

Impact of a Finite Shunt Resistance on the Dark Spectral Response of a-Si:H/μc-Si Thin-film Multi-junction Photovoltaic Devices

Mauro Pravettoni[1,2] and Alessandro Virtuani[1]
[1] University of Applied Science of Southern Switzerland, Institute of Applied Sustainability to the Built Environment (SUPSI-ISAAC), CH-6952 Canobbio, Switzerland
[2] Imperial College London, Blackett Laboratory, London, SW7 2BW, United Kingdom

ABSTRACT

Hydrogenated amorphous silicon (a-Si:H) multi-junction devices have demonstrated a way to increase the efficiency of a-Si:H thin-film photovoltaic (PV) modules, which is now well above 10%. Since the current–matching behaviour of all sub-cells is a critical aspect, the measurement of the spectral response (SR) of all junctions provides valuable information to optimize the device performance under a given spectral distribution. In this work the authors investigate the impact of low shunt resistances on the SR of a double-junction a-Si:H/μc-Si PV module. The origin of a low shunt resistance in a-Si:H multi-junction devices is revised. A simple theoretical approach is then used to describe the anomalous dark SR observed experimentally as a consequence of the presence of low shunt resistances. The dark SR allows therefore to detect the presence of shunts and discriminate the defective sub-cell.

INTRODUCTION

Hydrogenated amorphous silicon (a-Si:H) based single-junction thin film photovoltaic (PV) devices may be manufactured at low costs and have short energy pay-back times, but have relatively low conversion efficiencies. An easy approach to increase the efficiency of these devices is to realize multi-junction structures, resulting in a more efficient use of the solar spectrum. With this approach it has been possible to realize devices with efficiencies well above 10% also on module sizes (>1 m^2, [1,2]). Multi-junction PV devices consist of a stack of two or more semiconductive layers ("junctions"), each with different characteristic band gaps. In the case of the so-called "micromorph" (a-Si:H/μc-Si) device, the layer with the widest band gap – the top junction, absorbing the low wavelength radiation – is made of a-Si:H, and the bottom junction, absorbing longer wavelengths radiation, is made of microcrystalline Si (μc-Si).

The current–matching behaviour of all junctions is a critical aspect and the measurement of the spectral response (SR) of each junction provides valuable information to optimize the device performance under a given spectral distribution of the light [3].

In this work the authors investigate the impact that low shunt resistances have on the "dark SR" of a double-junction a-Si:H/μc-Si PV module. The dark SR is the SR measured according to the ASTM 2236-10 standard method [4] with no bias light. For some devices deviations from the expected triangular-shape dark SR have been observed (see for example Ref. [5]). A simple single-diode model for each junction [6] is analysed and the dark SR is theoretically investigated. Results show that low shunt resistances can be detected experimentally from the dark SR, which thus represents an interesting experimental tool for the quality control in multi-junction PV devices. Specifically, it becomes possible to detect the presence of shunts and discriminate the defective sub-cell.

SHUNT RESISTANCE AND ORIGIN OF SHUNTS IN a-Si:H/μc-Si PHOTOVOLTAIC DEVICES

The shunt resistance R_{sh} gives a macroscopic description of the shunting behaviour of a solar cell. In an ideal solar cell is $R_{sh} = \infty$ (i.e., no shunt). As described in Ref. [7], a low value of R_{sh} affects the low-light performance (usually at irradiances below 200 W/m^2) of the solar cell. However, for extremely low values of R_{sh} also the performance of the cell at 1000 W/m^2 (i.e., the standard test condition) may be affected, in terms of lower fill factor, open-circuit voltage and cell efficiency.

A *shunt* (i.e. an internal connection between the front and rear layers of the cell) acts as a low resistance in parallel to the ideal solar cell and may: (I) have a macroscopic origin or (II) be distributed in the solar cell. In the case of solar modules an additional origin of shunts may be related to (III) a poor structuring and bad interconnections of the solar cells. In the most severe cases the whole solar cell may be disabled from the module. Case III is beyond the objectives of this work.

In case I *macroscopic shunts* usually arise from local defects that can relatively easily be detected and in certain cases removed. Typical examples are the presence of dopants or impurities at the edges of c-Si solar cells and the presence of macroscopic *pin-holes* in thin-film solar cells (e.g. the presence of metal drops or segregations; the absence of the absorber material).

In case II *distributed shunts* may be related to some intrinsic properties of the absorber material or cell structure – as for example the conductivity for Cu(In,Ga)Se_2 solar cells (see Ref. [8]) – or to a distribution of microscopic pin-holes or defective portion of the absorber materials that act as "bad diodes".

In micromorph devices, as has been reported in Ref. [9], examples of macroscopic shunts (case I) may come from the presence of dust particles during the deposition of the absorber layers, which may then flake off during subsequent processing and create macroscopic shorts between back and front contacts. Examples of distributed shunts (case II) may occur in the top a-Si:H junction in the presence of a very rough transparent conductive oxide (TCO) in combination with a very thin a-Si:H intrinsic layer with the subsequent creation of distributed pin-holes. In case of the bottom μc-Si junction, the deposition of microcrystalline on rough substrates may as well lead to the formation of zones of non-dense material (commonly referred to as "cracks") with a shunting-like behavior.

DARK SPECTRAL RESPONSE OF A MULTI-JUNCTION PHOTOVOLTAIC DEVICE: A THEORICAL APPROACH

A basic model for a 2-junction PV cell is the equivalent circuit of Figure 1a: the electrical series of a resistor (*series resistance*: R_s) and two current generators representing the top and bottom junctions (*photogenerated current*, $I_{L,i}$, where *i* is either top or bottom), each one in parallel with a diode (*diode* or *dark current*, $I_{d,i}$) and a shunt resistor (*shunt resistance*, $R_{sh,i}$; *shunt cur-*

rent, $I_{sh,i}$). From Kirchhoff's Laws the following characteristic equation follows, giving the current I flowing through the circuit when an external voltage V is applied

$$I = I_{L,i} - I_{0,i}\left\{\exp\left[\frac{q(V-V_j+IR_s)}{nkT}\right] - 1\right\} - \frac{V-V_j+IR_s}{R_{sh,i}}, \tag{1}$$

where the second term on the right is the diode current through the i-th junction $I_{d,i}$ (q is the elementary charge, V_j is the voltage across the $j \neq i$ junction, n is the ideality factor, k is Boltzmann's constant and T is cell temperature) and the third term is the shunt current $I_{sh,i}$.

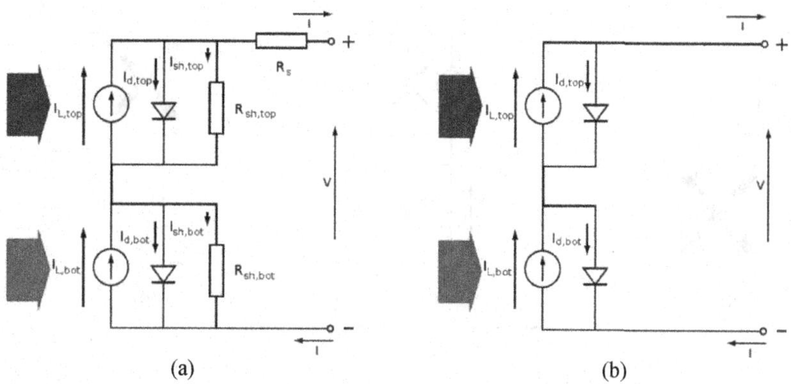

(a) (b)

Figure 1. (a) Basic equivalent circuit of a 2-junction PV device. (b) Infinite shunt resistance approximation.

In the following analysis the series resistance term will be neglected for simplicity (R_s = 0): this is a reasonable approximation when the current involved is small. Also, for simplicity and graphical purposes, "blue" (B) light means the short wavelength monochromatic light (which is not necessarily blue) that is absorbed by the top junction only; "green" (G) light means the intermediate wavelength component that is equally absorbed by the top and bottom junction; and "red" (R) is the long wavelength absorbed by the bottom junction only.

Only the case $V = 0$ will be considered (no external voltage bias: the overall device is in short-circuit).

Case 1: infinite shunt resistance approximation

Consider first the case of $R_{sh,i} = \infty$ (see Figure 1b). Equation (1) thus simplifies, giving

$$I = I_{L,i} - I_{0,i}\left\{\exp\left[\frac{q(V-V_j)}{nkT}\right] - 1\right\}. \tag{2}$$

According to equation (2), when B light is completely absorbed by the top junction (see Figure 2a), the bottom junction is in the dark and in reversed bias: its diode behaviour prevents the current flow through it and therefore the photogenerated current almost completely flows

through the diode of the top junction, which is in forward bias. No net current flow is observed at the cell terminals. Similar situation occurs when R light (to which the top cell is assumed to be totally transparent) is absorbed by the bottom junction (see Figure 2c).
In the intermediate case (G light, Figure 2b) a net current flow occurs. In this case both junctions are in short-circuit ($V_{top} = V_{bot} = 0$) and no current circulation through the diodes occurs.

Using the above results to simulate the dark SR gives the triangular shape shown in Figure 2d, that represents the modelled dark SR of a 2-junction PV device in the infinite shunt resistance approximation. The dark SR is given by the short wavelength tail of the bottom junction SR (450-625 nm in the example) and the long wavelength tail of the top junction SR (625-800 nm).

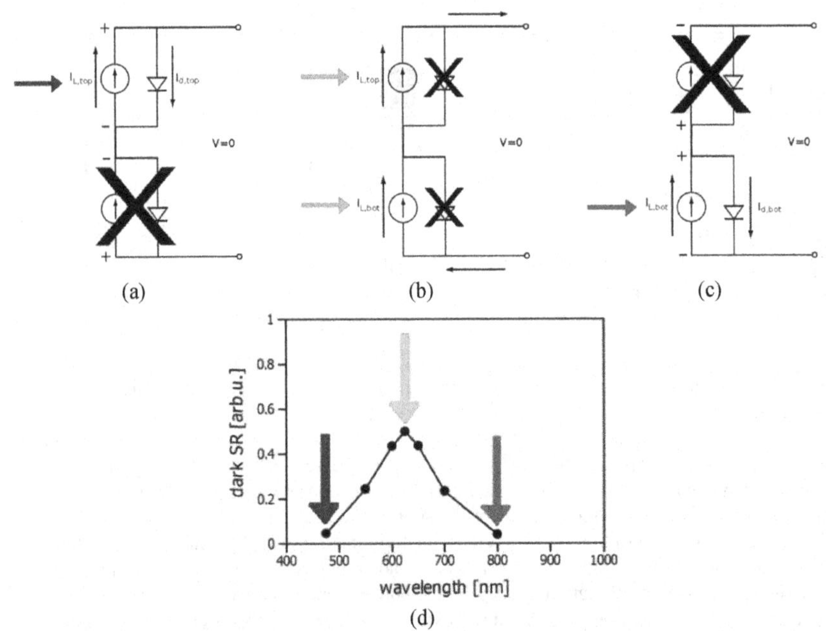

Figure 2. Dark SR measurement, infinite shunt resistance approximation: (a) B light ($I = 0$ and $V_{top} = -V_{bot}$); (b) G light ($I > 0$ and $V_{top} = V_{bot} = 0$); (c) R light ($I = 0$ and $V_{bot} = -V_{top}$); (d) the modelled dark SR as a function of wavelength for an a-Si:H/μc-Si PV cell with infinite shunt resistances. The device is kept at $V = 0$. The dark SR is non-zero only in the wavelength range where both junctions respond (500-800 nm for typical a-Si:H/μc-Si devices) and is maximum at the wavelength where the two junctions respond equally.

Case 2: finite shunt resistance

Without loss of generality, consider the case where $R_{sh,top} < \infty$. Under B light (see Figure 3a) the bottom junction blocks the current flow as in Case 1 and the dark SR at short wavelengths is zero again. Also under G light $R_{sh,top}$ does not show any effect (Figure 3b). At long wavelengths (R light, Figure 3c) instead, $R_{sh,top}$ represents a leakage through the top junction and the dark SR is therefore non-zero in this case. Figure 3d shows the different impacts that shunt resistances of various values (in arbitrary units) may have on the dark SR.

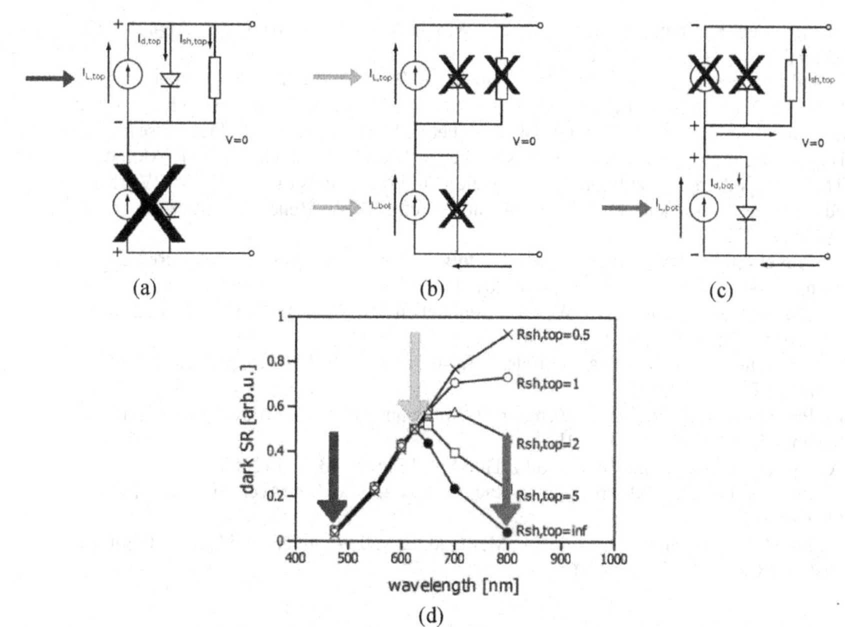

Figure 3. Dark SR measurement with a finite shunt resistance on the top junction (and infinite shunt resistance on the bottom): (a) B light ($I = 0$); (b) G light ($I > 0$); (c) R light ($I > 0$); (d) the modelled dark SR as a function of wavelength for an a-Si:H/μc-Si PV, taking into account a low shunt resistance in the top junction. The dark SR shows a current leakage above 800 nm, due to the flow through $R_{sh,top}$ at long wavelengths. Values of $R_{sh,top}$ are in arbitrary units (for a detailed analysis see Ref. [6]).

CONCLUSIONS

The origin of shunts in micromorph solar cells was revised in this work and two main sources have been taken into consideration: macroscopic shunts and distributed shunts.

In the simplest theoretical approach, shunts are described by resistors in parallel with the given junction in the multi-junction device. The effect of shunts on the dark SR of a micromorph cell has then been modelled, highlighting the difference with the case of infinite shunt resistance (i.e. no shunt).

Results suggest that low shunt resistances can be easily detected experimentally from the dark SR, representing thus an interesting experimental tool for the quality control in multi-junction PV devices.

REFERENCES

1. M. A. Green, K. Emery, Y. Hishikawa and W. Warta, Prog. Photovolt: Res. Appl. **19**, 84 (2011).
2. J. Bailat, L. Fesquet, J-B. Orhan, Y. Djeridane, B. Wolf, P. Madliger, J. Steinhauser, S. Benagli, D. Borrello, L. Castens, G. Monteduro, M. Marmelo, B. Dehbozorghi, E. Vallat-Sauvain, X. Multone, D. Romang, J-F. Boucher, J. Meier, U. Kroll, M. Despeisse, G. Bugnon, C. Ballif, S. Marjanovic, G. Kohnke, N. Borrelli, K. Koch, J. Liu, R. Modavis, D. Thelen, S. Vallon, A. Zakharian and D. Weidman, Proceedings of the 25th EU PVSEC, edited by G. F. De Santi, H. Ossenbrink and P. Helm (WIP – Renewable Energies, Munich, 2010) pp. 2720-2723.
3. M. Pravettoni, G. Tzamalis, K. Anika, D. Polverini and H. Müllejans, Mater. Res. Soc. Symp. Proc. Vol. 1245, 1245-A13-06 (2010).
4. ASTM Standard E2236, 2010, West Conshohocken, PA, 2003, DOI: 10.1520/E2236-10, www.astm.org.
5. F. A. Rubinelli, R. L. Stolk, A. Sturiale, J. K. Rath and R. E. I. Schropp, J. Non-Cryst. Solids **352**, 1876 (2006).
6. M. Pravettoni, R. Galleano, A. Virtuani, H. Müllejans and E. D. Dunlop, Meas. Sci. Technol. **22**, 045902 (10pp) (2011).
7. A. Virtuani, E. Lotter and M. Powalla, Thin Solid Films **443**, 431 (2003).
8. A. Virtuani, E. Lotter, M. Powalla, U. Rau, J. H. Werner and M. Acciarri, Jour. Appl. Phys. **99**, 014906 (2006).
9. M. Phyton, O. Madani, D. Dominé, F. Meillaud, E. Vallat-Sauvain and C. Ballif, Sol. En. Mat. Sol Cells **93**, 1714 (2009).

Mater. Res. Soc. Symp. Proc. Vol. 1321 © 2011 Materials Research Society
DOI: 10.1557/opl.2011.935

Investigation of local light scattering properties of thin-film silicon solar cells with subwavelength resolution

K. Bittkau, A. Hoffmann, J. Owen and R. Carius
IEK5-Photovoltaik, Forschungszentrum Jülich GmbH, 52425 Jülich, Germany

ABSTRACT

In order to obtain efficient light trapping within a thin-film silicon solar cell, randomly textured interfaces are used. The texture can be introduced by wet-chemical etching in diluted hyrdofluoric acid (HF). By varying of the HF concentration, a continuous transition to smaller surface structures can be achieved. Near-field scanning optical microscopy is applied to measure scattered light with sub-wavelength resolution. On those different surfaces, using Fourier high-pass filters on the measured near-field images, surface features with a high light trapping potential are identified. Finally, criteria for optimized scattering surfaces are obtained.

INTRODUCTION

Randomly textured transparent conducting oxide (TCO) surfaces are often used as front contact layer in thin-film silicon solar cells in order to scatter light into the absorber layer [1-3]. The sizes of the surface features are typically in the range of the wavelength of visible light or below. Far-field properties like haze and angular resolved scattering for TCO/air interfaces are typically used to characterize light trapping in solar cells [4-6]. These methods have a couple of major disadvantages. First, examining the TCO/air interfaces does not give sufficient information about scattering into thin-film solar cells. Second, the scattering properties of large areas are averaged, making it impossible to study the impact of different surface features on the scattering efficiency.

To overcome these disadvantages, near-field scanning optical microscopy (NSOM) is applied. NSOM is a well established tool used to investigate the local light intensities [7-9]. The optical resolution is about 50-80 nm. In a NSOM measurement, an optical fiber probe is scanned across a surface while being held 20 nm above. Due to the short distance, the probe collects the evanescent waves which result from light that is trapped in a solar cell by total internal reflection. These trapped light modes can only be studied by near-field experiments. The interference of counter-propagating light waves leads to periodic intensity patterns, where the spatial period is directly correlated to the scattering angle. By applying Fourier high-pass filters on the measured near-field images, the local distribution of trapped light modes can be extracted.

EXPERIMENT

In this work, different surface structures are investigated. Aluminum-doped zinc oxide (ZnO:Al) layers are frequency magnetron sputtered on glass substrates. The surface of the layers is textured by diluted hydrofluoric acid with different concentrations, resulting in a more or less continuous variation of the surface profile [10]. Scanning electron microscopy images of three of these surfaces are shown in Fig. 1. On top of the textured ZnO:Al films, hydrogenated amorphous silicon (a-Si:H) p-i-n solar cells are deposited by plasma enhanced chemical vapor deposition (PECVD). To allow optical experiments in transmission geometry, no back reflectors were deposited. Since the reflectivity at the a-Si:H/air interface is still high, this structure represents the complete solar cell with back reflector fairly well.

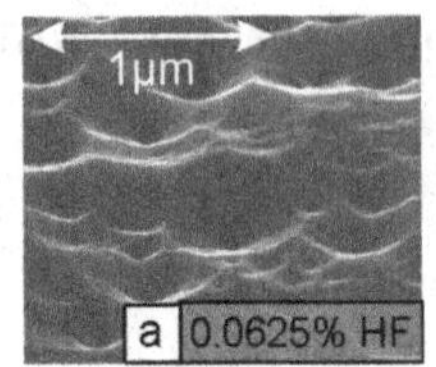

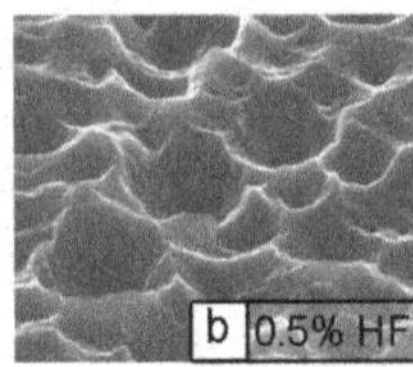

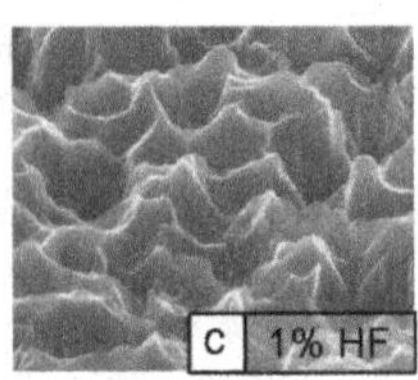

Figure 1. Scanning electron microscopy images of the investigated samples. The samples are textured by wet chemical etching in diluted hydrofluoric acid with 0.0625% (a), 0.5% (b) and 1% (c) concentration.

The experimental investigation is done by NSOM. The sample is illuminated from the glass side by coherent light source at a wavelength of λ=780 nm and the transmitted light intensity is collected at the surface by an Aluminum coated tapered fiber tip. By scanning the tip across the surface while holding it 20 nm above, images with the local light intensities and the topographic information are generated simultaneously. The tip has an aperture of about 80 nm, which defines the optical resolution.

Due to the small proximity between the tip and the surface, evanescent light modes are detected. These modes are directly correlated to light which is trapped inside the solar cell by total internal reflection. The intensity of evanescent light decreases exponentially with the distance from the surface and is, therefore, only detectable in the optical near-field.

Scanning areas are identified by laser scribing, which allows NSOM measurements of the same area for both before and after the deposition of a-Si:H on the ZnO:Al surface. By comparing the measurements, the impact of the a-Si:H layer on the local light scattering properties is investigated. Additionally, atomic force microscopy (AFM) measurements are performed, providing the surface topographies with a higher resolution than those by NSOM.

ANALYSIS TECHNIQUE

In the measured near-field intensity images, both, propagating and evanescent light modes are detected, since the total light intensity at the position of the aperture defines the signal. From these measurements, the entire scattering information can be obtained. In particular, the scattering into guided modes can be investigated. In this section, an important image processing technique is described which allows the extraction of angular resolved scattering (ARS) and the spatial distribution of trapped light.

The image processing is based on Fourier transform techniques combined with adapted filters. Due to the interference of counter-propagating waves, lateral light intensity oscillations occur. The spatial periods of these oscillations are defined by the scattering angle [Fig. 2(a)]. The light modes with spatial periods larger than half the wavelength are propagating while those with smaller spatial periods are evanescent. Since in a stochastically rough surface, a broad distribution of scattered light modes exists. The measured near-field intensity image is interpreted as a superposition of the interference patterns of light waves with different inclination angles. The distribution of the angles is given by the Fourier transform of the images.

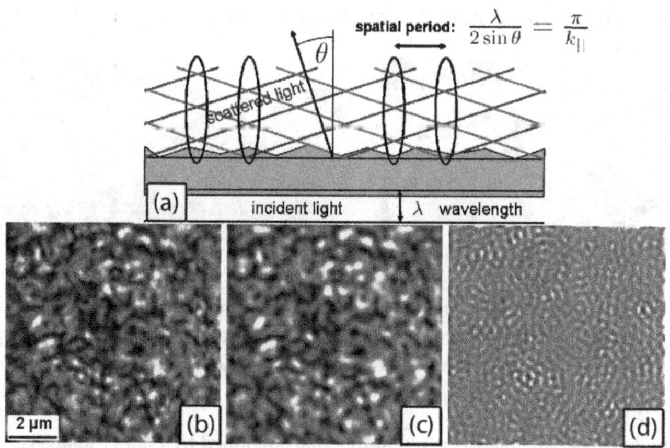

Figure 2. (a) Illustration of the interference of scattered wave fronts. Due to the interference of counter-propagating waves, lateral light intensity oscillations occur. The spatial periods of these oscillations are defined by the scattering angle θ. (b) shows the originally measured near-field image. By applying low-pass and high-pass filters to (b) the propagating and evanescent part of the light are extracted and shown in (c) and (d), respectively.

Various filters can be applied to the Fourier transform. For example, the propagating and evanescent waves are extracted by low-pass and high-pass step filters, respectively. The critical spatial frequency is given by $2/\lambda$ [Fig. 2(a) for $\theta = 90°$]. The filtered images are back-transformed to obtain the spatial distribution of the propagating and evanescent parts of the light. This is shown exemplarily in Fig. 2. The originally measured near-field image is shown in (b). In (c) and (d), the propagating and evanescent part of the near-field image is visualized, respectively, after back-transforming the filtered Fourier transform.

The ARS is given by the Fourier transform directly, since the spatial frequency can be directly translated into a specific scattering angle and the intensity in the Fourier transform defines the fraction of this scattering angle in the whole near-field image.

Since far-field experiments average over large sample areas, this information is difficult to obtain. Therefore, NSOM analysis is an essential tool for light trapping investigation and the further optimization of surface texture.

RESULTS & DISCUSSION

During deposition of a-Si:H, the surface morphology is changed. The changes depend significantly on the feature size and surface angles on the TCO surface. Figure 3 shows the AFM images of ZnO:Al etched in hydrofluoric acid with concentrations of 0.25% (a), (b) and 1% (c), (d) before and after the deposition of an a-Si:H layer.

Surface features in crater-like surface morphology of (a), referred to as surface I, are smoothed by deposition of a-Si:H. Especially at the slopes of craters, more silicon is deposited which leads to a smoothing and of the craters. Small features and thin crater rims are balanced by deposition and are not found on the surface of the later cell. The texture in (c), referred to as surface II, consists of mostly small sharp features which are reshaped to semi-spheres after deposition of a-Si:H (d).

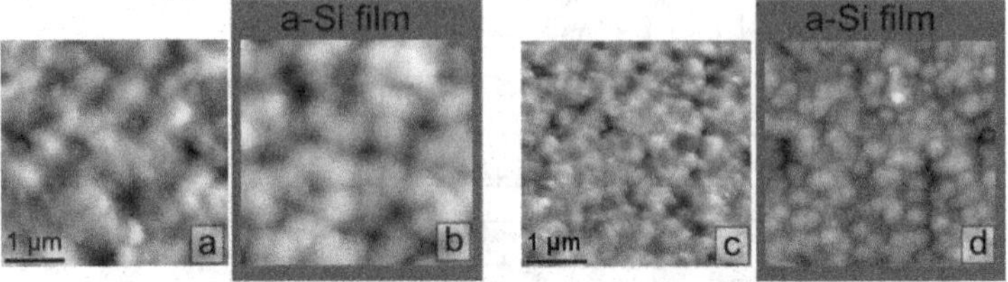

Figure 3. Atomic Force Microscopy pictures of two different TCO surfaces (0.25% and 1% HF concentration) before (a),(c) and after deposition of the a-Si:H layer (b),(d). Surface changes are slight in case of more crater-shaped features (b) (surface I) while the more pyramid-like peaks of (c) (surface II) are converted into semi-spheres after the deposition of a-Si:H.

ANGULAR RESOLVED SCATTERING

According to [9,11] angular distribution functions obtained by Fourier analysis adequately reproduce the behavior of angular resolved scattering experiments. Fig. 4 compares the angular scattering distributions of the ZnO:Al etched in HF concentrations of 0.0625% [Fig. 4(a)] and 1% [Fig. 4(b)]. The maximum of the angular distribution of surface I without silicon is found at 13°. By deposition of a-Si:H, the distribution is broadened and the maximum is shifted to higher angles. In case of the small sharper features on surface II, the maximum of the angular distribution is obtained at 21°. Depositing a-Si:H leads to a slight shift to lower scattering angles while the general shape of the distribution is maintained.

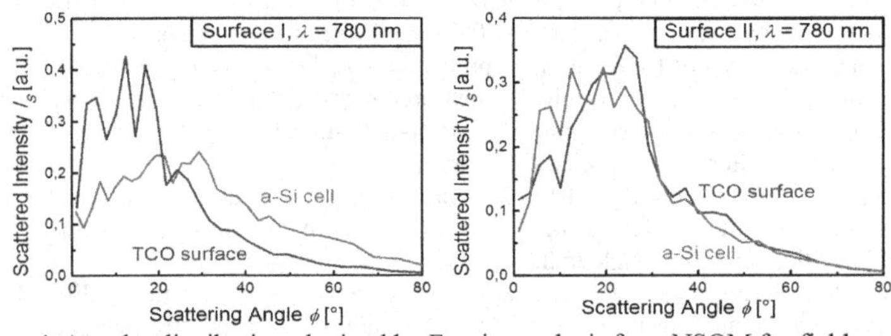

Figure 4. Angular distribution obtained by Fourier analysis from NSOM far-field measurements for surface I (a) and surface II (b). Scattering angles of surface I show a maximum at 13° and are shifted to higher angles after deposition of a-Si:H. In case of surface II, the maximum of angular distribution is found at 21° before and after deposition.

EVANESCENT FIELDS

In order to gain information about light trapping within the solar cell, near-field measurements performed by NSOM are analyzed using Fourier high-pass filters [9,12]. The localizations of evanescent modes shown in Fig. 5(c,f), are related to trapped light in the a-Si:H layer. Evanescent modes for the NSOM measurements of the ZnO:Al surfaces without a-Si:H (b,e) give information about light scattering in air with large transferred wave vectors. For a sufficient light trapping in the solar cell, these large wave vectors are necessary. Thus, regions where large wave vectors are transferred can be identified at the commonly investigated ZnO:Al/air interface.

Four different positions are picked in order to classify light trapping properties of the structure. Position A in Fig. 5 marks a typical crater of this surface type corresponding to a minimum of evanescent field intensity without (b) and with a-Si:H (c). Pos. B shows a small crater and its rim. The slope of the crater leads to evanescent modes in (b), as well as in (c). At the upper border of position B, a crater rim is found. With and without a-Si:H, there is a minimum of evanescent field intensity at this point. A crater rim (Pos. C) leads to evanescent modes for the ZnO:Al/air interface but not in case of the solar cell. Area D denotes a long rim that consists of a lot of vertices. Strong evanescent fields for the ZnO:Al/air as well as for the a-Si:H/air interface can be identified in this region. This structure is found to exhibit good light trapping properties.

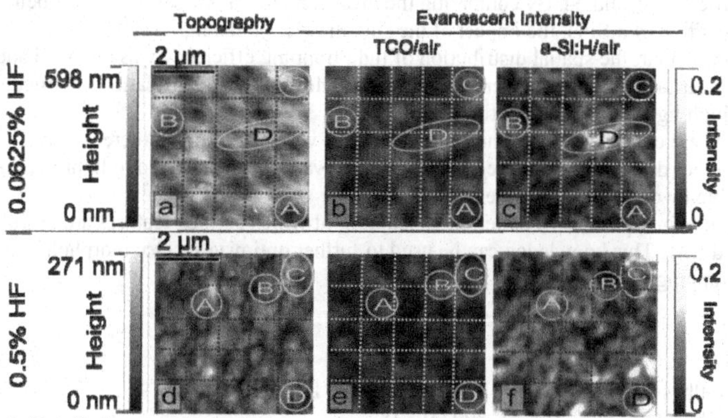

Figure 5. Detailed comparison of evanescent modes for 0.0625% [(a)-(c)] and 0.5% [(d)-(f)] HF concentration without (b),(e) and with a-Si:H (c),(f). Positions A-D were selected representing different light trapping behavior.

By increasing the concentration of HF, the crater-like features are replaced by sharper peak-like features. This increases the short circuit current density j_{sc} for a 350 nm thick a-Si:H solar cell by 0.6 mA/cm² to 16 mA/cm² which was measured using IV setup after preparing a silver back contact on top of the solar cells.

At the slope of a high peak (Pos. A), high evanescent field intensities can be detected for ZnO:Al and a-Si:H. Position B denotes several small features adjacent to a large crater. This area, though evanescent modes are found above the TCO surface, describes a minimum of evanescent fields above the a-Si:H layer.

Position C indicates a collection of several 300 nm surface peaks next to a large crater. Above ZnO:Al [Fig. 5(e)], high evanescent field intensities are found at the slopes of the crater. Above a-Si:H [Fig. 5(f)], the center of the crater gives a minimum of evanescent field intensities. Steep slopes and accumulations of small features lead to high evanescent field intensities and consequently to good light trapping in the a-Si:H layer.

Finally, position D with high surface angles is examined, and high evanescent field intensities can be identified with and without a-Si:H layer.

Generally, higher evanescent field intensities were observed above a-Si:H than without a-Si:H. This is due to the higher optical mode density of a-Si:H in comparison to ZnO:Al. While the crater-like texture in structure I only consists of certain structures that account to light trapping, the small steep features in structure II lead to more surface angle configurations that favor the total internal reflection of the light.

It should be emphasized that minima of evanescent fields above TCO can be related to crater centers and peaks in the surface morphology.

CONCLUSIONS

The light scattering at different surface textures of the ZnO:Al front contact of a-Si:H solar cells were investigated with NSOM. Using Fourier transform techniques, the local light scattering properties were studied. By comparing the measurements at the same location before and after the deposition of a-Si:H, the impact of the silicon layer on the light scattering was investigated. In particular, the spatial distribution of light trapping efficiency was obtained and correlations to the surface topography were found, which allowed the identification of structures which exhibit a high light trapping potential.

The results are consistent to the short circuit current which showed an increase for structures with higher density of evanescent waves. These waves are connected to light trapping and only detectable by optical near-field experiments.

NSOM, therefore, provides a better understanding of which surface features are useful to light trapping and why. This knowledge can be used to further optimize surface morphology with improved light trapping.

REFERENCES

1. J. Nelson, The Physics of Solar Cells, Imperial College, London (2003).
2. E. Yablonovitch, J. Opt. Soc. Am. **72**, 899 (1982).
3. P. Campbell, J. Opt. Soc. Am. B **10**, 2410 (1993).
4. J. Löffler, R. Groenen, J.L. Linden, M.C.M. van de Sanden, and R.E.I. Schropp, Thin Solid Films **392**, 315 (2001).
5. H. Stiebig, M. Schulte, C. Zahren, C. Haase, B. Rech, and P. Lechner, Proc. SPIE **6197**, 619701 (2006).
6. J. Krč, M. Zeman, O. Kluth, F. Smole, and M. Topič, Thin Solid Films **426**, 296 (2003)
7. K. Bittkau, T. Beckers, S. Fahr, C. Rockstuhl, F. Lederer, and R. Carius, Phys. Status Solidi (a) **205**, 2766 (2008).
8. K. Bittkau and T. Beckers, Phys. Status Solidi **207**, 661 (2010).
9. K. Bittkau, T. Beckers and R. Carius, Proceedings 25th European Photovoltaic Solar Energy Conference and Exhibition, 2996-2999 (2010).
10. J.I. Owen, J. Hüpkes, H. Zhu, E. Bunte, and S.E. Pust, Phys. Status Solidi (a) **208**, 109 (2011).
11. K. Bittkau, M. Schulte, T. Beckers, and R. Carius, Proc. SPIE **7725**, 77250N (2010).
12. T. Beckers, K. Bittkau, and R. Carius. J. Nonlin. Opt. Phys. Mat. **19**, 645 (2010).

Mater. Res. Soc. Symp. Proc. Vol. 1321 © 2011 Materials Research Society
DOI: 10.1557/opl.2011.812

Reflectance Improvement by Thermal Annealing of Sputtered Ag/ZnO Back Reflectors in a-Si:H Thin Film Silicon Solar Cells

Karin Söderström[1], Franz-Josef Haug[1], Céline Pahud[1], Rémi Biron[1], Jordi Escarré[1], Martial Duchamp[2], Rafal Dunin-Borkowski[3-2] and Christophe Ballif[1]
[1] Ecole Polytechnique Fédérale de Lausanne, Institute of Microengineering, Photovoltaics and Thin Film Electronics Laboratory, 2000 Neuchâtel, Switzerland
[2] Center for Electron Nanoscopy, Technical University of Denmark, 2800 Kongens Lyngby, Denmark
[3] Institute for Microstructure Research, Forschungszentrum Jülich, D-52425 Jülich, Germany

ABSTRACT

Silver can be used as the back contact and reflector in thin film silicon solar cells. When deposited on textured substrates, silver films often exhibit reduced reflectance due to absorption losses by the excitation of surface plasmon resonances. We show that thermal annealing of the silver back reflector increases its reflectance drastically. The process is performed at low temperature (150°C) to allow the use of plastic sheets such as polyethylene naphthalate and increases the efficiency of single junction amorphous solar cells dramatically. We present the best result obtained on a flexible substrate: a cell with 9.9% initial efficiency and 15.82 mA/cm^2 in short circuit current is realized in n-i-p configuration.

INTRODUCTION

Interest in thin film silicon solar cells has increased strongly in the last few years. Their development promises to help to produce abundant, low cost electricity at a time when mankind faces the challenge to move to alternative energy production processes. The low carrier lifetime in amorphous and microcrystalline silicon materials results in the need to reduce the thickness of the solar cells' active layer to well below the absorption length of light in the red part of the spectrum. Textured interfaces are used to scatter light within the solar cells to lengthen the light path and increase the generated photocurrent [1, 2, 3]. The n-i-p process is interesting because it can be used to produce light-weight, flexible and unbreakable modules, since the cell layers can be deposited on opaque substrates such as steel or on plastic sheets. The back contact and reflector in the n-i-p configuration is often realized by using metals having a high reflectivity like silver. Unfortunately, as the substrate itself needs to be rough to increase the light trapping, the silver layer has to be deposited onto a nano textured surface. In contrast to a silver layer deposited onto a flat surface, for which low parasitic absorptions are observed, silver deposited onto a rough substrate leads to parasitic absorption losses that take place within the metallic layer by the excitation of surface plasmon polaritons (SPPs) [4]. SPPs are known to be stronger when the metal quality is low [5]. In this contribution it will be shown that a proper thermal annealing of the silver layer leads to a large improvement of the film reflectance. The annealing is carried out at a moderate temperature, which allows the use of plastic substrates like polyethylene naphthalate (PEN). When the back reflectors are inserted into cells, the cell performance is improved significantly. In the first part, we present the effect of annealing on the bare reflectors, consisting only of the structured substrates and a silver layer. In the second part, the effect of annealing in the device is shown. By measuring the external quantum efficiency (EQE), and the total absorption of the whole cell we find a significant reduction in parasitic absorption losses. Finally, the best initial result demonstrating 9.9% efficiency on a plastic substrate textured by a nano imprinting replication process is presented.

EXPERIMENTAL DETAILS

We used the texture that develops naturally on zinc oxide when grown by low pressure chemical vapor deposition (LP-CVD ZnO:B) [6]. This texture consists of pyramidal shapes whose size is controlled by the film thickness. On our 2 μm thick films, the pyramids have base lengths of typically 200-300 nm. The V shaped valleys between the pyramids have been reported to reduce the open circuit voltage (Voc) and the fill factor (FF) because they lead to the nucleation and propagation of defective material [7]. We attempt to avoid this detrimental effect by using a plasma treatment that rounds out the valleys into U-shapes. On top of the ZnO substrates, a 300 nm thick silver layer was deposited by DC sputtering (Univex 450B, Leybold, deposited at room temperature). Half of the substrates were subjected to thermal annealing in ambient atmosphere at 150°C for 50 minutes. Total (TR) and diffuse (DR) reflectance of the bare substrates (ZnO + silver) were analyzed with a spectrophotometer equipped with an integrating sphere (Lambda 900, Perkin Elmer). To observe the morphology modification after annealing, scanning electron microscopy (SEM) was performed both from a standard top view (JSM-7500TFE, JEOL) and by using focused ion beam (FIB) milling for obtaining a cross-sectional profile for imaging (FEI Helios).

The remaining half of the substrates was used as the back reflector in single junction amorphous cells. For cell deposition, the back reflector was completed by adding an aluminum-doped ZnO barrier layer of around 80 nm by RF sputtering at room temperature (Univex 450B, Leybold, 2 wt % Al_2O_3 ceramic target). The cells were deposited in an n-i-p sequence with thick doped layers to avoid collection problems, the nominal thickness of the i-layer was 200 nm. The front contact was made of another LP-CVD ZnO:B layer. The cells were characterized by measuring their IV characteristics with a dual lamp solar simulator (Wacom WXS-220S-L2) under standard test conditions (STC, 25°C, AM 1.5 G spectrum, and 1000 W/m^2). Voc and FF were extracted from the IV curves, the short circuit current (Jsc) of the device was determined from the EQE after weighting it with the AM1.5 G spectrum. Correction of the EQE using the total cell reflection yields the internal quantum efficiency (IQE), i.e. the ratio of collected charge carriers per absorbed photon.

In a separate experiment, we applied the annealing process to produce a high efficiency device on a flexible substrate on polyethylene naphthalate (Goodfellow, 125 μm thick). The texture was made using a high fidelity process of replication using UV nano imprinting [8, 9]. The reproduced texture was a 2 μm thick LP-CVD ZnO plasma-treated slightly more than the texture used in the first experiments. The cell deposition was similar to that explained above except for the use of slightly thinner doped layers, which were used to improve Jsc.

Further results and details on the bare substrates analysis and cells can be found in [10].

RESULTS AND DISCUSSION

Analysis of bare substrates

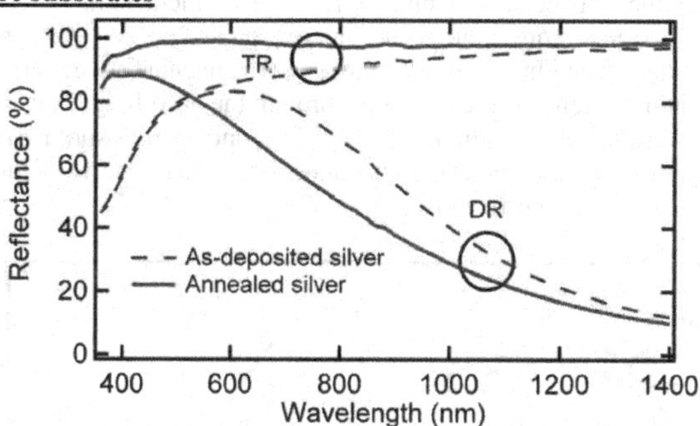

Figure 1: Bare substrate characterization: Total and diffuse reflectance of the silver samples.

Figure 1 compares the reflectance of the textured Ag films before and after annealing. In the as-deposited state, the total reflection is reduced by the strong absorption of short wavelengths. After annealing, the total reflectance is almost 100% over the entire wavelength range down to 360 nm, where the SPP resonance of the silver-air interface is expected in the presence of a rough surface. A reduction of the diffuse reflectance suggests reduced surface roughness after annealing.

Figure 2 shows SEM images of a top and a cross-sectional view. The SEM image of the annealed sample exhibits more rounded shapes than the as-deposited silver. The cross-sectional SEM images indicate that annealing results in a large reduction of the V shapes that can be observed in the as-deposited silver. It is interesting to note that the silver deposition itself brings out the pinched V shapes. The treated LP-CVD ZnO layer has few V shapes, but, as noted by the arrows on the right of the figure, a valley exhibiting a U shape develops into a V shape after silver deposition. This behavior suggests that a cell grown on top of the annealed silver, compared to a cell grown on the as-deposited silver, will not only benefit from improved optical properties but also from an improved Voc.

Figure 2. Bare substrate characterization: SEM images of the silver deposited on LP-CVD ZnO substrates; left) top view, right) cross sections made by FIB.

Cell analysis

Cells were co-deposited onto the same types of substrates as those describe above. Figure 3 shows that below 500 nm the cells behave identically because the incident light does not reach the back reflectors. Above 500 nm, more light reaches the rear part of the cell and is reflected by the textured silver back contact. The improved reflectance of the annealed silver film leads to more reflected light back into the cell for a second pass through the absorbing layer, resulting in higher EQE despite the lower total absorption of the cell. This conclusion is summarized in the IQE characteristics, which illustrate that the cell on the annealed reflector converts the absorbed light into collected carriers much more efficiently.

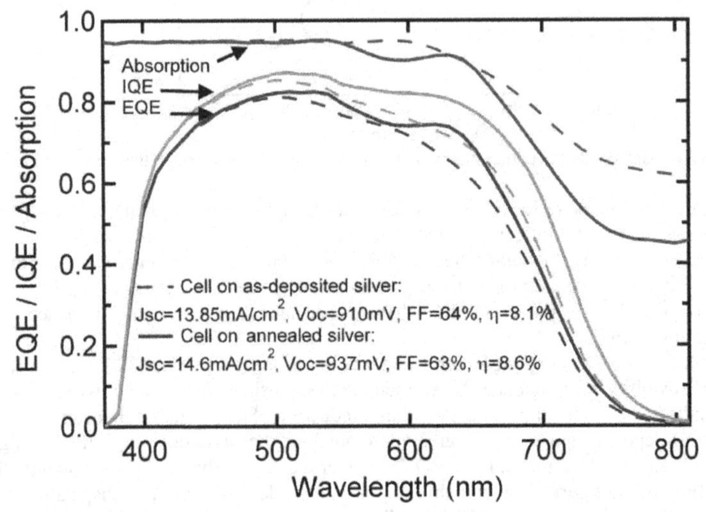

Figure 3: EQE, IQE and absorption measured on the cells with as-deposited silver and annealed silver back reflectors. The electrical parameters are also shown.

Interestingly, annealing decreases the surface roughness, as discussed above. Generally, a reduced roughness also yields cells with a higher Voc. The Voc of the cell on the annealed reflector is almost 30 mV higher and is well correlated with the reduction of the V shapes seen in the SEM images. However, a decreased substrate roughness also leads in general to lower light trapping. Based on the diffuse reflectance shown in Figure 1, we would expect less light scattering in the cell on the annealed reflector. The EQE in Figure 3 even shows interference effects in the cell grown on the annealed reflector, which is normally an indication of flat interfaces and poor light scattering. Nevertheless, the short circuit current density of the cell on the annealed reflector is almost 0.8 mA/cm^2 higher. Lower light scattering is therefore more than compensated by the better silver reflectance. In summary, a high efficiency improvement of 0.5% (from 8.1 to 8.6%) in absolute terms is observed, thanks to higher values of Jsc and Voc in the cell grown on annealed silver.

Best cell result on a flexible substrate using a replicated nano texture

Figure 4 shows the best cell result in terms of initial efficiency on a flexible substrate that has been textured with the replication of the LP-CVD ZnO surface texture via UV nano imprinting. An initial efficiency of 9.9% is obtained with a high current of 15.82 mA/cm^2. The silver annealing was performed as described above. This efficiency is better than that described above for several reasons. First, the cells shown in the previous section were fabricated with a robust base process, with relatively thick doped layers in order to avoid collection problems and to guarantee comparability across all cells of a series. Here, a compromise was achieved between a good Jsc for thin doped layers and good Voc and FF for thick doped layers. This improvement can be seen in the EQEs at short wavelengths; at 450 nm, the EQE of the cell on plastic is equal to 0.82, whereas at the same wavelength the EQE of the non-optimized cell on annealed silver presented above is equal to 0.76. Second, the deposition system cleaning history can explains the higher FF of this cell [11, 12].

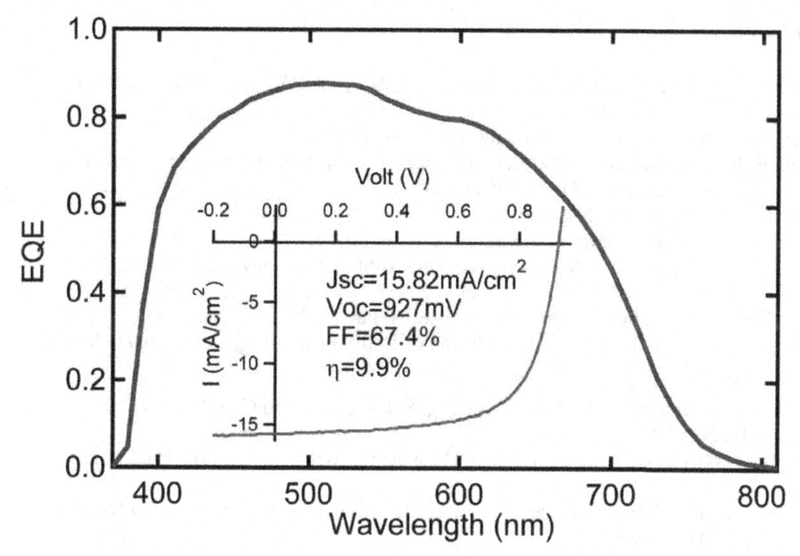

Figure 4: EQE and IV curve with corresponding electrical parameters for the best initial efficiency cell grown on a flexible substrate that was nano textured using UV nano imprinting.

CONCLUSIONS

Thermal annealing at moderate temperatures which allows the use of plastic substrates was performed on silver layers that serve as the back contact and reflector in thin film solar cells in n-i-p configuration. Several modifications to the silver film were observed; the reflectance increased substantially, the morphology changed towards a decreased substrate roughness and reduced V-shaped structures. The reflectance increase and the roughness decrease of the silver films are both beneficial for the cell device. The high reflectance reduces strongly the parasitic losses at the back reflector as can be observed in IQE curves. The decrease in roughness is beneficial in terms of Voc and also decreases the diffuse reflectance of the substrate, which in general leads to less light trapping. However, the high silver reflectance more than compensates for the light trapping losses, resulting in a higher Jsc value for the cell grown on the annealed silver reflector. An optimized cell on a flexible plastic substrate shows an initial efficiency of 9.9% and a current of 15.82 mA/cm^2.

ACKNOWLEDGMENTS

The authors acknowledge the support of the Swiss National Science Foundation under grant number 200021_12577/1 and European Union funding within the project Si-Light (contract no. 241277).

REFERENCES

1. H. Iida, N. Shiba, T. Mishuku, H. Karasawa, A. Ito, M. Yamanaka, and Y. Hayashi, *IEEE Electr. Device.* **L. 4**, 157-159 (1983).
2. A. Banerjee and S. Guha, *J. Appl. Phys.* **69**, 1030-1035 (1991).
3. O. Kluth, B. Rech, L. Houben, S. Wieder, G. Schöpe, C. Beneking, H. Wagner, A. Löffl, and H. W. Schock, *Thin Solid Films* **351**, 247-253 (1999).
4. R.H. Ritchie, *Phys. Rev.* **106**, 874 (1957).
5. K. Holst and H. Raether, *Optics Commun.* **2**, 312-316 (1970).
6. S. Faÿ, U. Kroll, C. Bücher, E. Vallat-Sauvain, and A. Shah, *Sol. Energy Mater. Sol. C.* **86**, 385-397 (2005).
7. J. Bailat, D. Dominé, R. Schlüchter, J. Steinhauser, S. Faÿ, F. Freitas, C. Bücher, L. Feitknecht, X. Niquille, T. Tscharner, A. Shah, and C. Ballif, *Proceedings of the Fourth WCPEC Conference*, 1533-1536 (2006).
8. K. Söderström, J. Escarré, O. Cubero, F.-J. Haug, S. Perregaux, and C. Ballif, *Prog. Photovoltaics* **19,** 202-210 (2011).
9. J. Escarré, K. Söderström, C. Battaglia, F.-J. Haug, and C. Ballif, *Sol. Energ. Mater. Sol. C.* **95,** 881-886 (2011).
10. K. Söderström, F.-J. Haug, J. Escarré, C. Pahud, R. Biron, C. Ballif, *submitted to Sol. Energ. Mater. Sol. C.* (2011).
11. O. Cubero, F.-J. Haug, Y. Ziegler, L. Sansonnens, P. Couty, D. Fischer and C. Ballif, *Sol. Energ. Mater. Sol. C.* **95,** 606-610 (2011).
12. S. Sweetnam, T. Söderström, F.-J. Haug, O. Cubero, X. Niquille, V. Terrazzoni-Daudrix and C. Ballif, *Thin Solid Films* (2011) **in press**.

Mater. Res. Soc. Symp. Proc. Vol. 1321 © 2011 Materials Research Society
DOI: 10.1557/opl.2011.815

11.0% Stable Efficiency on Large Area, Encapsulated a-Si:H and a-SiGe:H based Multijunction Solar Cells Using HF Technology

A. Banerjee, D. Beglau, T. Su, G. Pietka, G. Yue, B. Yan, J. Yang, and S. Guha
United Solar Ovonic LLC, 1100 West Maple Road, Troy, MI 48084, U.S.A.

ABSTRACT

We report on the investigation of large area a-Si:H/a-SiGe:H double-junction and a-Si:H/a-SiGe:H/a-SiGe:H triple-junction solar cells prepared by our proprietary High Frequency (HF) glow discharge technique. For investigative purposes, we initially used the simpler double-junction structure. We studied the effect of: (1) Ge content, (2) cell thickness, and (3) SiH_4 and GeH_4 gas flow on the light-induced degradation of the solar cells. Our results show that the double-junction cells with different Ge concentration have open-circuit voltage (V_{oc}) in the range of 1.62-1.75 V. V_{oc} exhibits a flat plateau in the range of 1.65-1.72 V for both initial and stabilized states. The light-induced degradation for cells in this range of V_{oc} is insensitive to the Ge content. In terms of thickness dependence of the intrinsic layers, we found that the initial efficiency increases with cell thickness in the thickness range 2000-4000 Å. However, light-induced degradation increases with increasing thickness. Consequently, the stabilized efficiency is invariant with cell thickness in the thickness range studied. The results of SiH_4 and GeH_4 gas flow on cell characteristics demonstrate that the deposition rate decreases by only 20% when the active gas flow is reduced to 0.25 times standard flow. The initial and stabilized efficiencies are similar. The information gleaned from the study was used to fabricate high efficiency, large area (~464 cm^2) double- and triple-junction solar cells. The highest stable efficiency, as measured by NREL, was 9.8% and 11.0% for the double- and triple-junction structures, respectively.

INTRODUCTION

We previously reported [1] a stabilized efficiency of 13.0% on a small-area (0.25 cm^2) a-Si:H/a-SiGe:H/a-SiGe:H triple-junction solar cell prepared using RF (13.56 MHz) Plasma Enhanced Chemical Vapor Deposition (PECVD) technique. Using low-rate RF process, we also attained [2] stable large-area (aperture area ~900 cm^2) efficiency of 10.5% using a similar triple-junction device structure. The RF technique, which has been widely used to deposit a-Si:H and a-SiGe:H films and thin film solar cells, is limited to a deposition rate of 1~3 Å/s in order to make high quality intrinsic layer materials for high efficiency solar cells. It is well known that higher deposition rates, using RF-excitation, usually results in lower material and cell quality [3-4].

In recent years, Very High Frequency (VHF) excitation has emerged as the preferred technique to deposit a-Si:H and a-SiGe:H based solar cells. Compared to conventional RF glow discharge methodology, VHF excitation has the advantages [5] of: (1) 2-3 times higher deposition rate, (2) provides solar cells exhibiting both superior light stability and high stabilized efficiency. We have developed a proprietary HF glow discharge technique to deposit device quality a-Si:H and a-SiGe:H films and solar cells at a high rate. In this paper, we present new work conducted on large-area a-Si:H/a-SiGe:H double-junction and a-Si:H/a-SiGe:H/a-SiGe:H triple-junction solar cells on Ag/ZnO back reflector coated stainless steel substrates, fabricated using the high-rate HF process. We conducted a series of experiments to investigate the effect of various deposition parameters on the light stability of the solar cells. For the sake of simplicity, we used an a-Si:H/a-SiGe:H double-junction device structure. The parameters investigated were: (1) Ge content in the a-SiGe:H layers, (2) cell thickness, and (3) SiH_4 and GeH_4 gas flow used for a-SiGe:H bottom cell *i* layer deposition. Using the optimized deposition parameters gleaned from the study, we fabricated large area (aperture are ~464 cm^2) double-junction and triple-junction solar cells. The highest stable efficiency, as confirmed by NREL, was 9.8% and 11.0% for the double- and triple-junction solar cells, respectively.

EXPERIMENT

The substrate used for the study was a stainless steel foil of thickness 125 μm and dimensions ~38 cm x 35 cm. A bi-layer of Ag/ZnO deposited onto the substrate served as the back reflector. The large-area multijunction devices were then deposited in a batch PECVD deposition reactor using our HF process. The deposition rate was ~6-8 Å/s. The film thickness uniformity and device characteristics were optimized by varying the process parameters, such as cathode-to-substrate spacing, cathode configuration, HF power, substrate temperature, and process gas mixture and pressure.

During the optimization phase, the coated substrate was cut into 4 cm x 4 cm pieces. Small-area (active area 0.25 cm^2) solar cells were delineated by depositing indium-tin-oxide (ITO) through an evaporation mask. Grid fingers were deposited on top for current collection. Current density-voltage (J-V) characteristics of the devices were measured under an AM1.5 solar simulator. Quantum efficiency (QE) as a function of wavelength was measured, and the integrated curve was used to determine the short-circuit current density (J_{sc}).

Large area (≥400 cm^2) solar cells were fabricated by depositing ITO over the large area, followed by the application of grid wires and bus bars for current collection. The cells were then encapsulated using our proprietary flexible, lightweight encapsulant. The light-stabilized J-V characteristics were obtained by exposing the solar cells (small-area and large-area modules) under an intensity of 100 mW/cm^2 white light at 50 °C for 1000 hours.

RESULTS AND DISCUSSION

Light stability vs. Ge content

Table I summarizes the initial and stable J-V characteristics of small-area double-junction a-Si:H/a-SiGe:H cells as a function of Ge content in the a-SiGe:H bottom cell. The Ge content is reflected in the V_{oc} of the cells, 1.62-1.75V. Higher Ge content leads to lower V_{oc}, and vice versa. The Ge content decreases from the sample shown in the first row to that in the last row in the table. J_{sc} contributions for the top and bottom component cells (QE_{top} and QE_{bot}), and the cumulative QE_{tot} for the initial and stable states are also listed. The initial and stable efficiency has been calculated using the QE value of the current-limiting cell. In order to separate the effect of device thickness on light degradation (discussed in the next section), we selected samples with similar thickness 2400-2700 Å.

Table I shows that for Ge content that corresponds to V_{oc} in the range of 1.65-1.72 V, there is a flat plateau for both initial efficiency 11.1-11.3% and stable efficiency 9.9-10.1%. In other words, the initial and stable efficiencies are independent of Ge concentration within this range.

Table I. Initial and stabilized J-V characteristics of small-area a-Si:H/a-SiGe:H double-junction cells for various Ge concentration.

Sample # 3D	State	V_{oc} (V)	FF	QE_{top} (mA/cm^2)	QE_{bot} (mA/cm^2)	QE_{total} (mA/cm^2)	Efficiency (%)
9489	Initial	1.624	0.681	9.95	11.13	21.08	11.0
	Stable	1.569	0.643	9.65	10.64	20.29	9.7
	Degradation	3.4%	5.5%	3.0%	4.4%	7.4%	11.4%
9260	Initial	1.654	0.677	9.94	10.68	20.62	11.1
	Stable	1.586	0.645	9.71	10.46	20.17	9.9
	Degradation	4.1%	4.6%	2.3%	2.1%	4.4%	10.6%
9305	Initial	1.683	0.688	9.75	10.32	20.07	11.3
	Stable	1.620	0.652	9.52	10.06	19.58	10.1
	Degradation	3.8%	5.1%	2.4%	2.5%	2.4%	10.9%
9121	Initial	1.695	0.699	9.51	10.17	19.68	11.3
	Stable	1.631	0.664	9.22	10.10	19.32	10.0
	Degradation	3.8%	5.0%	3.0%	0.7%	1.8%	11.4%
9228	Initial	1.718	0.703	9.3	9.88	19.18	11.2
	Stable	1.653	0.667	9.04	9.68	18.72	10.0
	Degradation	3.8%	5.2%	2.8%	2.0%	2.4%	11.3%
9541	Initial	1.748	0.710	8.71	9.47	18.18	10.8
	Stable	1.686	0.672	8.50	9.11	17.61	9.6
	Degradation	3.5%	5.3%	2.4%	3.8%	3.1%	10.9%

Figure 1 shows a plot of light-induced degradation of efficiency for the cells (shown in Table I) as a function of V_{oc}. The degradation value has been taken from the last column in Table I. The plot shows that the degradation is insensitive to Ge content for cells within this range of V_{oc}.

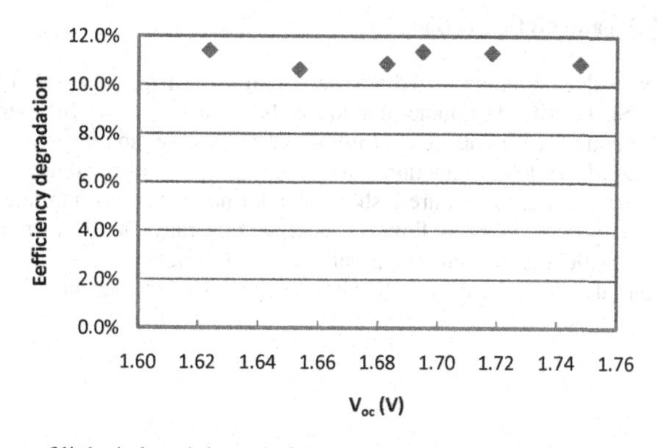

Figure 1. Extent of light-induced degradation as a function of V_{oc} for the cells in Table I.

Light stability vs. cell thickness

In this study, we selected four a-Si:H/a-SiGe:H double-junction cells with cell thickness in the range of 2290-3490 Å. In order to assure similar material quality for all the cells, we fixed all process parameters, and changed only the intrinsic layer deposition time. In each case, the top cell thickness was ~65% of the total device thickness. Table II summarizes the initial and stable J-V characteristics as a function of the total intrinsic layer thickness. As in Table I, the current contributions, QE_{top}, QE_{bot}, and QE_{tot} for the initial and stable states are also listed. The data shows that the initial efficiency increases with increasing intrinsic layer thickness. However, the extent of light-induced degradation also increases. Since the efficiency increase and stability decrease are similar, one can conclude that the stabilized efficiency is invariant with cell intrinsic layer thickness in this range of thickness.

Table II. Initial and stabilized J-V characteristics of small-area a-Si:H/a-SiGe:H double-junction cells for various intrinsic layer thickness. The top cell thickness is ~65% of the total device thickness.

Sample # 3D	State	V_{oc} (V)	FF	QE_{top} (mA/cm^2)	QE_{bot} (mA/cm^2)	QE_{total} (mA/cm^2)	Efficiency (%)	Thickness (Å)
9383	Initial	1.699	0.682	9.33	9.63	18.96	10.8	2290
	Stable	1.638	0.654	9.22	9.48	18.70	9.9	
9384	Initial	1.681	0.675	9.53	9.98	19.51	10.8	2540
	Stable	1.628	0.647	9.31	9.79	19.10	9.8	
9381	Initial	1.669	0.667	9.93	10.27	20.20	11.1	2880
	Stable	1.609	0.630	9.74	10.00	19.74	9.9	
9382	Initial	1.651	0.653	10.31	10.56	20.87	11.1	3490
	Stable	1.593	0.618	10.09	10.28	20.37	9.9	

Light stability vs. SiH_4 and GeH_4 gas flow

For this study, we selected an optimized a-SiGe:H bottom cell recipe as the reference from which to fabricate a-Si:H/a-SiGe:H double-junction cells. We varied the SiH_4 and GeH_4 gas flow in unison (in the bottom cell) with respect to the reference case, and investigated the initial and stable performance of the double-junction cells. The four different reference case gas flows were: 150%, 100%, 50%, and 25%. Figure 2 shows the normalized deposition rate of a-SiGe:H intrinsic layer as a function of total gas flow. The deposition rate initially increases with gas flow in the range of 25-100%, and then attains saturation up to 150%. It is interesting to note that the deposition rate decreases by only 20% when the gas flow was decreased from 100% to 25%.

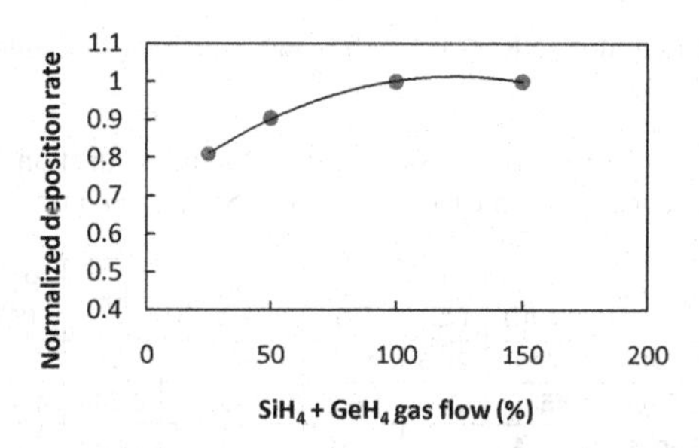

Figure 2. Normalized deposition rate of the a-SiGe:H intrinsic layer as a function of total gas flow of SiH_4 and GeH_4.

For the light stability study, we prepared three a-Si:H/a-SiGe:H double-junction cells with different SiH_4 and GeH_4 gas flows for depositing the bottom cell: 100%, 50% and 25%. Identical processing was used for the a-Si:H top cell. In order to exclude thickness effect, all three samples were made with the same cell thickness of 2400 Å by adjusting the a-SiGe:H bottom cell deposition time. The initial and light-soaked J-V characteristics of the cells are shown in Table III. Both the initial and stable efficiencies of the samples are similar: initial efficiency is ~11.5-11.7% and stable efficiency is ~9.9-10.2%. Thus, the cells prepared with gas flows differing by a factor of 4 all exhibit very similar film quality.

Table III. Initial and stabilized J-V characteristics of small-area a-Si:H/a-SiGe:H double-junction cells for various SiH_4 and GeH_4 gas flow.

Sample # 3D	Gas flow	State	V_{oc} (V)	FF	QE_{top} (mA/cm^2)	QE_{bot} (mA/cm^2)	QE_{total} (mA/cm^2)	Efficiency (%)
10034	100%	Initial	1.709	0.718	9.40	10.30	19.70	11.5
		Stable	1.644	0.665	9.20	9.91	19.11	10.1
10036	50%	Initial	1.704	0.710	9.64	10.10	19.74	11.7
		Stable	1.639	0.656	9.44	9.68	19.12	10.2
10023	25%	Initial	1.694	0.734	9.21	10.69	19.90	11.5
		Stable	1.627	0.683	8.93	10.29	19.22	9.9

High efficiency double- and triple-junction a-SiGe:H based cells

The optimized deposition parameters from the above studies were adopted to fabricate large-area (aperture area 464 cm^2), high efficiency, a-Si:H/a-SiGe:H double-junction and a-Si:H/a-SiGe:H/a-SiGe:H triple-junction solar cells. The cells were encapsulated using our proprietary flexible encapsulants and light soaked. Stable I-V characteristics were measured under a Spire solar simulator. Selected samples were sent to NREL for confirmation of cell efficiency. Table IV shows stable efficiency of the large-area cells, measured at United Solar and NREL. The highest stable efficiency as measured by NREL was 9.8% and 11.0% for the double- and triple-junction structures, respectively.

Table IV. Stable efficiency of large-area a-Si:H/a-SiGe:H double-junction and a-Si:H/a-SiGe:H/a-SiGe:H triple-junction solar cells, measured in United Solar and confirmed by NREL.

Cell # 3D	Cell structure	Measurements	Area (cm^2)	T (^oC)	V_{oc} (V)	J_{sc} (mA/cm^2)	FF	P_{max} (W)	Efficiency (%)
8489T	a-Si/ a-SiGe	USO Spire	464	24.3	1.67	8.79	0.647	4.393	9.47
		NREL Spire	453.7	25.5	1.685	8.93	0.648	4.421	**9.75**
5227T	a-Si/ a-SiGe/ a-SiGe	USO Spire	464	24.2	2.31	6.32	0.673	4.557	9.82
		NREL Spire	462.3	25.5	2.344	6.65	0.703	5.071	**11.0**

SUMMARY

We conducted a systematic study of the effect of 1) Ge content in the a-SiGe layers, 2) cell thickness and 3) SiH_4 and GeH_4 gas flow on a-SiGe:H cells prepared by using our proprietary HF technique. The results were used to fabricate large-area a-Si:H/a-SiGe:H double-junction and a-Si:H/a-SiGe:H/a-SiGe:H triple-junction solar cells. We attained conversion efficiency of 9.8% and 11.0% for the double- and triple-junction structures, respectively, as confirmed by NREL.

ACKNOWLEDGEMENTS

The authors thank X. Xu and J. Zhang for important contributions. The authors also thank D. Wolf, N. Jackett, L. Sivec, C. Worrel, J. Piner, Y. Zhou, S. Liu, T. Palmer, J. Owens, R. Caraway, and B. Hartman for sample preparation and measurements. The work was supported by US DOE under the Solar America Initiative Program Contract No. DE-FC36-07GO17053.

REFERENCES

1. J. Yang, A. Banerjee, and S. Guha, Appl. Phys. Lett. **70**, 2975 (1997).
2. A. Banerjee, J. Yang, and S. Guha, Mat. Res. Soc. Symp. Proc. **557**, 743 (1999)
3. S. Guha, J. Yang, S. Jones, Y. Chen, and D. Williamson, Appl. Phys. Lett. **61**, 1444 (1992).
4. S. Jones, Y. Chen, D. Williamson, X. Xu, J. Yang, and S. Guha, Mat. Res. Soc. Symp. Proc. **297**, 815 (1993).
5. X.Xu, D.Beglau, S.Ehlert, Y. Li, T.Su, G. Yue, B.Yan, K. Lord, A. Banerjee, J. Yang, S. Guha, P. Hugger, and J.D. Cohen, Mat. Res. Soc. Symp. Proc. **1153**, 99 (2009).

Mater. Res. Soc. Symp. Proc. Vol. 1321 © 2011 Materials Research Society
DOI: 10.1557/opl.2011.939

Calibration of multi-junction (tandem) thin film photovoltaic modules under natural sunlight

Georgios Tzamalis, Harald Müllejans
European Commission, DG JRC, IE, Renewable Energy Unit, 21027 Ispra (Va), Italy.

ABSTRACT

We present a procedure for calibrating tandem thin film photovoltaic modules under natural sunlight. The distinct steps involve pre-conditioning via light soaking, spectral response measurements and I-V measurements under natural sunlight. The measurements were done using a crystalline silicon reference cell as well as two filtered reference cells designed to spectrally match the top and bottom junctions in the tandem module.

INTRODUCTION

Thin film technology has secured its place as a commercially viable alternative to crystalline silicon cells due to its lower material and manufacturing cost and increasing lifetime and stability. Although the efficiencies of thin film solar modules are still lower than crystalline, they can drastically improve by the ability to deposit different type of materials and alloys.

One method of enhancing the efficiency of a thin film module is by creating a multi-junction (e.g. tandem) thin film module. It consists of a stack of two or more semi-conductive layers with different band gaps so that the gap energy decreases from the top. The top layer, usually a-Si, converts the shorter wavelength of the solar spectrum, while the bottom layer, usually microcrystalline Si, converts the longer wavelengths more efficiently. Electrically they are connected in series, which means that the current of the module is limited by the sub-cell with the lowest photocurrent under the prevailing spectral and irradiance conditions. Usually the limiting junction is not known a-priori, thus providing an additional challenge for the measurement and final calibration of a multi-junction module. The increased complexity of a multi-junction thin film module is reflected by the fact that although several type of multi-junction modules are now available in the PV market, there is no IEC standard for their calibration (even though ASTM E2236 addresses some aspects).

Based on our experience at the European Solar Test Installation (ESTI) laboratory, the calibration procedure of a tandem module consists of the following steps: stabilization of the module by successive light soakings, spectral response (SR) measurements, I-V measurements under natural sunlight (with simultaneous measurement of the spectral irradiance) and final correction to standard test conditions (STC). Based on the calculation of the current limiting junction from the SR a spectral mismatch correction for the limiting junction is calculated and applied to obtain the electrical parameters at STC, i.e. at total irradiance equal to 1000 W/m^2 with a reference spectral irradiance for AM1.5G (as defined by IEC 60904-3) and at device temperature equal to 25°C.

We present a detailed calibration study of two multijunction PV modules of the latest technology. The measurements involved all the aforementioned steps until the final calibration values. Issues regarding pre-conditioning and stability are discussed together with aspects of the reliability and uncertainty of the final calibration parameters.

EXPERIMENTAL DETAILS

The performance of tandem modules, as that of other thin-film technologies, is affected considerably by their previous temperature and irradiation history. Effects such as long-term degradation under light soaking and in situ variations due to reversible degradation and annealing can be observed, making the application of an appropriate pre-conditioning treatment necessary in order to ensure that the reported performance values are representative of those expected in normal operation. The IEC 61646 standard provides a procedure to stabilize the electrical parameters of the module via controlled light soaking under load. The modules were light-soaked in a temperature controlled facility at ESTI [1]. The total irradiation of each cycle was approximately 48kWh/m^2 at an average module temperature of 50°C.

Before and after each light soaking cycle, I-V measurements were performed on each module using an indoor large area pulsed solar simulator (PASAN) for monitoring the relative performance. During light-soaking the modules were under load, using a resistance chosen for module operation near their maximum power point. According to IEC 61646 the modules are considered stable when measurements from three consecutive light soaking cycles meet the criterion $(P_{max}-P_{min})/P_{average}<2\%$.

After stabilization, the modules were calibrated under natural sunlight on a sun tracker at clear sky conditions and at air mass values close to AM1.5. For the IV measurements three reference cells were used. First a crystalline Si, which is the standard reference for single junction crystalline Si PV devices. Furthermore two filtered crystalline reference cells were used to establish whether they could substitute the standard crystalline silicon reference cell. These filtered reference cells are also made of crystalline Si for stability but in addition have a filter mounted in front of them so that the combined SR of the package matches closely the SR of the top and bottom junctions respectively in the module. During the measurements, the module temperature was kept in the range (25±2) °C. The irradiance of the natural sunlight was within the range (1000±25)W/m^2 and its spectral distribution was measured with a spectroradiometer OL750.

The spectral response measurements were performed using a setup which has been described in detail previously [2, 3]. A crystalline Si reference cell was used for the spectral response measurements since it has a broad spectral response covering that of both sub-cells in the tandem module.

RESULTS AND DISCUSSION

Pre-conditioning and stabilization.

Two tandem modules (laboratory codes NH801 and NH804) from the same technology (a-Si/μ-Si) but with differences in their deposition parameters have been stabilized. Their maximum power variation is within 2% after 4 light soaking cycles (Figure 1). Therefore the modules were considered stable and ready for calibration measurements.

Spectral Response Measurements

The spectral response of the tandem modules was measured for each junction separately using colored biased light to saturate the response of the other junction respectively [2,3]. For the measurements, mini-modules of smaller dimensions but manufactured in the same production batch were used (NH802 in Figure 2 corresponds to module NH801). Due to their identical manufacturing conditions, the mini-modules can be assumed to have the same SR characteristics as the full size modules. Figure 2 plots the SR of one of the tandem mini-modules, together with the relative spectral response of the crystalline reference cell (PX303c) and the two filtered crystalline-Si reference cell (asp002 with top filter and asp003 with bottom filter). From Figure 2 it is apparent that the SR of the filtered reference cells matches closely the spectral response of the top and bottom junction of the module respectively.

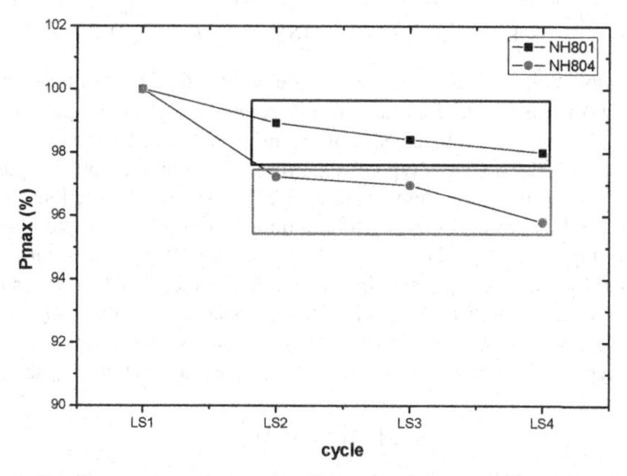

Figure 1 The normalized power measurement of two tandem modules. After four light-soaking cycles the stability criterion of 2% is satisfied, as indicated by the boxes.

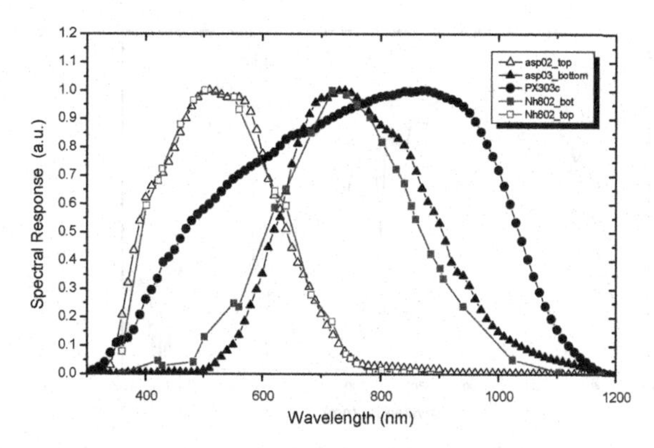

Figure 2 The normalized spectral response measurements of the top and bottom junction of the tandem module and the spectral responses of the three reference cells used for I-V measurements.

Spectral Mismatch Factor Calculations

The spectral response measurements are necessary in order to correct the spectral mismatch between the test spectrum and the reference spectrum and for the mismatch between the spectral responses (SR) of the reference cell and the device under test. This is done by calculating the mismatch factor (MMF)

$$MMF = \frac{\int SR(\lambda) E_{AM1.5}(\lambda) d\lambda}{\int SR(\lambda) E_{L}(\lambda) d\lambda} \cdot \frac{\int SR_{Ref}(\lambda) E_{L}(\lambda) d\lambda}{\int SR_{Ref}(\lambda) E_{AM1.5}(\lambda) d\lambda} \tag{1}$$

where $SR_{Ref}(\lambda)$ is the spectral response of the reference cell used, $E_L(\lambda)$ is the spectral irradiance of the solar simulator in use, while $E_{AM1.5}(\lambda)$ is the reference AM1.5g spectral irradiance as defined in IEC 60904-3. For well matched SRs of reference cell and device under test the MMF would be close to a value of 1 even for a typical spectrum of a solar simulator (Figure 3).

In the case of a multijunction device the spectral mismatch correction should be made with the MMF value for the current limiting junction under the irradiance conditions used for the I-V measurement. Using Equation (1), the MMF values for each of the junctions, top and bottom, were obtained by using the spectral irradiance distribution $E_L(\lambda)$ of the natural sunlight at the time of the IV measurement (Figure 3), and the spectral response $SR(\lambda)$ of the limiting junction. The limiting junction was established from the spectral response measurements by calculating first the hypothetical short-circuit current I_{sc} for each junction, top and bottom, from the following equation

$$I_{sc} = A_{test} \int SR(\lambda) E_L(\lambda) d\lambda \tag{2}$$

where A_{test} is the module area. The limiting junction corresponds to the lower current value. The results indicate that the module NH804 is clearly top limited, whereas module NH801 is also top limited but closer to current matching, both under these irradiance conditions and the reference spectrum.

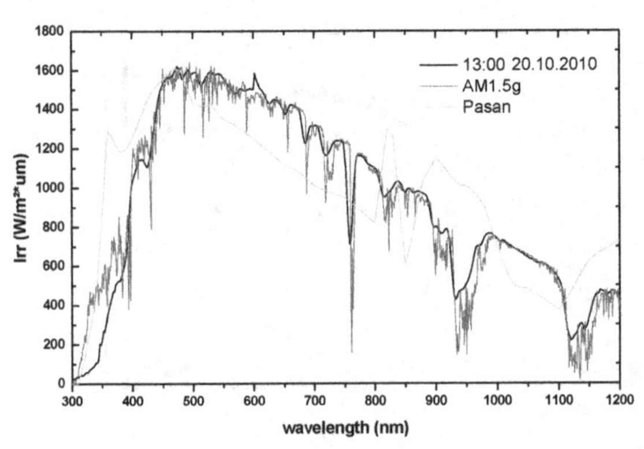

Figure 3: Spectral irradiance as defined in IC60904-3 (AM1.5g) and of natural sunlight at the time of calibration. For comparison the spectrum of an indoor solar simulator (PASAN) is also shown.

Calibration Values obtained under natural sunlight

Table 1 shows the final results for the two modules under test using the aforementioned reference cells. The spectral irradiance of the natural sunlight was close to AM1.5 (Fig. 3), which is reflected by the mismatch factors all being close to unity (i.e. no spectral mismatch). For completeness each limiting junction was considered separately and therefore the MMF was calculated for each.

The standard calibration is done with the crystalline silicon (c-Si) reference cell. As the calculation showed that both modules are top limited, the calibration values of short circuit current I_{sc}, open circuit voltage V_{oc} and maximum power P_{max} of this case (top junction limiting) are used as reference values (100%) and all other values are expressed with respect to them.

Table 1 The results at STC of the outdoor calibration

Module	Ref_Cell	MMF		Isc (A)		Voc (V)		Pmax (W)	
		top	bottom	top	bottom	top	bottom	top	bottom
NH801	c-Si	1.0099	0.9989	100.0%	98.9%	100.0%	99.9%	100.0%	99.9%
	aspire (top)	0.9964	0.9958	102.9%	102.8%	100.9%	100.9%	104.5%	104.4%
	aspire (bottom)	1.0059	1.0009	100.5%	100.0%	100.5%	100.5%	100.4%	99.8%
NH804	c-Si	1.0050	1.0010	100.0%	99.6%	100.0%	99.9%	100.0%	99.5%
	aspire (top)	0.9956	0.9934	103.6%	103.3%	100.1%	100.0%	102.6%	102.3%
	aspire (bottom)	1.0075	0.9997	101.6%	100.8%	99.5%	99.3%	99.7%	98.8%

The measurement uncertainty of the parameters are determined based on our experience with calibration of single junction PV modules and the following considerations [4]. The expanded combined measurement uncertainty (U95%) for the short circuit current I_{sc} is ±1.3%, the same as for single junction devices. This value already includes the uncertainty of the mismatch factors of ±1 %, which is the variation observed in the calculated mismatch factor between different reference cells and limiting junctions. The uncertainty of the open circuit voltage V_{oc} is ±1.4%, as the measurements were made close to 25°C. This uncertainty includes a deviation of up to 2°C and an additional inhomogeneity inside the module up to 3°C. Further contributions are from cabling and possible influence from a current mismatch between top and bottom junctions were considered to obtain a final value of combined expanded uncertainty of ±2.8% for the maximum power, P_{max}.

The following observations can be made:

a) The calibration values for the c-Si reference cell agree well with each other independently of the assumption which junction is limiting.
b) Under the natural sunlight on a clear sky day the choice of reference cell does not significantly affect the mismatch factor for outdoor measurements. This is due to the fact that when $E_L \approx E_{AM1.5}$, equation (1) gives MMF≈1.
c) For both modules the results using the c-Si or the bottom filtered reference cell are well within experimental uncertainty.
d) For both modules and the top filtered reference cell the uncertainty bands for I_{sc} do not overlap and those for P_{max} are just overlapping. This can be attributed to a different

irradiance reading in comparison to the other two reference cells (V_{oc} is less sensitive to this, because it depends logarithmically on irradiance.).

The observations a) – c) are in excellent agreement, as expected given the fact that the spectrum of the natural sunlight was a good match for the defined reference spectrum. However, consistent results were also expected for the case of the top filtered reference cell. As this was not the case, in order to check for possible errors the top filtered reference cell was recalibrated. However, the original calibration value was confirmed. Nevertheless, a further investigation is under way, including possible systematic improvements in the calibration procedure for filtered reference cells. This investigation will in particular consider the accuracy of determining the near UV light component, which might be significant due to instrumentation limits. Once this aspect is solved, it is planned to repeat the calibration using both filtered reference cells and two modules, one top and one bottom limited. In the current work, despite the different manufacturing conditions, both modules resulted in being top limited. Nevertheless, the calibration procedure is independent of the current limiting junction.
Therefore currently from the two filtered reference cells only the one with a bottom filter can substitute the crystalline silicon reference cell for calibration purposes.

CONCLUSIONS

After suitable pre-conditioning two tandem PV modules were calibrated under natural sunlight. The outdoor conditions were close to standard test conditions. Consistent results were obtained using both a standard c-Si reference cell as well as a bottom filtered reference cell. Results with an analogous top filtered reference cell were not fully consistent and require further investigation. Measurement uncertainties of less than 3% for maximum power were reached.

ACKNOWLEDGMENTS

The authors gratefully thank Komlan Anika for his invaluable help in the experimental setup and with the measurements.

REFERENCES

1. Anatoli I. Chatzipanagi, Robert P. Kenny and Tony Sample, "Preconditioning of various thin film PV module technologies by light soaking", Proc. PHOTOVOLTAIC TECHNICAL CONFERENCE - THIN FILM 2010, Aix-en-Provence, 27-28 May 2010

2. M. Pravettoni, G. Tzamalis, K. Anika, D. Polverini and H. Müllejans: "Standard Characterization of Multi-junction Thin-film Photovoltaic Modules: Spectral Mismatch Correction to Standard Test Conditions and Comparison with Outdoor Measurements", in Amorphous and Polycrystalline Thin-Film Silicon Science and Technology — 2010, edited by Q. Wang, B. Yan, S. Higashi, C.C. Tsai, A. Flewitt (Mater. Res. Soc. Symp. Proc. Volume 1245, Warrendale, PA, 2010), 1245-A13-06.

3. M. Pravettoni, K. Anika, R. Galleano, H. Müllejans and E. D. Dunlop: " An alternative method for spectral response measurements of large-area thin film photovoltaic modules" *Progress in Photovoltaics: accepted*

4. H. Müllejans, W. Zaaiman, R. Galleano "Analysis and mitigation of measurement uncertainties in the traceability chain for the calibration of photovoltaic devices" Measurement Science and Technology 20 (2009) 075101 1-12

Mater. Res. Soc. Symp. Proc. Vol. 1321 © 2011 Materials Research Society
DOI: 10.1557/opl.2011.941

Flexible, Lightweight, Amorphous Silicon Based Solar Cells on Polymer Substrate for Space and Near-Space Applications

K. Beernink, A. Banerjee, J. Yang, K. Lord, F. Liu, G. DeMaggio, G. Pietka, C. Worrel, and S. Guha

United Solar Ovonic LLC, 1100 West Maple Road, Troy, MI 48084-5352, USA

ABSTRACT

United Solar Ovonic has leveraged its history of making amorphous silicon solar cells on stainless steel substrates to develop amorphous silicon alloy (a-Si:H)-based solar cells and modules on ~25 μm thick polymer substrate using high-throughput roll-to-roll deposition technology for space and near-space applications. The solar cells have a triple-junction a-Si:H/a-SiGe:H/a-SiGe:H structure deposited by conventional plasma enhanced CVD (PECVD) using roll-to-roll processing. The cells have distinct advantages in terms of high specific power (W/kg), high flexibility, ruggedness, rollability for stowage, and irradiation resistance. The large area (23.9 cm x 32.1 cm) individual cells manufactured in large quantity can be readily connected into modules and have achieved initial, 25 °C, AM0 aperture-area efficiency of 9.8% and initial specific power of 1200 W/kg. We have conducted light-soak studies and measured the temperature coefficient of the current-voltage characteristics to determine the stable values at an expected operating temperature of 60 °C. The stable total-area efficiency and specific power at 60 °C are 7.2% and 950 W/kg, respectively. In this paper, we review the challenges and progress made in development of the cells, highlight some applications, and discuss current efforts aimed at improving performance.

INTRODUCTION

Flexible, light, low-cost reliable photovoltaic (PV) modules are desirable for both space and stratospheric solar arrays. Future high-power spacecraft will require low-cost, lightweight PV arrays with reduced stowage volume. Flexible PV incorporated into lightweight roll-out array designs can meet these requirements. In terms of the requirements for low mass, reduced stowage volume, and the harsh space environment, thin film amorphous silicon (a-Si) alloy cells have several advantages over some other technologies, such as a low temperature coefficient, resistance to degradation from electron and proton irradiation, and the ability to deposit on large-area flexible substrates [1].

For stratospheric applications, such as the high altitude airship, the required PV arrays must provide considerably higher power than current space arrays. Airships typically have a large area available for the PV, but weight is of critical importance. As a result, low cost and high specific power (W/kg) are key factors for airship PV arrays. Again, thin-film a-Si alloy solar cell technology is well suited to such applications. Another stratospheric application is the solar-powered unmanned aerial vehicle. One design utilizes extremely lightweight materials, and is enabled by very lightweight, flexible, high specific power (W/kg), thin-film a-Si based solar cells.

United Solar Ovonic (USO), in collaboration with space companies and government laboratories, has been working on optimization of a-Si alloy solar cells for space and stratospheric applications for a number of years. United Solar has leveraged its decades of experience in providing a-Si alloy solar cells deposited on stainless steel for the terrestrial market to develop solar cells and strings for space and stratospheric markets [2-6]. Considerable progress has been made over the last several years in demonstrating the use of a-Si alloy solar cells for space and stratospheric use, including the optimization of solar cells for high efficiency under the AM0 spectrum and qualification of the cells and modules under harsh space and stratospheric conditions. The culmination of this work has been the transition from R&D demonstrations to the availability in production volumes of lightweight a-Si alloy solar cells on polymer substrate for space and stratospheric applications.

In the following, we present details on the cell structure, design, and performance characteristics. We also review some of the challenges and progress made in development of the cells, highlight some applications, and discuss current efforts aimed at improving performance.

EARLY CELL DEVELOPMENT

Batch deposition on free-standing polymer

USO's development of amorphous silicon alloy solar cells on polymer started with depositions in batch machines on free-standing polymer. These depositions were on polyimide with a thickness of 1 or 2 mil in large-area batch machines, and resulted in cells with aperture area of 412 cm². A schematic view of the cross-section of a cell is shown in Figure 1. The cells consisted of a textured Ag/ZnO back reflector (BR), triple-junction a-Si/a-SiGe/a-SiGe *pin* layers, and a top layer of Indium Tin Oxide (ITO). The Ag/ZnO back reflector layer was sputtered onto a polyimide substrate. The deposition parameters for the Ag/ZnO BR were such that the back reflector was textured to facilitate light scattering and multiple reflections. The a-Si alloy layers were deposited by radio-frequency (RF) PECVD. The sputtered ITO served as a transparent conductor and anti-reflective coating. A stress-balancing layer was sputtered onto the back of the polyimide to balance the stress in the front side layers. The cells were next passivated to remove any shunts and shorts existing in the large area cells. Wire grids and bus bars were added to the top of the structure. A top view of a cell showing the layout of the wires and bus bars is shown in Figure 2. The bus bars at either side serve as the positive cell contacts. At that time, a portion of the substrate and back reflector extended out past the bus bars to serve as the negative cell contact. In later designs, a negative bus bar was added under the positive bus bar.

Early cells made in batch mode using this design achieved initial AM0 efficiencies of 9.0% and specific powers of >1200 W/kg, for an aperture area of 412 cm² [7, 8]. The I-V measured at NASA Glenn for such a cell is shown in Figure 3.

Optimization for high specific power

For space and stratospheric applications targeted for these cells, specific power (W/kg) is an important metric. Reference 8 details analyses to increase specific power by optimizing for high efficiency and low mass. Modifications to reduce the mass included variation in the substrate material and thickness and changes to various module components. In some cases, modifications to reduce mass are accompanied by increased electrical loss and reduced efficiency. The goal of the effort was to explore reductions in mass to maximize specific power, while maintaining high efficiency.

A model which included the mass, shadow loss, and electrical losses was used to analyze specific power of the modules. The model started with a measured or assumed efficiency for a module with standard substrate dimensions, bus bars, and wire spacing. Then, changes were made to module components, and the mass and total loss were used to calculate efficiency and specific power of the modified structure. Variations in bus bar materials and thickness, substrate thickness, wire spacing, and cell aspect ratio were examined to evaluate their effects on specific power. Figure 4 shows the effects of positive bus bar material and thickness on specific power for substrate thicknesses of 1 mil and 2 mil.

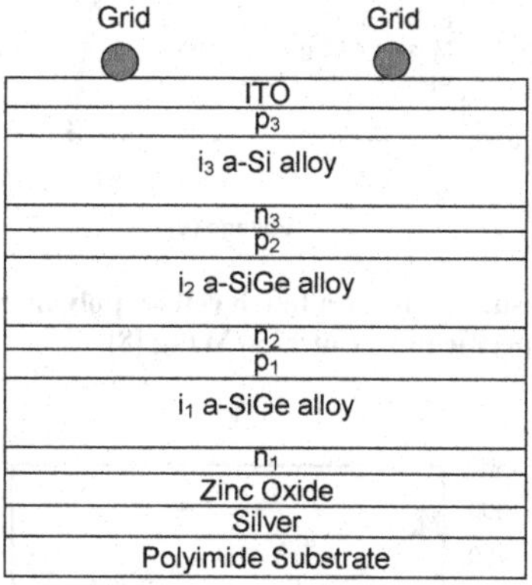

Grid Grid
ITO
p_3
i_3 a-Si alloy
n_3
p_2
i_2 a-SiGe alloy
n_2
p_1
i_1 a-SiGe alloy
n_1
Zinc Oxide
Silver
Polyimide Substrate

Figure 1. Cross section of triple-junction cell on polymer substrate deposited on 2 mil free-standing polymer in large-area batch machines.

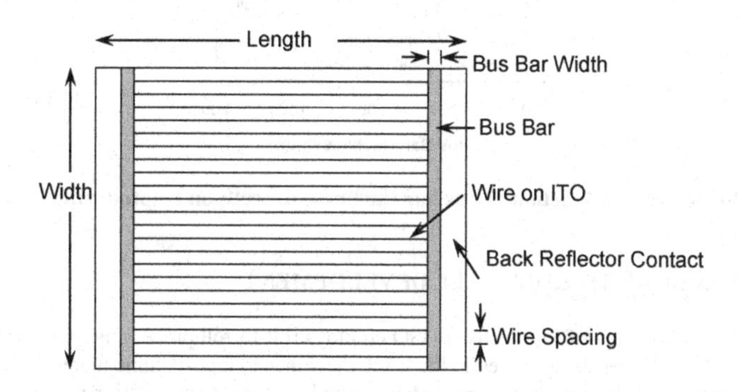

Figure 2. Top view of an early batch cell design.

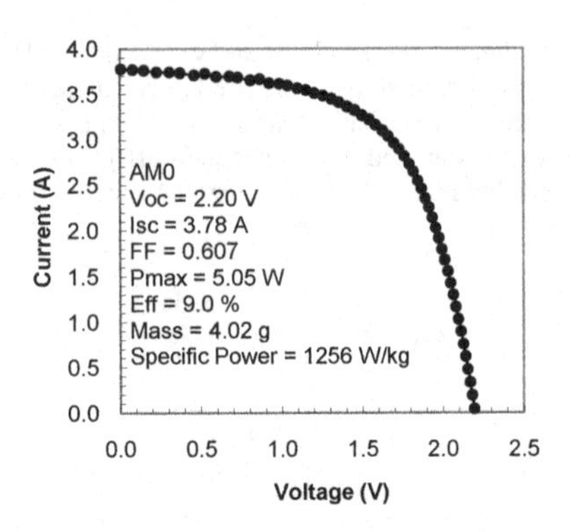

Figure 3. Current-voltage measurement for a batch cell on polymer substrate with 9.0 % AM0 aperture area efficiency and specific power of 1256 W/kg [8].

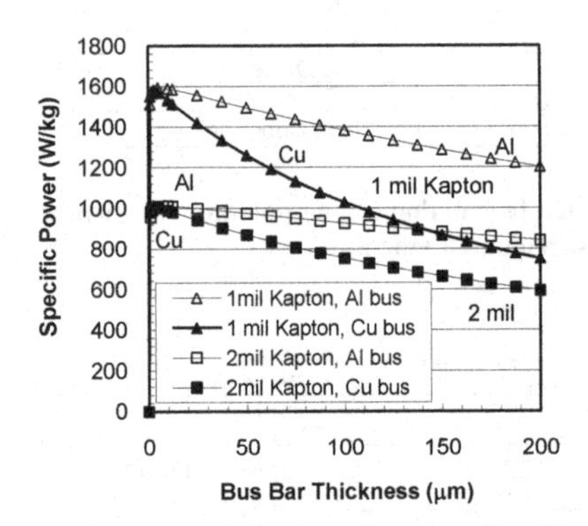

Figure 4. Specific power as a function of bus bar thickness for cells on Kapton with Cu or Al top bus bars [8].

CHALLENGES IN ROLL-TO-ROLL CELL DEVELOPMENT

The large-scale production of solar cells at USO employs roll-to-roll processing onto stainless steel substrates. Taking the results achieved in the batch machines to high-volume manufacturing required transitioning from the low-throughput batch research systems to production roll-to-roll equipment. As with any technology, several challenges arose in moving from lab to production scale machines. Some of these challenges are discussed in this section.

Roll-to-roll Polymer

USO's production roll-to-roll machines are designed to run rolls of stainless steel. Since the expected market for cells on polymer substrate was low, it was necessary to find a way to run the polymer through existing equipment designed for stainless steel. Because of the length and design of the machines, it was not possible to run free-standing polymer through them. We therefore first attached the polymer to the stainless steel, then processed the sandwiched rolls through the systems.

Bonding the polymer in a manner that is compatible with a vacuum deposition process at temperatures up to about 300 °C was difficult. Most adhesives were quickly eliminated because of outgassing or temperature limits. Although it was very desirable, we were unsuccessful in finding an adhesive that could serve the requirement of attaching the polymer to the steel for the deposition, but that could later be removed from the back of the polymer. As a result, the adhesive would remain on the back surface of the polymer substrate. Of course, since an over-riding goal was to make the cells light, a thick, heavy adhesive was not acceptable. In the end, we were able to find an adhesive that could be as little as 0.1 mil thick and was compatible with high-temperature vacuum processing.

In-house lamination processes were developed to bond the polymer and adhesive to the stainless steel substrate, allowing for transport through the roll-to-roll machines.

Stress Balancing

Another major issue was the stress in the deposited films. When the polymer was removed from the steel, the stress in the films caused curling of the polymer towards the back. In the batch systems, a stress-balancing layer was deposited onto the back of the polymer. In the roll-to-roll system, with the polymer bonded to the steel, there was no straightforward way to deposit the stress-balancing layer on the back of the polymer. The result was cells that rolled up to a diameter less than 1 cm, or, depending on the adhesive, cells that curled backwards along all edges, such as the cell in Figure 5.

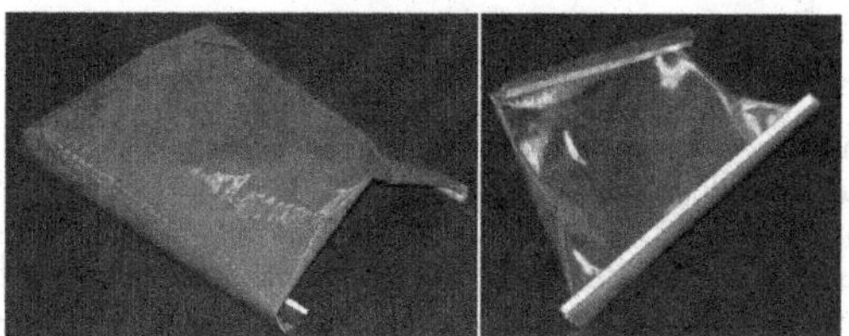

Figure 5. Front (left) and back (right) pictures of a cell without stress-balancing layers on the back.

The solution to the stress-balancing problem is to transfer the stress-balancing layers from the steel to the back of the polymer, as in Figure 6. A chemical etch-stop layer is deposited on the steel, followed by the stress balancing layers. The polymer with adhesive is then bonded to the coated side of the stainless steel. Following depositions onto the front side of the polymer and further cell processing, the steel is chemically etched from the back. The etching stops at the etch-stop layer, leaving the stress-balancing layer and the etch-stop layer on the back of the cell. Figure 7 shows a series of polymer substrates coated with the BR on the front, and with stress-balancing layers of increasing thickness on the back. The samples were prepared following the method of Figure 6.

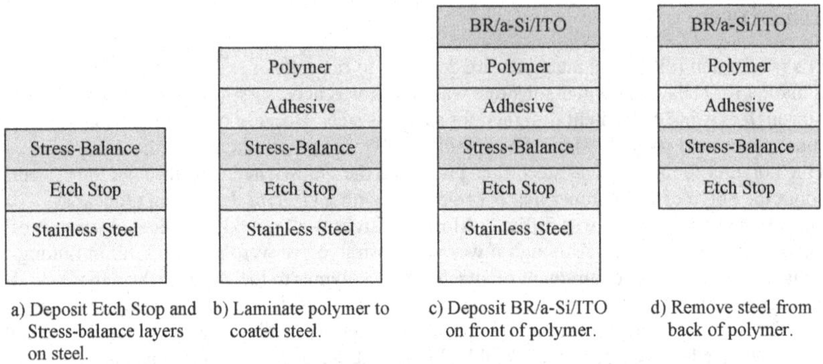

a)	b)	c)	d)
		BR/a-Si/ITO	BR/a-Si/ITO
	Polymer	Polymer	Polymer
	Adhesive	Adhesive	Adhesive
Stress-Balance	Stress-Balance	Stress-Balance	Stress-Balance
Etch Stop	Etch Stop	Etch Stop	Etch Stop
Stainless Steel	Stainless Steel	Stainless Steel	
a) Deposit Etch Stop and Stress-balance layers on steel.	b) Laminate polymer to coated steel.	c) Deposit BR/a-Si/ITO on front of polymer.	d) Remove steel from back of polymer.

Figure 6. Process for putting stress-balancing layers on the back of the polymer cells.

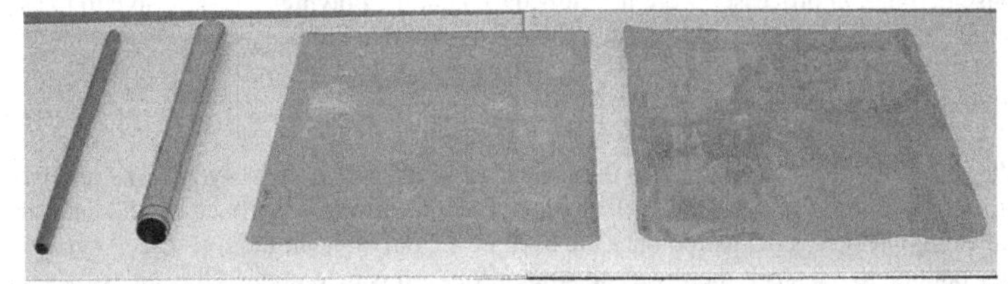

Figure 7. Samples with 4 different thicknesses of stress-balancing layer on the back of 1 mil polyimide with Ag/ ZnO on the front. Thickness of the stress-balancing layer on the back ranges from ~110nm (left) to ~500 nm (right). Samples were made using roll-to-roll processes.

CURRENT STATUS OF CELLS FROM ROLL-TO-ROLL DEPOSITION

a-Si:H/a-SiGe:H/a-SiGe:H triple-junction cells have been fabricated using a roll-to-roll deposition process. The substrate is a roll of stainless steel, 35.6 cm wide and ~125 μm thick. It is laminated onto a roll of polymer ~25 μm thick. A layer of Ag/ZnO back reflector is deposited on the polymer substrate followed by an RF PECVD deposition of the triple-junction stack. A layer of ITO is finally deposited on top to complete the device. The roll is then cut into smaller size pieces, wires and bus bars are applied for current collection, and coated with a proprietary USO top coating [6]. The stainless steel substrate is removed from the back, resulting in a completed solar cell on a polymer substrate. The overall cell dimensions are 23.9 cm x 32.1 cm, and a schematic of the device is shown in Figure 8.

The AM1.5 I-V characteristics were measured using a pulsed solar simulator, and the initial AM0 values were extracted using spectral mismatch and intensity correction factors. In order to obtain light-stabilized AM0 performance, the cells were light soaked under AM0 intensity at 60 °C for 1000 hours. I-V measurements were made at 25 °C and 60 °C. The cells can be interconnected in series to fabricate a string. Pictures of a completed cell and an interconnected string are shown in Figure 9.

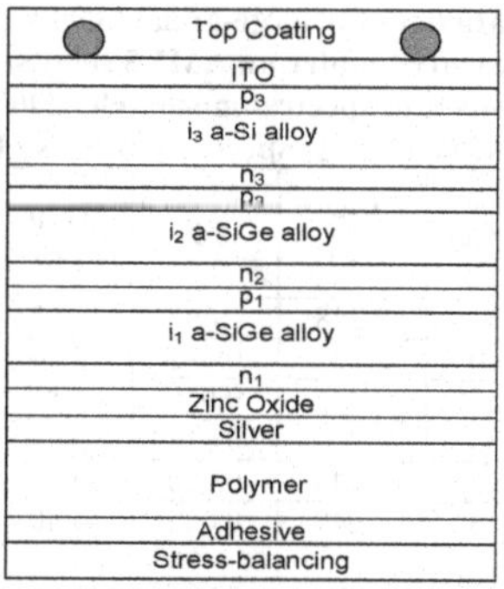

Top Coating
ITO
p_3
i_3 a-Si alloy
n_3
p_2
i_2 a-SiGe alloy
n_2
p_1
i_1 a-SiGe alloy
n_1
Zinc Oxide
Silver
Polymer
Adhesive
Stress-balancing

Figure 8. Cross section of triple-junction cell on polymer substrate.

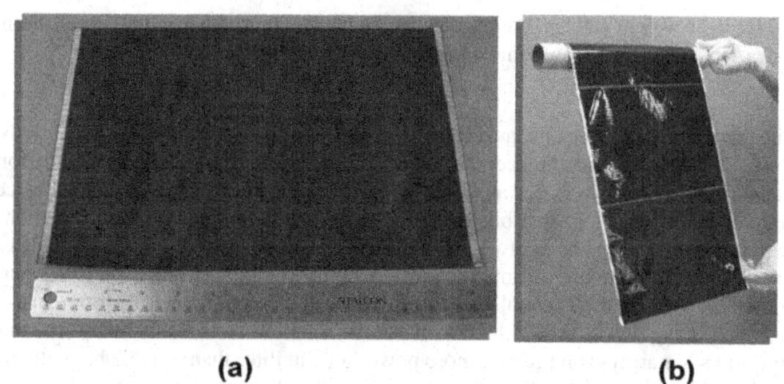

(a) (b)

Figure 9. (a) Solar cell on polymer substrate and (b) interconnected string of solar cells.

The extrapolated initial AM0 characteristics of six large area (23.9 cm x 32.1 cm) solar cells on ~25 μm polymer substrate and measured at 25 °C are summarized in Table I. The AM0 aperture-area efficiency of the cells is in the range of 9.7-9.8%. These correspond to total-area efficiency of 9.2-9.3%. The mass of the cells is ~8.0 g, which corresponds to a specific power of ~1200 W/kg.

In order to obtain stable performance, representative cells have been light soaked at 60 °C for 1000 hours and measured at 25 °C and 60 °C. The typical Staebler-Wronski degradation is ~15%. I-V measurements of these cells in the temperature range 25-100 °C show that the power derating from 25 °C to 60 °C is ~8%. We used these derating factors to determine the light-stable performance of the cells. The stable aperture-area efficiency measured at 25 °C is in the range of 8.3-8.4%. The stable, total-area efficiency values are 7.8-7.9% measured at 25 °C, and 7.2% measured at 60 °C. The corresponding stable specific power at 60 °C is ~950 W/kg.

Table I. Initial, aperture-area AM0 data for large-area (23.9 cm x 32.1 cm) triple-junction cells on ~25 μm polymer substrate measured under an AM1.5 solar simulator and extrapolated to AM0 using corrections for intensity and spectral mismatch. The cells have a USO top coating of ~6 μm. The estimated initial AM0 average specific power is ~1200 W/kg [9].

Cell ID (5MW-)	Area (cm^2)	Temp (°C)	Isc (A)	Voc (V)	FF	Imp (A)	Vmp (V)	Pmax (W)	Aperture Area Eff (%)
2235-277	721.8	24.7	6.64	2.248	0.647	5.57	1.734	9.66	9.80
2235-278	721.8	24.5	6.60	2.251	0.652	5.44	1.779	9.68	9.82
2235-280	721.8	24.7	6.58	2.251	0.655	5.56	1.747	9.70	9.84
2235-743	721.8	24.9	6.53	2.251	0.654	5.45	1.765	9.62	9.76
2235-744	721.8	24.7	6.56	2.243	0.650	5.42	1.765	9.57	9.71
2235-745	721.8	24.7	6.53	2.249	0.652	5.40	1.775	9.59	9.73

EXAMPLE APPLICATIONS

The flexible, light-weight cells on polymer substrate have potential or existing applications in space and in near-space/stratosphere. Some applications are highlighted in this section.

Space

For use in space, the benefits of the a-Si based cells on polymer substrates compared to those traditionally used are lower temperature coefficient, greater resistance to damage from electron and proton irradiation, flexibility, lower cost, and high specific power. These attributes make the cells particularly interesting for missions in orbits having significant electron and proton irradiation, such as MEO (Medium Earth Orbit). The lightweight and flexible features also allow for a new class of array designs based on the ability to roll-out the array, such as Boeing's High Power Solar Array (HPSA) concept [10]. Composite Technology Development is developing their Roll-out and Passively Deployed Array (RAPDAR), which incorporates their elastic memory composite materials [11]. These roll-out solar arrays can provide more power for satellites than is available with current array designs. A picture of a string of cells designed for a roll-out space solar array is shown in Figure 10.

Near-Space/Stratosphere

The cells also have found use in near-space/stratospheric applications. Currently, there is considerable interest in lighter-than-air airships and other stratospheric craft that can act as communications relays or provide continuous surveillance or remote sensing and have the ability to stay in the stratosphere for lengthy periods of time. For large airships, with relatively high surface area available for PV, one of the most important metrics for the solar array is high specific power. The array must be able to provide power to the airship without adding too much mass. USO's high-specific-power cells on polymer are well-suited for this type of application.

High-altitude, long-endurance, unmanned aerial vehicles (HALE UAVs) are another example application of these cells. The QinetiQ Zephyr is a solar-powered HALE UAV with an extremely lightweight design. In 2010, the Zephyr flew continuously for over 2 weeks, reaching altitudes over 20 km, setting three world records, while being powered by USO's lightweight, flexible cells on polymer [12].

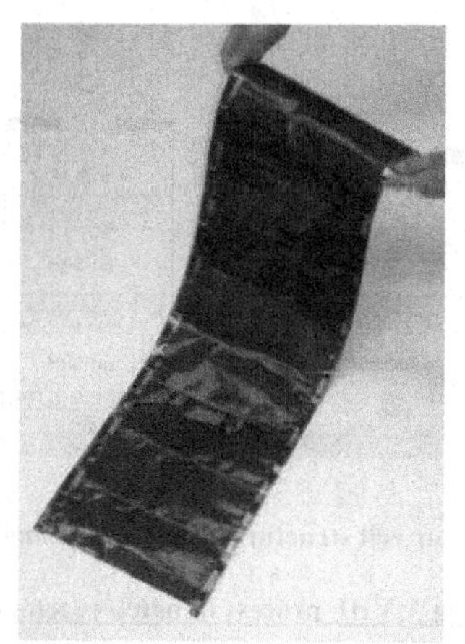

Figure 10. A string of cells on polymer designed for use in a roll-out space solar array.

EFFORTS TO IMPROVE EFFICIENCY

In order to increase the efficiency and specific power of the cells further, we have developed a roadmap that invokes two new technologies: (i) Modified Very High Frequency (MVHF) deposition [13] and (ii) nc-Si:H based solar cells [14]. nc-Si:H solar cells have superior long wavelength response and lower light-induced degradation compared to their a-SiGe:H counterparts. The work was done using a large-area batch reactor on 38.1 cm x 35.6 cm substrates.

Cells made using MVHF process in batch reactors

We developed a high rate MVHF process for the deposition of nc-Si:H, a-Si:H and a-SiGe:H solar cells using a batch reactor on ~25 μm thick polymer substrate of dimensions 38.1 cm x 35.6 cm. The deposition rate is typically 2-4 times that of conventional RF PECVD. We fabricated two multi-junction structures, a-Si:H/a-SiGe:H/a-SiGe:H and a-Si:H/nc-Si:H/nc-Si:H triple-junction solar cells, as shown in Figure 11. The substrate was cut into smaller 5 cm x 5 cm pieces and ITO dots were evaporated through a mask to define an array of individual cells of active area 0.25 cm^2. The AM0 I-V characteristics were measured using a continuous source solar simulator.

We also fabricated large area cells for the two cell structures with aperture areas of 464 cm^2 (21.5 cm x 21.5 cm). All intrinsic layers were deposited using MVHF PECVD.

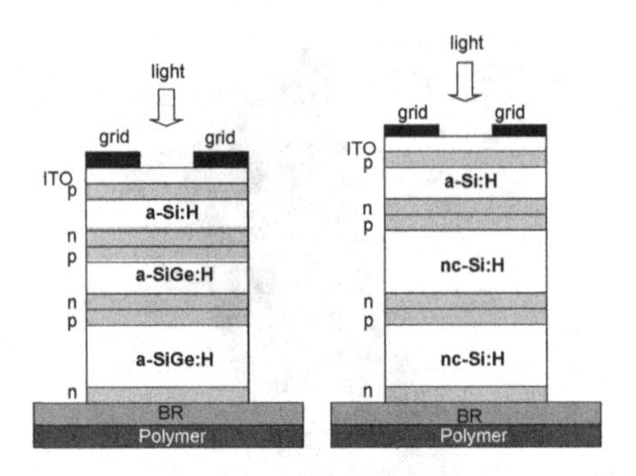

Figure 11. Schematic of two solar cell structures developed using MVHF plasma processes.

Performance of cells made using MVHF process in batch reactors

Initial AM0 I-V characteristics of small area (0.25 cm^2) cells, cut from 38.1 cm x 35.6 cm deposition on ~25 μm polymer substrates, are summarized in Table II. The initial AM0 efficiency for the a-Si:H/nc-Si:H/nc-Si:H cell structure, 10.5-10.6% is slightly higher than that for the a-Si:H/a-SiGe:H/a-SiGe:H cell structure, 10.0-10.2%. The QE response of the a-Si:H/nc-Si:H/nc-Si:H cell is shown in Figure 12.

Table II. Initial AM0 values of small-area (0.25 cm^2) a-Si:H/a-SiGe:H/a-SiGe:H and a-Si:H/nc-Si:H/nc-Si:H triple-junction cells made with MVHF on ~25 μm polymer substrate.

Sample No.	Triple-junction cell structure	V_{oc} (V)	FF	AM0 QE (mA/cm^2)				AM0 Efficiency (%)
				top	middle	bottom	total	
8182	a-Si/a-SiGe/a-SiGe	2.33	0.71	8.38	8.34	8.54	25.3	10.1
8191	a-Si/a-SiGe/a-SiGe	2.36	0.73	8.06	8.22	8.90	25.2	10.2
15951	a-Si/nc-Si/nc-Si	1.91	0.74	11.24	10.17	10.11	31.5	10.5
15954	a-Si/nc-Si/nc-Si	1.94	0.75	11.08	10.03	10.35	31.5	10.6

In order to attain superior stabilized solar cell performance, one needs to attain 1) high quality nc-Si:H, a-Si:H, and a-SiGe:H component cells, 2) optimized current match between the component cells, and 3) optimized cell structure with the least light-induced degradation. We have optimized the MVHF process for nc-Si:H, a-Si:H, and a-SiGe:H film deposition and obtained equivalent or slightly better performance for the high deposition rate, MVHF high rate cells, as compared to comparable (low-rate) RF cells.

We conducted a light soaking study to compare a-Si:H/a-SiGe:H/a-SiGe:H triple-junction cells with a-Si:H/nc-Si:H/nc-Si:H triple-junction cells. The light-soaking experiments were conducted under AM0 light intensity for up to 1000 hours (open circuit, at 60 °C). As illustrated in Figure 13, both type triple-junction cells deposited by MVHF PECVD only degraded about 6-8% while RF deposited a-Si:H/a-SiGe:H/a-SiGe:H triple-junction cells usually degrade ~15%.

For large-area (464 cm²) cells, the nc-Si:H based triple-junction cell shows 8.7% stabilized AM0 efficiency, while the a-SiGe:H based triple junction cell yields 8.5%. In comparison, the best stabilized AM0 efficiency for RF a-SiGe:H based triple-junction cells is 8.2-8.4% [9].

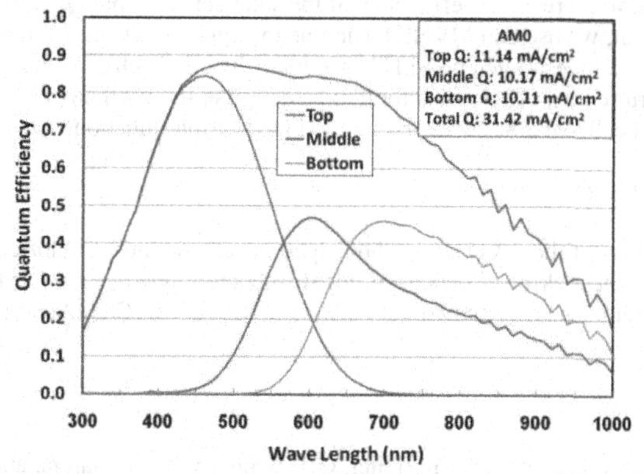

Figure 12. Quantum efficiency of an a-Si:H/nc-Si:H/nc-Si:H triple-junction cell. The active-area of the cell is 0.25 cm^2, fabricated from a 38.1 cm x 35.6 cm deposition [9].

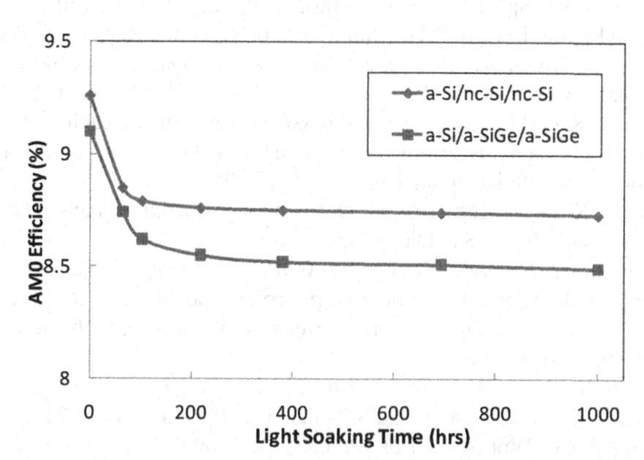

Figure 13. AM0 efficiency of a-Si:H/a-SiGe:H/a-SiGe:H and a-Si:H/nc-Si:H/nc-Si:H triple-junction cells, deposited by using MVHF PECVD process, plotted against light soaking time. Light soaking conditions are one sun AM0 light intensity, 60 °C, at open circuit [9].

CONCLUSIONS

We have developed lightweight a-Si:H/a-SiGe:H/a-SiGe:H triple-junction solar cells on ~25 μm thick polymer substrate using high-throughput roll-to-roll deposition processing. The stable aperture-area efficiency measured at 25 °C is 8.4%. The corresponding stable specific power at 60 °C is ~950 W/kg. These cells are suitable for space and near-space/stratospheric applications. In order to increase the efficiency of the solar cells, we pursued two new approaches. In the first, we used an MVHF technique to deposit multi-junction a-SiGe:H based cells. In the second, we investigated nc-Si:H based multi-junction cells. We have achieved stabilized AM0 aperture area efficiencies for large area cells (464 cm^2) on polymer of 8.5% for the MVHF a-SiGe:H cells, and 8.7% for the nc-Si:H based triple-junction cell.

ACKNOWLEDGEMENTS

We thank members of the R&D team who helped in cell and module fabrication, measurements, and light soaking. Portions of this work were funded by the Air Force Research Laboratory, Space Vehicles Directorate under contracts P29601-03-C0122 and F29601-00-C-0024.

REFERENCES

1. S. Guha, J. Yang, A. Banerjee, T. Glatfelter, G.J. Vendura, Jr., A. Garcia, and M. Kruer, 2nd World Conf. on Photo. Solar Ener. Conv. Proc., Vienna, 3609 (1998).
2. A. Banerjee, F. Liu, K. Beernink, K. Lord, G. DeMaggio, B. Yan, G. Pietka, C. Worrel, X. Xu, J. Yang, and S. Guha, Space Power Workshop, Los Angeles, CA (2009).
3. X. Xu, K. Lord, G. Pietka, F. Liu, K. Beernink, C. Worrel, G. DeMaggio, A. Banerjee, J. Yang, and S. Guha, 33rd IEEE Photov. Spec. Conf. Proc., San Diego, CA (2008).
4. A. Banerjee, K. Beernink, X. Xu, B. Yan, K. Lord, F. Liu, G. DeMaggio, G. Pietka, C. Worrel, J. Yang, and S. Guha, *Space Power Workshop*, Manhattan Beach, CA (2008).
5. F. Liu, J. Owens, G. Pietka, K. Beernink, A. Banerjee, J. Yang, and S. Guha, 34th IEEE Photov. Spec. Conf. Proc., Philadelphia, PA, 1370-1373 (2009).
6. F. Liu, K. Beernink, X. Xu, A. Banerjee, G. DeMaggio, G. Pietka, J. Yang, and S. Guha, 33rd IEEE Photov. Spec. Conf. Proc., San Diego, CA (2008).
7. K. Beernink, G. Pietka, J. Noch, K. Younan, D. Wolf, A. Banerjee, J. Yang, S. Jones, and S. Guha, Proceedings of the Mater. Res. Soc. Symp. Proc., paper V6.2, San Francisco (2002).
8. K.J. Beernink, G. Pietka, J. Noch, D. Wolf, A. Banerjee, J. Yang, S. Guha, and S.J. Jones, 29th IEEE PVSC, 998-1001 (2002).
9. A. Banerjee, X. Xu, K. Beernink, F. Liu, K. Lord, G. DeMaggio, B. Yan, T. Su, G. Pietka, C. Worrel, S. Ehlert, D. Beglau, J. Yang, and S. Guha, 35th IEEE PVSC, 2651-2655 (2010).
10. R. Stribling, Space Power Workshop, Los Angeles, CA (2007).
11. R. Barret and R. Taylor, Nanotech Conference and Expo 2009, Houston (2009).
12. QinetiQ Group PLC, www.qinetiq.com.
13. X. Xu et al., 34th IEEE PVSC, pp. 2159-2164 (2009).
14. J. Meier et al., Appl. Phys. Lett. **65**, 860-862 (1994).

Mater. Res. Soc. Symp. Proc. Vol. 1321 © 2011 Materials Research Society
DOI: 10.1557/opl.2011.942

Properties of amorphous silicon passivation layers for all back contact c-Si heterojunction solar cells

Lulu Zhang[1,2], Ujjwal Das[1], Jesse Appel[1], Steve Hegedus[1], Robert Birkmire[1,2]
[1]Institute of Energy Conversion, University of Delaware, Newark DE 19716
[2]Department of Physics and Astronomy, University of Delaware, Newark DE 19716

ABSTRACT

Low temperature deposited Interdigitated All Back Contact a-Si:H/c-Si Heterojunction (IBC-SHJ) devices are a promising approach for high efficiency, low cost solar cells on thin wafers. Thin intrinsic a-Si:H films (i-a-Si:H) deposited below 300°C provide excellent surface passivation and high Voc. However, the optical properties of a-Si:H layers and electronic band alignment at the heterointerface are critical to reduce optical losses and transport barriers in IBC-SHJ solar cells. At the front illumination surface, a wide band gap (E_g) i-a-Si:H layer with good passivation is desirable for high Voc and Jsc while at the rear surface a narrower E_g i-a-Si:H layer with good passivation is required for higher FF and Voc as seen in 2D numerical simulation. Various substrate temperature, H_2/SiH_4 dilution ratio and plasma power conditions were explored to obtain i-a-Si:H with good passivation and desired E_g. All the deposited films are characterized by Variable Angle Spectroscopic Ellipsometry (VASE) to determine E_g and thickness and by Fourier Transform Infrared spectroscopy (FTIR) to estimate hydrogen content and microstructure factor. Passivation qualities are examined by quasi-steady state photoconductance (QSS-PC) measurement. The i-layer E_g, was varied in the range from ~1.65eV to 1.91eV with lifetime >1 ms. Lowest E_g is obtained just prior to the structure transition from amorphous to epitaxial-like growth. The FF of IBC-SHJ devices improved from 20% to 70% as E_g of the a-Si:H rear passivation layer decreased from 1.78 to 1.65 eV.

INTRODUCTION

Interdigitated All Back Contact Silicon Hetero-Junction (IBC-SHJ) solar cells combine the advantages of back contact designs with c-Si/a-Si:H heterojunction technology, which have potentials to reach 26% efficiency at lower cost [1]. The all back contact design eliminates front grid shading loss resulting in high short circuit current (J_{sc}), reduces grid resistance leading to improve fill factor (FF) by increasing contact coverage and thickness, and simplifies interconnection between cells when integrated into a module. Additionally, silicon heterojunction technology utilizes a low-temperature, <300°C, continuous processing with excellent surface passivation and high open circuit voltage (V_{oc}), which is compatible with thin, <50 micron wafers. The use of an a-Si:H structure for both the emitter and contact results in excellent surface passivation and high open circuit voltage (V_{oc}). Thus IBC-SHJ solar cells have advantages in performance, and manufacturability with a low temperature process.

It is well known that a thin (<10nm) i-a-Si:H buffer layer deposited on both surfaces of a c-Si wafer reduces surface recombination and emitter saturation current by reducing structural and electronic defect at c-Si/a-Si:H interfaces and therefore improves Voc of SHJ solar cells. Without i-a-Si:H buffer layers, IBC-SHJ solar cells suffer low Voc and Jsc due to poor surface passivation quality [2]. The carrier transport in IBC-SHJ solar cells is governed by interface band structure and alignment. An i-a-Si:H buffer layer without the appropriate properties causes a very low FF, and an anomalous "S" shaped J-V curve, which has been seen in experiments and simulations [1]. Reducing rear surface buffer layer thickness and/or reducing buffer layer E_g are the effective methods to improve FF in IBC-SHJ solar cells as has been predicted by 2D simulation. In this paper, we determine conditions for a lower E_g rear buffer layer with good passivation quality to apply to IBC-SHJ devices. Varying substrate temperature, H2 dilution ratio, and plasma power yields a wide range of buffer layer E_g with high quality passivation, which allows the buffer layer conditions for the back surface to be optimized.

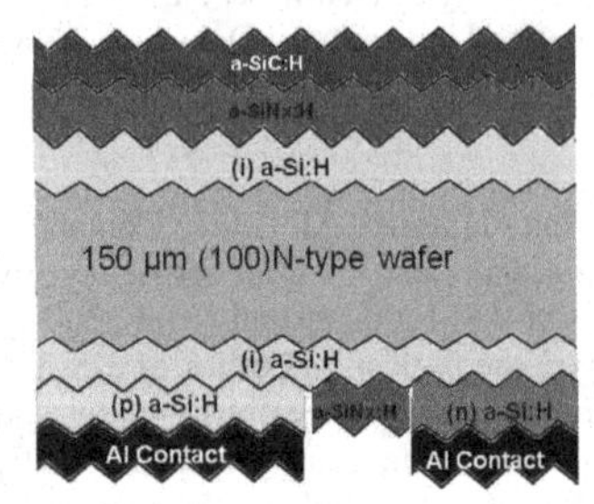

Figure 1 Cross section of an IBC-SHJ device on 150um n-type textured Cz wafer

EXPERIMENT

The cross section view of an IBC-SHJ solar cell structure is shown in Figure 1. Substrates were 150µm thick n-type wafers, with resistivity of ~2.5Ω•cm and the device area was 1.25 cm2. All the a-Si:H and its alloy layers were deposited in a multi-chamber plasma enhanced chemical vapor deposition (PECVD) system in low temperature (<300°C) using either direct current (DC) or radio frequency (RF) plasma. All the c-Si surfaces were passivated with i-a-Si:H buffer layers. The emitters and contact structures were an interdigitated finger-like configuration on the rear surface and the gaps between the fingers were capped by a-SiN_x:H[3]. A multilayer stack AR coating was applied on the top of the front buffer layer [4] to minimize optical loss. The i-a-Si:H buffer layer conditions were varied to optimize surface passivation and the device performance. Prior to deposition, the solvent cleaned silicon wafers were oxidized for 5 min in a mixture of H_2SO_4/H_2O_2 (2:1) followed by a 5 min DI water rinsing. The oxide layer was then removed by a dip in 10% HF for 1 min. Two step photolithography processes were used to form the rear finger-like interdigitated pattern. Back contact layer was formed by 500nm thick aluminum deposited by electron beam evaporation.

To characterize ~10nm i-a-Si:H film, symmetric structure were prepared where i-a-Si:H films were deposited on both sides of 150um polished c-Si wafers, and thermally annealed at 300°C for 25mins. These symmetric structures were used for FTIR measurements to characterize the Si-H bonding configuration. From the FTIR spectra, hydrogen content C_H and microstructure factor $R_{mf}=I_{SiH_2}/[I_{SiH}+I_{SiH_2}]$ of the films were determined, which reflected the structure and quality of the films [5]. Passivation quality of the i-a-Si:H layers was also evaluated using the symmetric structures where the effective minority carrier lifetime was determined from QSS-PCD measurements. Variable Angle Spectroscopy Ellipsometry (VASE) was used to determine i-layer thickness and optical properties. The optical band gap (E_g) was determined from Tauc's plot.

RESULTS AND DISCUSSION

Three critical deposition parameters of i-a-Si:H layer investigated in this work were : (1) Substrate temperature; (2) H2 Dilution ratio; (3) DC or RF plasma power. It is shown that all three parameters can change the E_g of i layer but maintain good quality of passivation until the film develops epitaxial-like growth. Optical band gap was found to increase with increase in hydrogen content in the films seen in Figure 5.

Very thin low band gap i-Si:H films

A series of ~10nm intrinsic buffer layers were deposited at different substrate temperatures using DC plasma and H2 dilution ratio R=H2/SiH4 of 2. With increasing substrate temperature from 150°C to 250°C, the optical E_g of i-a-Si:H films gradually decreased from 1.90eV to 1.65eV as shown in Figure 2-a. At 300°C, structure in the imaginary portion of dielectric function obtained from VASE clearly suggested appearance of crystalline phase (Figure 2-b). In Figure 3, hydrogen content and microstructure factor are shown to decrease with increasing substrate temperature where microstructure factor is an indication of the defect level of the films: the larger Rmf, the more defective the film. The hydrogen content in the <10nm i-a-Si:H films varied from 5% to 30% which is a much larger range then has been reported for a-Si:H films with thickness >0.5μm. The thin a-Si films with very low hydrogen content, ~5%, usually showed epitaxial-like growth.

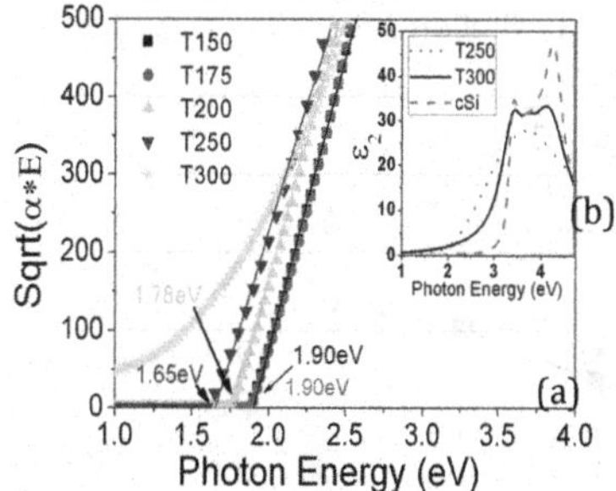

Figure 2 (a) Tauc's plot of various i-a-Si:H layers deposited at different temperatures; (b) Imaginary part of dielectric function of a-Si:H layers at various temperature obtained from spectroscopic Ellipsometry. The c-Si spectra is shown for comparison.

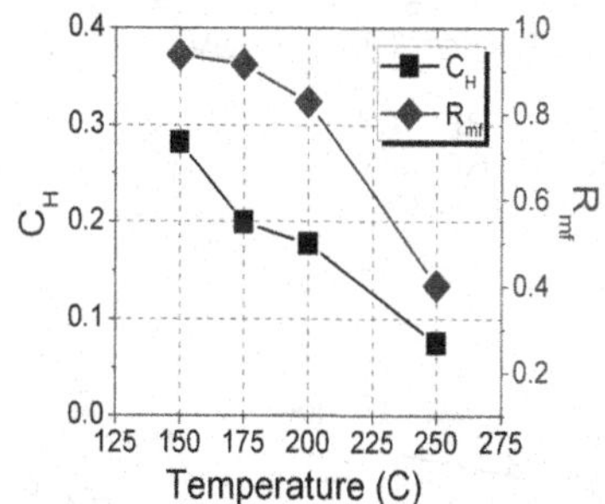

Figure 3 Hydrogen content (C_H) and microstructure factor (R_{mf}) decreasing with substrate temperate increasing.

In second series, H2 dilution ratio, R, was varied with a constant substrate temperature of 200°C and DC plasma. E_g of i-a-Si:H also showed monotonic decrease from 1.91eV to 1.69eV, as seen in Figure 4-a with increasing R. When R was larger than 5 at 200°C, the films enter the crystalline transition region where epitaxial-like growth occurred. Relations between R value and C_H and R_{mf} shown in Figure 4-b are similar to relations seen with varying temperature.

Low E_g i-a-Si:H films were also deposited using RF plasma where the power was varied from 20W to 60W, at a fixed 250°C substrate temperature without H_2 dilution. Optical E_g of RF deposited films was only slightly influenced by RF power between 40-60W as shown in Figure 4-c while decreasing the RF plasma power to 20W allowed microcrystal structure to form. Hydrogen content and microstructure factor decreased with decrease of RF plasma powers as shown in Figure 4-d.

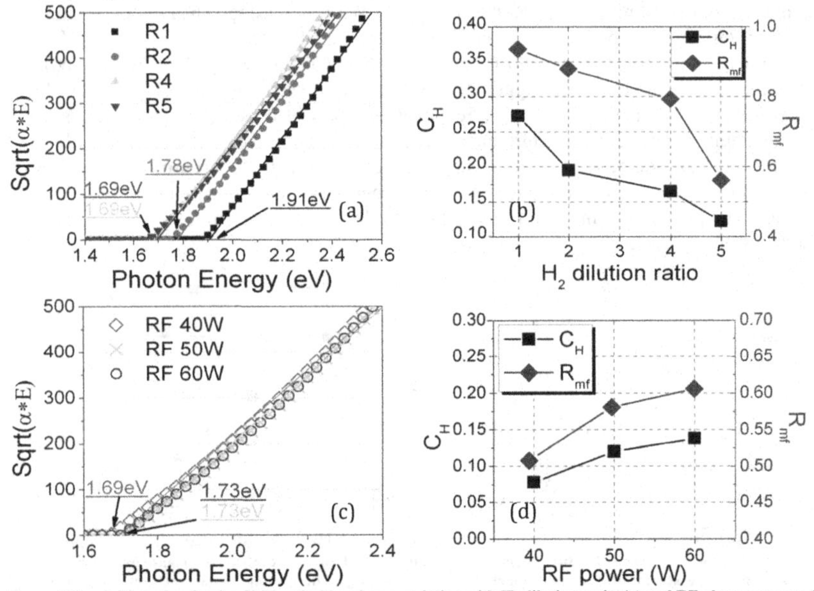

Figure 4 Tauc's Plots showing i-a-Si:H optical band gap variation with H_2 dilution ratios(a) and RF plasma powers (c), respectively. Hydrogen Content (C_H) and Microstructure Factor (R_{mf}) decreased with increase of H_2 dilution ratios (b) and decrease of RF plasma powers (d).

Figure 5 shows that E_g increased with the hydrogen content, regardless of which deposition condition was varied. Since the goal was to obtain films with low E_g and good passivation, we have a wide range of conditions to independently optimize the passivation while maintaining low E_g. This finding also provided option for two separate intrinsic buffer layers for passivation at the front (higher E_g for reduced absorption loss) and rear (lower E_g for improved carrier transport) surfaces of IBC-SHJ devices.

Effect of a-Si:H conditions on passivation quality

Minority carrier effective lifetimes of these symmetric structures were measured by QSS-PCD technique. Figure 6 shows the variation of effective minority carrier lifetime as a function of hydrogen content in the intrinsic buffer layer. Based on results of others, we assume that high hydrogen content and high microstructure factor are suggestive of a more defectively structured film and poor passivation quality. Figure 6 shows that this is generally true but poor passivation can also occur with very low hydrogen content, when an epitaxial-like growth was observed from VASE measurements. However, the relation between lifetime and hydrogen content, hence the band gap still offered a wide range of conditions with good passivation. Within this range, appropriate i-a-Si:H layers can be chosen for IBC-SHJ devices.

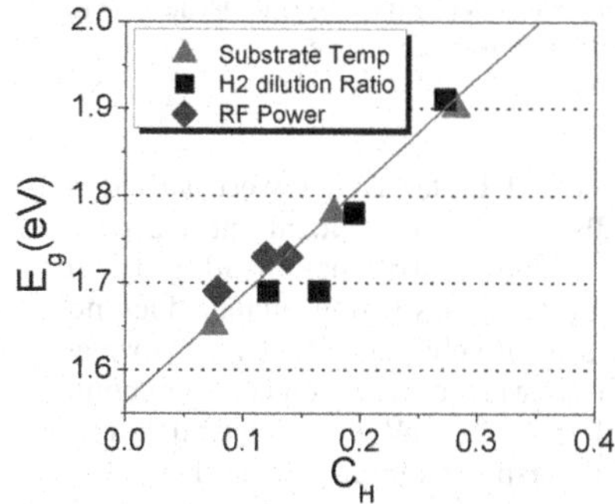

Figure 5 E_g has linear relation with C_H of films.

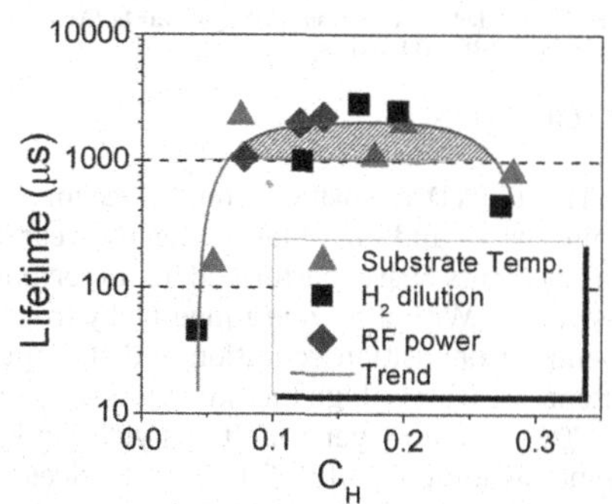

Figure 6 Relation between minority carrier lifetime and hydrogen content of films.

IBC-SHJ device results

Results of 2D simulation seen in Figure 7 indicated that reducing the E_g of i-layer under the p-strip (emitter) from 1.72eV to 1.65eV will increase the FF from 55% to 78%, assuming the same passivation quality and electron affinity of i-layer. IBC-SHJ devices were fabricated with 5 nm thick i-layers having three different bandgap for the rear side passivation: (1) 1.78eV at 200°C and R=2 ratio; (2) 1.72eV at 250°C and R=2.5 ratio; (3) 1.65eVat 250°C and R=2 ratios. Figure 8 shows that the fill factor was significantly improved from an S-shape curve to 70% as E_g decreased from 1.78eV to 1.65eV consistent with simulation results. The variance in FF of the experiments and simulation may be caused by other factos [1, 6].

However, the Voc of the devices were still not optimum. The initial lifetimes on passivated c-Si wafers were all well over 1ms with an implied V_{oc} over 700mV. However, the front surface passivation quality had degraded during the subsequent processing of the back surface, as evidenced by a change in reflection and a visual non-uniformity of the front surface. This may be due to the etching of the front protective passivation stack. Further investigation has been currently carried out.

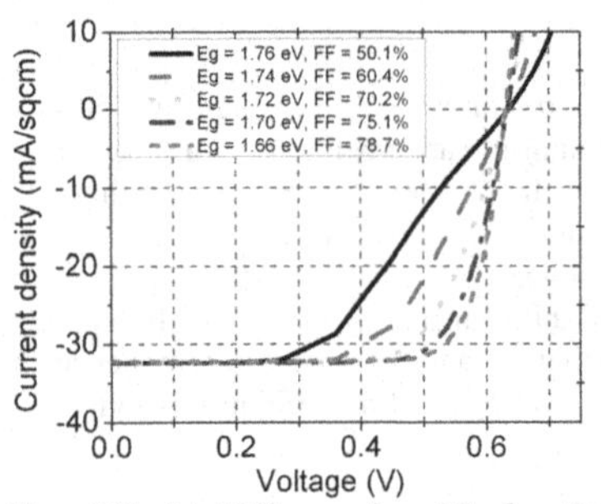

Figure 7 Simulated J-V curves showed E_g of rear buffer layer effect on IBC-SHJ devices.

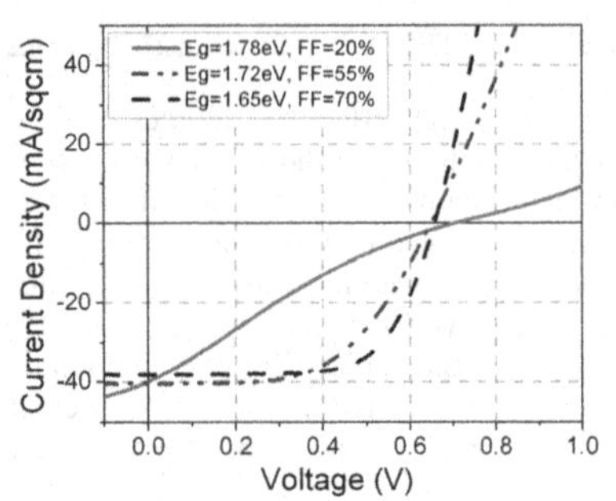

Figure 8 J-V curves for IBC-SHJ cells with different E_g i-a-Si:H on the rear.

CONCLUSION

The PECVD deposition conditions for ~10nm i-a-Si:H buffer layers were varied to obtain low E_g and high lifetime for the rear side of IBC-SHJ device. Optical and electrical measurements were performed to determine film thickness, structure, band gap, and passivation. Within a wide range of hydrogen content, film passivation quality does not depend on deposition condition and the optical E_g is positively dependent on hydrogen content. To achieve high lifetime but low E_g intrinsic buffer layer, hydrogen content should be <30%. Three i-layer conditions yielding E_g from 1.78 to 1.65 eV were selected as rear side passivation of an IBC-SHJ. FF of devices showed the expected improvement from 20% to 70% consistent with 2D simulation.

ACKNOWLEDGMENTS

Authors thank K. Hart for skilled processing contributions. This work is funded by US Department Energy SETP program under contract number DE-FG36-08GO18077.

REFERENCES

1. M.Lu, U.Das, S.Bowden, S.Hegedus and R.Birkmire, Prog. Photovoltaics: Res & Appl. 19,326 (2011)
2. M.Lu, U.Das, S.Bowden and R.Birkmire, 33rd IEEE PVSC , 1-5 (2008)
3. J. Appel, L.Zhang, U.Das, S.Hegedus, S. Mudigonda, R.Birkmire and J.Rand, 35th IEEE Photovoltaic Specialists Conference (PVSC) 001295-001298 (2010)
4. B.Shu,U. Das, J.Appel, B.McCandless, S.Hegedus and R.Birkmire, 35th IEEE, Photovoltaic Specialists Conference (PVSC), 003223-003228 (2010)
5. A.A Langford, M.L.Fleet, B.P.Nelson,W.A.Landford and N.Maley, Phys. Rev.B 45, 13367 (1992)
6. J. Allen, B. Shu and S. Hegedus, presented at 37th IEEE PVSC, Seattle, WA 2011. (to be published)

Mater. Res. Soc. Symp. Proc. Vol. 1321 © 2011 Materials Research Society
DOI: 10.1557/opl.2011.943

Thin Film Si Photovoltaic Devices on Photonic Structures Fabricated on Steel and Polymer Substrates.

S. Pattnaik[1],N. Chakravarty[1], J. Bhattacharya[1], R. Biswas[1], D. Slafer[2],V.L. Dalal[1]
[1]Electrical and Comp. Engr., Iowa State University, Ames, Iowa;
[2]Lightwave Power, Cambridge, Massachusetts.

ABSTRACT

In this paper, we report on the growth and fabrication of thin film Si photovoltaic devices on photonic structures which were fabricated on steel and PEN and Kapton substrates. Both amorphous Si and thin film nanocrystalline Si devices were fabricated. The 2 dimensional photonic reflector structures were designed using a scattering matrix theory and consisted of appropriately designed holes/pillars which were imprinted into a polymer layer coated onto PEN, Kapton and stainless steel substrates. The photonic structures were coated with a thin layer of Ag and ZnO. Both single junction and tandem junction (amorphous/amorphous and amorphous/nanocrystalline) cells were fabricated on the photonic layers. It was observed that the greatest increase in short circuit current and efficiency in these cells due to the use of photonic reflectors was in nanocrystalline Si cells, where an increase in current approaching 30% (compared to devices fabricated on flat substrates) was obtained for thin (~ 1 micrometer thick i layers) films of nano Si deposited on steel structures. The photonic structures (which were nano-imprinted into a polymer) were shown to stand up to temperatures as large as 300 C, thereby making such structures practical when a steel (or glass) of kapton substrate is used. Detailed measurements and discussion of quantum efficiency and device performance for various photonic back reflector structures on steel, kapton and PEN substrates will be presented in the paper.

INTRODUCTION

Thin film silicon solar cells are a very important part of the solar industry, by decreasing the amount of material used as well as using inferior so cheaper quality material they pursue the ability to decrease the cost of manufacturing [1]. But, using lower quality material does not allow making devices thick as they lead to poor device quality. On the other hand, absorption is a strong function of thickness of the absorber layer, which creates a dilemma. Many groups try to overcome this issue by using increasing the light path in the absorber by using light trapping methods by introducing randomly roughened back reflectors namely annealed silver or hot silver(Ag), etched Zinc oxide(ZnO) and silica spheres[2-5]. All these structures use a metallic reflector which is mostly silver and sometimes aluminum in conjunction with a layer separating the active device to prevent diffusion of metal into the active device. ZnO is the most general choice for this layer. Most of these textured reflectors lead to light path length enhancement close to ~10, whereas theoretical limit is close to 49, which is calculated for a perfect lambertian diffuser which should then give an enhancement of $4n^2$, where n is the refractive index of Silicon. This clearly indicates that we have room for further improvement.

Therefore, a more organized approach has been studied extensively in recent years [6-9]. This approach relies on modeling to incorporate increasing the path length of light, albeit in a specific target wavelength range and also get rid of the detrimental effects of randomly oriented structures which suffer from intrinsic losses from generated at the metallic reflector interface [10]. This design is based on a 2 d photonic structure which can act as a 2d diffraction grating and hence diffract light sideways which could enhance the path length of the light travelling in the absorber. The design is also modified as to have a plasmonic effect to decrease losses at the metal interface.

The development of the photonic structure was based on modeling using scattering matrix simulations, where in Maxwell equations are solved in Fourier space [8,9]. For the case of proof of concept, we earlier fabricated the above structures on a c-Si using photolithography and reactive ion etching. Although this was a costly process, it did show that the increase in current density was significant for the photonic structured substrates compared to their planar counterpart [9].

To take this process to the next level, cheaper substrates had to be used. Plastics are a very formidable alternative as they are cheap, flexible, easily processable and can be applied to roll to roll processing [11,12]. In this study, two plastics namely polyethylenenapthalate (PEN) and Polyimide (Kapton) have been used in conjunction with PEN coated on a thin sheet of Stainless steel (SS). The plastics are treated to be able to withstand higher deposition temperatures like 200°C, 250°C and 300°C respectively. The photonic structures were then nano imprinted on the above structure using nano-imprint lithography. Nano-imprint technology is an inexpensive method for fabricating nanoscale patterning and is also compatible with commercial mass production methods using roll to roll production techniques [12].

EXPERIMENTAL DETAILS

The substrates mentioned above were fabricated by Lightwave Power inc. in their in-house nano-imprinting unit. The structures fabricated with a hexagonal lattice with a pitch of ~760nm and distance between each structure of ~450nm and the initial depth of as received structures being ~180nm. The substrates are then covered with a layer of Ag of 200nm by thermal evaporation and then 80nm layer of ZnO is sputtered on top providing a spacer layer between the device and silver. This layer acts as a spacer for silver not to diffuse into the device as well as is necessary for optical matching. An important thing to note is that the layer of Ag/ZnO smoothens out the structure made on the plastic substrates.

The active layers are then deposited on the substrate at p-i-n configuration using our in-house PECVD system at 45MHz from a mixture of Silane, hydrogen and other dopant gases. The maximum temperature of deposition is controlled by the substrate used, although temperature is also varied during the fabrication according to the layer being deposited. Both single cells of a-Si and μc-Si of similar thickness respectively are deposited on different types of substrates to see the effectiveness of the photonic structures; finally tandem devices were also fabricated on them. The last step is to deposit an anti-reflecting top contact Indium tin oxide (ITO) of ~70nm with resistivity of 4×10^{-4} ohm.cm is deposited on the device. Figure 1 shows a schematic of the final device structure.

Current density voltage measurements was done using Oriel sun simulator, and external quantum efficiency(EQE) measurements were done using a monochromatic light source in conjunction with a chopper and a Stanford system lock in amplifier through a preamp.. To estimate the current density we integrate the signals from the desired range using the photon flux of global air mass 1.5(AM1.5). To measure EQE on tandem devices a secondary light source was used to saturate the top and bottom cells by external blue and red bias light illumination respectively.

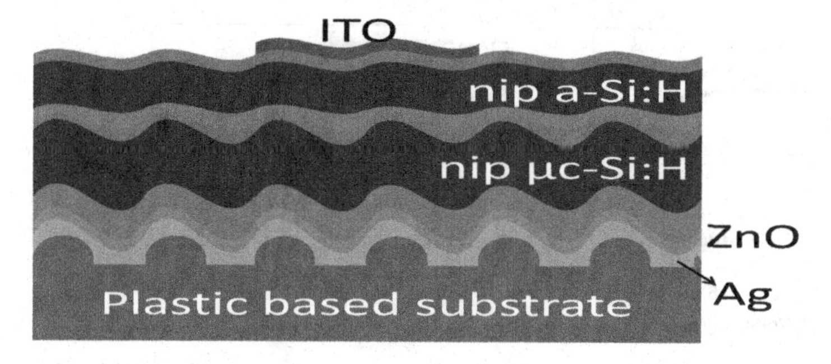

Figure 1: Schematic of the micromorph tandem cell after deposition of all the layers

RESULTS AND DISCUSSION

As we had expected from the earlier work, there was increase in the current density of a-Si single cell devices made on PEN. In this case device deposited at a temperature not exceeding 200˚C. Figure 2(a) shows the increase in current density on the photonic structure, quantitatively a 52% increase in comparison to SS and 25% in comparison planar reflector. The EQE measurement shown in Figure 2(c) clearly shows the better absorption in the longer wavelength (i.e. >550nm). This clearly proves the better quality of substrates with photonic structures. Similar experiments were also done on other substrates namely polymer on SS giving us similar enhancement in current density (not shown here).

Single cell µc-Si devices were fabricated on different substrates with similar i-layer thicknesses, a significant increase in the current densities was observed 49% increase from SS and 30.5% increase from planar counterparts as shown by JV plot in Figure 2(b) and EQE on Figure 2(d). In this case, the data shown here has devices made at 250˚C on polymer on SS; again devices were also made on PEN at 200˚C which showed similar enhancement in currents although had lesser currents as deposition temperature was lower (data not shown). In Figure 2(d) the EQE on the photonic structures shows a very clear change of the longer wavelengths being absorbed much better on the photonic structures.

It has been proven by many groups that if thin film solar cells need to compete with their crystalline counter parts, there is need to increase their efficiencies. A very standard method used by many groups all over the world is to make tandem devices which use two or more materials of different band gaps stacked together to efficiently use the spectrum of sunlight. Tandems made with microcrystalline (1.1eV) and amorphous (1.75eV) silicon also known as micromorphous is one of the best bandgap matches available to create bilayer tandem devices. Here we show, tandem devices made on two different types of plastic substrates namely PEN and Kapton. Devices made on PEN have been limited to a deposition temperature of 200˚C and Kapton substrates upto 250˚C.

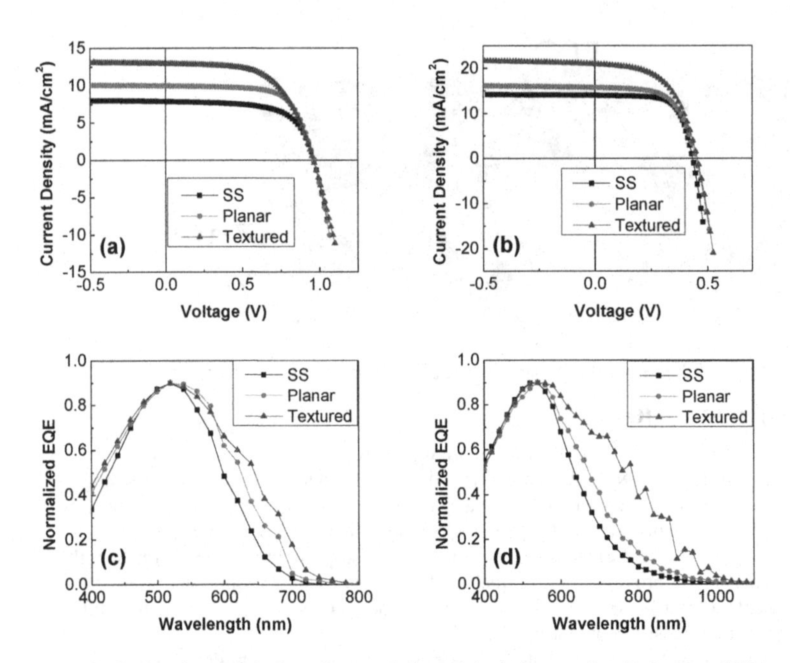

Figure 2: Comparison of device characteristics of single layer p-i-n devices on different substrates (a) J-V plot of a-Si device (b) J-V plot of μc-Si device (c) EQE of single layer a-Si device (d) EQE of single layer μc-Si device

Figure 3(a) and 3(b) shows JV and EQE plots for a micromorphous tandem device made on a photonic structured PEN substrate and Figure 3(c) and 3(d) shows a similar device on Kapton substrates. These devices perform better due to higher currents obtained due to the textured substrates. We have taken appropriate care to change the thickness of the top and bottom cells to match their currents. As the temperatures of deposition are different the quality of device also varies a little. Devices made on PEN need thorough care to get a good junction owing to the low temperature leading to barriers in the μc-Si/a-Si junction. Kapton can bear higher temperatures but care has to be taken as difference coefficient of thermal expansion of Kapton and deposited materials may lead to problems in the devices.

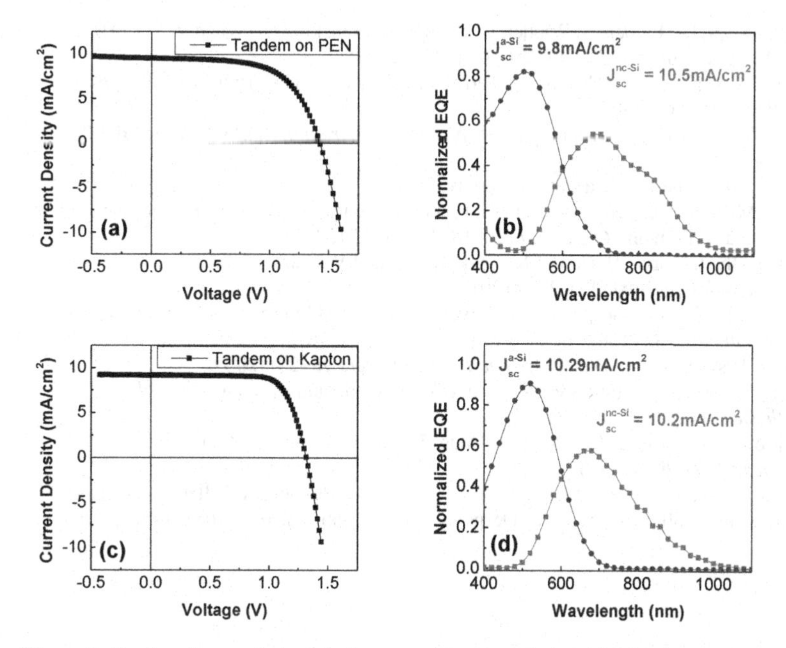

Figure 3: Device characteristics of micro morph tandem device (a) J-V characteristics and (b) EQE of the tandem device fabricated on textured PEN substrate (c) J-V characteristics and (d) EQE of the tandem device fabricated on textured Kapton substrate

CONCLUSIONS

We were successfully able to fabricate solar cells which show high current enhancement due to photonic structures on different types of substrates. All the fabrication procedure has been kept accountable to meet commercial needs to get high through put and easy process ability. Both single cells of a-Si and µc-Si show significant increase in current densities when proper care is taken during deposition depending on the substrate type. Finally tandem devices have been shown on textured plastic substrates.

ACKNOWLEDGEMENTS

This work was supported in part by a subcontract from NREL and a grant from NSF. We thank our colleagues at Iowa State, particularly K Han, M Noack, A Shyam, P Webster, S. Konduri ,S Kajjam and D Congreve for their experimental help in this project.

REFERENCES

1. A. Shah, P. Torres, R. Tscharner, N. Wyrsch, and H. Keppner, *Science* **285** (5428), 692 (1999).

2. O. Kluth, B. Rech, L. Houben, S. Wieder, G. Schope, C. Beneking, H. Wagner, A. Loffl, and H. W. Schock, *Thin Solid Films* **351** (1-2), 247 (1999);
3. F. J. Haug, T. Soderstrom, O. Cubero, V. Terrazzoni-Daudrix, and C. Ballif, *Journal of Applied Physics* **104** (6) (2008);
4. S. Pillai, K. R. Catchpole, T. Trupke, and M. A. Green, *Journal of Applied Physics* **101** (9) (2007).
5. B. W. Lewis, M.S. Thesis, Iowa state university, 2010.
6. V. E. Ferry, M. A. Verschuuren, H. B. T. Li, E. Verhagen, R. J. Walters, R. E. I. Schropp, H. A. Atwater, and A. Polman, *Optics Express* **18** (13), A237 (2010);
7. R. Biswas, J. Bhattacharya, B. Lewis, N. Chakravarty, and V. Dalal, *Solar Energy Materials and Solar Cells* **94** (12), 2337 (2010);
8. R. Biswas and D. Zhou, Amorphous and Polycrystalline *Thin-Film Silicon Science and Technology 2007* **989**, 35 (2007).
9. B. Curtin, R. Biswas, and V. Dalal, *Applied Physics Letters* **95** (23) (2009).
10. J. Springer, A. Poruba, L. Mullerova, M. Vanecek, O. Kluth, and B. Rech, *Journal of Applied Physics* **95** (3), 1427 (2004).
11. K. Soderstrom, J. Escarre, O. Cubero, F. J. Haug, S. Perregaux, and C. Ballif, *Progress in Photovoltaics* **19** (2), 202 (2011).
12. C. Battaglia, J. Escarre, K. Soderstrom, L. Erni, L. Ding, G. Bugnon, A. Billet, M. Boccard, L. Barraud, S. De Wolf, F. J. Haug, M. Despeisse, and C. Ballif, *Nano Letters* **11** (2), 661 (2011).

Mater. Res. Soc. Symp. Proc. Vol. 1321 © 2011 Materials Research Society
DOI: 10.1557/opl.2011.944

PERFORMANCE of HYDROGENATED a-Si:H SOLAR CELLS with DOWNSHIFTING COATING

Bill Nemeth[1], Yueqin Xu[1], Haorong Wang[2], Ted Sun[2], Benjamin G. Lee[1], Anna Duda[1], and Qi Wang[1]
[1] National Renewable Energy Laboratory, Golden, CO, 80401
[2] Sun Innovations, Inc, Fremont, CA 94539

ABSTRACT

We apply a thin luminescent downshifting (LDS) coating to a hydrogenated amorphous Si (a-Si:H) solar cell and study the mechanism of possible current enhancement. The conversion material used in this study converts wavelengths below 400 nm to a narrow line around 615 nm. This material is coated on the front of the glass of the a-Si:H solar cell with a glass/TCO/p/i/n/Ag superstrate configuration. The initial efficiency of the solar cell without the LDS coating is above 9.0 % with open circuit voltage of 0.84 V. Typically, the spectral response below 400 nm of an a-Si:H solar cell is weaker than that at 615 nm. By converting ultraviolet (UV) light to red light, the solar cell will receive more red photons; therefore, solar cell performance is expected to improve. We observe evidence of downshifting in reflectance spectra. The cell J_{sc} decreases by 0.13 mA/cm^2, and loss mechanisms are identified.

INTRODUCTION

Researchers continue to develop novel approaches to drive the economics of solar cells to be more competitive with traditional fossil fuel power generation. One of the primary limitations of performance for all solar cells is the response to the spectral output of the sun, and many tactics have been utilized or theorized in an effort to circumvent this. Multi-junction cells are tailored to absorb specific wavelength ranges of sunlight, whereas wavelength conversion layers can be used to tailor the sunlight to the particular device. By doing so, lattice thermalization and material transmission losses can be minimized [1].

Light conversion techniques can utilize quantum dots, rare earth ions, as well as various organic dye materials [2], and can be broadly placed into downconversion (high energy to lower energy) and upconversion (vice versa) processes. A subcategory of downconversion is downshifting (or photoluminescence [3]), which occurs at sub-unity quantum efficiencies. The narrow emission lines of rare earth ion light converters [4] lend their utility to single junction solar cells; however, narrow absorption bands limit the likelihood that broadband conversion is likely to occur for a single rare earth ion type. Many solutions have been implemented to films and phosphors with a few applications to solar cells giving mixed results in device performance [5]. Amorphous silicon (a-Si:H) solar cells show peak quantum efficiencies in wavelengths between 500 and 600 nm with sharp declines approaching 350 and 750 nm. The highest laboratory scale stable amorphous silicon (a-Si:H) single junction cell efficiency is in excess of 10% [6] with a theoretical efficiency limit between 15% [7] to 22% [8]. In this paper, we address a luminescent downshifting (LDS) layer applied to a single junction amorphous silicon solar cell to determine performance changes.

EXPERIMENTAL

Solar cells with a *p-i-n* structure were grown on Asahi U-type SnO:F substrates. Substrates were ultrasonically cleaned in DI water, rinsed with acetone and isopropanol, and dried with N_2. All a-Si:H layers were grown by PECVD in a multi-chamber cluster tool manufactured by MVSystems, Inc. at a substrate temperature of approximately 200 °C. The 5000 Å intrinsic layer was grown using SiH_4 without hydrogen dilution with $E_{Tauc} = 1.78$ eV and $\sigma_{dark} \sim 2x10^{-10}$ S/cm. The *n*-layer was grown using SiH_4 and PH_3/H_2 with $E_{Tauc} = 1.75$ eV and $\sigma_{dark} \sim 2x10^{-2}$ S/cm. Two p-layers were grown using SiH_4, CH_4, and TMB/He with $E_{Tauc} = 2.0$ eV and $\sigma_{dark} \sim 4x10^{-8}$ S/cm and $E_{Tauc} = 1.8$ eV and $\sigma_{dark} \sim 2x10^{-5}$ S/cm. All layers were deposited with 9.5 mW/cm^2 13.56 MHz RF power, and 1 cm^2 3000 Å thick Ag back contacts were deposited by electron beam evaporation [9]. The LDS layer was synthesized and applied on the glass front surface of the device as well as on a standard microscope slide by Sun Innovations, Inc. It consists of dissolving 30 mg Eu phosphor material (Red1) and Lucite Elvacite 2042 resin in methyl ethyl ketone, and applying a 100 μm thick layer using a scraper, allowing the film to dry naturally.

Indium lines were soldered to the front TCO and used as the front contact for current density voltage (J-V) measurements. The J-V measurements via a Keithley model 2400 were made using an ELH projector lamp light source calibrated with an AM1.5 standard reference solar cell. The resulting measurements reported are averaged over device areas of 1 cm^2. Layer thicknesses, reflection, and transmission measurements were measured using an n&k 1700 R-T analyzer from n&k Technology, Inc. Photoconductivity measurements were made using standard current-voltage (I-V) measurement under illumination utilizing 1 cm bars spaced 1 mm apart under a 100 V bias. Global reflectance measurements were taken with a Cary-6 spectrometer in an integrating sphere. Device external quantum efficiency (EQE) measurements were made using an inhouse QE measurement system calibrated with a reference cell. Photoluminescence (PL) quantum efficiency is measured in a LabSphere integrating sphere, with excitation of 365 nm (selected from a xenon lamp passed through a monochromator). The excitation and emission spectra are fiber coupled to an LN_2-cooled silicon CCD spectrometer. All spectra are corrected for grating, fiber, sphere, and detector efficiencies using a calibrated lamp.

RESULTS AND DISCUSSIONS

We calculated J_{sc} values to determine prospective enhancement due to the LDS coating by integrating the AM1.5 flux values (Figure 1) with respect to weighting factors in wavelengths between 350 and 400nm. An ideal solar cell with 100% EQE would yield 1.02 mA. Our cell shows a 75% EQE at 615nm, which is the emission wavelength of the conversion. The J_{sc} resulting from a 75% EQE would be 0.77 mA, and our cell would yield 0.49 mA. Since this is the light that is being converted, this represents a loss. Therefore, the maximum theoretical gain that we could expect from our device with this LDS coating is 0.28 mA/cm^2.

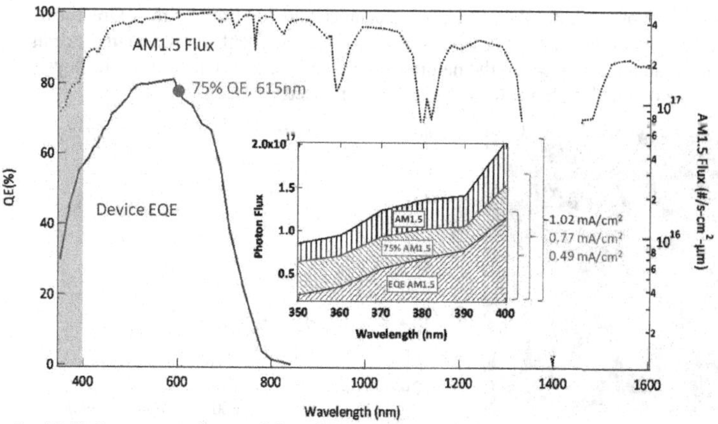

Figure 1. AM1.5 spectrum (log scale) and EQE measurement of a-Si:H device without LDS coating. Inset shows photon flux with weighting factors to calculate theoretical J_{sc} values.

To characterize the LDS coating on glass, we extracted the quantum yield of the conversion using PL data (Figure 2). The spectrum on the left shows the amount of excitation light absorbed by the LDS coating. The spectrum on the right shows the amount of light emitted by the LDS coating. The ratio of the emission integral (emitted photon flux) to the excitation integral (absorbed photon flux) results in a quantum yield of 88% for the LDS coating. We take the approach that the LDS coating is not part of the device, but a separate component to the system. As such, it can be conceptualized that the AM1.5 spectrum has been altered by the LDS coating incident to the device as depicted in the figure inset.

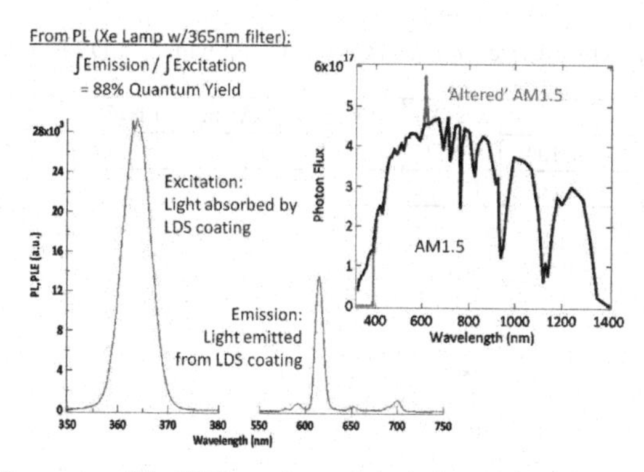

Figure 2. PL spectra of the LDS coating and depiction of altered AM1.5 spectrum.

Figure 3 shows devices with (left) and without (right) LDS coating illuminated with a 400 nm filter on a Xe lamp. The reflectance data on the right shows an additional 2% in reflectance. This is presumably the 10% UV reflectance and additionally the extra 2% of converted red light that has met the escape cone criteria for incident surface. This is designated in the UV, but is actually red due to the nature of the measurement, where discrete steps in excitation source coupled with a solid state detector are used.

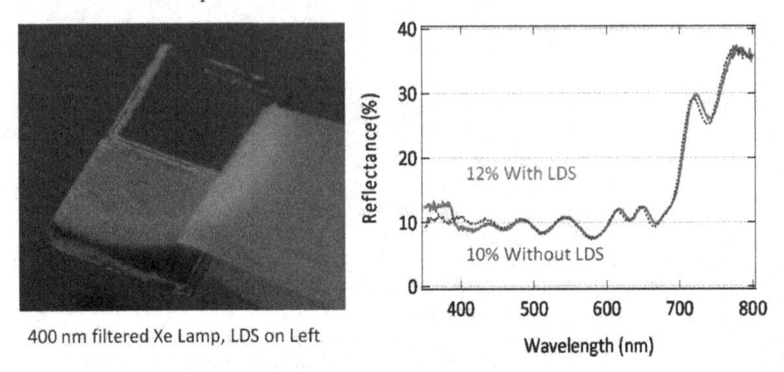

Figure 3. Device with and without LDS coating under Xe illumination and reflectance data of these devices.

The following table (Table I) shows the device performance compared with and without LDS coating application. Our results for 1 cm^2 cells show that 0.13 mA was lost with the LDS coating, while V_{oc} and FF remained the same. Any gain experienced by the cell's increased quantum efficiency (75% at 615 nm versus 50% at 365 nm) was negated by losses in conversion (88% quantum yield) and scattering (12% reflection and 43% waveguiding). Figure 4 shows the loss mechanisms responsible for the device performance.

Table I. 1cm^2 Device Characteristics with and without LDS Layer

	V_{oc} (V)	FF	J_{sc} (mA/cm^2)	Eff (%)
Without LDS	0.843	0.607	17.728	9.066
With LDS	0.843	0.606	17.598	8.975

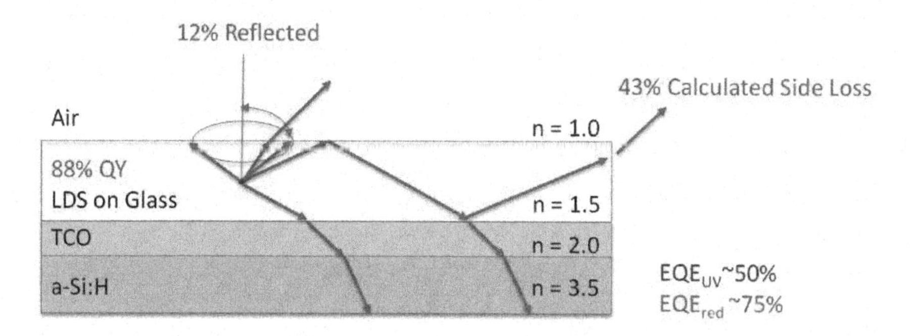

Figure 4. Loss mechanisms of LDS coating on device.

CONCLUSIONS

Light downshifting has marginal application for use in single junction amorphous silicon solar cells as evidenced by theoretical calculation (possible 0.28 mA/cm^2 net gain) as well as by application to solar cells (0.13 mA/cm^2 loss). Downshifting materials may be applied with better results to other materials systems with lower blue response. Downconversion, where quantum yields of greater than 100% occur, shows more promise for application to amorphous silicon.

ACKNOWLEDGEMENTS

The authors would like to thank Pauls Stradins and Eugene Iwaniczko for helpful discussion and dedicated effort to the world of photovoltaics. This work was supported by the U.S. Department of Energy under Contract DE-AC36-08-GO28308 to NREL.

REFERENCES

1. B.S. Richards, *Solar En. Matls. and Sol. Cells* **90**, 2329, 2006.
2. B.M. van der Ende, L. Aarts, A. Meijerink, *Phys. Chem. Chem. Phys.* **11**, 11081, 2009.
3. C. Strumpel, M. McCann, G. Beaucarne, V. Arkhipov, A. Slaoui,V. Svrcek, C. del Canizo, and I. Tobias, *Sol. En. Matls. and Sol. Cells* **91**, 238–249, 2007.
4. G.H.Dieke and H.M. Crosswhite **2**, *Appl. Opt.*, 1963.
5. E. Klampaftis, D. Ross, K.R. McIntosh, B.S.Richards, , *Solar En. Matls. and Solar Cells* **93**, 1182, 2009.
6. M.A. Green, K. Emery, Y. Hishikawa, and W. Warta, *Prog. Photovolt: Res. Appl.* **19**, 84, 2009.
7. D.E.Carlson, *IEEE Trans. Elec. Dev.* **24**, 449, 1977.
8. F. Meillaud, A. Shah, C. Droz, E. Vallat-Sauvain and C. Miazza, *Solar En. Matls. and Solar Cells* **90**, 2952, 2006.
9. Y. Xu, B. Nemeth, F. Hasoon, L. Hong, A. Duda, Q. Wang, *MRS Proc. Spring 2010*, 1245-A07-15, 2010.

Mater. Res. Soc. Symp. Proc. Vol. 1321 © 2011 Materials Research Society
DOI: 10.1557/opl.2011.816

Light-induced Open-circuit Voltage Increase in Amorphous Silicon/Microcrystalline Silicon Tandem Solar Cells

Xiaodan Zhang, Guanghong Wang, Shengzhi Xu, Shaozhen Xiong, Xinhua Geng and Ying Zhao
Institute of Photo-electronic Thin Film Devices and Technology of Nankai University, Weijin Road 94#, Nankai District, Tianjin 300071, P.R.China

ABSTRACT

Light-induced metastability of amorphous/microcrystalline (micromorph) silicon tandem solar cell, in which the microcrystalline bottom cell was deposited in a single-chamber system, has been studied under a white light for more than 1000 hours. Two different light-induced metastable behaviors were observed. The first type was the conventional light-induced degradation, where the open-circuit voltage (V_{oc}), fill factor (FF), and short-circuit current density (J_{sc}) were degraded, hence the efficiency was degraded as well. This phenomenon was observed mainly in the tandem cells with a bottom cell limited current mismatch. The second type was with a light-induced increase in V_{oc}, which sometimes resulted in an increase in efficiency. The second type of light-induced metastability was observed in the tandem cells with a top cell limited current mismatch. The possible mechanisms for these phenomena are discussed.

INTRODUCTION

Hydrogenated amorphous silicon (a-Si: H)/microcrystalline (micromorph) silicon (μc-Si: H) tandem solar cell structure, which can utilize more solar spectrum [1], has been considered as one of the next generation thin film silicon based solar cell structures. In order to make this technique more cost effective and competitive than other techniques, one need to further improve the cell efficiency, decrease the equipment cost, and minimize the light-induced degradation in cell efficiency. Single-chamber plasma-enhanced-chemical-vapor deposition (PECVD) technique, even though it has a cross contamination problem [2], is considered to be a low cost technique compared to multi-chamber systems.

Light-induced defect generation in a-Si: H, known as the Staebler-Wronski effect, causes the degradation in a-Si: H solar cells [3]. The metastable defects are not only induced by prolonged illumination [4, 5] but also by charge accumulation [6] or carrier injection [7, 8]. The light-induced defects reduce the mobility-lifetime product of the carriers, thereby degrading solar cell performance. The μc-Si: H solar cells have attracted a remarkable attention due to its ability to absorb long wavelength light. Previously, several groups reported that no light-induced degradation in μc-Si: H solar cells were observed. [9-11]. Recently, several groups reported that μc-Si: H solar cells also showed light-induced degradations, where the light-induced efficiency degradations are in the range of 2%-15% [12-17]. It was found that the light-induced degradation is dominated by the amount of amorphous component in the μc-Si: H intrinsic layer [16, 17]. However, it is interesting that Yue et al. [18] found that the volume fraction of the amorphous component is not necessarily the determining factor for the light-induced degradation in μc-Si: H solar cell. They suggested that smaller grains and intermediate range orders may provide a better grain boundary passivation and hence improve the solar cell stability.

The stability study of micromorph solar cells is essential for the improvement of stable solar cell performance. Yan et al. [19] have reported that a bottom cell limited current mismatch for a-Si: H/a-SiGe: H/μc-Si: H triple junction solar cell, prepared in a multi-chamber system, is desirable for high efficiency in the light soaked state. They suggested that the light-induced degradation in μc-Si: H solar cells occur only in the amorphous tissue phase. In this paper, we report on light-induced metastable phenomena in micromorph solar cells, where the μc-Si: H bottom cell was prepared in a single chamber system. We founded an abnormal phenomenon of light-induced increases in V_{oc} and FF for some micromorph tandem solar cells.

EXPERIMENT

All a-Si: H top cells were deposited in a multi-chamber system, where radio frequency glow discharge was used. The finished top cells were transferred to another single chamber system to deposit μc-Si: H bottom cells in the micromorph structures; very high frequency (75 MHz) glow discharge was used. Individual solar cells were defined by an aluminum back contact (about 1.0 cm^2). Light soaking was carried out under a white light with a spectrum and density close to AM 1.5 light at 50 °C for more than 1000 hours. Current-Voltage (I-V) measurements were used to monitor the solar cell performance after different periods of light soaking time. Quantum efficiency (QE) was measured from 300-1100 nm.

RESULTS AND DISCUSSION

Figures 1 (a), (b), (c) and (d) show the kinetics of V_{oc}, FF, J_{sc}, and conversion efficiency, respectively, where the three samples showed different behaviors. In sample NT 18, the efficiency was not degraded, instead, it was increased by 2% by light soaking, which was mainly from a 6% increase in V_{oc}. However, sample NT 16 showed a conventional degradation behavior, where a light-induced degradation of 10% in conversion efficiency was mainly from the decrease of FF and J_{sc}. Sample NT 2 also showed a significant light-induced enhancement in V_{oc}, but the light-induced degradations in the FF and J_{sc} were so large that the overall efficiency was reduced by the light soaking.

In order to obtain a better understanding of the mechanisms of the light-induced degradations for the tandem solar cells and their relation to the single chamber deposition process, the QE measurements were carried out directly on the sample NT 16 and NT 2. Figures 2 (a) and (b) show the QE curves of samples NT 16 and NT 2, where a red light filter (RG 695) and a blue light filter (IF 450) were used for the a-Si: H top cell and μc-Si: H bottom cell, respectively. NT 16 shows the 9.12 mA/cm^2 of current density from the μc-Si: H bottom cell is less than the a-Si: H top cell's current density of 9.78 mA/cm^2. The overall efficiency degradation for this solar cell is only 10%, which is relatively low and benefit from the improved a-Si: H top cell deposited with high hydrogen dilution and with an intrinsic layer thinner than 300 nm. [20, 21]

It is noticed that samples NT 18 and NT 2 show light-induced increases in V_{oc}, which contributes to the low light-induced degradation in the efficiency of NT 2 and the light-induced increase in efficiency of NT 18. In fact, the QE curves of NT 18 and NT 2 show that the μc-Si: H bottom cell current is larger than a-Si: H top cell. Figure 2(b) shows that the current from the bottom cell is 11.72 mA/cm^2 and 9.49 mA/cm^2 from the a-Si: H top cell. Here, we believe that the light-induced degradation in the cell performance of this tandem solar cell is controlled by the μc-Si: H bottom cell because μc-Si: H single-junction solar cells prepared the same deposition conditions also showed increases in V_{oc} and FF after prolonged light soaking. For the above phenomenon, we have suggested that a light-induced grain boundary decrease, which is replaced by amorphous content. As a result, the V_{oc} and FF of the μc-Si: H cells are improved because of the decrease of defect density [22].

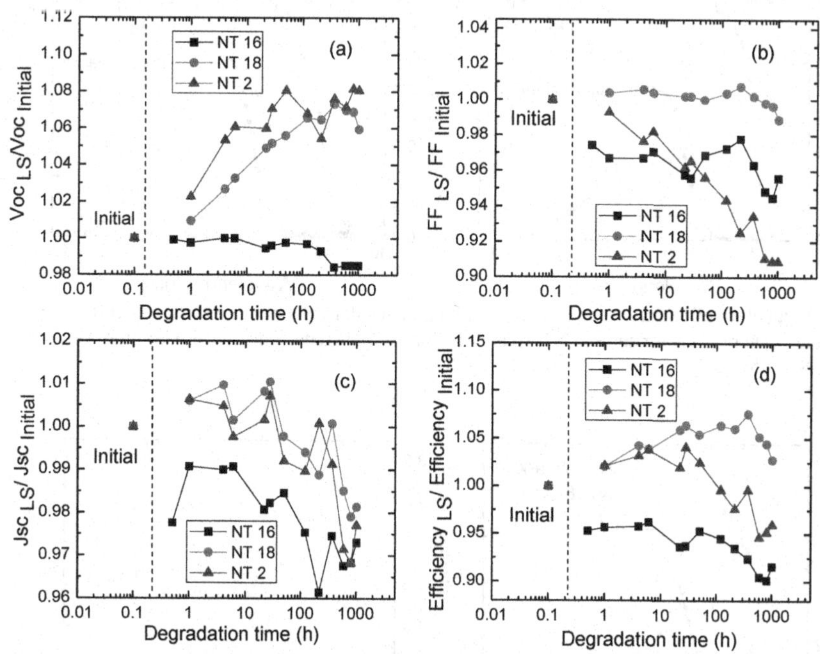

Figure 1 Relative performance changes as a function of light soaking time. The solar cells were measured under a white illumination with a spectrum and intensity close to AM 1.5. The plots of (a), (b), (c), and (d) represent the kinetics of V_{oc}, FF, J_{sc}, and Efficiency, respectively.

In order to investigate the mechanisms of the different meta-salability phenomena further, we conducted QE measurements in the dark. From Figure 3, one can see that the QE curves of samples NT 18 and NT 2 in the dark are similar to the top cell QE curves measured under the red bias light, which means that the bottom cell does not effectively block the top cell current generated in the top cell by the chopped light. Thus, there must be a leaking path through the bottom cell for such low reverse currents. J. Löffler et al. [23] have reported that this current leaking arises from the tunnel recombination junction (TRJ). As we know, TRJ between the sub-cells plays a key role in limiting the efficiency of multi-junction devices. It has to have an ohmic and low resistive electrical connection between the two adjacent cells with a low parasitic optical absorption. So, one or two microcrystalline doped layers were usually used to fabricate the high performance tandem solar cells. However, if there is no protecting layer to inhibit the current leaking for one of sub-cells, the tandem solar cells may demonstrate low V_{oc} and FF. The

J-V performance for NT 18 and NT 2 have shown in Table I, which is evidently shown that the cells with lower V_{oc} and *FF*.

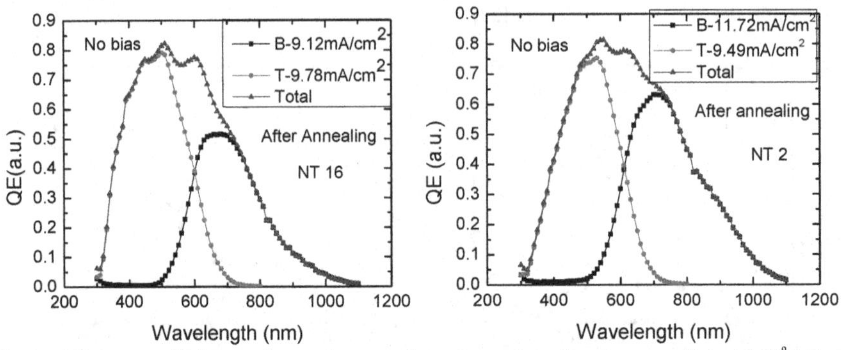

Figure 2 Quantum efficiency of two micromorph tandem solar cells after annealing at 160 °C for 30 minutes, where (a) is for sample NT 16 and (b) for NT 2. B represents μc-Si: H bottom cell and T a-Si: H top cell.

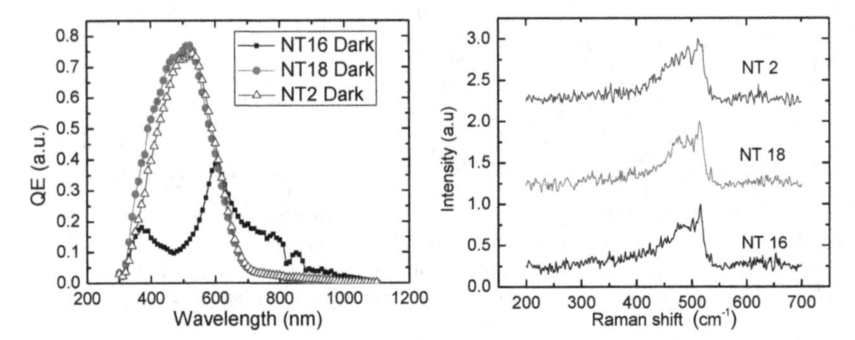

Figure 3 External QE curves of three a-Si: H/μc-Si: H tandem solar cells measured without optical bias.

Figure 4 Raman spectra of a-Si: H/μc-Si: H tandem solar cells excited by a 325-nm laser on the film side.

It is noticed that the dark QE of NT 16 basically followed the expected curve along the lower response of the individual cells at all wavelengths. That means NT 16 has a low shunting current through the bottom cell during the dark QE measurements, and corresponding to the slightly higher V_{oc} (1.248 V) than samples NT 18 (1.173 V) and NT2 (1.217 V) (in Table I).

Table I also shows that annealing at 160 °C for 30 minutes increased the V_{oc} and *FF*. It is noticed that samples were performed annealing before the light-degradation. We speculate that the annealing might change the properties of doping layer for TRJ and then improved the performance of tandem solar cells.

As shown in Figure 2, the J_{sc} values are 9.12 mA/cm^2 and 9.49 mA/cm^2 for samples NT 16 and NT 2, which are very different from the values of 12.96 mA/cm^2 and 12.93 mA/cm^2 listed in Table I. This phenomenon primarily arises from the coplanar current collections from the n layer. Raman spectra of the a-Si: H/μc-Si: H tandem solar cells measured on the top surface using a 325-nm wavelength laser showed microcrystalline characteristics, which proved the surface coplanar current collection (shown in Figure 4). As a result, the J_{sc} values obtained from J-V measurements do not reflect the real current density. However, the V_{oc} increase after prolonged light soaking cannot be attributed to this coplanar current collection. Because single-junction μc-Si: H solar cells prepared under this condition also showed an enhancement in V_{oc} even though the surface side Raman spectra of μc-Si: H single-junction solar cell measured using the 325-nm wavelength laser showed amorphous characteristics.

Table I Initial and annealing characteristics of three a-Si: H/μc-Si: H tandem solar cells.

Sample	State	J_{sc} (mA/cm^2)	V_{oc} (V)	FF (%)	η (%)
NT 18	Initial	13.64	1.173	54.5	8.73
	Annealing	13.87	1.278	57.2	10.22
NT 2	Initial	12.42	1.217	56.0	8.49
	Annealing	12.93	1.246	58.1	9.36
NT 16	Initial	12.96	1.248	52.4	8.48
	Annealing	12.96	1.27	54.3	8.93

CONCLUSIONS

We find that the micromorph silicon tandem solar cells, in which the μc-Si: H bottom cell was prepared in a single-chamber system, showed different light-induced metastable phenomena during prolonged light soaking. The micromorph solar cells with an a-Si: H top cell limited current mismatch showed a light-induced increase in V_{oc}, which was also observed in single-junction μc-Si: H solar cell prepared under the same condition as the bottom cells in the micromorph solar cells. Even though the np tunnel junction of tandem cell causes coplanar current collections and leakage through the bottom cell, we still ascribe the light-induced increase of Voc in the tandem cells from the changes in the μc-Si: H bottom cells.

ACKNOWLEDGEMENTS

The authors greatly thank W. Reetz in Forschungszentrum Juelich for the help in the measurements. The authors also gratefully acknowledge B. Yan of United Solar Ovonic LLC and F. Finger and A. Gerber of Forschungszentrum Juelich for helpful discussion and proofreading The work was supported by Hi-Tech Research and Development Program of China (Grant nos.2007AA05Z436, 2009AA050602), Science and technology support project of Tianjin (08ZCKFGX03500), National Basic Research Program of China (Grant nos.2011CBA00705, 2011CBA00706, and 2011CBA00707), National Natural Science Foundation of China (60976051), International Cooperation Project between China-Greece Government

(2009DFA62580), and Program for New Century Excellent Talents in University of China (NCET-08-0295).

REFERENCES

1. J. Meier, S. Dubail, R. Platz, P. Torres, U. Kroll, J. A. Anna Selvan, N. Pellaton Vaucher, Ch. Hof, D. Fischer, H. Keppner, R. Flückiger, A. Shah, V. Shklover, K. –D. Ufert, Sol. Energy Mater. Sol. Cells 49, 35(1997)
2. U. Kroll, C. Bucher, S. Benagli, I. Schönbächler, J. Meier, A. Shah, J. Ballutaud, A. Howling, Ch. Hollenstein, A. Büchel, M. Poppeller, Thin Solid Films 451, 525(2004)
3. D. L. Staebler, C. R. Wronski, Appl. Phys. Lett. 31, 292(1976)
4. S. Guha, J. Yang, D. L. Williamson, Y. Lubianiker, J. D. Cohen, A. H. Mahan, Appl. Phys. Lett. 74, 1860 (1999)
5. J. K. Kim, J. Y. Lee, K. S. Nam, J. Appl. Phys. 77, 95(1995)
6. J. Meier, P. Torres, R. Platz, S. Dubail, U. Kroll, J. A. A. Selvan, N. Pellaton-Vaucher, C. Hof, D. Fischer, H. Keppner, A. Shah, K.-D. Ufert, P. Giannoulès, J. Köehler, Mater. Res. Soc. Symp. Proc. 420, 3(1996)
7. G. Yue, J. D. Lorentzen, J. Lin, Q. Wang, D. Han, Appl. Phys. Lett. 75, 492(1999)
8. G. Yue, D. Han, D. L. Williamson, J. Yang, K. Lord, S. Guha, Appl. Phys. Lett. 77, 3185 (2000)
9. J. Meier, R. Flückiger, H. Keppner, and A. Shah, Appl. Phys. Lett. 65, 860(1994)
10. K. Yamamoto, IEEE Trans. Electron Devices 46, 2041(1999)
11. O. Vetterl, F. Finger, R. Carius, P. Hapke, L. Houben, O. Kluth, A. Lambertz, A. Mück, B. Rech, and H. Wagner, Sol. Energy Mater. Sol. Cells 62, 97(2000)
12. A. Chowdhury, S. Mukhopadhyay, S. Ray, Sol. Energy Mater. Sol. Cells 93, 597(2009)
13. V. Smirnov, S. Reynolds, F. Finger, C. Main, and R. Carius, Mater. Res. Soc. Symp. Proc. 808, 47(2004)
14. Y. Wang, X. Geng, H. Stiebig, F. Finger, Thin Solid Films 516, 733(2008)
15. M. Sendiva- Vassileva, S. Klein, F. Finger, Thin Solid Films 501, 252(2006)
16. S. Klein, F. Finger, R. Carius, T. Dylla, B. Rech, M. Grimm, L. Houben, and M. Stutzmann, Thin Solid Films 430, 202(2003)
17. F. Meillaud, E. Vallat-Sauvain, X. Niquille, M. Dubey, J. Bailat, A. Shah, and C. Balif, Proceedings of the 31st IEEE Photovoltaic Specialists Conference (IEEE, New York, 2005). p. 1412
18. G. Z. Yue, B. J. Yan, G. Ganguly, J. Yang, and S. Guha, Appl. Phys. Lett. 88, 263507(2006)
19. B. J. Yan, G. Z. Yue, J. M. Owens, J. Yang, and S. Guha, Appl. Phys. Lett. 85, 1925(2004)
20. R. J. Koval, J. Koh, Z. Lu, L. Jiao, R. W. Collins, C. R. Wronski, Appl. Phys. Lett. 71, 1317(1997)
21. S. Shimizu, A. Matsuda, M. Kondo, Sol. Energy Mater. Sol. Cells 92, 1241(2008)
22. A. Gordijn, L. hodakova, J. K. Rath, R. E. I. Schropp, J. Non-Cryst. Solids 352, 1868(2006)
23. J. Löffler, A. Gordijn, R. L. Stolk, H. Li, J. K. Rath, R. E. I. Schropp, Sol. Energy Mater. Sol. Cells 87, 251(2005)

Mater. Res. Soc. Symp. Proc. Vol. 1321 © 2011 Materials Research Society
DOI: 10.1557/opl.2011.945

Modulated surface-textured substrates with high haze for thin-film silicon solar cells

O. Isabella[1], P. Liu[1], B. Bolman[1], J. Krč[2], M. Zeman[1]
[1] Delft University of Technology, EEC Unit / DIMES, 2600 GB Delft, The Netherlands
[2] University of Ljubljana, Faculty of Electrical Engineering, SI-1000 Ljubljana, Slovenia

ABSTRACT

Modulated surface-textured substrates for thin-film silicon solar cells exhibiting high haze in a broad range of wavelengths were fabricated. Glass substrates coated with different thicknesses of a sacrificial layer were wet-etched allowing the manipulation of the surface morphology with surface roughness ranging from 200 nm up to 1000 nm. Subsequently, zinc-oxide layers were sputtered and then wet-etched constituting the final modulated textures. The morphological analysis of the substrates demonstrated the surface modulation, and the optical analysis revealed broad angle intensity distributions and high hazes. A small anti-reflective effect with respect to untreated glass was found for etched glass samples. The performance of solar cells on high-haze substrates was evaluated. The solar cells outperformed the reference cell fabricated on a randomly-textured zinc-oxide-coated flat glass. The trend in the efficiency resembled the increased surface roughness and the anti-reflective effect was confirmed also in solar cell devices.

INTRODUCTION

Thin-film silicon solar cells is one of the major photovoltaic technologies. The short energy payback time and a small amount of needed materials offer a market profitable performance-price ratio. To sustain the growth of this technology an increase in the efficiency is mandatory. One way to approach this challenge is to enhance the photocurrent. Photocurrent can be increased by applying different light management techniques. One of them is light scattering at internal interfaces, which prolongs the light path and consequently increases the absorption in the absorber layers.

Nowadays thin-film silicon solar cells are formed by two junctions stacked on top of each other with rough internal interfaces for light scattering. Amorphous silicon (a-Si:H) and microcrystalline silicon (μc-Si:H) are used as absorber layers in the *top* and *bottom* component cells, respectively, forming a *tandem* device. To evaluate scattering properties of a surface texture the wavelength-dependent haze parameter is commonly used [1]. When the surface morphologies exhibit features smaller than the wavelength of incident light, the amount of scattered light will exponentially decrease with the haze parameter approaching zero (scalar scattering approach) [2], leading to poor absorption in the long wavelength region. The bottom cell that is optically active up to 1100 nm, especially suffers from this problem, as the photons passing through the μc-Si:H absorber are not efficiently scattered. To boost the absorption in the bottom cell a different family of interface morphology is needed that ensures high scattering also at long wavelengths ($\lambda > 700$nm).

Advanced surface morphologies have been recently proposed to tackle this requirement. Most of them are characterized by the superposition of two or more different types of features, using one [3] or more materials [4, 5]. The larger features that constitute the background on which the final morphology is developed, have geometric dimensions in the same order of magnitude or bigger than the wavelengths of incoming light. The smaller features are present on top of the final surface morphology and modulate the background. With this modulated surface texturing approach, the haze parameter still decreases with the increasing wavelengths but saturates at a minimal value higher than zero.

One way to realize the concept of modulated surface texturing is to combine an etched glass carrier with wet-etched aluminum-doped zinc-oxide (ZnO:Al) that results in an advanced texture on the front surface of the transparent conductive oxide (TCO). The glass carrier provides the large background textures, while the etched ZnO:Al layer modulates the glass morphology and forms a transparent and conductive substrate suitable for thin-film silicon solar cells. In this contribution we describe a technique to manipulate the background textures on glass. Examples of five broadband high-haze substrates will be presented. The morphological analysis of the substrates confirms the concept of modulated surface texturing, while the optical analysis reveals broad angle intensity distributions (AID) and high hazes. An anti-reflective effect caused by the etched glass substrates is presented. Thin-film single-junction silicon solar cells were deposited on the advanced textured substrates. The electrical and spectral behavior of the cells is reported and compared with a reference solar cell deposited on flat glass / etched ZnO:Al substrate.

THEORY

The concept of modulated surface texturing was developed to unveil the physics underlying the strong optical response of advanced textures for thin-film silicon solar cells. It was found that combining together different textures in one morphology results in the superposition of the scattering mechanisms activated by each type of the surface features [6]. The optical outcome from nano-scale features is characterized by scalar scattering theory (exponential decay of the haze with wavelengths) and by Mie solution of Maxwell equations for micro-scale features (weakly wavelength-dependent haze). For all combinations between these boundaries a superposition of the two effects is expected, with one dominating the other according to the geometrical size of the background texture. Particularly the minimal value to which the haze saturates (*offset effect*) is strongly determined by the background texture. This effect can be manipulated by changing the correlation length and the peak-to-peak height of the background features.

EXPERIMENT

Five sets of textured substrates were investigated. Each set comprised three samples: (i) etched glass (EG), (ii) etched glass coated with non-treated ZnO:Al (EG/AZON), and (iii) etched glass coated with surface wet-etched (rough) ZnO:Al (EG/AZOR). The latter is the example of the modulated surface texture (MST). The etched glass carriers were prepared by depositing five increasing thicknesses of a sacrificial layer on Corning Eagle XG glass sheets which were subsequently wet-etched in a mix of HF and H_2O_2 resulting in five different textures EG1 to EG5. The ZnO:Al layers were rf-magnetron sputtered. The non-treated ZnO:Al layer was 1.2 μm thick; the thickness of wet-etched ZnO:Al layer was initially 1.6 μm in order to reach 1.2 μm of thickness after the acidic bath in 0.5% HCl. An AZOR layer was fabricated also on a flat glass substrate as a reference. The morphological analysis was carried out with both NT-MDT nTegra AFM and Philips XL-50 SEM. Digital data from AFM were used to carry out the calculations on modulated surface textures. The measurement of AID and hazes were done with a Perkin-Elmer Lambda spectrophotometer equipped with ARTA and TIS accessories [7]. The advanced surface-textured substrates presented here are designed for tandem devices, however, since the deposition process for the tandem cell has not been optimized yet, we deposited a-Si:H single junction solar cells on the substrates. The thickness of the intrinsic a-Si:H layer was 300 nm,

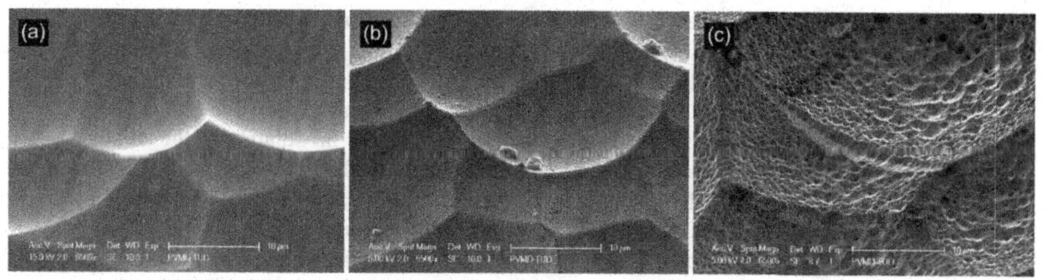

Figure 1: SEM images of (a) EG4, (b) EG4/AZON, and EG4/AZOR. Scale bars denote 10 μm.

the substrates. The thickness of the intrinsic a-Si:H layer was 300 nm, while the thickness of the back ZnO:Al was optimized for high reflection at λ_{air} = 700nm. The solar cells were characterized with a Pasan Flash Simulator and with the external quantum efficiency (EQE) setup.

RESULTS AND DISCUSSION

Morphological analysis

A selection of SEM images of the fabricated samples is presented in Fig. 1. The EG4 substrate shows large microscopic carved features. The AZON layer coats the EG4 background, smoothing the surface features, especially the ridges, because of different ZnO:Al growth orientations. Finally, the AZOR layer with typical craters covers the big carved features of EG4 forming the MST. It is noticeable that the wet-etching of ZnO:Al affected the ridges differently than the rest of the morphology resulting in further horizontal widening of small craters. These three effects could be recognized also in the other sets of textured substrates (not shown here). For each fabricated sample, AFM scans were performed in four different spots. Particularly, in Fig. 2(a) the average correlation length (**L_C**) and the surface roughness (**σ_{RMS}**) with relative error bars are reported for all MST substrates. It was found that increasing the thickness of the sacrificial layer resulted in a strong roughening of the glass surface morphology in both horizontal and vertical directions. However, EG5-based MST demonstrated a decrease in statistical surface parameters. Similar trend (not reported here) was found also for the substrates EG and EG/AZON. The inclination of the features was evaluated with respect to the direction perpendicular to an

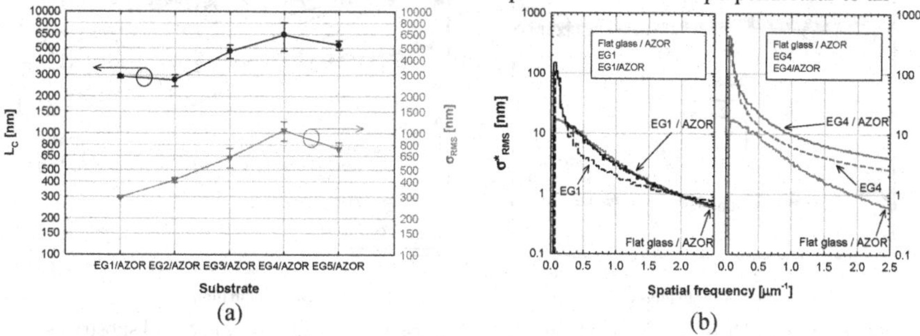

Figure 2: (a) **L_C** and **σ_{RMS}** of the MST substrates, (b) spatial frequency surface representation of EG1-based and EG4-based substrates. All y-axes are in logarithmic scale.

inclination of the features was evaluated with respect to the direction perpendicular to an ideal flat surface, which is defined as zero degrees. In all sets of textured substrates the mean inclination was 13 ± 3 degrees. Despite the increasing roughening of the glass surface with the increased thickness of the sacrificial layer, such small angle deviation indicates that the features become bigger without changing shape. To demonstrate the surface modulation of the etched glass substrates the spatial frequency surface representation [6] of the EG1-based and EG4-based substrates is presented in in Fig. 2(b) along with the reference flat glass / AZOR substrate. In the first case, the surface modulation is clearly the envelope between the background (EG1) and the reference AZOR. In the second case, the EG4 representation is always higher than the reference AZOR, nevertheless the MST representation is shifted up by the reference AZOR spectral components, thus demonstrating the surface modulation of the etched glass background.

Optical analysis

The total transmittance (T) and reflectance (R) of the EG substrates was compared with flat glass. As presented in Fig. 3 the etched glass substrates have similar R and T as the flat counterpart and exhibit a slight anti-reflective (AR) effect in the visible range. The haze of transmitted light (H_T) was measured for both EG and MST substrates and compared to the reference flat glass / AZOR substrate (see Fig. 4). Even the least rough MST of the series (EG1/AZOR) has H_T higher than the reference wet-etched ZnO:Al, taking advantage of the *offset effect* of its background (EG1). The same effect is also reported for EG3-based substrates performing a stronger offset (~ 0.7). In accordance with the statistical analysis, the EG4/AZOR substrate resulted in the highest haze (average $H_T = 0.89$), while the H_T of EG5/AZOR decreased following the observed trend. In contrast to the reference textured TCO the angle intensity distribution of transmitted light (AID_T) of the advanced textured substrate exhibited a very weak dependency on wavelengths In Fig. 5 the measurements at 700 nm are reported. Increasing the roughness of the MST substrates resulted in a decreased specular component of transmitted light and broader AID_T than that of the reference TCO. In the same plot also the *offset effect* is reported, illustrating how much the EG5/AZOR substrate benefits from the AID_T of its background (EG5).

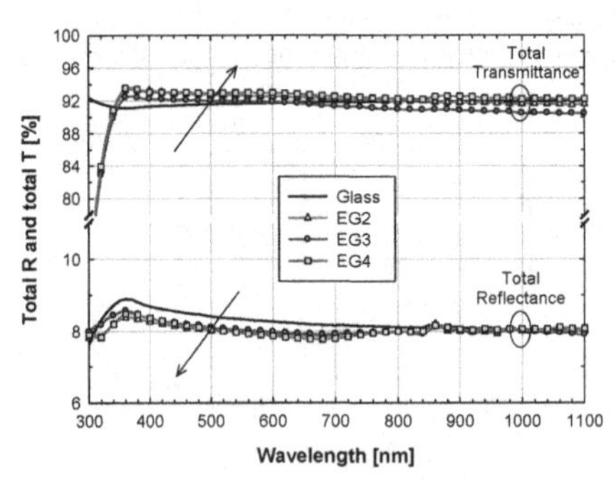

Figure 3: Total R and T of different etched glass carriers compared to flat glass substrate.

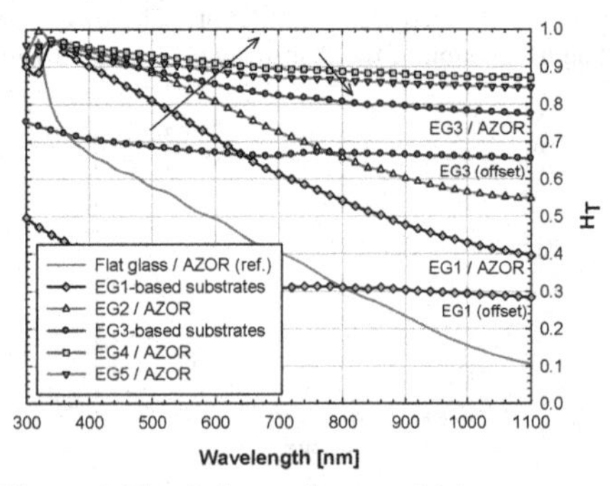

Figure 4: H_T of advanced textured substrates compared to reference textured TCO.

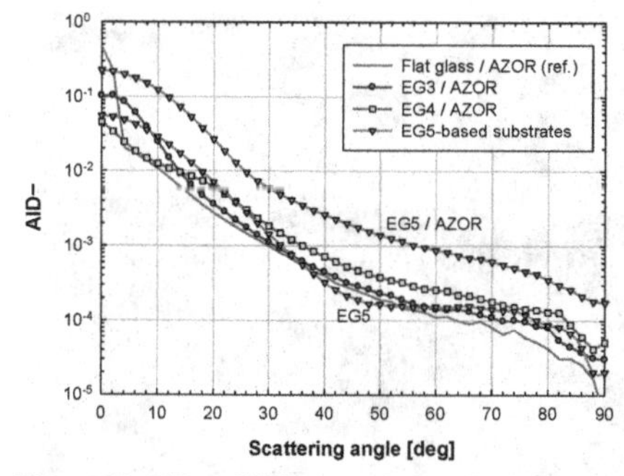

Figure 5: AID$_T$ of different advanced substrates compared to reference textured TCO (reported trends are for λ_{air} = 700 nm).

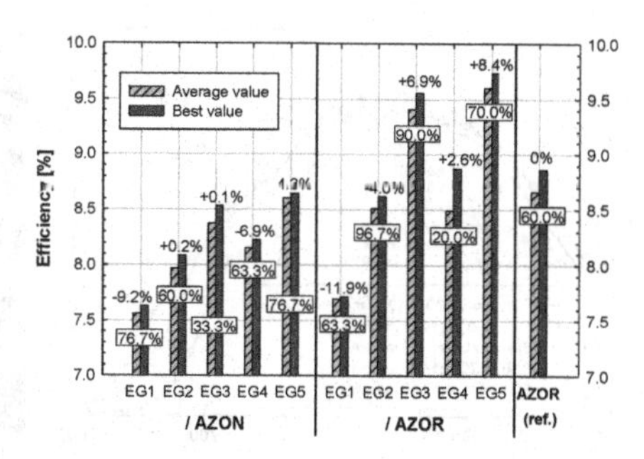

Figure 6: Initial efficiency of the solar cells deposited on advanced textured substrates. Numbers on bars are ΔJ_{SC} with respect to the reference, framed numbers are Y_{90} for each substrate.

Solar cells

The fabricated MST were used as substrates in a-Si:H single junction solar cells. Dealing with highly textured morphologies, the yield (Y_{90}) for each substrate was calculated considering the number of cells with an efficiency higher or equal than 90% of the best one. In Fig. 6 the measured efficiencies are shown with the short-circuit current density percentage increasing (ΔJ_{SC}) calculated with respect to the reference. The efficiency increased accordingly to the roughening of the surface morphology. Only the EG4-based substrates did not follow the trend because of too large roughness and/or problems during the deposition of the cells (low yield for EG4/AZOR cells). The solar cells deposited on EG/AZON substrates performed less than the solar cells on the reference TCO, because the p-i-n thin-films on the non-treated front ZnO:Al were locally flat (see Fig. 7). It is evident from the EQE measurements (EG1/AZON curve in Fig. 8) that the solar cells on non-treated front ZnO:Al deposited on etched glass behave like flat cells. The only gain of these solar cells with respect to the reference one is observed in the range 400 nm – 550 nm that confirms the AR effect. The solar cells deposited on EG/AZOR substrates demonstrated efficiencies higher than the reference TCO-based solar cells with peak performance on EG5/AZOR substrate. Such enhancement depends on the increased lateral and vertical features of the MST substrates resulting in the high H_T and the broad AID$_T$ previously reported. In Fig. 8 the EQE of the best solar cells on MST substrates are presented. One can observe a widened AR effect (up to 600 nm) and additional scattering that was triggered by

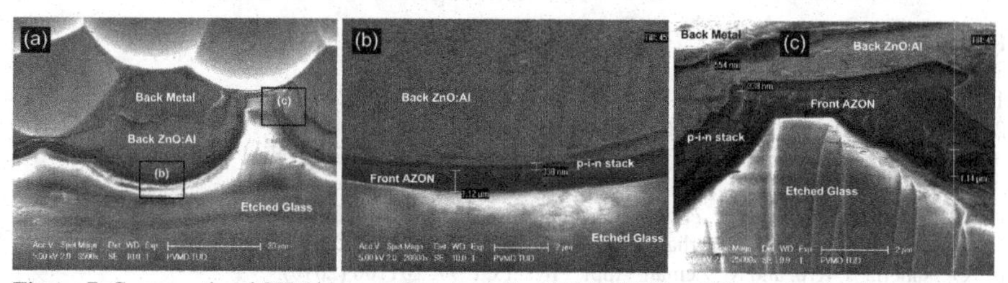

Figure 7: Cross-sectional SEM images of a solar cell on EG4/AZON at different levels of zoom.

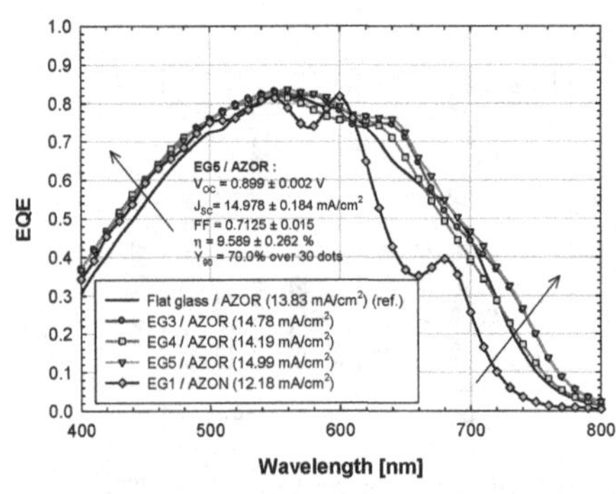

Figure 8: EQE of the best advanced textured substrates. In the inset the external parameters for EG5 / AZOR are reported.

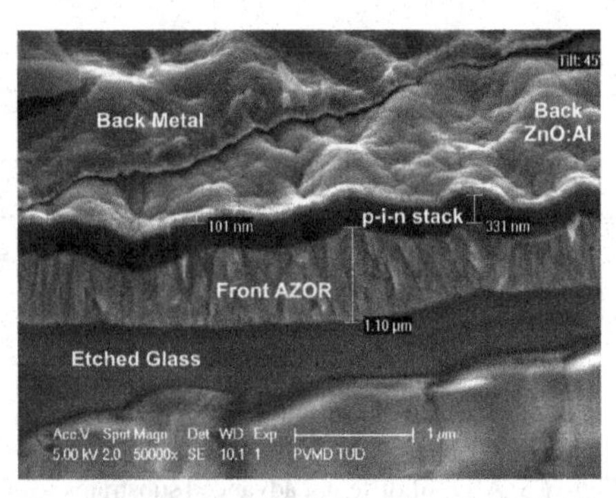

Figure 9: Cross-sectional SEM image of a single junction solar cell on EG4 / AZOR advanced substrate.

the superposition of large and small features. Evidence of this superposition is shown in a cross-sectional SEM image of the device deposited on EG4/AZOR in Fig. 9.

CONCLUSIONS

Five high haze transparent conductive MST substrates for thin-film silicon solar cells were presented. SEM and AFM analysis demonstrated the controlled manipulation of the etched glass and the modulated surface texturing. The developed MST showed broad AID_T and very high H_T due to strong and controllable *offset effect*. The etched glass substrates exhibit a small AR effect. The developed etched glass-based substrates were used in single junction cells. EG/AZON substrates do not result in efficient solar cell devices since the a-Si:H p-i-n films were found to be locally flat. The best performance was found on a-Si:H single junction devices deposited on EG5/AZOR substrate. Also solar cells on the EG3/AZOR substrate performed well with a very high Y_{90}.

ACKNOWLEDGEMENTS

This work was carried out with a subsidy of the Dutch Ministry of Economic Affairs under EOS-LT program (project number EOSLT04029).

REFERENCES

1. J. Krč, M. Zeman, O. Kluth, *et al.*, Thin Solid Films 426 (2003) 296–304.
2. C. K. Carniglia, Opt. Eng. 18, 104 (1979).
3. M. Kambe, A. Takahashi, N. Taneda, *et al.*, Proceedings of the 33rd IEEE PVSC, San Diego (IEEE, New York, 2008), pp. 1–4.
4. O. Isabella, F. Moll, J. Krč, *et al.*, Phys. Status Solidi A, 10.1002/pssa.200982828.
5. J. Bailat, L. Fesquet, J.-B. Orhan, *et al.*, 10.4229/25thEUPVSEC2010-3BO.11.5
6. O. Isabella, J. Krč, and M. Zeman, Appl. Phys. Lett. 97, 101106 (2010);
7. K. Jäger, O. Isabella, L. Zhao, et al., Phys. Status Solidi C, 10.1002/pssc.200982695.

Mater. Res. Soc. Symp. Proc. Vol. 1321 © 2011 Materials Research Society
DOI: 10.1557/opl.2011.946

Excitation of guided-mode resonances in thin film silicon solar cells

F.-J. Haug, K. Söderström, A. Naqavi, C. Ballif
Ecole Polytechnique Fédérale de Lausanne (EPFL), Institute of Microengineering (IMT),
Photovoltaics and Thin Film Electronics Laboratory,
Rue A.-L. Breguet 2, CH-2000 Neuchâtel, Switzerland

ABSTRACT

Thin film silicon solar cells are attractive for photovoltaics; however, the poor charge transport in this material requires that the devices are thinner than the absorption length. Adequate absorption can nevertheless be achieved by light scattering at textured interfaces because light can get trapped inside the absorber layer if it is scattered into angles above the critical angle of total internal reflection. This situation can be identified with the propagation of a guided mode in a waveguide where silicon plays the role of the high index guiding medium and the interface texture serves to couple the incident light to modes via grating coupling. We present an experimental realization of a solar cell structure on a line grating where the enhanced photocurrent can be clearly related to resonant excitation of waveguide modes.

INTRODUCTION

Thin film silicon is an interesting option for low cost solar energy production because its manufacturing technology is well mastered and it is an abundant material. Its most fundamental limitation against large scale production is related to its poor electronic quality compared to its crystalline counterpart. Poor charge transport poses several fundamental limitations on the device design. For example, the absorber layer thickness is normally much smaller than the absorption length, particularly for light with energy close to the band gap [1]; therefore, state-of-the-art devices require additional elements that enhance the absorption within the absorber layer. The earliest, and so far experimentally most successful approach employs light scattering at textured interfaces [2]. Light that is scattered beyond the critical angle for total internal reflection is thus trapped into the absorber film. Obviously, the condition of total internal reflection is not perfect in the presence of interface texture; light trapping is therefore an equilibrium between in-coupling which results in the desired absorption enhancement in the silicon film, and the undesired out-coupling.

In the terminology of the optics community, the silicon absorber layer can be identified with the high index guiding medium, the metal film at the back that serves as reflector and electric contact translates to the cladding, and the interface texture establishes the grating coupling between the external radiation field and eigen-modes of the multilayer stack. Different from standard waveguides, typical dimensions of a solar cell are likely to support multiple modes, the guiding medium is absorbing and the requirement of broad-band coupling means a spectral range between 600 and 750 nm for amorphous solar cells and an even larger range from 700 to 1000 nm for microcrystalline solar cells.

In this contribution, we describe an amorphous solar cell as planar waveguide and discuss the main features of the dispersion relation, using realistic thicknesses for all component films and including the dispersion of the respective refractive indices. The theoretical simplification of flat interfaces is checked by measuring resonant absorption enhancement in the external quantum efficiency of a solar cell on a line grating. Thanks to the one-dimensionality of the grating, polarization phenomena are distinguished.

EXPERIMENTAL

We used a commercially available grating with a period of 560 nm and blazed shape (Thorlabs, GR18). In order to be compatible with the subsequent fabrication steps, the grating was replicated on glass substrates, using a UV-curable lacquer [3]. The cross section in Figure 1 shows a peak-to-valley depth of 140 nm and a saw-tooth with slightly rounded tip. A. The grating was covered with a back reflector consisting of a sputtered double layer of silver (120 nm) and zinc-oxide (60 nm). The silicon film is grown in an n-i-p sequence with a total thickness of 200 nm and the front contact consists of ITO (In_2O_3-SnO_2) with a thickness of 65 nm which serves as anti-reflection coating [4].

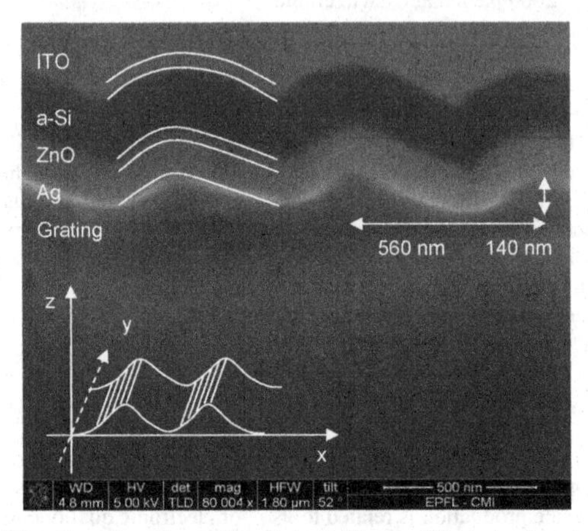

Figure 1: Cross section of the solar cell on the grating substrate. The inset illustrates the orientation of the coordinates.

The polarization resolved external quantum efficiency (EQE) was measured under perpendicular incidence with a broad-band wire polarizer (ProFlux, Moxtek). Figure 1 illustrates the coordinate system; in s-polarization, the E-field of is parallel to the y-axis, in p-polarization the E-field is contained in the x-z plane. The EQE is the probability of creating and collecting an electron-hole pair per incident photon at a specific wavelength in the illumination spectrum.

THEORY

Eigen-modes of a multilayer system are easily found with the pole method [5]. Based on the assumption that a resonance corresponds to a singularity of the system, the pole method simply looks for zeros in the denominator of the Fresnel coefficient. Fresnel coefficients of multilayer stacks are defined recursively, starting from the coefficients for the interface between two component media i and j. For s- and p-polarization they are defined as follwos:

$$r_{ij}^{s} = \frac{S_j - S_i}{S_j + S_i} \text{ and } r_{ij}^{p} = \frac{\varepsilon_i S_j - \varepsilon_j S_i}{\varepsilon_i S_j + \varepsilon_j S_i} \qquad (1)$$

The first and the last layers of the stack assumed to be semi-infinite, the propagation constants S_i are defined for every individual layer i in terms of the complex dielectric constant ε_i and $k_0=2\pi/\lambda$, the modulus of the wave vector in air:

$$S_i^2 = k_0^2 - \varepsilon_i k_{\|}^2 \qquad (2)$$

For more than two layers, Fresnel coefficients are defined via the single interface coefficients of eq. (1), using coherent propagation across the film thicknesses d_i:

$$r_{1...n} = \frac{r_{12} \cdot r_{2...n} \cdot \exp\{-2S_2 d_2\}}{r_{12} + r_{2...n} \cdot \exp\{-2S_2 d_2\}} \quad (3)$$

Eqns. (1) to (3) were used to determine the dispersion relation for a solar cell stack using five layers [6]; the silver layer is assumed semi-infinite and we considered a 60 nm thick ZnO buffer layer, a 200 nm thick amorphous silicon film, a 65 nm thick ITO front contact and an air space above.

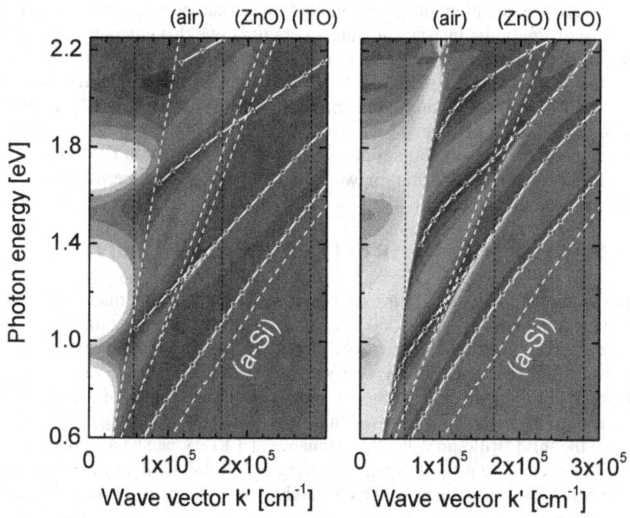

Figure 2: Dispersion diagrams in s-(left) and p-polarization (right). The symbols denote the individual modes of the solar cell stack, the dashed lines represent light lines of the respective materials. Vertical lines denote the Brillouin zones for a grating with period of 560 nm. The background is a contour-plot of Abs(D($r_{1...5}$))^2.

Before discussing the dispersion relation of the solar cell stack, we would like to point out a few general aspects of dispersion diagrams like the one shown in Figure 2. Propagation of light in an unbounded medium corresponds to straight lines according to the dispersion relation of photons which is given by $\omega(k) = (ck)/n$. In Figure 2, the lines are not exactly straight because of the refractive index dispersion. Guided modes can be defined approximately by requiring nodes at the boundaries of the guiding medium; in case of the fundamental or zero-order mode, half of the wavelength can be fit into the thickness of the high-index layer. Likewise, higher order modes can be defined by fitting integer multiples. This definition explains the cut-off condition because modes with wavelengths larger than twice the guiding film thickness cannot be guided.

Figure 2 shows the guided mode structure of the solar cell stack. The left and right panels correspond to s- and p-polarization, respectively, and their backgrounds show contour-plots of the squared modulus of the denominator of $r_{1...5}$. Dark shades correspond to values close to zero. The symbols represent numeric solutions. The cut-off condition of the zero-order mode (squares) is not observed, but with increasing energy first order (circles), second order (up-triangles) and third order (down-triangles) modes appear alternatingly in the two polarization directions. At their onset, all modes lie close to the light line of air. This observation is explained by the dielectric nature of the ITO front contact; compared to the silver back contact, the confinement at the front of the device is relatively weak. At their onset, the modes resemble standing waves in z-direction rather than guided modes in x-direction. With increasing energy, all modes approach the light line of amorphous silicon.

The region towards the left of the air-line represents a continuum of radiation modes. The radiation modes are straight lines that emerge from the origin, their slope is related to their angle of incidence with respect to the average cell surface. The shown light line of air represents grazing incidence, the light line of perpendicular incidence coincides with the vertical energy axis. Note that none of the continuum modes intersects with the guided modes, conservation of energy and momentum forbids an interaction of radiation modes with Eigen-modes. Darker shades in the contour-plot reveal nevertheless that a certain amount of resonant behaviour does exist in this region because of multiple reflections into angles below the critical angle.

Interactions between the radiation modes and guided modes can be established by grating coupling. In the presence of a periodic interface texture, every light line that emerges from the origin can pick up multiples of the reciprocal lattice vector that translates it into the centre of higher order Brillouin zones. Note that this approach also applies to random interface textures by means of their spatial Fourier expansion. For the case of sinusoidal textures, relations for the strength of out-coupling from a guided mode to the radiation field have been worked out in ref. [7]. In a future contribution, we plan to adapt the theory to the inverse situation of in-coupling from the radiation field to a (lossy) guided mode. In the remainder of this contribution, we would like to present experimental evidence on grating coupling in a real solar cell device.

ABSORPTION ENHANCEMENT IN SOLAR CELLS BY GRATING COUPLING

The excitation of guided modes by grating coupling was investigated by manufacturing a complete solar cell on a grating with period of 560 nm. Figure 3 shows the polarization resolved EQE of the weakly absorbing region; in addition to the Fabri-Perot interferences that are also observed on flat cells, the external quantum efficiency shows clearly resolved resonances [8]. The energies of the resonances are transferred to Figure 4 which shows the weakly absorbing region of Figure 2. Positions of the resonances are tentatively assigned by arrows; in s-polarization, the coupling at 1.63 eV is most likely to the s1 mode in the third Brillouin zone, the resonance at 1.81 eV can be attributed to the s0 mode in the fourth Brillouin zone. In p-polarization, coupling at 1.91 eV involves the p2 mode in the third Brillouin zone, whereas the coupling at 1.79 eV is likely to be with the p0 mode in the fourth Brillouin zone.

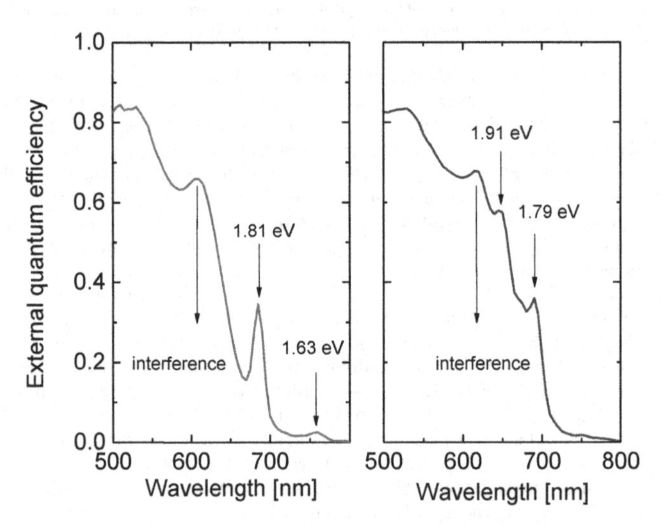

Figure 3: Polarization resolved external quantum efficiency (EQE) of a solar cell on a line grating with a period of 560 nm in s-(left) and p-polarization (right).

We note that the energies of the dispersion diagram of Figure 4 would suggest that the s-coupling in the fourth Brillouin zone occurs at lower energy than the p-coupling. This discrepancy can possibly be explained by the fact that the theoretical dispersion diagram was determined on the basis of flat interfaces but did not take into account the true nature of the interface texture.

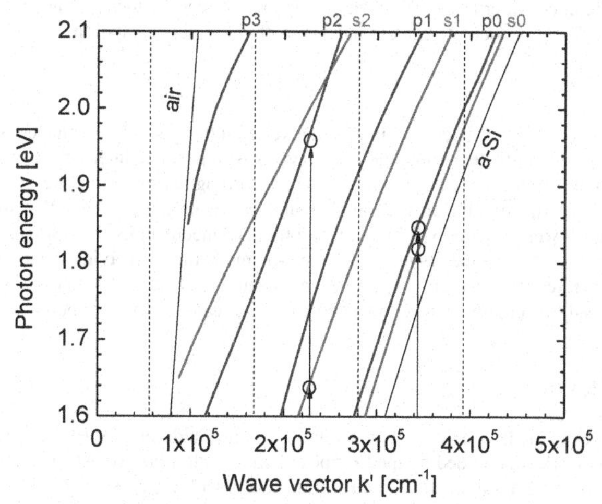

Figure 4: Detail of the dispersion diagram of Figure 2, only the weakly absorbing region is shown; grating coupling for the four resonances of Figure 3 is indicated tentatively.

FIELD DISTRIBUTION

In order to assess the amount of absorption enhancement, the amplitude of the electromagnetic field throughout the structure must be known. Figure 5 shows the s-polarized field intensity for the wavelength of the dominant resonance at 690 nm calculated with rigorous coupled wave analysis (RCWA) for the exact interface texture shown in Figure 1 [9]. For the shown 1D problem on the line grating, the required computing power is still reasonable. However, for more complicated scenarios it would be desirable to revert to analytic approximations, provided they can provide predictive power.

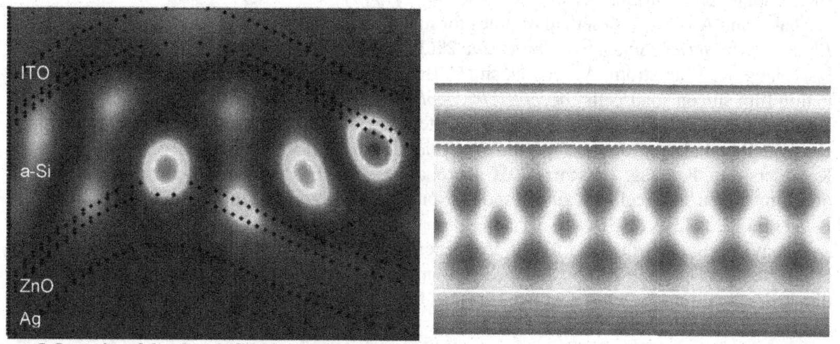

Figure 5: Intensity of the electric field in s-polarization at the resonance wavelength of 690 nm. The left panel shows the rigorous solution, the right panel illustrates the flat waveguide approximation. The horizontal scale corresponds to one period of 560 nm.

The rather complicated field profile of the exact solution in Figure 5 can be understood by superimposing the field intensity of the s0 guided at $k_{\|} = 3.43 \times 10^5$ cm^{-1} with the interference pattern of the Fabry-Perot resonance. The former shows 6 clearly resolved maxima in x-direction because this value of the wave vector accommodates three full wavelengths within the grating period, and one single maximum in z-direction because the order number zero means that there are no nodes within the guiding film. The interference pattern adds three maxima in the z-direction; the resulting superposition of the fields yields a surprisingly accurate correspondence with the exact calculation.

CONCLUSIONS

We investigated the effect of a periodic interface texture on the absorption enhancement in amorphous silicon solar cells. In the experimental part, devices with periodic interfaces were found to exhibit clearly resolved absorption signatures in the weakly absorbing region. In the theoretical part, we treated the solar cell in terms of a planar waveguide that supports multiple modes. With the help of the dispersion diagram, we were able to explain the absorption enhancement by the resonant excitation of waveguide modes and to attribute the resonances to specific mode orders. Having identified the modes that participate in a resonance, their coupling strength can be determined with existing theories; ultimately this should lead to a parameter or a few parameters that can be tuned for better absorption enhancement.

ACKNOWLEDGEMENTS

We thankfully acknowledge Dr Aïcha Hessler and Caroline Calderone for FIB cross sectioning and SEM imaging. This work was funded by the European Union within the project Si-Light (Contract No 241277) and by the Swiss National Science Foundation (Contract No. 2000021_125177/1).

REFERENCES

1. D. E. Carlson and C. R. Wronski, "Amorphous Si Solar Cell" *Applied Physics Letters* **28(11)**, p. 671-673 (1976)
2. H. W. Deckman, C. R. Wronski, H. Witzke, and E. Yablonovitch, "Optically enhanced amorphous silicon solar cells" *Applied Physics Letters* **42(11)**, p. 968-970 (1983)
3. K. Söderström, J. Escarré, O. Cubero, F.-J. Haug, S. Perregaux, and C. Ballif, "UV nano imprint lithography technique for the replication of back reflectors for n-i-p thin film silicon solar cells" *Progress in Photovoltaics: Research and Applications* **19(2)**, p. 202-210 (2011)
4. H. Okamoto, Y. Nitta, T. Adachi, and Y. Hamakawa, "Glow discharge produced amorphous silicon solar cells" *Surface Science* **86**, p. 486-491 (1979)
5. S. Shakir and A. Turner, "Method of poles for multilayer thin-film waveguides" *Applied Physics A: Materials Science & Processing* **29(3)**, p. 151-155 (1982)
6. F.-J. Haug, K. Söderström, A. Naqavi, and C. Ballif, "Resonances and absorption enhancement in thin film silicon solar cells" *accepted for application in Journal of Applied Physics* (2011)
7. I. Avrutsky, A. Svakhin, and V. Sychugov, "Interference phenomena in waveguides with two corrugated boundaries" *Journal of Modern Optics* **36(10)**, p. 1303-1320 (1989)
8. K. Söderström, F. Haug, J. Escarré, O. Cubero, and C. Ballif, "Photocurrent increase in n-ip-thin film silicon solar cells by guided mode excitation via grating coupler" *Applied Physics Letters* **96**, p. 213508 (2010)
9. A. Naqavi, K. Söderström, F. J. Haug, V. Paeder, T. Scharf, H. P. Herzig, and C. Ballif, "Understanding of photocurrent enhancement in real thin film solar cells: towards optimal one-dimensional gratings" *Optics Express* **19(1)**, p. 128-140 (2011)

Mater. Res. Soc. Symp. Proc. Vol. 1321 © 2011 Materials Research Society
DOI: 10.1557/opl.2011.947

Effect of Buffer Structure on the Performance of a-Si:H/a-Si:H Tandem Solar Cells

C.H. Hsu, C.Y. Lee, P.H. Cheng, C.K. Chuang, C.C. Tsai
Department of Photonics, National Chiao Tung University, Hsinchu, Taiwan

ABSTRACT

The study focuses on the influence of the hydrogenated amorphous silicon carbide (a-SiC:H) buffer layer in hydrogenated amorphous silicon (a-Si:H) single-junction and tandem thin-film solar cells. By increasing the undoped a-SiC:H buffer layer thickness from 6nm to 12nm, the J_{SC} in single-junction cell was significantly improved, and the efficiency was increased by 4.5%. The buffer layer also effectively improves the efficiency of the a-Si:H/a-Si:H tandem cells by 7% as a result of the increase in open-circuit voltage (V_{OC}) and short-circuit current (J_{SC}). Although the bottom cell absorbs less short-wavelength photons, the wider-bandgap doped and buffer layers were still necessary for improving the cell efficiency. Presumably, this is because these wider-bandgap layers allow more photons to reach the bottom cell. Also, they can reduce interface recombination.

INTRODUCTION

Hydrogenated amorphous silicon (a-Si:H) has received much attention due to its superior properties for the thin-film solar cells. The high absorption coefficient of a-Si:H allows it to absorb light with less material. The bandgap around 1.75eV also makes it suitable for the effective absorption of the solar spectrum. However, the Staebler-Wronski effect (SWE) [1] influences the long-term stability of the solar cells, resulting in a decrease in the cell efficiency [2]. The amount of the degradation depends on the material quality and the film thickness. By the use of a-Si:H/a-Si:H tandem structure, the absorber is divided into two devices [3]. The thinner absorber not only reduces the degree of SWE, but also creates a stronger build-in electric field to assist the carrier transport. Wieder et al. [4] had reported an initial cell efficiency of 9.2% with a relative decrease of 8% after 900 hours light soaking due to a more effective carrier extraction in thinner undoped layers. Furthermore, other materials with lower bandgaps, such as hydrogenated amorphous silicon germanium (a-SiGe:H) or hydrogenated microcrystalline silicon (μc-Si:H) can be used in the bottom cells. However, a-SiGe:H uses costly germane and μc-Si:H requires longer deposition time. As a comparison, the a-Si:H/a-Si:H solar cell has a lower production cost.

A high efficiency a-Si:H single-junction thin-film solar cell normally contains a wide-bandgap window layer to allow more high energy photons to be absorbed in the undoped layer. However, the bandgap difference between doped and undoped a-Si:H at the p/i interface hinders the hole transportation [5]. A wide-gap undoped layer was therefore used to accommodate the band offset. Such a buffer layer also significantly reduces the p/i interface recombination, prohibits the electron from moving into p-layer [6] and may prevent the boron diffusion into absorber. As a result, the buffer layer can significantly improve the V_{OC}, accompanied by the increases in J_{SC}, fill factor (FF) and efficiency. In an a-Si:H/a-Si:H tandem solar cell, although the two absorber have identical or similar bandgaps, the bottom cell still absorbs less high energy photons than the top cell. Replacing the material of the window layer (in the bottom cell) to doped a-Si:H may have a better electrical property due to more effective doping. To investigate whether the wide-bandgap material was needed in the bottom cell, the study focused on the properties of the a-SiC:H material and the structure of the tandem solar cell.

EXPERIMENT

In this work, the devices were prepared in a 27.12MHz plasma-enhanced chemical vapor deposition (PECVD) system on a SnO:F glass in a superstrate configuration. A gas mixture of SiH_4 and H_2 was used to deposit undoped a-Si:H. Undoped a-Si:H film with thicknesses ranging from 40nm to 80nm were used in the top cell. The p-type and n-type layers were achieved by introducing B_2H_6 and PH_3, respectively, during deposition. With the addition of CH_4, doped or undoped a-SiC:H thin-film were obtained. The p-type a-SiC:H thin-film had a conductivity of $1.48x10^{-5}$S/cm with an optical bandgap of 1.95eV. By modulating the CH_4 flow rate, the optical bandgap of the undoped a-SiC:H film can be altered. Amorphous a-SiC:H p-layer and undoped buffer layer were used in the top or the bottom cell, as can be seen in Fig.1. For the study of replacing the a-SiC:H p- and buffer layer by a-Si:H p-layer in the bottom cell, the thin-film with a conductivity of $1.12x10^{-5}$S/cm with an optical bandgap of 1.75eV was used. Considering the light-induced-degradation, a fixed absorber thickness of 300nm was used in the bottom cell. No particular tunneling junction between top and bottom cell was used. A back reflector consisted of TCO and silver was also used to enhance the reflection of long-wavelength photons.

The optical bandgap of the a-SiC:H thin-film was derived from the UV/VIS/IR spectrometer. The electrical properties of the thin-film and the devices were characterized with an AM1.5G illuminated I-V measurement system and a quantum efficiency (QE) instrument.

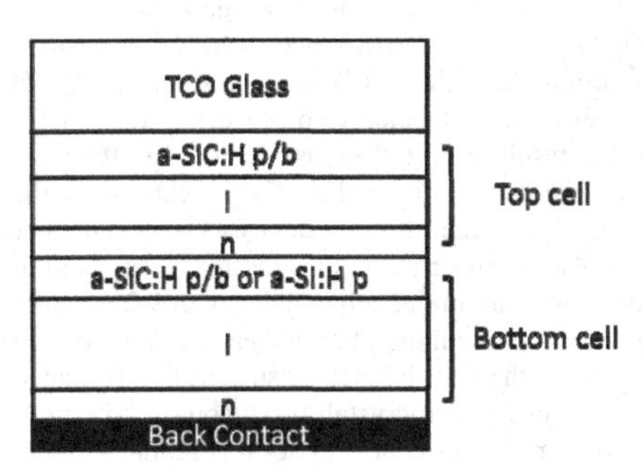

Figure 1. Schematic diagram of the a-Si:H/a-Si:H solar cell structure

RESULTS AND DISCUSSION

We first studied the single-junction cells from the optimization and characterization of the undoped a-SiC:H thin-films. The bandgap of the a-SiC:H film can be changed by varying the carbon content in the film. Instead of stoichiometric composition of the crystalline SiC, the a-

SiC:H material used here has a much lower carbon content, due to the consideration of electrical and optical properties. Fig.2 shows the effect of methane-to-silane flow rate ratio on the conductivity and the bandgap. As the flow ratio in the gas phase increased from 0 to 1.5, the bandgap increased from 1.80eV to 2.04eV, indicating an increased incorporation of C in the film. Accompanied with the increasing bandgap, the photo-conductivity significantly decreased mainly due to the increasing defects. Besides, fewer photons were able to be absorbed by the materials with the higher bandgap. The dark-conductivity did not show obvious decrease which can also be due to the increasing defect density induced by the carbon [7]. The selection of the a-SiC:H material should consider the trade-off between electrical and optical properties based on conductivity and bandgap, respectively. In this study we use the a-SiC:H with methane-to-silane flow ratio of 0.5.

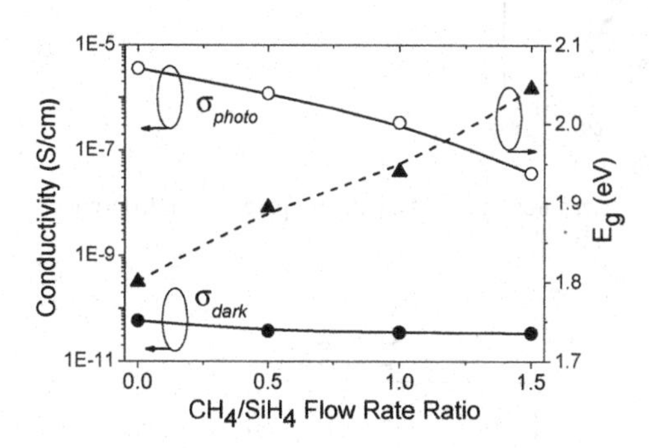

Figure 2. Effect of the methane-to-silane flow rate ratio on the conductivity and the bandgap

The buffer layer with thickness of 6nm has a bandgap of 1.89eV, slightly smaller than the p-type layer which was 1.95eV. The 12nm buffer layer had changed bandgaps which were achieved by decreasing the methane flow rate during deposition. Although the a-SiC:H layer was located at the front side of the device, a total enhancement was observed in the range from 350nm to 700nm, as shown in Fig. 3. This indicates that the wide-bandgap material not only allows more short-wavelength photons be absorbed, but also assisted the transport of carriers generated near the n/i interface. The increase in the EQE measurement reflected the increase in J_{SC} and FF while V_{OC} shows no significant change, compared to the study which has significant improvement in V_{OC} [6]. By optimizing the thickness of the buffer layer with compromised bandgap and conductivity, a relative increase of 4.5% in single-junction cell efficiency from 7.95% to 8.31% was obtained. In this study we achieved a higher efficiency with buffer layer thickness of 12nm. Further optimization was carried out to achieved a single-junction a-Si:H solar cell with efficiency of 9.45%, with J_{SC}=14.39mA/cm^2, V_{OC}=0.90V and FF of 73.33% [8].

On the next we fabricated the tandem device based on the previous single-junction cell structure. The a-Si/a-Si tandem structure was fabricated based on the structure of the single-junction solar cell. In order to obtain better solar cell efficiency, the current generated from the top cell should match to that from the bottom cell because of the limitation of a series connection. With the increasing thickness, the generated photo-current in the top cell should increase, while the current generated in the bottom cell decreased. As the two current are perfectly the same, the whole device would have the highest efficiency.

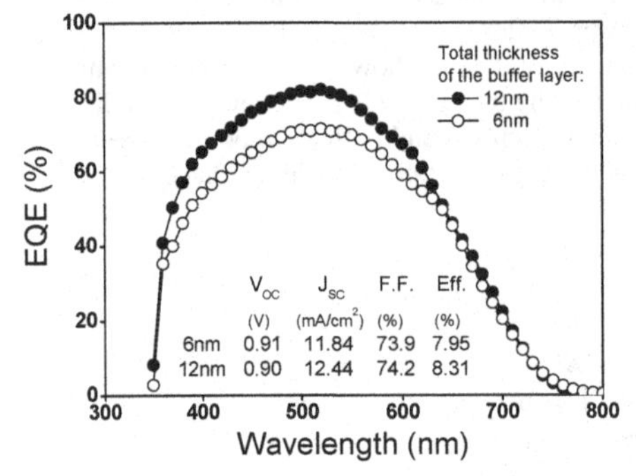

Figure 3. External quantum efficiency measurement of the cells prepared with different thicknesses of a-SiC:H buffer layer.

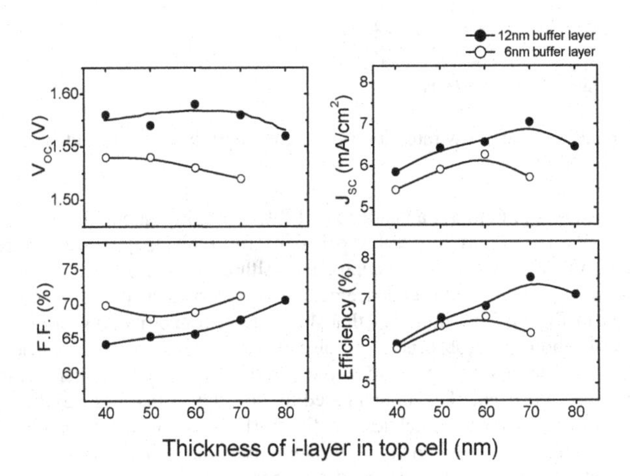

Figure 4. Effect of the i-layer thickness in the top cell on the performance of the tandem cells with different buffer layer thicknesses.

Fig. 4 shows the effect of the i-layer thickness in top cell on the cell performance with 6nm and 12nm buffer layer thicknesses, while the thickness of undoped a-Si:H in the bottom cell was kept at 300nm under the consideration of degradation. Although the V_{OC} has no improvement in the single-junction cell, significant improvement of V_{OC} in the tandem cell was observed. There was decrease in fill factor due to the absence of tunneling recombination junction [9], but the increase of J_{SC} accompanied with increased V_{OC} improved the overall efficiency by 7.0%. The cell with 12nm-thick buffer layer in both top and bottom cell has an efficiency of 7.54% with i-layer thickness of 70nm in top cell, compared to the cell with 6nm buffer layer and 60nm i layer having an efficiency of 7.05%. The thicker i-layer required for the top cell having 12nm buffer layer may be due to the reduction of build-in field affected by the insertion of the undoped a-SiC:H, while the increased thickness of a-SiC:H in the bottom cell shows less influence due to small portion in the 300nm undoped a-Si:H.

In Fig. 4, the FF gradually increased even if the thickness of the top cell was thicker than the matched one. This behavior in FF is only observed in cell without the tunnel junctions. The normal behavior in FF was observed in other experiments where tunnel junctions were employed. Although we do not understand the nature of this phenomenon, we believe this is related to the absence of the tunnel junctions. Further experiments are needed to investigate such effect.

The comparison of EQE measurement from the cells having different buffer structures is shown in Fig. 5. The i-layer thickness in the bottom cell was also fixed at 300nm with 60nm or 70nm i-layer thickness in the top cell, depending on a better current matching. As the buffer thickness increased from 6nm to 12nm for the cell having a-SiC:H in both top and bottom cell, the current density increased. As a result the total quantum efficiency increased with a maximum of over 80%.

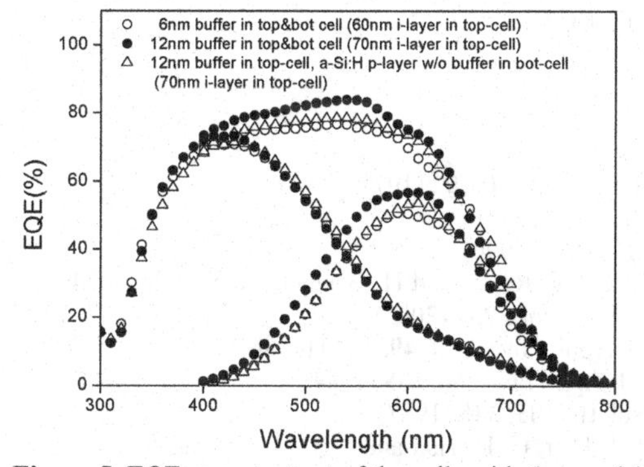

Figure 5. EQE measurement of the cells with 6nm and 12nm a-SiC:H buffer layer in the top and the bottom cell, and the cell without buffer layer in the bottom cell.

We have also made a cell with bottom cell contained p-type a-Si:H only (also without a-SiC:H buffer layer) to see if the a-SiC:H was still necessary in the bottom cell, since the wide-bandgap a-SiC:H mainly increases the short-wavelength incident light and the buffer layer may act as a resistor in such device. The result shows that the bottom cell with a-SiC:H p-layer and 12nm buffer layer still generated a higher current. Compare to the cell with 6nm a-SiC:H buffer, the cell composed of no a-SiC:H in the bottom cell had no apparent improvement. The carrier generated in the bottom cell with shorter wavelength decreased and showed similar result with cell having 6nm buffer. The result indicates the wide-bandgap window layer can still benefit the absorption of the bottom cell to have a higher current density. Part of the contribution may also arise from the reduced recombination at p/i interface by the insertion of high quality a-SiC:H.

CONCLUSIONS

We have found by increasing the undoped a-SiC:H buffer layer thickness from 6nm to 12nm, the single-junction cell performance improved significantly in J_{SC}, and the overall efficiency by 4.5%. The buffer layer also effectively improves the efficiency of the a-Si:H/a-Si:H tandem cell by 7% because of the increased V_{OC} and J_{SC}. Although the bottom cell absorbs less short-wavelength photons, the wide-bandgap doped and buffer layers were still necessary for improving efficiency which may be due to the absorption of short-wavelength photons and the advantage of reducing interface recombination.

ACKNOWLEDGMENTS

This work was sponsored by the Center for Green Energy Technology at the National Chiao Tung University and the National Science Technology Program-Energy of National Science Council under contract no.100-3113-E-009-007-CC2.

REFERENCES

1. D. L. Staebler and C. R. Wronski, Appl. Phys. Lett. **31**, 292 (1977).
2. J. Yang and S. Guha, Appl. Phys. Lett. **61**, 2917 (1992).
3. M. Bennett and K. Rajan, J. Appl. Phys. **67**, 4161 (1990).
4. S. Wieder, B. Rech, C. Beneking, F. Siebke, W. Reetz, and H. Wagner, Proceedings of the 13th European Photovoltaic Solar Energy Conference 234 (2005).
5. R. R. Arya, A. Catalano and R. S. Oswald, Appl. Phys. Lett. **49**, 1089 (1986).
6. K. S. Lim, M. Konagai and K. Tajahashi, J. Appl. Phys. **56**, 538 (1984).
7. J. Bullot and M. P. Schmidt, Phys. Stat. Sol. (b) **143**, 345 (1987).
8. P. H. Cheng, S. W. Liang, Y. P. Lin, H. J. Hsu, C. H. Hsu and C. C. Tsai, Mat. Res. Soc. Proc. of 2011 Spring Meeting, Symposium A (to be published).
9. D. S. Shen, R.E.I. Schropp, H. Chatham, R. E. Hoilingsworth, P.K. Shat and J. Xi, Appl. Phys. Lett. **56**, 1871 (1990).

Mater. Res. Soc. Symp. Proc. Vol. 1321 © 2011 Materials Research Society
DOI: 10.1557/opl.2011.817

Annealing Effects of Microstructure in Thin-film Silicon Solar Cell Materials Measured by Effusion of Implanted Rare Gas Atoms

W. Beyer[1,2], D. Lennartz[2], P. Prunici[1], H. Stiebig[1]
[1]Malibu GmbH & Co. KG, Böttcherstr. 7, D-33609 Bielefeld, Germany
[2]IEK5-Photovoltaik, Forschungszentrum Jülich, D-52425 Jülich, Germany

ABSTRACT

In thin film silicon solar cell technology, annealing (heat treatment) effects are of interest since (i) annealing of underlying films often cannot be avoided during deposition and (ii) heat treatment (e.g. by laser) may be actively used for improvement of as-deposited material. Changes in the microstructure of several thin film silicon solar cell materials like hydrogenated amorphous silicon, microcrystalline silicon and zinc oxide by heat treatment were investigated by effusion measurements of hydrogen and implanted helium. Densification is observed for all materials studied, i.e. interconnected voids disappear or are transformed to isolated voids. We attribute the observed annealing effects primarily to an incomplete polymerization during growth. Important for solar cell processing is the result that the annealing effects involving structural changes set in at temperatures close to the temperature of deposition.

INTRODUCTION

Knowledge on temperature treatment (annealing) processes in thin film silicon solar cell materials is desirable by various reasons. On one hand, during processing of thin film silicon related devices, annealing of underlying material often cannot be avoided. Annealing processes, on the other hand, may be of interest to change and improve individual layers during processing of devices like solar cells. In particular, the application of large area laser annealing and laser crystallization may serve as a tool for improvement of thin film silicon related devices. Various annealing processes are conceivable, like (i) diffusion of hydrogen, (ii) structural changes of amorphous material and (iii) partial and complete crystallization of the amorphous material. Improvement of individual thin film layers applied in silicon solar cell technology by annealing have been reported, like the reduction of light-induced defect generation for hydrogenated amorphous silicon (a-Si:H) films [1], enhanced passivation properties of thin film silicon materials for crystalline silicon processing [2] and the improvement of zinc oxide (ZnO) films used as transparent contacts [3]. Here, we focus on microstructural changes in thin film materials caused by annealing. For characterization of microstructure, we apply effusion measurements of hydrogen and in particular of implanted helium. Since the presence of interconnected voids or of compact material causes quite different hydrogen effusion processes, hydrogen effusion is sensitive to microstructure [4-6]. Effusion of rare gas atoms like helium is highly sensitive to material density and voids as these gas atoms do not bind to the host material [6,7].

EXPERIMENT

Thin films of 1 – 2.5 μm thickness deposited on c-Si wafers were investigated. The a-Si:H films were grown in a single chamber system at a frequency of 13.56 MHz, a power of 10W, a pressure of 0.5 mbar and a SiH_4 flow of 3 sccm. Note that some microstructure data of this set of samples was reported before [6,7]. For the microcrystalline silicon (μc-Si:H) films a three chamber (10 x10 cm^2) system was applied. The SiH_4 and H_2 flows were 4 sccm and 360 sccm, respectively, and the rf power was 80 W. The zinc oxide films were grown using low pressure chemical vapor deposition (LPCVD) employing diethylzinc and water vapor as precursor substances. Annealing was done in high vacuum. The annealing time was 5-10 min after rapidly ramping up. Blister and pinhole formation in the films was avoided by thin SiO_2 layers between film and substrate [6]. The effusion measurements were performed as described elsewhere using a heating rate of 20°C/min [6]. Helium was implanted after annealing at an energy of typically 40 keV and at rather low doses of $3x10^{15}$ to $10^{16}/cm^2$ which are not expected to modify greatly microstructure [7]. According to the TRIM routine, the maxima of the implanted He distributions are at depths of ~0.4 μm for a-Si:H and of ~0.2 μm for ZnO.

RESULTS AND DISCUSSION

In Fig. 1, hydrogen and helium effusion spectra are shown for a-Si:H, deposited at a substrate temperature of 200°C and annealed at various temperatures T_A. The deposition conditions were those often used for device-grade material. The hydrogen content is 11-12 at.% and the infrared microstructure parameter is approximately 0.3. Hydrogen effusion proceeds by a single effusion peak with a maximum at $T \approx 650°C$ as expected for a film of thickness $d > 1.5$ μm [5,6]. He effusion for this type of material shows two and more effusion peaks, in particular a major one near 430°C and minor one at $T > 800°C$. In addition, there is a small structure near 750°C attributed to crystallization. The low temperature effusion peak is attributed to a diffusion-limited release of the implanted He. For helium implanted in crystalline silicon, we get a similar effusion temperature of about 400°C [6] which is also in agreement with the (extrapolated) helium diffusion coefficient in c-Si from literature [8]. This low temperature (LT) He effusion apparently can be influenced by interconnected voids and/or the presence of lower density material causing an effective increase of the (average) diffusion coefficient and, consequently, a shift of the effusion temperature to lower temperature. We characterize this helium effusion process by the peak temperature T_M. The high temperature (HT) He effusion process takes place

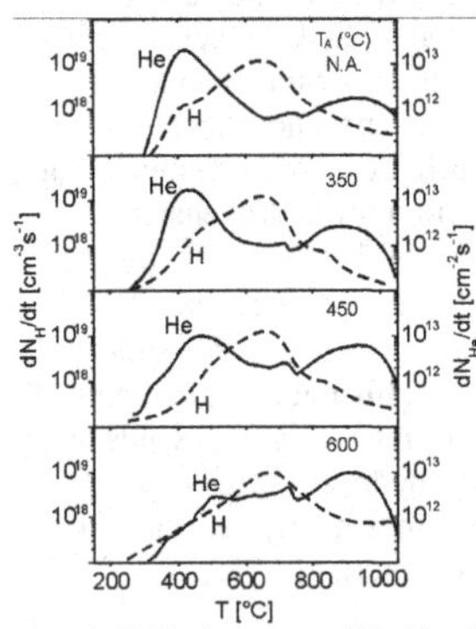

Figure 1. Effusion rates dN_H/dt of hydrogen and dN_{He}/dt of He versus temperature for a-Si:H (T_S =200°C) annealed at T_A.

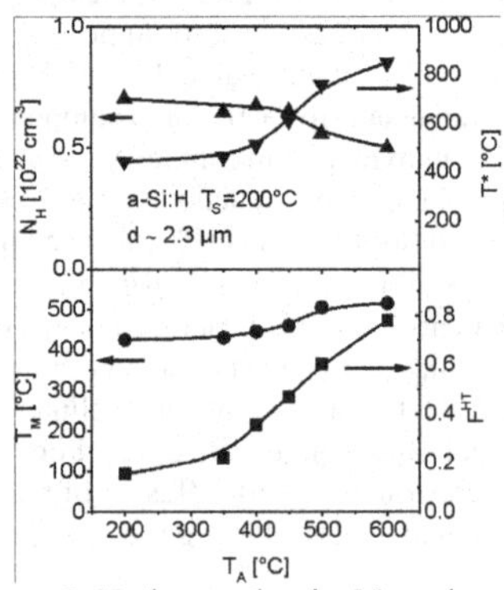

Figure 2. Hydrogen density N_H and helium effusion parameters T*, T_M and F^{HT} versus T_A for film of Figure 1.

in a temperature range (T =600-1000°C) where all He should have diffused out according to its diffusion coefficient in c-Si. However, such high helium effusion temperatures are possible if there are isolated voids embedded in the material where diffusing helium can precipitate. In this case, helium may leave these voids only if according to the gas equation (ideal gas law) the pressure within the voids gets high enough that He is driven to the actual film surface by a permeation process [7,8]. Since implanted He on its diffusion path to the film surface will either reach the surface directly or will get trapped in voids, the fraction F^{HT} of He effusing out in the HT peak (in relation to the total implanted He) can be used as a measure of the void concentration in the material. Note that this is the void concentration at temperatures when He is diffusing in the amorphous material, i.e. at $T \approx 300$-400°C. Furthermore, the overall stability of He in the material may be characterized by the mean helium effusion temperature which is defined by the temperature T* at which half of the implanted He has effused. Finally, another characteristic quantity of the annealed material is its total density of hydrogen N_H, obtained (e.g.) from the H effusion measurements. The results in Fig. 1 show that the hydrogen effusion spectra are only slightly affected by annealing up to $T_A = 600$°C, apart from a decrease in signal height. The He effusion, on the other hand, shows a clear re-distribution of He atoms effusing in the LT and HT processes. While for not annealed (N.A.) material the maximum effusion rate in the LT helium effusion peak exceeds that of the HT peak, after annealing at 600°C the maximum effusion rate in the HT He peak exceeds the LT one. Moreover, the peak temperature T_M of the LT helium effusion peak shifts to higher temperature by annealing.

The detailed results from Fig. 1 are compiled in Fig. 2. Plotted are the H density N_H, the mean helium effusion temperature T*, the temperature T_M of the LT He effusion peak and the fraction of HT helium effusion F^{HT}. It is seen that at temperatures > 300°C the hydrogen content starts to decrease. This is caused by H out-diffusion and will, of course, depend on film thickness. As the hydrogen density starts to decrease, the quantity T* starts to increase, thus indicating a rising He stability. Two processes apparently contribute to this increased stability: the quantities T_M and F^{HT} also shown in Fig. 2. As is seen, the concentration of isolated voids (characterized by F^{HT}) is increasing and the density of the amorphous network (characterized by T_M), is rising, too. We estimate the amount of isolated voids to 10^{18} -10^{19} /cm^3 for equal signal height of LT and HT helium effusion [7]. We note that by SAXS measurements the appearance of (spherical) voids after annealing of device-grade a-Si:H was also demonstrated [9].

Densification effects by annealing are in particular expected for material which grows at much reduced density compared to crystalline material. Indeed, we previously reported a significant shrinkage of a-Si:H films deposited near room temperature [4,6]. In Fig. 3, we show for this type of material the results of He and H effusion for various annealing steps. As deposited, the hydrogen content was about 23 at.% according to H effusion. The microstructure parameter was 0.55. Hydrogen effusion (see Fig.3) shows two effusion peaks, one near 400 °C and the other at T> 600°C. The 400°C peak has been explained by release of molecular hydrogen (H_2) from internal void surfaces and rapid effusion of H_2. Indeed, for this type of material, molecular hydrogen has been identified as the major effusing species [4-6]. From the result that in the HT hydrogen effusion peak atomic H was identified as the major diffusing species [4-6], a densification of the material, in particular near annealing temperatures of 400°C was concluded [4-6]. The present results confirm this picture. In the as-deposited state, the material has a rather open microstructure with He effusion peaked near 300°C. The lack of high temperature He effusion indicates a low concentration of isolated voids in agreement with the concept that the presence of interconnected voids makes the occurrence of isolated voids unlikely [6,7]. Changes

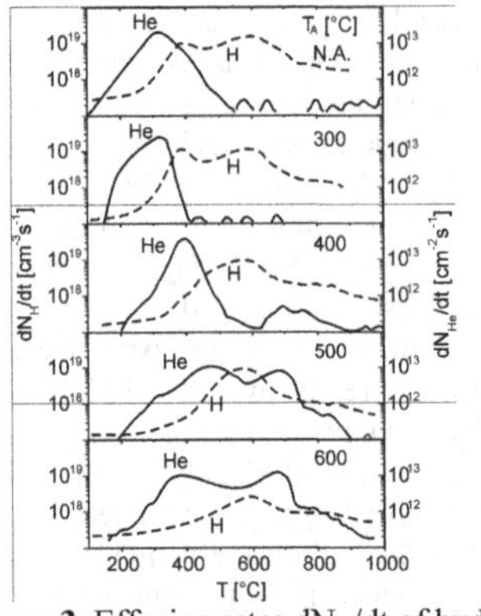

Figure 3. Effusion rates dN_H/dt of hydrogen and dN_{He}/dt of He versus temperature for a-Si:H (T_S =50°C) annealed at T_A.

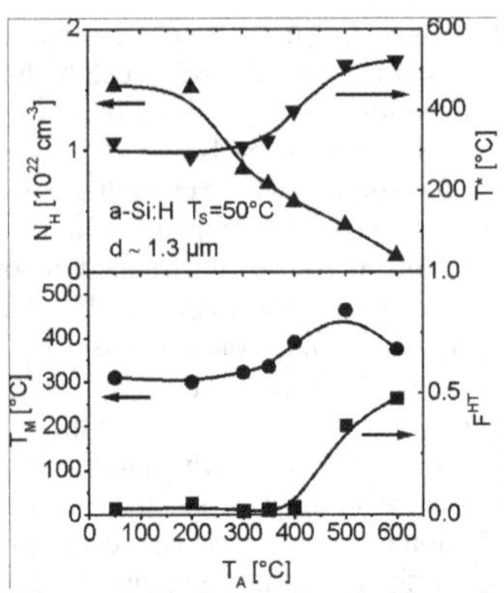

Figure 4. Hydrogen density N_H and helium effusion parameters T*, T_M and F^{HT} versus T_A for film of Figure 3.

in the shape of the He effusion peak indicate microstructural changes and densification already at rather low annealing temperatures ($T_A \approx 300$°C). The LT hydrogen effusion peak attributed to the presence of interconnected voids disappears after annealing at $T_A \geq 400$°C. At the highest annealing temperatures of 500 and 600°C, the material apparently is so much densified that isolated voids appear. The compiled results of Fig. 3 are shown in Fig. 4. Plotted are the quantities T*, T_M and F^{HT}, along with N_H as a function of T_A. As is seen from the annealing dependences of N_H, T* and T_M, densification starts when the first hydrogen release takes place at $T_A > 200$°C. However, while the hydrogen release is strongest between $T_A = 250$ and 400°C (i.e. in the temperature range of the LT H effusion peak), densification according to T_M is strongest at somewhat higher temperatures between about 350 and 500°C. This result confirms the concept that material reconstruction is not directly involved in low temperature H release but occurs retarded by the collapse of interconnected voids [4,5]. Presumably, when the interconnected voids are depleted of hydrogen the resulting dangling bonds will tend to reconstruct. In the annealing range where densification according to T_M is strongest, F^{HT} rises significantly. However, between $T_A = 500$ and 600°C, the (low temperature) helium effusion temperature T_M decreases again and F^{HT} increases less strongly. We attribute this latter effect to newly generated interconnected voids (cracks), presumably caused by the shrinkage of the material.

We note that densification effects by annealing were also reported for a-Si alloys like silicon nitride [10]. The results of Figs. 5 and 6 demonstrate the presence of densification effects by annealing also for microcrystalline Si:H films. However, due to problems with film peeling from substrate we can present only one annealing step and the results must be considered preliminary. The material was deposited at a substrate temperature of 100°C and had a hydrogen content of about 10 at.%. The IR microstructure parameter was 0.21, the Raman crystallinity was somewhat below 70%. It is seen that T* and F^{HT} rise strongly by annealing, i.e. isolated voids form. On the other hand, T_M is rather high and changes only slightly. This may be attributed to the presumably different nature of interconnected voids in a-Si:H and µc-Si:H. While in (plasma-deposited) a-Si:H these voids likely arise from extensive H incorporation during growth ("polysilane material") interrupting potential Si-Si bonds, in microcrystalline material

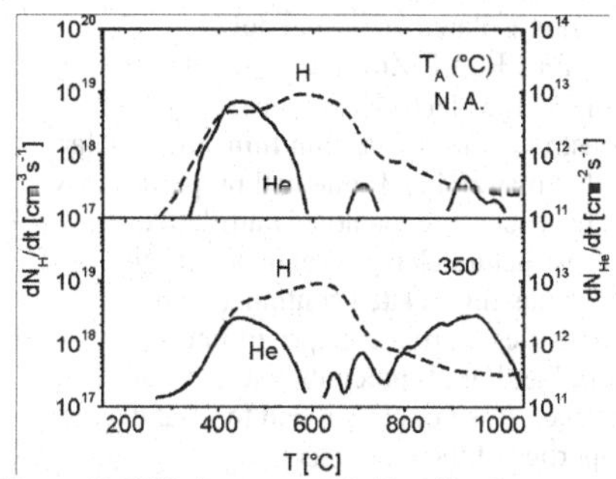

Figure 5. Effusion rates dN_H/dt of hydrogen and dN_{He}/dt of He versus temperature for µc-Si:H (T_S =100°C) not annealed and T_A=350°C.

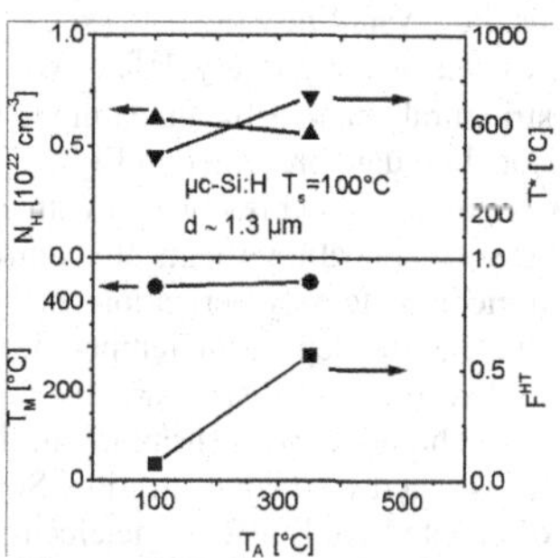

Figure 6. Hydrogen density N_H and helium effusion parameters T*, T_M and F^{HT} versus T_A for film of Figure 5.

interconnected voids (seen by 400°C H effusion) are likely related to grain boundaries. Thus, while in a-Si:H voids may collapse upon annealing related H release and Si-Si reconstruction, this cannot be expected for µc-Si:H as the grain boundaries are unlikely to disappear. However, long range H_2 and He diffusion along these grain boundaries may get modified by annealing.

The same is presumably true for annealing effects in LPCVD grown polycrystalline zinc oxide material, as shown in Figs. 7 and 8 [3]. The material shown in Fig. 7 was deposited at a substrate temperature of 150 °C. We focus here on the effusion of implanted helium which shows in the as-deposited state an effusion peak near 300°C, while after annealing at 350-400°C an effusion peak near 700°C becomes dominant. By comparison with helium effusion from single crystalline ZnO which shows an effusion peak near 550 °C [3] associated with He out-diffusion, we attribute the low helium effusion temperature of the as-deposited material to the presence of interconnected voids at grain boundaries. Upon annealing, these long range helium pathways seem to disappear and isolated voids, causing the He effusion peak near 700°C are increasing in concentration. In Fig. 8, T* for this type of material shows a strong increase upon annealing between $T_A \approx$ 200°C and 400 °C. Also depicted in Fig. 8 are data for T* of ZnO material deposited at the higher substrate temperatures of 175 and 200°C. In this case, T*

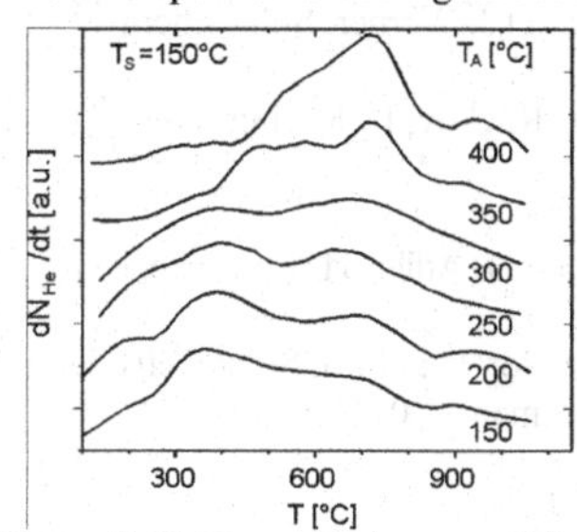

Figure 7. Helium effusion rate dN_{He}/dt versus temperature for LPCVD ZnO film (T_S = 150°C) for various annealing temperatures T_A.

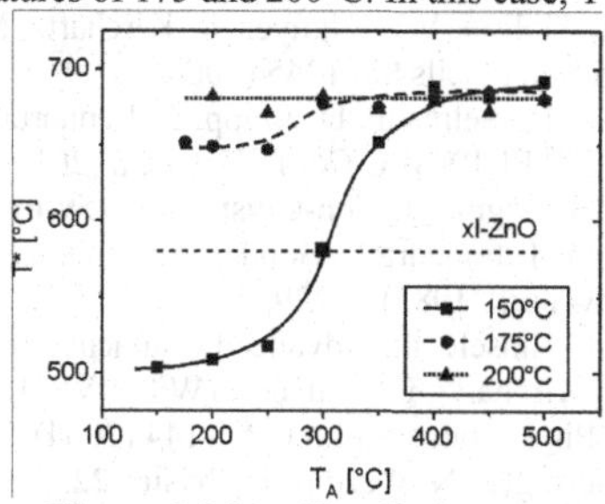

Figure 8. Mean He effusion temperature T* versus annealing temperature T_A for ZnO films of different T_S. Dashed line indicates T* for crystalline ZnO.

always exceeds the crystalline ZnO data (xl-ZnO) indicating isolated voids and not interconnected voids as the primary defect. We note that for LPCVD ZnO we were able to correlate the structural changes by annealing to variations in the electrical properties [3].

The annealing data presented in Figs. 1-8 demonstrate that various thin film silicon solar cell materials are rather unstable to heat treatment and that material and solar cell properties may get improved if such annealing effects are controlled. The effects are tentatively attributed to a growth related incomplete polymerization, as these microstructure changes set in at annealing temperatures close to the deposition temperature. In this annealing related continuation of polymerization, hydrogen is released and the material densifies. Partial collapse of inter-connected voids, inhomogeneous shrinking of the material and/or H_2 precipitation may cause the enhancement of isolated void concentration. So far however, it is not clear if and to what degree the presence of isolated voids affects the electronic properties of the materials.

CONCLUSIONS

Using effusion measurements of hydrogen and of implanted helium, annealing related changes of microstructure are demonstrated to occur for various thin film silicon solar cell materials. In case of a-Si:H films, the onset of annealing effects seems to be related to the onset of hydrogen effusion. In most cases, the materials densify by annealing, i.e. interconnected voids disappear or are transformed into isolated ones. We attribute these effects primarily to an incomplete polymerization during film growth.

ACKNOWLEDGEMENTS

The authors wish to thank A. Dahmen, F. Hamelmann, M. Hülsbeck, F. Pennartz and R. Schmitz for technical support. Part of the work was financed by the state of Nordrhein-Westfalen (project EN/1008B "TRISO").

REFERENCES

1. M. Ohsawa, T. Hama, T. Akasaka, T. Ichimura, H. Sakai, S. Ishida, Y. Uchida, Jpn. J. Appl. Phys. **24**, L838 (1985).
2. P.J. Rostan, U. Rau, V.X. Nguyen, T. Kirchartz, M.B. Schubert , H.J. Werner, Solar Energy Materials and Solar Cells **90**, 1345 (2006).
3. W. Beyer, F. Hamelmann, D. Knipp, D. Lennartz, P. Prunici, A. Raykov, H. Stiebig, Proceedings 25th EUPVSEC Conf., Valencia (2010) p. 3094.
4. W. Beyer, H.Wagner, J. Non-Cryst. Solids **59-60**, 161 (1983).
5. W. Beyer, in: Tetrahedrally-Bonded Amorphous Semiconductors, D. Adler, H. Fritzsche, eds. (Plenum, New York, 1985) p. 129.
6. W. Beyer, F. Einsele, in: Advanced Characterization Techniques for Thin Film Solar Cells, D. Abou-Ras, T. Kirchartz, U. Rau, eds. (Wiley-VCH, Weinheim, Germany, 2011) p. 449.
7. W. Beyer, Phys. Status Solidi C **1**, 1144 (2004).
8. A. Van Wieringen, N. Warmoltz, Physica **22**, 849 (1956).
9. S. Acco, D.L. Williamson, W.G.J.H.M. van Sark, W.C. Sinke, W.F. van der Weg, A. Polman, S. Roorda, Phys. Rev. B**58**, 12853 (1998).
10. W. Beyer, H.F.W. Dekkers, MRS Symp. Proc. **910**, A06-05 (2006).

Mater. Res. Soc. Symp. Proc. Vol. 1321 © 2011 Materials Research Society
DOI: 10.1557/opl.2011.818

Room Temperature Fabricated ZnO:Al with Elevated and Unique Light-Trapping Performance

E. V. Johnson[a], C. Charpentier[a,d], T. Emeraud[b], J.F. Lerat[b], C. Boniface[c], K. Huet[c], P. Prod'homme[d], and P. Roca i Cabarrocas[a]

[a] LPICM-CNRS, Ecole Polytechnique, 91128 Palaiseau, France
[b] PV BU, EXCICO Group NV, Kempischesteenweg 305 bus 2, B-3500 Hasselt, Belgium
[c] Process & Application Team, EXCICO France SAS, 13-21 Quai des Gresillons, F-92230 Gennevilliers, France
[d] TOTAL S.A. - Gas & Power, R&D Division, Tour La Fayette - 2 Place des Vosges - La Défense 6, 92 400 Courbevoie, France

ABSTRACT

We present a novel ZnO:Al fabrication process consisting of room-temperature vacuum sputtering followed by an excimer laser annealing (ELA). The ELA treatment improves the optical transmission of the films, and the film resistivities (<1 mΩ·cm) remain stable or improve with increasing laser fluence up to 0.6 J/cm^2, as the carrier density increases but the carrier mobility is degraded. This process is followed by a standard dilute HCl chemical texturing step, and produces substrates with suitable texture, conductivity, and transparency properties for thin-film photovoltaic applications. Substrates resulting from this process display elevated haze levels (80% at 600 nm and 50% at 800 nm) after the wet-chemical etching step. Such substrates have been used to make single junction hydrogenated nanocrystalline silicon solar cells, and an increase in the short-circuit current of up to 2.2 mA/cm^2 is observed compared to a substrate deposited by a standard room-temperature sputtering + wet-etch process. This gain is primarily due to increased photo-response in the red due to improved light-scattering, as at wavelengths greater than 600 nm, a gain in photocurrent of up to 1.7 mA/cm^2 is observed.

INTRODUCTION

An important technology for the production of doped ZnO for use in photovoltaics consists of a vacuum sputtering technique using a ceramic (ZnO:Al_2O_3) or metallic (Zn:Al) target to deposit films onto a glass substrate at optimized temperatures between 250-375 °C. This method produces high quality but smooth films [1,2], and the texturing necessary for light scattering is obtained through a subsequent wet-etching step. The texture morphology produced has been related to the deposition conditions [3], but room temperature deposited films [4] typically do not show the chemical texturing (CT) effect necessary for PV applications. Post-deposition processing consisting of rapid thermal annealing treatments have been shown to be effective in improving the electrical properties of rf-magnetron sputtered ZnO:Al films [5], primarily through the activation of Al dopants.

In this work, we show the positive effect of excimer laser annealing of room temperature sputtered ZnO:Al thin films. Other groups have investigated changes in ZnO:Al thin films deposited by various means and upon laser exposure from various sources [6,7] including examination of undoped ZnO films deposited by magnetron sputtering at 400 °C and then subjected to different post-treatment techniques: laser annealing (up to 200 mJ/cm^2), furnace annealing, and rapid thermal annealing [8], and showed that the laser annealed films demonstrated an increase in PL intensity and a decrease in defect density.

In this paper, we describe the use of Excimer Laser Annealing (ELA) to modify the properties of ZnO:Al thin films deposited by sputtering at room temperature. The as-deposited films showed sheet resistances suitable for large-area thin film silicon solar cell applications (<1 mΩ cm), but transparencies that were insufficient for photovoltaic applications and a low level of haze when etched. It is demonstrated that this ELA technique can significantly improve the optical properties of such films compared to their as-deposited state. In addition, by combining the use of ELA of certain pulse fluences and a well-known CT technique, such films can provide a broad range of unique optical scattering properties, suitable for thin film silicon photovoltaic devices. The promise of this technique is additionally demonstrated through the deposition of hydrogenated microcrystalline silicon (μc-Si:H) solar cells on ELA + CT ZnO:Al thin films.

EXPERIMENTAL DETAILS

In this work, the ZnO:Al samples were deposited on Corning Eagle glass at room temperature (RT) by sputtering from a ZnO:Al_2O_3 (1 wt% Al) target using pure Ar as a carrier gas. The sputtering process results in a deposition rate of ~30 nm/min. All samples in this study are approximately 1 μm thick after deposition. Although actual deposition conditions varied slightly between sample sets, all samples were deposited at ambient temperature.

The ZnO:Al thin films were excimer laser annealed (ELA) using a pulsed UV exciplex XeCl laser (EXCICO LTA 15 series) with an emission wavelength of 308 nm. The duration of the pulses generated by this source ranges from 130-185 ns, and the energy per pulse was varied. Each energy dose was delivered in a single shot over a constant area of 1 cm^2 with a cross beam non-uniformity of less than +/-5%.

Some ZnO:Al samples were subsequently wet-etched in a dilute (0.5%) aqueous HCl solution to induce a textured roughness on the surface of the ZnO and thus enhance its light scattering properties. This CT step is similar to the one reported by the Juelich group [2] to produce textured substrates from ZnO:Al deposited at high temperature (250-400 °C). The texturing method used in this study differs in that the HCl etching step was shorter (20 s) than is typically employed (40 s). The optical transmission measurements were acquired using a Perkin-Elmer 950 Spectrometer with a 150 mm integrating sphere for the hemispherical (diffuse + specular) transmission measurements. The Hall Effect measurements were performed using a Microworld Ecopia HMS-5000.

The μc-Si:H photovoltaic cells used to study the effect of the ZnO:Al texturing on device performance were composed of the following layer stack: Corning Eagle / ZnO:Al / p-μc-Si:H / μc-Si:H / n-a-Si:H / Ag. All silicon layers were deposited by plasma enhanced chemical vapour deposition (PECVD) with excitation at the standard frequency of 13.56 MHz.

RESULTS

Electrical and optical measurements before chemical texturing

The optical and electrical properties of two sets of films were characterized after ELA but before CT. The optical transmission of a sample set of films is presented in Fig 1(a), wherein it can be seen that a clear modification in the transmission is induced by the ELA process. A slight increase of 0.13 eV can be observed in the bandgap, along with an increased transmission in the blue and a decreased transmission in the infrared. The average transmission, averaged from 400-1000 nm, is presented in Fig 1(b), as this is the range of interest for thin film silicon solar cells. Two samples are presented, that underwent ELA treatments with slightly different pulse length durations (A=140 ns and B=185 ns). Up to fluences of 0.6 J/cm^2, a net positive effect (~ 1% absolute) is seen in the transmission for both samples.

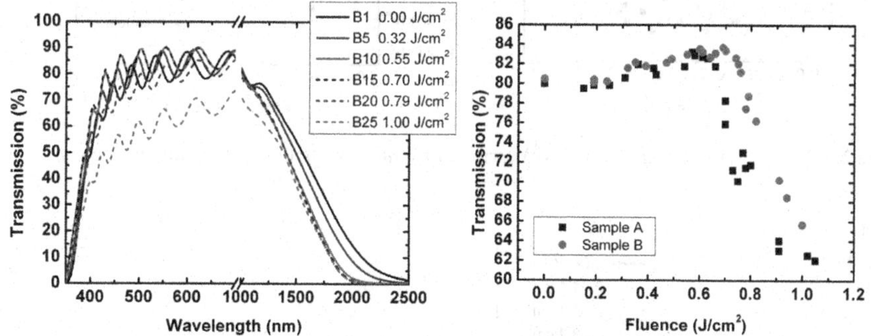

Figure 1. (a) Transmission spectra and (b) average transmission (averaged between 400 – 1000nm) for two ZnO:Al samples, each treated by different ELA fluences.

Conductivity and Hall Effect measurements were performed on 1 cm x 1 cm samples irradiated with fluences of up to 1.05 J/cm^2, and the resistivities measured are presented in Figure 2(a). It can be seen that from an initial value of 0.8-0.9 mΩ·cm, the resistivity varies between values of 0.6-0.9 mΩ·cm with increasing fluence up to a value of 0.5 J/cm^2, at which the resistivity begins to increase unpredictably for some samples. Using Hall Effect measurements, the carrier concentration and mobility can be determined independently, and the carrier concentration is shown in Figure 2(b). A clear increase can be noted; however, the increase in carrier concentration is offset by a decrease in mobility (not shown), and so no dramatic decrease in resistivity is seen. However, the as-deposited values of sheet resistance were sufficient for photovoltaic applications, and so no improvements were necessary. Importantly, the threshold fluence for a resistivity increase were above that for the improved transmittance.

Optical measurements after chemical texturing

After undergoing ELA, a selection of samples were chemically textured as described above and their specular and total optical transmissions (T_{spec} and T_{total}) were measured. From these values, the optical haze was calculated (as 100% x (T_{total}-T_{spec})/ T_{total}) and is shown in Figure 3. It can be seen that for the samples annealed with more than 0.52 J/cm^2, a dramatic increase in optical haze is observed. A haze value of 50% at a wavelength of 800 nm is obtained for the sample annealed with 0.86 J/cm^2, whereas the unannealed but CT sample shows negligible light scattering.

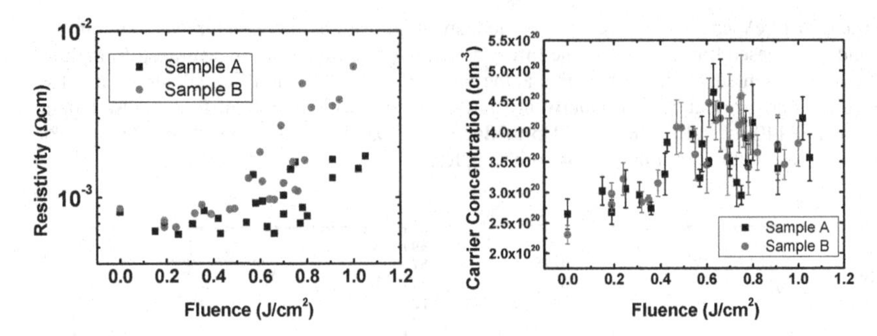

Figure 2. (a) Resistivity of samples, and (b) carrier concentration, both as a function of fluence.

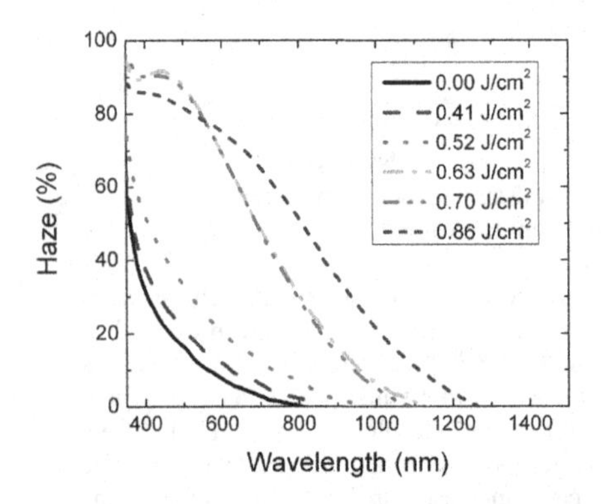

Figure 3. Optical haze for ZnO:Al samples deposited at room temperature after ELA and chemical texturing. Unannealed area is indicated as 0 J/cm^2.

Microcrystalline silicon cell deposition results

The elevated values of haze measured for the substrates suggest that they may be useful as a front electrode in thin film silicon photovoltaic devices, as scattering TCO layers can increase the light trapping in such cells. To explore this possibility, we have selected the use of μc-Si:H cells, due to their broad wavelength response but lower absorption coefficient compared to a-Si:H, which bring forward the necessity of light-trapping for such devices [9]. The μc-Si:H cells were deposited (as described above) on the ZnO:Al substrates after ELA and CT.

Typical external quantum efficiency (EQE) measurements of a set of devices (as measured at -1V to minimize the effects of collection) are depicted in Figure 4(a). An enhancement can be seen in the EQE for wavelengths above 650 nm for all levels of ELA, compared to the non-annealed substrate (0 J/cm^2). However, at the highest values of ELA fluence (0.86 and 0.88 J/cm^2) the increase in EQE above 650 nm is accompanied by a decrease below 600 nm.

Using the results of these external quantum efficiency (EQE) measurements and the AM1.5 standardized solar spectrum, values of photocurrent have been calculated for these devices, showing a maximal increase in photocurrent of 2.2 mA/cm^2 compared to an unannealed substrate. Focusing on the wavelength range above 600 nm, we plot the integrated photocurrent (again measured at -1V) as a function of ELA fluence in Figure 4(b). Results are shown for cells from three co-deposited substrates (results from Figure 4(a) are from substrate 4). This figure quantifies what is visible from the EQE graphs – an increase in photocurrent of up to 1.7 mA/cm^2 for this long wavelength region on samples having undergone ELA before CT.

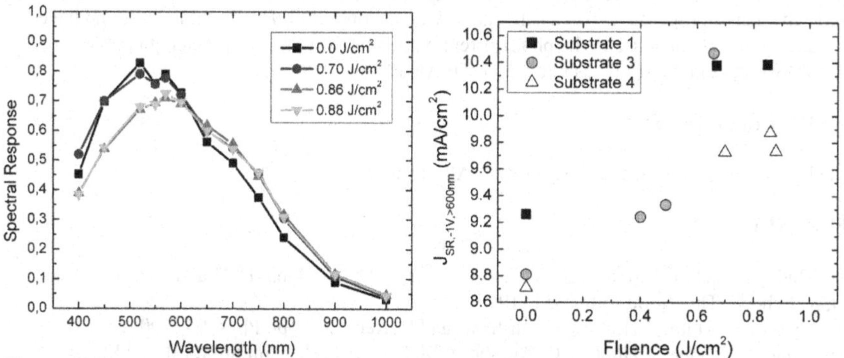

Figure 4. (a) EQE measurements taken at -1V for μc-Si:H cells deposited on ELA and CT ZnO:Al substrates, and (b) integrated photocurrent using portion of AM1.5 spectrum above 600 nm.

DISCUSSION

The optical and electrical measurements taken on the samples before etching show that an increased carrier density is responsible for the optical changes: notably an increased free-carrier absorption in the infrared, and an increase in the optical band-gap, presumably due to the Burstein-Moss shift, as previously seen in epitaxial ZnO:Al thin films due to doping effects [10]. The increased carrier density translates to a slightly lower resistivity, but only for ELA fluences up to 0.5 J/cm², however, as a decrease in mobility at higher laser fluence offsets the increased carrier density. This decrease of mobility for highly doped ZnO:Al is well known and can be explained by an effect of impurity clustering and the non-parabolicity of the conduction band. For higher laser fluences, the decreased mobility may be due to the formation of "missing-puzzle piece" morphology in the material, as observed by AFM (not shown).

Upon chemical etching, the areas having undergone ELA showed a significantly higher haze, despite the original layers being deposited at room temperature. To take advantage of the potentially increased light-trapping properties of such substrates, μc-Si:H PIN devices were fabricated. Although the devices showed an improvement from a light-trapping point of view, it should be noted that the devices were far from state of the art (~ 4 %). This may be attributed to the fact that only a single trial was performed and with a new deposition process conditions, one must also keep in mind that the texture of a substrate can significantly impact the quality of the active layers of a μc-Si:H cell [11,12,13]. Further work is necessary to optimize the interplay between such substrates and the devices deposited on them.

CONCLUSIONS

We have shown that an Excimer Laser Annealing and subsequent Chemical Texturing process is effective in producing highly scattering ZnO:Al thin films despite their deposition by sputtering having been done at room temperature. The optical haze value at 800 nm is increased from ~0% to 50% when samples have undergone ELA and CT. We have investigated the light-trapping enhancing properties of such substrates by depositing μc-Si:H PIN solar cells, and have shown an appreciable increase in the photocurrent, particularly in the red wavelength range (>600 nm) where it increases by as much as 1.7 mA/cm^2.

ACKNOWLEDGMENTS

The authors wish to thank J. Charliac for ZnO:Al preparation.

REFERENCES

1. O. Kluth, B. Rech, L. Houben, S. Wieder, G. Schöpe, C. Beneking, H. Wagner, A. Löffl, and H.W. Schock, Thin Solid Films 351 (1999) 247.
2. C. Agashe, O. Kluth, J. Hüpkes, U. Zastrow, and B. Rech, J. Appl. Phys. 95 (2004) 1911.
3. O. Kluth, G. Schöpe, J. Hüpkes, C. Agashe, J. Müller, B. Rech, Thin Solid Films 442 (2003) 80.
4. O. Kluth, A. Loffl, S. Wieder, et al., Proc. 26th, IEEE Photovoltaic Specialists Conf., Anaheim (1997) 715.
5. K.K.Kim, H. Tampo, J.O. Song, T.Y. Seong, S.J. Park, J.M. Lee S.W. Kim, S. Fujita, and S. Niki, Jpn. J. Appl. Phys. (2005) 4776.
6. I. Ozerova, M. Araba, V.I. Safarova, W. Marinea, S. Giorgiob, M. Sentisc, L. Nanaid, Appl. Surf. Sci. 226 (2004) 242.
7. G.K. Bhaumik, A.K. Nath, S. Basu, Mater. Sci. Eng. B52 (1998) 25.
8. T. Yen, D. Strome, Sung Jin Kim, A. N. Cartwright and W. A. Anderson, J. Elec. Mater. 37 (2007) 764.
9. J. Meier, R. Fluckiger, H. Keppner, and A. Shah, Appl. Phys. Lett. 65 (1994) 860.
10. K. Postava, H. Sueki, M. Aoyama, T. Yamaguchi, K. Murakami, and Y. Isasaki, Appl. Surf. Sci. 175-176 (2001) 543.
11. G. Yue, L. Sivec, J. M. Owens, B. Yan, J. Yang, and S. Guha, Appl. Phys. Lett. 95 (2009) 263501.
12. H. Li, R. H. Franken, J. Rath, and R. E. I. Schropp, Sol. Energy Mater. Sol. Cells 93, (2009) 338.
13. M. Python, O. Madani, D. Dominé, F. Meillaud, E. Vallat-Sauvain, and C.Ballif, Sol. Energy Mater. Sol. Cells 93 (2009) 1714.

Mater. Res. Soc. Symp. Proc. Vol. 1321 © 2011 Materials Research Society
DOI: 10.1557/opl.2011.948

N-Type Hydrogenated Microcrystalline Silicon Oxide Films and Their Applications in Micromorph Silicon Solar Cells

Amornrat Limmanee, Songkiate Kittisontirak, Channarong Piromjit, Jaran Sritharathikhun and Kobsak Sriprapha
Solar Energy Technology Laboratory, National Electronics and Computer Technology Center, National Science and Technology Development Agency, 112 Thailand Science Park, Phahonyothin Road, Klong 1, Klong Luang, Pathumthani 12120, Thailand.

ABSTRACT

We have prepared n-type hydrogenated microcrystalline silicon oxide films (n μc-SiO:H) and investigated their structural, electrical and optical properties. Raman spectra shows that, amorphous phase of the n μc-SiO:H films tends to increase when the CO_2/SiH_4 ratio increases from 0 to 0.28 resulting in a reduction of the crystalline volume fraction (X_c) from 70 to 12%. Optical bandgap (E_{04}) becomes gradually wider while dark conductivity and refractive index (*n*) continuously drop with increasing CO_2/SiH_4 ratio. The n μc-SiO:H films have been practically applied as a n layer in top cell of a-SiO:H/μc-Si:H micromorph silicon solar cells. We found that, open circuit voltage (V_{oc}) and fill factor (FF) of the cells gradually increased, while short circuit current density (J_{sc}) remained almost the same value with increasing CO_2/SiH_4 ratio for n top layer deposition up to 0.23. The highest initial cell efficiency of 10.7% is achieved at the CO_2/SiH_4 ratio of 0.23. The enhancement of the V_{oc} is supposed to be due to a reduction of reverse bias at sub cell connection (n top/p bottom interface). An increase of shunt resistance (R_{sh}) which is caused by a better tunnel recombination junction contributes to the improvement in the FF. Quantum efficiency (QE) results indicate no difference between the cells using n top μc-SiO:H and the cells with n top μc-Si:H layers. These results reveal that, the n μc-SiO:H films in this study do not work as an intermediate reflector to enhance light scattering inside the solar cells, but mainly play a key role to allow ohmic and low resistive electrical connection between the two adjacent cells in the micromorph silicon solar cells.

INTRODUCTION

Wide bandgap silicon oxide based materials have been widely studied for thin film silicon solar cell applications [1-4]. Properties of boron doped hydrogenated amorphous and microcrystalline silicon oxide films (p a-SiO:H and p μc-SiO:H) and their applications as window layer of solar cells have been reported by many research groups [5-6]. On the other hand , there are a few works regarding n type SiO:H films. Our group has been investigated wide bandgap SiO:H based materials and also fabricated a-SiO:H based solar cells with single junction and multi-junction structures [7]. In this work, we focus on properties of n μc-SiO:H films and their appropriateness for the use in micromorph silicon solar cells. Properties of the n μc-SiO:H films are presented along with the performance of a-SiO:H/ μc-Si:H micromorph silicon solar cells.

EXPERIMENT

Preparation of n μc-SiO:H films

N μc-SiO:H films have been prepared by very high frequency plasma enhanced chemical vapor deposition (60 MHz VHF-PECVD) technique. The gas sources are SiH_4, H_2 and CO_2 , and PH_3 is employed as a doping source. For film characterizations, the n μc-SiO:H films are deposited on Corning glass substrates at the deposition temperature of 180°C, a plasma power of 70 mW/cm^2, deposition pressure of 0.5 Torr, H_2/SiH_4 ratio of 35, PH_3/SiH_4 ratio of 0.38 and CO_2/SiH_4 ratio in the range of 0~0.28. The thickness of the films is about 350 nm. The crystalline volume fraction (X_c) of the n μc-SiO:H films is estimated by Raman scattering experiment. The Raman scattering spectra of the n μc-SiO:H films in the 400-600 cm^{-1} region can be deconvoluted into three spectra. A peak distribution around 470-475 cm^{-1} is assigned to the transverse optical (TO) mode of amorphous silicon. The corresponding integrated area is identified as I(*a*). A sharp peak arising at around 519-522 cm^{-1} corresponds to the transverse optical vibrational mode of crystalline silicon and the associated integrated area is identified as I(*c*). The intermediate component corresponding to a peak at around 506-510 cm^{-1} is identified as I(*b*). The crystalline volume fraction is calculated by using the simplified empirical relation as follows [8];

$$X_c = [I(c) + I(b)]/[I(a) + I(b) + I(c)] \quad (1)$$

We have measured the absorption data (α) of the films at visible range by UV/Visible spectrophotometer. Due to the varying structure of the films from microcrystalline to amorphous phase, we avoid Tauc's plots and to give a numerical presentation of the shift in the absorption spectra we determine E_{04}, i.e, the energy corresponding to $\alpha = 10^4$ cm^{-1} as an indicator of relative optical bandgap (E_{op}). Refractive index (*n*) spectra of the films is estimated by Spectroscopic Ellipsometry (SE) using Tauc-Lorentz model [9]. The dark conductivity (σ_d) of the films is measured in a coplanar configuration with Al electrode at room temperature.

Fabrication of a-SiO:H/μc-Si:H micromorph silicon solar cells

We have applied the n μc-SiO:H films as a n top layer of the a-SiO:H/μc-Si:H micromorph silicon solar cells. Cell structure is TCO glass/ ZnO/ p-μc-SiO:H/ i-a-SiO:H/ n-μc-SiO:H/ p-μc-SiO:H/ i-μc-Si:H/ n-μc-Si:H/ ZnO/ Ag. Cell active area is 0.75 cm^2. Note that absorber layer of the top cell is a wide bandgap a-SiO:H film and p type μc-SiO:H films are used as p layer in both top and bottom cells. There is no intermediate layer at the junction connection between top and bottom cells. Thicknesses of the i top a-SiO:H and i bottom μc-Si:H layers are 400 and 1500 nm, respectively. The CO_2/SiH_4 ratio for n top layer deposition is varied from 0 to 0.28, while other conditions in cell fabrication are kept as the same. The thickness of the n top layer is approximately 30 nm. The current-voltage (I-V) characteristics of the solar cells have been investigated under standard conditions; AM1.5, 100 mW/cm^2, 25°C in a Wacom solar simulator. Quantum efficiency (QE) of the solar cells also has been evaluated by spectral response measurements.

DISCUSSIONS

Properties of n μc-SiO:H films

Figure 1 shows Raman spectra of n μc-SiO:H films deposited with different CO_2/SiH_4 ratio. It is obviously shown that, a peak corresponding to crystalline phase, peak (*c*), gradually decreases with increasing CO_2/SiH_4 ratio, and amorphous silicon (*a*) becomes a dominant phase at the ratio above 0.23.

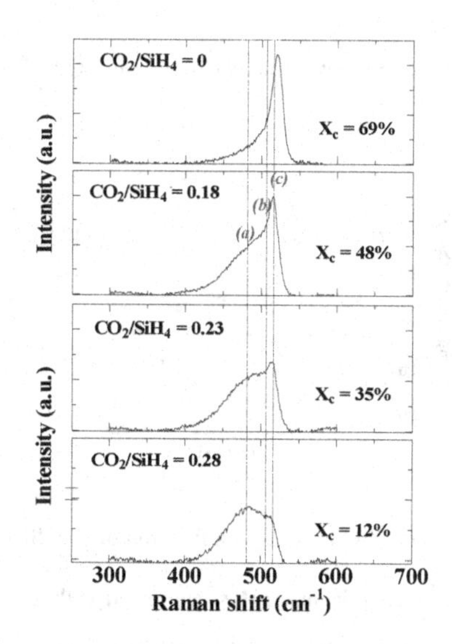

Figure 1. Raman spectra of n μc-SiO:H films deposited with different CO_2/SiH_2 ratio.

Optical bandgap of the films tends to increase while refractive index measured at the wavelength of 550 nm shows opposite behavior when the CO_2/SiH_4 ratio becomes higher, as shown in figure 2. Incorporation of oxygen into the Si:H network has a direct consequence on optical gap widening. A component of the increase in optical bandgap is associated with the Si-O bonds because of the stronger bond energy of Si-O compared to those of Si-Si and Si-H [10]. Addition of oxygen atoms to Si:H films can widen optical bandgap, however, the more participation of oxygen atoms the lower conductivity of the films as indicated in figure 3.

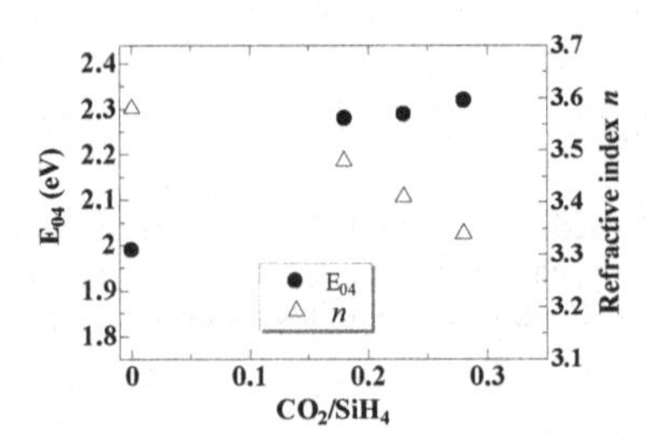

Figure 2. Optical bandgap (E_{04}) and refractive index (*n*) of n μc-SiO:H films as a function of CO_2/SiH_2 ratio.

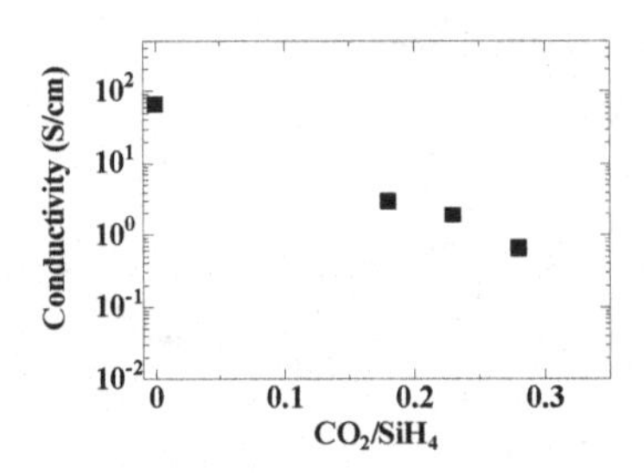

Figure 3. Dark conductivity of n μc-SiO:H films as a function of CO_2/SiH_2 ratio.

Characteristics of a-SiO:H/μc-Si:H micromorph silicon solar cells

As shown in figure 4, open circuit voltage (V_{oc}) and fill factor (FF) of the solar cells obviously improve when the n μc-SiO:H film is applied as n top layer instead of n μc-Si:H film ($CO_2/SiH_4 = 0$). The best cell with initial conversion efficiency of 10.7% is achieved at the CO_2/SiH_4 ratio of 0.23, where the X_c of the film is approximately 35%. At the higher ratio, short circuit current density (J_{sc}) of the cell begins to drop resulting in a decrease in cell efficiency. Both series resistance (R_s) and shunt resistance (R_{sh}) of the solar cells increase with increasing CO_2/SiH_4 ratio. Since the n μc-SiO:H films possess wide optical bandgap of about 2.3 eV and higher defect density compared to n μc-Si:H film, these are supposed to allow a better continuity of band diagram and also a better tunnel recombination junction at the connection between top and bottom cells. As mentioned previously, the i top and p bottom layers in these solar cells are wide bandgap SiO:H based materials. The E_{04} of the p bottom μc-SiO:H layer is estimated to be about 2.25 eV, so the n μc-SiO:H film with the E_{04} of 2.3 eV probably better suites for applying as n top layer in these solar cells. The enhancement of the V_{oc} is supposed to be due to a reduction of reverse bias at sub cell connection (n top/p bottom interface). The R_s slightly increases while the R_{sh} significantly enhances from 1500 to 3200 Ω when the CO_2/SiH_4 ratio increases from 0 to 0.28. The increase of R_{sh} is supposed to be caused by a better tunnel recombination junction, and this contributes to the improvement in the FF. No obvious difference is found in the QE of the micromorph solar cells with different CO_2/SiH_4 ratio for n top layer deposition. This suggests that the n μc-SiO:H films with relatively low *n* in this study do not work as an intermediate reflector to enhance light scattering inside the solar cells, but mainly play a key role to allow ohmic and low resistive electrical connection between the two adjacent cells in the micromorph silicon solar cells. After 1000 h of light soaking at 50°C, our best cell shows a stabilized efficiency of 8.8%.

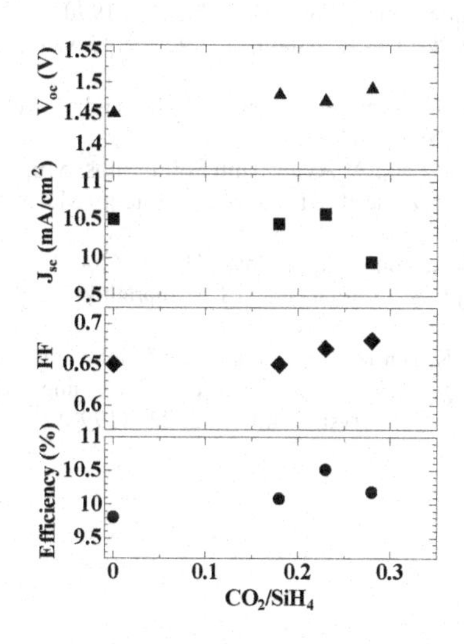

Figure 4. Photovoltaic parameters of a-SiO:H/μc-Si:H micromorph silicon solar cells using n top μc-Si(O):H layer deposited with various CO_2/SiH_2 ratio.

CONCLUSIONS

We have developed n type μc-SiO:H films and applied them as n top layer in the a-SiO:H/μc-Si:H micromorph silicon solar cells. With increasing CO_2/SiH_4 ratio, the X_c and conductivity of the films gradually decrease while the optical bandgap increases. Solar cells using the n top μc-SiO:H layer show higher V_{oc} and FF compared to the cell with n top μc-Si:H layer. The best SiO:H/μc-Si:H micromorph silicon solar cell is achieved with 10.7% initial and 8.8% stabilized values at the CO_2/SiH_4 ratio of 0.23. Enhancement in the cell performance is supposed to be due to a better tunnel recombination junction and a better continuity of band diagram at the sub cell connection, which is mainly owing to the developed n μc-SiO:H films in this study.

ACKNOWLEDGMENTS

This work was supported by Cluster and Program Management Office (CPM) of NSTDA, Thailand (P-00-10470).

REFERENCES

1. K. Haga and H. Watanabe, Jpn. J. Appl. Phys. 29, 636 (1990).
2. Y. Matsumoto, F. Melendez and R. Asomoza, Sol. Energy Mater. Sol. Cells. 52, 251 (1998).
3. P. Buehlmann, J. Bailat, D. Domine, A. Billet, F. Meillaud, A. Feltrin and C. Ballif, Appl. Phys. Lett. 91, 143505 (2007).
4. Y. Matsumoto, V. Sanchez and A. Avila, Thin Solid Films. 516, 593 (2008).
5. Y. Matsumoto, F. Melendez and R. Aromoza, Sol. Energy Mater. Sol. Cells. 66, 163 (2001).
6. A. Sarker and A. K. Barua, Jpn. J. Appl. Phys. 41, 765 (2002).
7. K. Sriprapha, C. Piromjit, A. Limmanee and J. Sritharathikhun, Sol. Energy Mater. Sol. Cells. 95, 115 (2011).
8. D. Das, M. Jana and A. K. Barua, Sol. Energy Mater. Sol. Cells. 63, 285 (2000).
9. H. Fujiwara, *Spectroscopic Ellipsometry*. (Maruzen Publishing, Tokyo, 2003) p. 141.
10. A. Singh and E. Davis, J. Non-Cryst. Solids. 122, 223 (1990).

Mater. Res. Soc. Symp. Proc. Vol. 1321 © 2011 Materials Research Society
DOI: 10.1557/opl.2011.955

Modeling of Advanced Light Trapping Approaches in Thin-Film Silicon Solar Cells.

Miro Zeman, Olindo Isabella, Klaus Jäger, Pavel Babal, Serge Solntsev, Rudi Santbergen
Delft University of Technology, PVMD/DIMES, P.O. Box 5053, 2628 CD Delft, Netherlands

ABSTRACT

Due to the increasing complexity of thin-film silicon solar cells, the role of computer modeling for analyzing and designing these devices becomes increasingly important. The ***ASA*** program was used to study two of these advanced devices. The simulations of an amorphous silicon solar cell with silver nanoparticles embedded in a zinc oxide back reflector demonstrated the negative effect of the parasitic absorption in the particles. When using optical properties of perfectly spherical particles a modest enhancement in the external quantum efficiency was found. The simulations of a tandem micromorph solar cell, in which a zinc oxide based photonic crystal-like multilayer was incorporated as an intermediate reflector (IR), demonstrated that the IR resulted in an enhanced photocurrent in the top cell and could be used to optimize the current matching of the top and bottom cell.

INTRODUCTION

Thin-film silicon solar cell technology is a promising photovoltaic (PV) technology for delivering low-cost solar electricity. However, the efficiency of thin-film silicon solar cells has to achieve a level of more than 20% in order to stay competitive with bulk crystalline silicon solar cells and other thin-film solar cell technologies. Light management [1] is one of the key areas for improving the performance of thin-film silicon solar cells and decreasing the production costs by using less material for an absorber layer and shortening its deposition time.

Performance improvement of a tandem micromorph silicon solar cell has introduced new challenges for light management. In the micromorph solar cell hydrogenated amorphous silicon (*a*-Si:H) is incorporated as the absorber layer in the top cell and hydrogenated micro-crystalline silicon (μc-Si:H) in the bottom cell. An intermediate reflector (IR) is applied between the top and bottom cells and is optimized to reflect a particular part of the solar spectrum back into the top *a*-Si:H absorber. In this way the IR makes it possible to use thinner absorber layers which results in a higher stability of the double-junction cell under light exposure and thus a higher stabilized efficiency. Due to the use of μc-Si:H absorber in the bottom cell a high level of scattering caused by rough interfaces is required for a broad wavelength range up to 1100 nm. Additional light scattering that increases the absorption in the absorber layers is expected from the implementation of metal nanoparticles in the solar cell structure due to plasmonic effects. On the other hand, in order to eliminate the parasitic plasmonic absorption in the rough back metal electrode [2], layers of transparent conductive oxides (TCO), distributed Bragg reflectors based on dielectric materials and/or white paint are developed and tested in solar cell structures.

Modeling of these advanced light-trapping techniques in thin-film solar cells is an important tool for analyzing and understanding their effect on the performance of solar cells. The ***ASA*** computer program has become a standard simulation tool in the thin-film PV community for analyzing and optimizing both optical and electrical performance of multi-junction silicon-based solar cells. The ***ASA*** program has been recently extended with features that enable the user to simulate effects of advanced light trapping techniques. In this article we demonstrate the novel features of the ***ASA*** simulator by presenting results of integrated optical and electrical simulations on two types of advanced silicon based solar cell devices: (i) an amorphous silicon solar cell with embedded silver nanoparticles and (ii) a tandem micromorph solar cell with 1-D photonic crystal-like-structure (PC) as an intermediate reflector.

THE ASA DEVICE SIMULATOR

The ***ASA*** program is designed for the simulation of devices based on amorphous and crystalline semiconductors. It solves the basic semiconductor equations (the Poisson equation and two continuity equations for electrons and holes) in one dimension and uses the electrostatic potential and the free electron and hole concentrations as variables. The program meets the requirements for a suitable thin-film solar cell simulation program as listed in Table I [3, 4]. The comparison with other available programs, such as AMPS or SCAPS, is to be found in [3].

Table I. Model requirements for a thin-film solar cell simulation program

- Multiple layers
- Band discontinuities in the CB and VB
- Large band gap materials: $E_g > 2.0 - 3.7$ eV
- Grading of material parameters
- Recombination and charge in the localized states
- Simulation of non-routine measurements: J(V), QE, C(V), etc., all as a function of T
- Fast and easy to use

Further, the ***ASA*** program uses several advanced physical models which describe specific device operation and material opto-electronic properties [5]. The main features of the ***ASA*** program are:

- Modeling of multilayer amorphous and/or crystalline semiconductor devices
- Optical simulation of scattering at rough interfaces using an optical model that takes into account both the specular and the scattered part of the incident light.
- Models describing a complete density of states (DOS) as function of energy, which include both the extended and localized (tail and defect) states.
- Calculation of the defect-states distribution in *a*-Si:H using the defect-pool models
- Recombination-generation statistics for the acceptor-like, donor-like and ambipolar states
- Change (grading) of all input parameters as a function of position in the device or energy level in the gap
- Models for the tunnel-recombination junction in multi-junction solar cells
- Modeling of degradation of *a*-Si:H solar cells
- High computational speed due to the native C++ implementation
- Command line support for automation and scripting

MODELING OF PLASMONIC SOLAR CELLS

We use the ***ASA*** program to simulate the performance of an *a*-Si:H solar cell with silver nanoparticles that are embedded in the rear TCO layer as shown in figure 1. The nanoparticles can be very efficient light scattering elements that due to plasmon resonance help to trap weakly absorbed light in the *a*-Si:H absorber layer [6]. The optical properties of the nanoparticles are strongly affected by their size, shape and position. The ***ASA*** program uses a complex refractive index N to characterize optical properties of layers constituting the solar cell. We use the Bergman effective medium theory [7] to represent the layer containing the silver nanoparticles by an effective medium layer with similar macroscopic optical properties. First we explain how the effective medium layer's complex refractive index N_{EM} is determined. Then we will use ***ASA*** to perform solar cell simulations using the obtained N_{EM} as input.

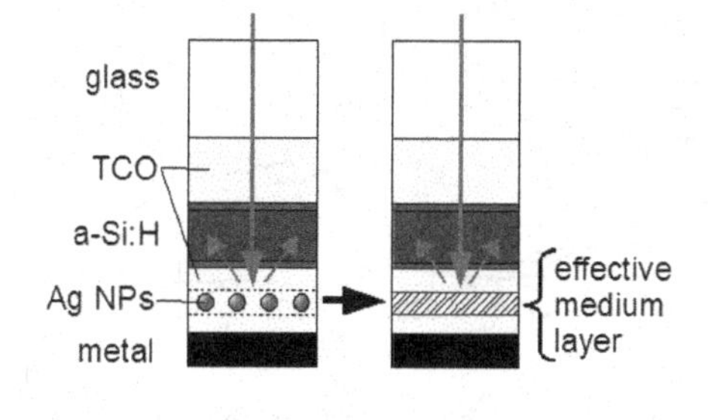

Figure 1: Schematic cross-section of *a*-Si:H solar cell with silver nanoparticles embedded in the rear TCO layer.

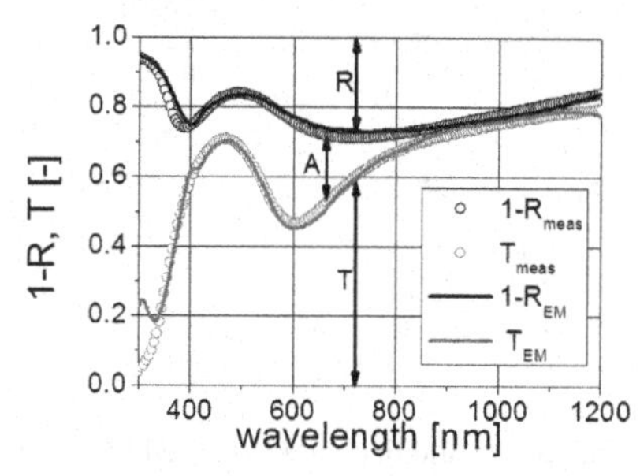

Figure 2: R and T of glass/TCO/NP/TCO layer stack. Circles: measured. Lines: eff. medium simulation.

Effective complex refractive index

Silver nanoparticles were formed on a glass/TCO substrate by a commonly used self-assembly technique [8]. This resulted in a film of irregular silver nanoparticles with distribution of particles sizes. Three samples were prepared. SEM inspection showed that average nanoparticle sizes are approximately 20, 40 and 60 nm. The surface coverage was 30 to 40% in all cases. The nanoparticle films were then over coated with a second layer of TCO. The reflectance and transmittance of the layer stacks were measured using an integrating sphere setup. The dip in transmittance at 600 nm (see figure 2), which would not occur in a continuous silver film, indicates the localized surface plasmon resonance of the silver nanoparticles. Next, the reflectance R and transmittance T of this layer stack were simulated with the nanoparticles replaced by an effective medium layer. The parameters in the Bergman effective medium theory were used as fit parameters. A good agreement was obtained with the experimental results, as illustrated in figure 2. The resulting N_{EM} (i.e. the real part n and imaginary part k) for all three fabricated samples are shown in the left panel of figure 3.

We applied this approach not only to experimental layers of nanoparticles, but also to a virtual layer of ideal nanoparticles obtained by simulations. Using finite difference time domain (FDTD) simulations we obtained the R and T of a square two-dimensional array of spherical silver particles embedded in TCO. Keeping the surface coverage fixed at 20%, the particle diameter was varied to be 20, 40 and 60 nm. Again a good fit could be obtained using the Bergman effective medium theory. The obtained N_{EM} are shown in the right panel of figure 3. As can be seen in figure 3, the n and k curves of nanoparticles obtained from the FDTD simulations have sharper features compared to those fabricated experimentally. This clearly illustrates the effect of different particle geometries on the optical properties. The fabricated particles have a size distribution and somewhat irregular shapes while in the FDTD simulation all particles have the same size and perfectly spherical shape that results in a sharper plasmon resonance.

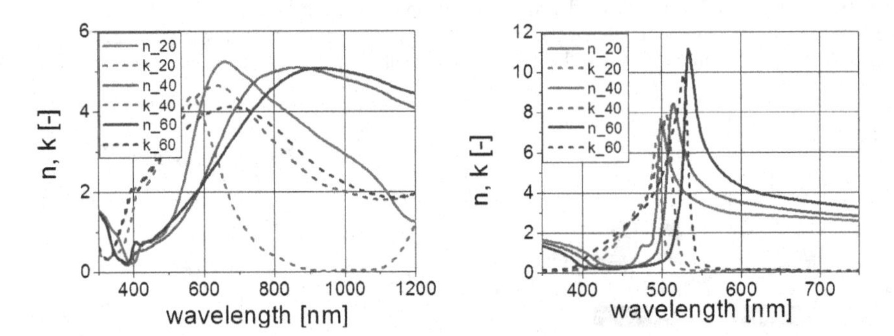

Figure 3: *n* (solid line) and *k* (dashed line) of effective medium of layers with 20, 40 or 60 nm nanoparticle layers for the fabricated (left) and simulated (right) silver nanoparticles.

Solar cell simulations

We used the obtained N_{EM} in the ***ASA*** program to simulate the solar cell structure shown in figure 1. We assume that all light interacting with the particles is scattered with a Lambertian angular distribution. As a reference, a solar cell shown in figure 1 without nanoparticles was simulated. The external quantum efficiency (*EQE*) of the reference cell is shown as the black line in figure 4. The *EQE* of the solar cell with either the fabricated or the perfectly spherical nanoparticles, are shown in the left and right panel of figure 4, respectively. The *a*-Si:H layer is only transparent for wavelengths larger than 500 nm and only in this wavelength region the silver nanoparticles have an effect. These simulations suggest that the *EQE* is *reduced* when the fabricated nanoparticles are used (left panel of figure 4). We attribute this to the parasitic absorption in the silver that dominates over light scattering. This absorption is most likely caused by the wide size distribution (i.e. the presence of much smaller particles), the somewhat irregular shape or the rather high surface coverage inherent to the used particle fabrication technique. The results are in line with the previously obtained experimental results [9,10]. When the perfectly spherical particles are used instead, a modest enhancement is found (right panel of figure 4). Analysis shows that parasitic absorption plays a less prominent role in this case. This illustrates that nanoparticle fabrication techniques with good control over nanoparticle size, shape and surface coverage are essential for enhancing solar cell performance.

1-D PHOTONIC CRYSTAL-LIKE INTERMEDIATE REFLECTOR

One approach to improve the tandem micromorph silicon solar cells is to apply wavelength-selective IR between the top and bottom cells. Single films that have been used for IR, such as zinc oxide or silicon oxide, do not allow to manipulate the wavelength region and/or the level of high *R*. In order to improve the *R* a stack of several pairs of layers with different refractive indices, referred to as 1-D PC, can be applied [11].

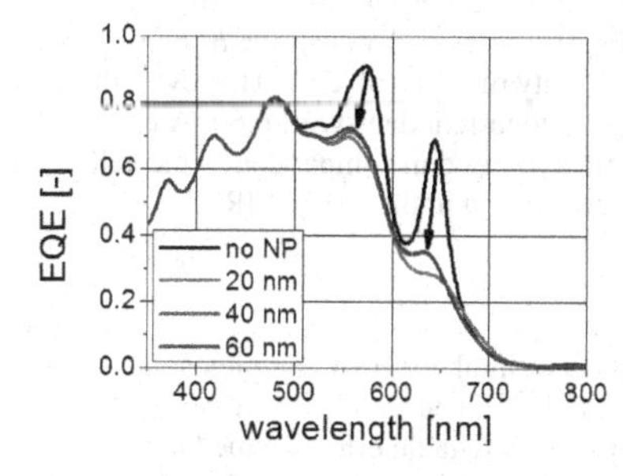

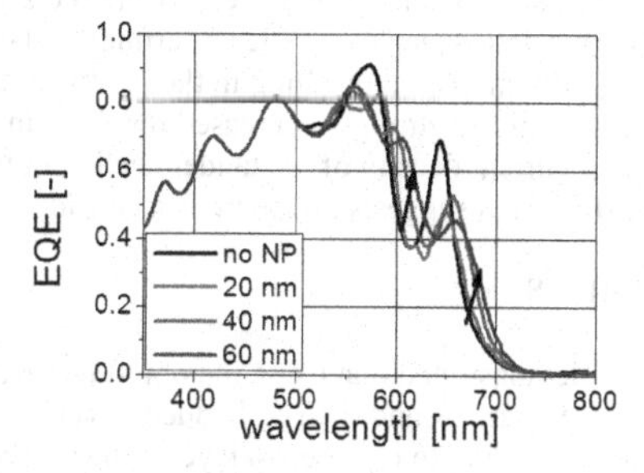

Figure 4: Simulated external quantum efficiency of solar cells with embedded silver nanoparticles. Either the optical properties of irregular (left) or perfectly spherical (right) nanoparticles with a diameter of 20, 40 and 60 nm were used.

We used the ***ASA*** program to design a 1-D PC based on n-type *a*-Si:H and zinc oxide (ZnO) and simulate the performance of the micromorph solar cell with the IR based on the designed 1-D PC. We have chosen the maximum reflectance to peak around the Bragg's wavelength of 600 nm and calculated the optimal thickness of the n-type *a*-Si:H and ZnO layers to be 30 nm and 63 nm, respectively [12]. In order to verify the design of the 1-D PC we fabricated the 1-D PC containing 2 pairs of *a*-Si:H and ZnO layers on a glass substrate and measured the total *R* and *T* of this stack. In figure 5 we show that we obtained a good matching between the measured and simulated *R* of the stack on the glass substrate. Also in figure 5 we present the *R* of the designed IR inside the solar cell which is not possible to determine experimentally.

The ***ASA*** program was used to simulate the performance of the tandem micromorph silicon solar cell with the 1-D PC IR. The thickness of *a*-Si:H and μc-Si:H absorbers was 250 nm and 1000 nm, respectively. Figure 6 presents the integrated optical and electrical simulation of the *EQE* of the tandem solar cell with flat interfaces without and with the IR. The input parameters

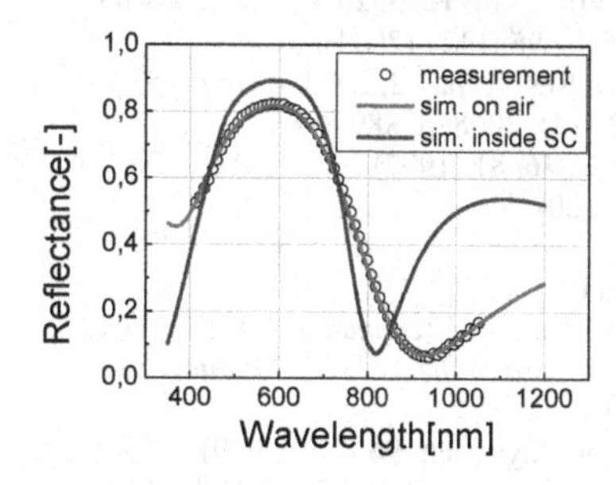

Figure 5: Reflectance of the 1-D PC inside a solar cell designed by the ***ASA*** program and of measured and simulated 1-D PC on glass.

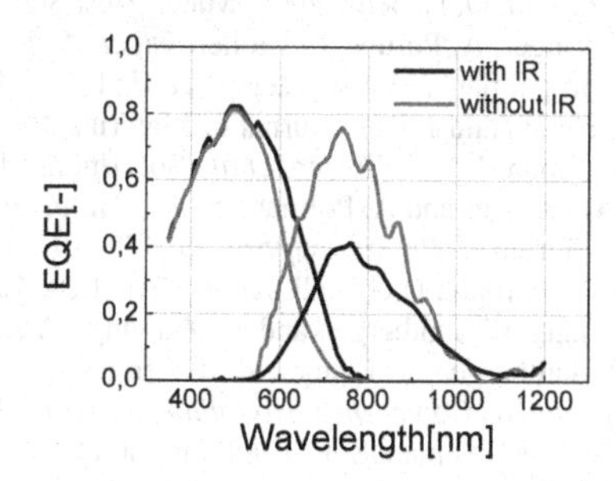

Figure 6: The *EQE* of the micromorph solar cell with and without the IR.

for the simulations were reported elsewhere [4]. The thickness of the top *a*-Si:H and bottom *μc*-Si:H absorber was chosen to obtain the current matching in the tandem cell without the IR. The photocurrent density calculated in this case from the *EQE* was 10.4 and 10.9 mA cm^{-2}, for the top and bottom cell, respectively. After inserting the 1-D PC IR in the tandem cell the *EQE* of the top cell clearly increased resulting in the photocurrent density of 12.1 mA cm^{-2}. However, the *EQE* of the bottom cell strongly decreased resulting in the photocurrent density of 6.6 mA cm^{-2}. The total photocurrent density of the tandem cell with the IR decreased in comparison to the cell without the IR, which suggests that there is a considerable absorption in the 1-D PC IR.

CONCLUSIONS

We have demonstrated that using the opto-electrical device simulator ***ASA*** complex solar cell structures with an IR and with embedded silver nanoparticles can be simulated. We described the procedure to represent a layer of metal nanoparticles with an effective medium layer and to use the optical constants of the effective medium layer as input for the ***ASA*** simulator. When using the optical constants reflecting the experimentally fabricated nanoparticles the simulations of *a*-Si:H solar cell show a decrease in the *EQE* for the wavelengths < 500 nm. When the optical constants of perfectly spherical particles are used, a modest enhancement in the *EQE* is found.

The ***ASA*** program was used for design of 1-D PC as an IR in the tandem micromorph solar cell and the optical and electrical simulation of the performance and current matching in the tandem cell. The simulation results demonstrate the effect of reflecting light back into the top cell due to the implementation of the IR.

ACKNOWLEDGEMENTS

This work was carried out with subsidies of the Dutch Ministry of Economic Affairs, Nuon Helianthos company, and the NMP-Energy Joint Call FP7 Solamon Project (www.solamon.eu).

REFERENCES

1. M. Zeman, O. Isabella, *et al.*, Mater. Res. Soc. Symp. Proc. Vol. **1245**, 2010, 1245-A03-03.
2. J. Springer, A. Poruba, L. Mullerova, *et al.*, J. Appl. Phys. **95**, 1427 (2004).
3. M. Burgelman, J. Verschraegen, *et al.*, Prog. Photovolt: Res. Appl. 12, 2004, p. 143–153.
4. M. Zeman and J. Krc, Journal of Materials Research **23** (4), 2008, p. 889-898.
5. M. Zeman, J.A. Willemen, *et al.*, Sol. Ener.. Mat. Sol. C. **46**, 81 (1997).
6. H.A. Atwater and A. Polman, *Nature Materials* **9**, 205 (2010).
7. D.J. Bergman, *Physics Reports* **43**, 377 (1978).
8. H.R. Stuart and D.G. Hall, Appl. Phys. Lett. **73**, 3815 (1998).
9. R. Liang, R. Santbergen and M. Zeman, *Advances in Science and Technology* **47**, 182 (2010).
10. R. Santbergen, R. Liang and M. Zeman, *Conference Record of the 35th IEEE Photovoltiac Specialists Conference, Honolulu, Hawaii,* 748 (2010).
11. J. Krc, M. Zeman, S. Luxembourg and M. Topic, Appl. Phys. Lett. **94** (15) (2009) 153501.
12. O. Isabella, B. Lipovšek, J. Krc, and M. Zeman, Mater. Res. Soc. Symp. Proc. Vol. **1153**, 2009, 1153-A03-05.

Polycrystalline Films

Mater. Res. Soc. Symp. Proc. Vol. 1321 © 2011 Materials Research Society
DOI: 10.1557/opl.2011.930

Flash-Lamp-Induced Lateral Solidification of Thin Si Films

K. Omori[1,2], G.S. Ganot[1], U.J. Chung[1], A.M. Chitu[1], A.B. Limanov[1], and James S. Im[1,3]

[1] Program in Materials Science and Engineering, Department of Applied Physics and Applied Mathematics, Columbia University, New York, NY, USA

[2] Technical Development Department, The Japan Steel Works, LTD., Yokohama, Japan

[3] Department of Materials Science and Engineering, College of Engineering, Korea Advanced Institute of Science and Technology, Korea

ABSTRACT

In this paper we show that a flash lamp can be employed to induce controlled lateral solidification of a-Si thin films. Specifically, a dual xenon-arc-lamp-based system was utilized to induce location-controlled complete melting by shaping the incident beam using a contact mask. The resulting laterally solidified microstructure consisted of exceptionally long grains (~10s to ~100s of μm) that were relatively free of intragrain-defects. With further development and optimization, the approach may lead to cost-effective/high-throughput processes and systems that can capture and enhance the advantages of laser-based/melt-mediated crystallization techniques.

INTRODUCTION

Using a flash lamp to heat and crystallize a-Si films can be recognized as an interesting and noteworthy technical procedure for a number of reasons: (1) it was demonstrated as a viable crystallization method nearly thirty years ago [1,2], (2) it is an extremely flexible technique capable of being used for solid-phase as well as melt-mediated crystallization of a-Si films [3-6], and (3) the irradiation-system-related components are well developed as a consequence of the "flash lamp annealing" method being evaluated and developed for the semiconductor manufacturing industry [7].

In this paper, we demonstrate the controlled lateral solidification (referred to as controlled super-lateral-growth (C-SLG)) of a-Si films using a flash-lamp as the source of crystallization, and we show that low-defect-density Si films are created in the process. Controlled super-lateral growth of a-Si films is a melt-mediated thin film crystallization approach using pulsed lasers which has been studied extensively in the past [e.g. 8,9]. By systematically manipulating and controlling the locations, shapes, and extent of melting induced by the incident beam, the C-SLG approach can generate grains with low structural defect densities. Specifically, the method takes advantage of the fact that when one controls the locations of the silicon film within which complete melting is induced, the resulting solidification will proceed via lateral growth to generate a polycrystalline material with large and elongated grains.

EXPERIMENTAL DETAILS

Our custom-built flash lamp irradiation system comprises two independently controllable xenon lamps (Figure 1). Each lamp is capable of delivering a pulse (~50μsec to 15msec pulse duration), with sufficient energy density to induce melting over the entire pulse duration range. A copper grid with periodically arranged square patterns (with the sides measuring either 37, 90, or 200μm) was placed on the film to act as a contact mask in order to induce complete melting in the selectively irradiated areas. The results presented here correspond to simultaneous irradiation by both lamps each with a 500μsec pulse. The samples used were 50nm-thick amorphous Si films, prepared on SiO_2-buffer-layer-coated glass substrates using a standard PECVD method. Analysis of the irradiated samples was performed using plan-view optical microscopy, TEM, and EBSD techniques.

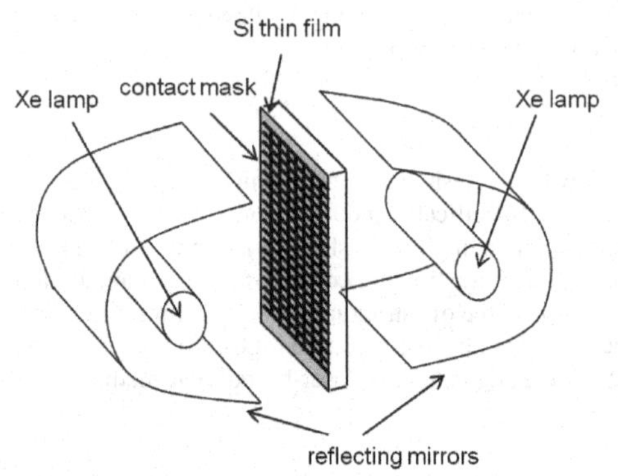

Figure 1. A schematic representation of the xenon-flash-lamp system used in this work. The pulse duration, voltage, and ignition time of each lamp can be fully adjusted to achieve a wide range of conditions. A focusing mirror is positioned around each lamp in order to direct the optical energy towards the sample, which is mounted in between the two lamps. In the present work, the lamp on the back-side of the sample was used to uniformly irradiate the film, while the lamp on the front-side was used to induce localized melting of the film (achieved via a contact mask).

RESULTS AND DISCUSSION

We have chosen to utilize a xenon-arc-lamp-based approach as it can potentially lead to cost-effective and high-throughput processes and systems; these lamps can deliver large amounts of optical power over an extremely wide range of C-SLG-suitable pulse durations (10s of μsec to msec). These technical specifications mean that a definite possibility exists here for developing a non-laser crystallization process that can capture the material-quality-related advantages that are typically associated with laser-based techniques, while avoiding the associated cost-related disadvantages.

From the crystallization perspective, the most salient characteristic of the process is the exceptionally long lateral-growth distance (~10s to 100s of μm, as can be seen in Figure 2) that can be achieved compared to previously demonstrated pulsed-laser-based C-SLG processes. This result, which is fully expected from thermal and kinetic considerations associated with the encountered experimental conditions (i.e., ~50μsec to ~10msec pulse duration range), endows the approach with an unprecedented level of flexibility for generating various high-performance-device-enabling low-defect-density materials. The TEM micrograph of the grains, shown in Figure 3, reveals that these laterally solidified grains contain very few intragrain defects.

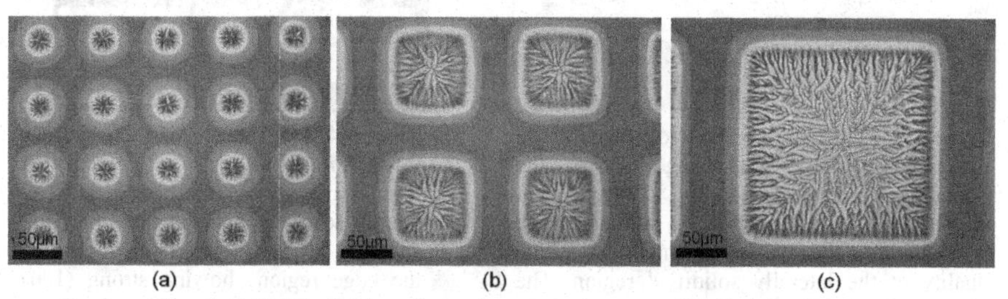

Figure 2. Bright-field optical microscope images of flash-lamp-induced lateral solidification of a-Si thin films using a contact mask with varying square openings, with square side length of (a) 37μm, (b) 90μm, and (c) 200 μm.

The analysis of the crystallographic orientation of the laterally solidified grains (Figure 4) shows that there is a systematic trend associated with the surface-normal direction of the grains. It can be seen that many of the grains at the edge of the completely melted region start with {100} surface orientation, but as the growth proceeds inward, the (100) texture is eventually and gradually lost. The prevalence of the initial {100} orientation is attributed to the fact that the condition at the boundary corresponds to the condition that is needed to induce mixed-phase solidification of Si films; the sudden increase in the value of reflectivity as the film melts will lead to the selection and stabilization of the grains that possess thermodynamically favorable characteristics [6,10].

CONCLUSIONS

Based on the preliminary experimental results that are presented in this paper, that xenon-arc-lamp-based irradiation can be configured to readily satisfy the conditions that are needed to perform C-SLG. With further development, the approach may provide engineers with a robust crystallization technique, which can be utilized for making high-performance-enabling Si films for various macroelectronic and microelectronic applications ranging from flat-panel displays to 3D-ICs.

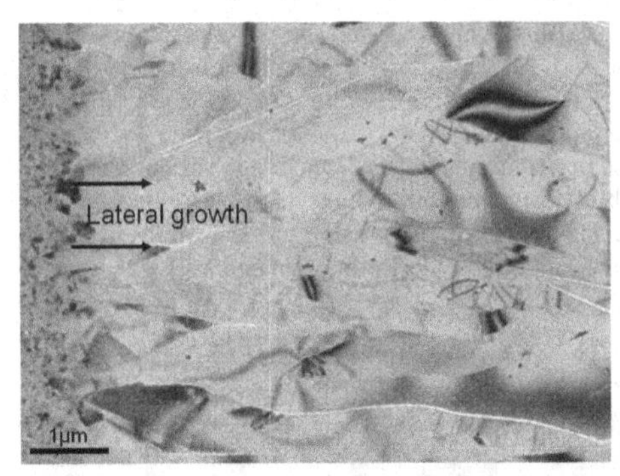

Figure 3. TEM micrograph of the C-SLG processed region. The image shows the boundary region corresponding to the point at which lateral solidification is initiated. The micrograph also reveals the high crystalline quality of the laterally solidified region. The film has been defect etched to reveal grain boundaries.

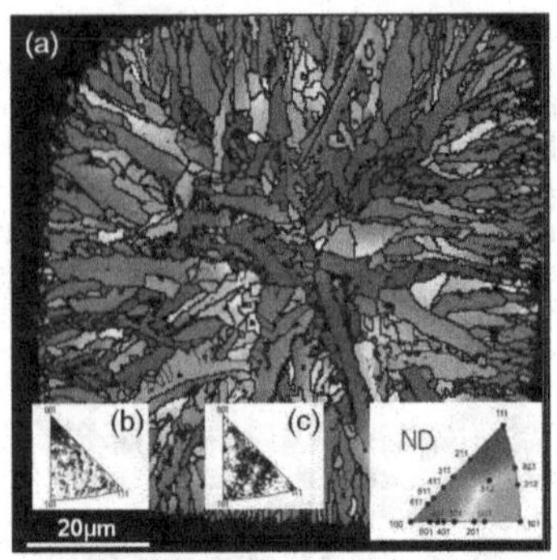

Figure 4. (a) The electron backscatter diffraction (EBSD) map of a region that has undergone flash-lamp-induced lateral solidification (normal direction shown). (b) The inverse pole figure of the edge region showing strong (100) orientation. (c) The inverse pole figure of the inner region showing that as growth proceeds, the (100) texture is lost.

ACKNOWLEDGEMENTS

We would like to acknowledge the support by The Japan Steel Works, LTD., the Lenfest Center for Sustainable Energy at Columbia University, and the WCU Flexible Signage Program at KAIST.

REFERENCES

1. H.A. Bomke, H.L. Berkowitz, M. Harmatz, S. Kronenberg, R. Lux, Appl. Phys. Lett. **33**, 955 (1978).
2. L. Correra and L. Pedulli, Appl. Phys. Lett. **37**, 55 (1980).
3. H. Wirth, D. Panknin, W. Skorupa, E. Niemann, Appl. Phys. Lett. **74**, 979 (1999).
4. B. Pécz, L. Dobos, D. Panknin, W. Skorupa, C. Lioutas, N. Vouroutzis, Appl. Surf. Sci. **242**, 185 (2005).
5. F. Terai, S. Matunaka, A. Tauchi, C. Ichimura, T. Nagatomo, T. Homma, J. Electrochem. Soc. **153,** H147 (2006).
6. James S. Im, M. Chahal, P.C. van der Wilt, U.J. Chung, G.S. Ganot, A.M. Chitu, N. Kobayashi, K. Omori, A.B. Limanov, J. Cryst. Growth **312**, 2775 (2010).
7. T. Gebel, M. Voelskow, W. Skorupa, G. Mannino, V. Privitera, F. Priolo, E. Napolitani, A. Carnera, Nucl. Instrum. Methods Phys. Res., Sect. B **186**, 287 (2002).
8. H.J. Kim and James S. Im, Appl. Phys. Lett. **64**, 2303 (1994).

9. James S. Im, M.A. Crowder, R.S. Sposili, J.P. Leonard, H.J. Kim, J.H. Yoon, V.V. Gupta, H. Jin Song, H.S. Cho, Phys. Stat. Sol. (a) **166**, 603 (1998).
10. W.G. Hawkins and D.K. Biegelsen, Appl. Phys. Lett. **42**, 358 (1983).

Mater. Res. Soc. Symp. Proc. Vol. 1321 © 2011 Materials Research Society
DOI: 10.1557/opl.2011.804

Poly-Si Thin Film Formation Using a Novel Low Thermal Budget Process

Minghao Zhu[1], Yue Kuo[1], Chen-Han Lin[1] and Qi Wang[2]
[1] Thin Film Nano & Microelectronics Research Laboratory, Texas A&M University, College Station, TX 77843-3122, U.S.A.
[2] National Renewable Energy Laboratory, Golden, Colorado 80401, U.S.A.

ABSTRACT

Polycrystalline silicon thin films were formed from the amorphous silicon thin film by the pulsed rapid thermal annealing process enhanced with a thin nickel seed layer through the vertical crystallization mechanism. In this paper, authors presented the results on the material properties of the crystallized film. The dopant and film thickness effects were also investigated. It has been demonstrated that a 2 μm thick amorphous silicon n^+-i-p^+ diode structure could be transformed into polycrystalline stack with a 4-pulse 1 sec 850°C heating and 5 sec cooling cycle process.

INTRODUCTION

Thin film amorphous silicon (a-Si:H) solar cells have advantages over the single crystal silicon solar cells, such as requiring a very small amount of composing materials, low temperature fabrication process, unlimited supply of low-cost materials, and large-area substrate capability [1,2]. Compared with a-Si:H thin film solar cells, the polycrystalline silicon (poly-Si) thin film cells have additional advantages of high conversion efficiency and improved stability [3]. J.-D Hwang *et al.* reported a high conversion efficiency poly-Si solar cell formed from a-Si without passivation or antireflection layers, with a conversion efficiency of 10.4% [4]. However, the fabrication of the poly-Si solar cells is limited to processes that take a long time or require a high temperature, such as chemical vapor deposition (CVD), solid phase crystallization, or metal induced crystallization [5-10]. A novel, low thermal-budget pulsed rapid thermal annealing (PRTA) process enhanced with a Ni seed layer has been introduced to transform the a-Si thin film into a poly-Si thin film in the horizontal direction, i.e., parallel to the substrate surface [11]. Recently, this method has been used to form the poly-Si film in the vertical growth mechanism, i.e., perpendicular to the substrate surface [1]. In this paper, detailed material properties and process effects on this kind of poly-Si thin film have been investigated.

EXPERIMENTAL

The intrinsic or doped a-Si:H thin film was deposited in a PECVD system, which has the parallel-plate electrode arrangement driven by a 13.56 MHz RF generator, on the corning 1737 glass substrate pre-coated with the sputter deposited Mo (150 nm) and Ni (1nm) films. The intrinsic a-Si:H film was deposited at a rate of 6 nm/min from SiH_4 50 sccm, 80 W, 150 mT at 250°C. The n^+ film was deposited at a rate of 4 nm/min from SiH_4/PH_3 (6.88% in H_2)/H_2 10 sccm /10 sccm/1,000 sccm, 500 W, 750 mT at 250°C. The p^+ film was deposited at a rate of 10 nm/min from SiH_4/B_2H_6 (2% in H_2)/H_2 10 sccm /20sccm/1,000 sccm, 500W, 750mT at 250°C.

The n^+/intrinsic/p^+ a-Si:H thin film stack was deposited sequentially in one-pump down in the same chamber without breaking the vacuum.

Before PRTA, the PECVD a-Si:H thin film was dehydrogenated to form a-Si by annealing at 500°C under N_2 for 3 to 5 hours depending on the layer thickness. Then, the sample was exposed to the PRTA condition for several cycles to transform the a-Si film into a poly-Si film [1]. Each cycle was composed of 1 sec of 850°C heating and 5 sec of cooling. The programmed temperature ramp in the heating cycle was 180°C/sec for the first pulse and 140°C/sec for the rest pulses. The actual temperature ramp, as a result, was about 190°C/sec for the first pulse and 140°C/sec for the rest pulses. The cooling cycle was done by turning the heater. The actual minimum temperature in the cooling cycle was about 530°C after the first cycle and slightly higher in subsequent cycles. After all cycles, the sample was cooled down to room temperature under the N_2 ambient for approximately 20 minutes.

RESULTS AND DISCUSSION

Crystallization of a-Si:H films

Figure 1 shows the cross-sectional transmission electron microscopy (TEM) views of the p^+ a-Si film, formed from the a-Si film after the 10-pulse PRTA process. Lattice fringes in TEM figures are clearly observable. An interface layer between Si and Mo/Ni is detected. Since no lattice fringe is detected at this interface layer, it is probably an amorphous oxide layer formed from the Mo native oxide [12]. The existence of this interface layer does not stop the formation of the poly-Si film. Separately, it was confirmed that this native oxide layer could be eliminated using various surface treatment methods.

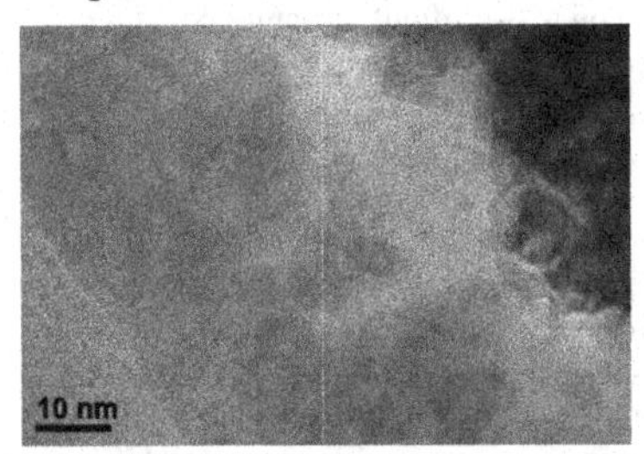

Figure 1. Cross-sectional TEM view of the p^+ a-Si film after a 10-pulse PRTA process.

Dopant effects of PRTA crystallization

Dopant effects on the PRTA crystallized poly-Si thin film formation process have also been investigated using XRD and Raman spectra. Figure 2 shows the XRD patterns of 100 nm thick p^+, n^+, and intrinsic poly-Si films formed by PRTA. All films contain (111) and (220) peaks, with a very small (311) minor peak. The average crystalline size is estimated to be between 200Å and 400 Å using the Scherrer equation [13]. The grain size in these films decreases in the order of intrinsic > n^+> p^+, i.e., 395 Å >332 Å >257 Å

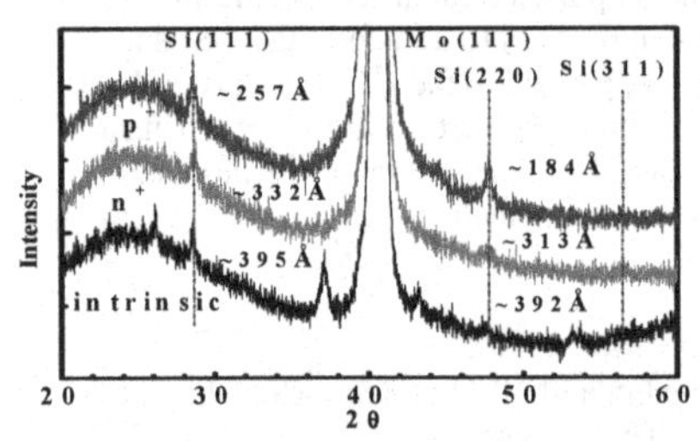

Figure 2. XRD of PECVD n^+, p^+ and intrinsic a-Si films after 10-pulse PRTA.

at the (111) orientation and 392 Å >313 Å >184 Å at the (220) orientation. Therefore, the dopant in the original a-Si:H film suppresses the grain growth. In addition, the intrinsic poly-Si film contains several minor peaks, which could be related to other Ni-silicide phase, such as Ni_3Si and NiSi [14].

Figure 3 shows the Raman spectra of the p^+, n^+ and intrinsic poly-Si films. Compared with the single crystalline Si peak at 520 cm^{-1}, all peaks are located in the lower frequency region, which can be contributed to the finite size effect. It was reported that the Raman peak of the crystalline Si film shifted from 520 cm^{-1} to about 512 cm^{-1} when the grain size decreased from >10 nm to 3 nm [15-17]. With the further decrease of the grain size, a third intermediate peak centered at around 500 ± 10 cm^{-1} could be detected [18,19]. The Raman peak location in Fig. 3 increases in the order of $\nu_{p\text{-type}}$(514.5 cm^{-1}) < $\nu_{n\text{-type}}$(516.9 cm^{-1}) < $\nu_{intrinsic}$(517.7 cm^{-1}). This is consistent with the XPD result that the dopant in the original a-Si:H film suppresses the crystallization process. The suppression effect of the boron dopant is much more significant than that of the phosphorus dopant.

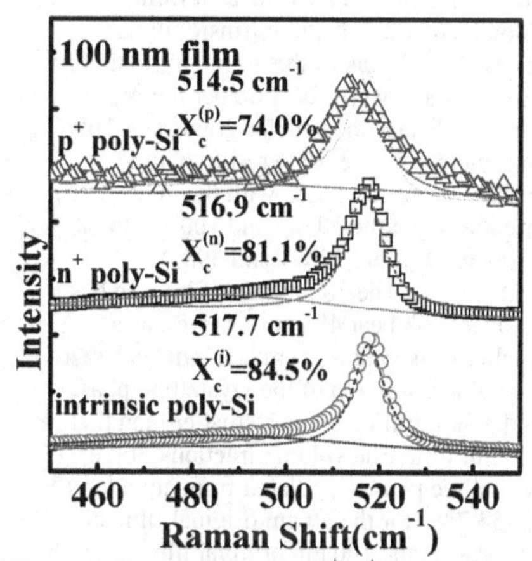

Figure 3. Raman spectra of p^+, n^+ and intrinsic poly-Si thin films formed after the 10-pulse PRTA process.

The volume fraction of the crystalline phase could be calculated by the following equation [20]:

$$X_C = \frac{I_C + I_{GB}}{I_C + I_{GB} + y(L) I_a} \qquad (1)$$

where I_C, I_{GB} and I_a are integrated intensities of decovoluted peaks of crystalline, grain boundary, and amorphous phases, separately, and y(L) is the ratio of the cross-section area between the amorphous and the crystalline phases. Bustarret *et al.* [21] reported that when the grain size is > 3 nm, y(L) could be expressed with the following equation:

$$y(L) = 0.1 + \exp\left[-\left(\frac{L}{250}\right)\right] \qquad (2)$$

Assuming that L is the average (111) grain size estimated from the XRD analysis, i.e., 395 Å for intrinsic Si, 332 Å for n^+ Si and 257 Å for p^+ Si, the volume fraction of the crystalline phase X_c decreases in the order of 84.5% (intrinsic) > 81.1% (n^+) > 74.0% (p^+). The suppression effect of the boron and phosphorous dopants could be related to the breakdown of dangling bond between silicon and hydrogen and the bond formation between Si and the dopant [22]. It was reported that in the a-Si:H film, the Si-H_2 and Si-H bonds broke at 400°C and 550~600°C [23,24], separately, which is a primary factor to deter the MILC growth. Johnson *et al.* [25], reported the dopant enhanced H diffusion effect in the solid phase epitaxial growth process, which could be responsible for the suppression effect of dopants. The boron dopant could enhance the H diffusion more effectively than the phosphorous dopant. In addition, since dopants compete with Ni atoms to form bonds with Si, it becomes difficult for Ni to form Ni-silicide. Phosphorous was also known to trap $NiSi_2$, which suppresses the diffusion of Ni-silicide through the a-Si film and raises the formation energy of $NiSi_2$ [26].

Original film thickness effects of poly-Si characteristics

T. Ma *et al.* [27] and T. Chang *et al.* [28] reported that the a-Si film thickness could affect the Ni induced lateral crystallization result. As the thickness of the original a-Si film increases, the volume fraction of the crystalline phase increases because of the reduced geometry effect on the horizontal transport of Ni. In this study, the volume fraction of crystalline phase also increases with the increase of the original a-Si film thickness. Figure 4 shows the thickness dependence on the Raman spectrum on the PRTA crystallized intrinsic a-Si film. Compared with 100 nm intrinsic films, the crystalline Si peak of the 30 nm intrinsic film shifts to the lower wave number region, which indicates the smaller grain size. This is consistent with the XRD result that the average grain sizes, i.e., with the (111) orientation, in the 30 nm and 100 nm thick intrinsic films are 228 Å and 395 Å, respectively. The 30 nm thick film also has a broader peak bear 480 cm^{-1} as well as a smaller peak near 520 cm^{-1}. It indicates that the volume fraction of the crystalline phase in the 30 nm film is much smaller than that in 100 nm film. The volume fractions of the crystalline phase calculated from equation (2) are 54.73% for the 30 nm original film and 84.52% for the 100 nm original film.

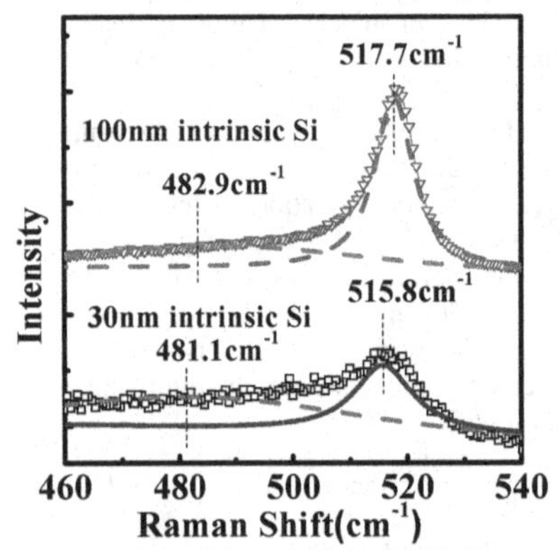

Figure 4. Raman spectra of poly-Si films formed from 30 nm and 100 nm thick intrinsic a-Si films after 10-pulse PRTA process.

PRTA pulse number effects and crystallization of n^+-i-p^+ stacks

The thermal budget effect of the PRTA process was studied by varying the number of heating-cooling cycles. Figure 5 shows the Raman spectra of an n^+ (30 nm)-i (2 μm)-p^+ (30 nm) stack after 4,7, and 10 pulses of 1 sec 850°C heating and 5 sec cooling cycles. The complete stack could be crystallized after only 4 pulses. In the crystal volume fraction calculation, if L is arbitrarily set as 300Å, and the value of y(L) is 0.4, the volume fractions of the crystalline phase would be 89.9%, 89.2% and 89.3%, after 4, 7 and 10 pulses.

Figure 6 shows the influence of the PRTA pulse number on (a) the Raman

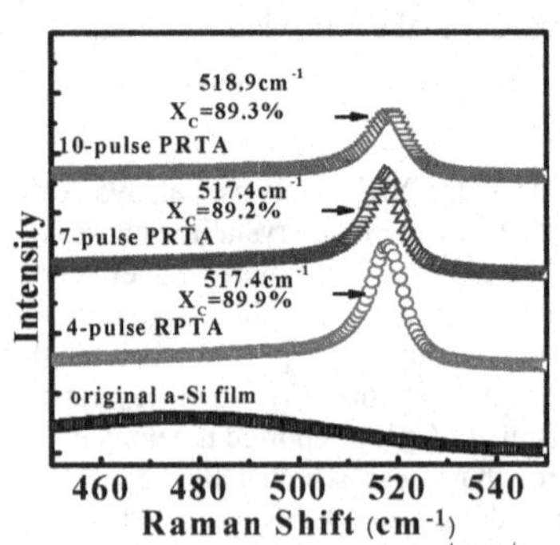

Figure 5. Raman spectra for a-Si n^+-i-p^+ stacks before and after 4-, 7-, and 10-pulse PRTAs.

spectra and (b) the normalized intensities of the amorphous, crystalline and grain boundary phases in the n^+-i-p^+ stacks. Each Raman spectrum can be deconvoluted into three peaks representing the crystalline (520 cm^{-1}), amorphous (480 cm^{-1}), and grain boundary (510cm^{-1}) phases. The intensity of the amorphous phase is almost steady, i.e., independent of the number of pulses. However, the volume fraction of the grain boundary phase decreases and the volume fraction of the crystalline phase increase with the increase of the number of pulses. After 4 pulses, about 90% of the film is composed of the crystalline phase and grain boundary phases. The further increase of the pulse numbers enhances the grain size but slightly reduces the volume fraction of the grain boundary size.

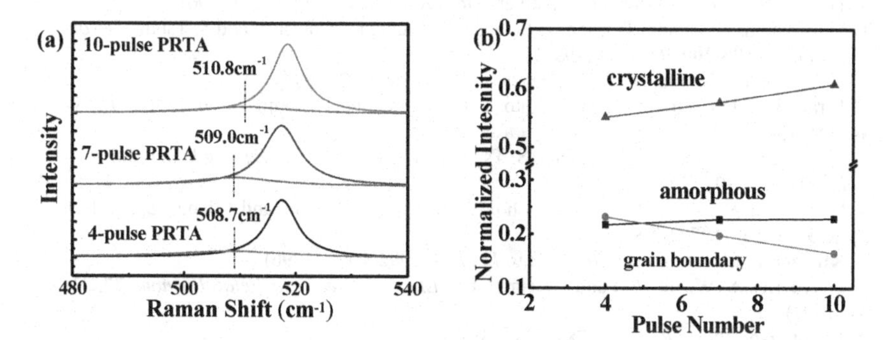

Figure 6. PRTA pulse number effect on (a) Raman spectra of poly-Si and (b) normalized intensity of crystalline, grain boundary, and amorphous phases.

SUMMARY

It has been demonstrated that the a-Si film could be transformed into the poly-Si film with a low thermal budget Ni-enhanced PRTA process in the vertical crystallization mechanism. Material properties of the poly-Si film under various process conditions have been studied. The existence of the boron and phosphorous dopant in the original a-Si film suppressed the crystallization process. The influence of the boron dopant is more pronounced than that of the phosphorous dopant. The volume fraction of the crystalline phase in the poly-Si film increases with the original a-Si layer thickness. The 2-μm thick a-Si n^+-i-p^+ stack was crystallized with only 4 pulses of the 1 sec 850°C and 5 sec cooling cycle. The increase of the pulse number slightly increases the grain size but does not increase the total volume fractions of the crystalline and grain boundary phases.

ACKNOWLEDGEMENTS

Authors acknowledge NSF CMMI project 0968862 for supporting this research.

REFERENCES

1. Y. Kuo, C.-H. Lin and M. Zhu, *Conf. Rec. IEEE Photovoltaic Spec. Conf.* 3698-3701 (2010)
2. W. G. J. H. M. van Sark, G. W. Brandsen, M. Fleuster and M. P. Hekkert, *Energ. Policy* **35**, 3121-3125 (2007)
3. R. E. I. Schropp, R. H. Franken, H. D. Goldbach, Z. S. Houweling, H. Li, J. K. Rath, J. W. A. Schüttauf, R. L. Stolk, V.Verlaan and C. H. M. van der Welf, *Thin Solid Films* **516**, 496-499 (2008)
4. J.-D Hwang, T.-Y. Chi, J.-C Liu, C.-Y Kung and I.-C Hsein, *Jpn. J. Appl. Phys.*45, 7675-7676 (2006)
5. N. H. Nickel, W. B. Jackson and J. Walker, *Phys. Rev. B* **53**, 7750-7761 (1996)
6. T. Matsuyama, N. Terada, T. Baba, T. Sawada, S. Tsuge, K. Wakisaka and S. Tsuda, *J. Non-Cryst. Solids* **198-200**, 940-944 (1996)
7. S.-W. Lee and S.-K. Joo, *IEEE T. Electron Dev.* **17**, 160-162 (1996)
8. F. Terai, H. Kobayashi, S. Katsui, Y. Sato, T. Nagatomo and T. Homma, *Jpn. J. Appl. Phys.* **44**, 125-130 (2005)
9. T. Matsuyama, T. Baba, T. Takahama, S. Tsuda and S. Nakano, *Sol. Energy Mater. Sol. Cells* **34,** 285-289 (1994)
10. S. Y. Yoon, S. K. Kim, J. Y. Oh, Y. J. Choi, W. S. Shon, C. O. Kim and J. Jang, *Jpn. J. Appl. Phys.* **37**, 7193-7197 (1998)
11. Y. Kuo and P. M. Kozlowski, *Appl. Phys. Lett.* **69**, 1092-1094 (1996)
12. A. Galyayries, S. Wisniewski and J. Grimblot, *J. Electron Spectrosc. Relat. Phenom.* **87**, 31-44 (1997)
13. A. L. Patterson, *Phys. Rev.* 56, 978-982 (1939)
14. W. Huang, L. Zhang, Y. Gao and H. Jin, *Microelectron. Eng.* **84**, 678-683 (2007)
15. S. Vepřek and Z. Iqbal, *J.Phys. C* **15** 377-392 (1982)
16. S. Vepřek, Z. Iqbal and F. -A. Sarott, *Philos Mag. B* ***45***,137-145 (1982)
17. Y. He, C. Yin, G. Cheng, L. Wang, X. Liu and G. Y. Hu, *J. Appl. Phys.* **75**, 797-803 (1994)
18. S. Vepřek, F. –A. Sarott and Z. Iqbal, *Phys. Rev. B* **36**, 3344-3350 (1987)
19. G. Yue, J. D. Lorentzen, J. Lin, D. Han and Q. Wang, *Appl. Phys. Lett.* **75**, 492-494 (1999)
20. D. Han, J. D. Lorentzen, J. Weinberg-Wolf, L. E. McNeil and Q. Wang, *Appl. Phys. Lett.* **94**, 2930 -2936 (2003)
21. E. Bustarret, M. A. Hachicha and M. Brunel, *Appl. Phys. Lett.* **52,** 1675-1677 (1988)
22. C.-W. Byun, Y.-W. Lee, J.-H. Park, C.-K. Seok, Y.-S. Kim and S.-K. Joo, *Electron. Mater. Lett.* **4**, 79-83 (2008)
23. H.-Y. Kim, J.-B. Choi and J.-Y. Lee, *J. Vac. Sci. Technol. A* **17**, 3240-3245 (1999)
24. J. R. Woodyard and D. R. Bowen, *J. Appl. Phys.* **57**, 2243-2248 (1985)
25. B. C. Johnson, P. Caradonna and J. C. McCallum, *Mater. Sci. Eng. B* **157**, 6-10 (2009)
26. J.-D Hwang, J.-Y. Chang and C.-Y. Wu, *Appl. Surf. Sci.* **249**, 65-70 (2005)
27. T. Ma and M. Wong, *J. Appl. Phys.* **91,** 1236-1241 (2002)
28. T.-K. Chang, C.-W Lin, Y.-H Chang, C.-H. Tseng, F.-T. Chu, H.-C Cheng and L.-J. Chou, *J. Electrochem. Soc.* **150**, G494-G497 (2003)

Mater. Res. Soc. Symp. Proc. Vol. 1321 © 2011 Materials Research Society
DOI: 10.1557/opl.2011.932

Impact of rapid thermal annealing and hydrogenation on the doping concentration and carrier mobility in solid phase crystallized poly-Si thin films

A. Kumar, [1,2] P.I. Widenborg, [1,2] H. Hidayat, [1,2] Qiu Zixuan[1] and A.G. Aberle[1,2]
1. Solar Energy Research Institute of Singapore, National University of Singapore, Singapore
2. Department of Electrical and Computer Engineering, National University of Singapore, Singapore

ABSTRACT

The effect of the rapid thermal annealing (RTA) and hydrogenation step on the electronic properties of the n^+ and p^+ solid phase crystallized (SPC) poly-crystalline silicon (poly-Si) thin films was investigated using Hall effect measurements and four-point-probe measurements. Both the RTA and hydrogenation step were found to affect the electronic properties of doped poly-Si thin films. The RTA step was found to have the largest impact on the dopant activation and majority carrier mobility of the p^+ SPC poly-Si thin films. A very high Hall mobility of 71 cm^2/Vs for n^+ poly-Si and 35 cm^2/Vs for p^+ poly-Si at the carrier concentration of 2×10^{19} cm^{-3} and 4.5×10^{19} cm^{-3}, respectively, were obtained.

INTRODUCTION

With the increase in natural disasters and serious environmental problems such as global warming, it has been never more apt to look into alternative clean energy sources which are not detrimental to the environment. Out of the various renewable energy sources available, solar energy has attracted significant attention in the past two decades and has grown at a substantial rate of about 30-40% per annum. More than 90% of today's photovoltaic market is dominated by crystalline silicon (c-Si) wafer solar cells [1]. However this technology is material and energy intensive, which acts as bottleneck in further cost reduction. High efficiency thin film module is an effective alternative technology.

Thin-film polycrystalline silicon (poly-Si) is a promising semiconductor material for photovoltaic (PV) devices. The thin-film poly-Si PV technology was pioneered by Sanyo in the 1990s. The company achieved an efficiency of 9.2% for small-area (1 cm^2) solar cells fabricated on metal substrates [2]. Pacific Solar (now CSG Solar) then developed poly-Si on glass and achieved a record efficiency of 10.4% for a 94-cm^2 mini-module [3]. It is possible to improve the efficiency of poly-Si thin film solar cell modules further by optimizing the doping concentration and improving the thin film electronic properties such as majority carrier mobility and minority carrier diffusion length. There are various fabrication techniques to fabricate the poly-Si thin films, such as liquid phase crystallization [4], vapor phase processing [5] and laser induced crystallization from amorphous silicon [6]. However, these processes require high process temperatures and, hence, limit the substrate choice.

In this work we investigate poly-Si thin film prepared by solid phase crystallization (SPC) [7] of hydrogenated amorphous silicon (a-Si:H). The SPC method has many features suited towards low cost fabrication process of solar cell materials such as large area, simple process and low process temperature. Also, the Si deposition rate in the SPC method is not restricted by the need for excellent electronic properties of the a-Si:H material as in the case for a-Si:H solar cell. The electronic properties of the poly-Si are mainly ruled by the post-SPC thermal processing steps [8].

In this paper, we investigate the effect of rapid thermal annealing (RTA) and hydrogenation on the active doping concentration and majority carrier mobility of n^+ and p^+ SPC poly silicon thin films. These n^+ and p^+ layers form the emitter and back surface field of a poly-Si thin film solar cell. Thus, it is important to optimize the doping concentration of these two layers and understand the impact of thermal processing and hydrogen passivation to further enhance the efficiency of the poly-Si thin film solar cell.

EXPERIMENT

n^+ and p^+ a-Si:H layers were deposited by plasma-enhanced chemical vapor deposition (PECVD, MV Systems, USA) onto 400 mm × 300 mm Borofloat glass sheets with a thickness of 3 mm. The glass substrate temperature during PECVD was ~380°C for all depositions. The experimental conditions of n^+ and p^+ a-Si film deposition are summarized in Table I. Prior to the a-Si:H deposition, an antireflection coating of 70-nm-thick silicon nitride was deposited by PECVD onto the glass substrate. This silicon nitride film acts as an antireflection coating and a diffusion barrier for impurities from glass.

Table I. Experimental details used for the PECVD of the n^+ and p^+ a-Si:H films.

Process condition	*n^+ a-Si layer*	*p^+ a-Si layer*
SiH_4 (sccm)	10	8-20
2% PH_3:H_2 (sccm)	0.1-1.5	0
2% B_2H_6:H_2 (sccm)	0	0.2-0.3
Substrate temperature (^{0}C)	380	380
Pressure (mTorr)	600	600
RF Power (W)	12	12-40

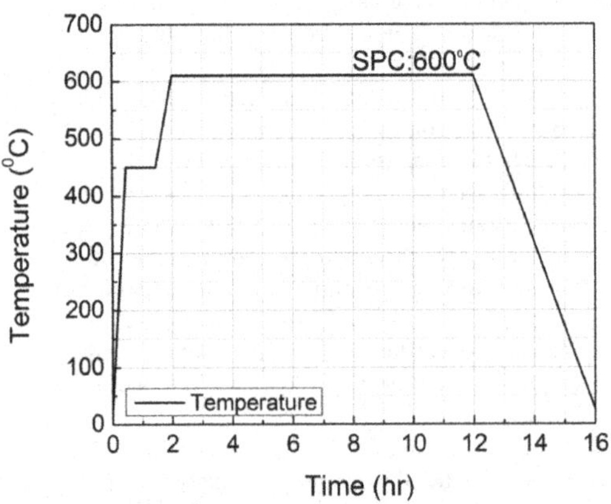

Figure1. Temperature profile used for the SPC crystallization of the a-Si:H films.

The a-Si:H coated glass sheets were subsequently annealed (Nabertherm, N 120/65HAC furnace, Germany) in a N_2 atmosphere for 12 hours. Figure 1 shows the temperature profile used to achieve the solid phase cysratallization (SPC) of the a-Si:H film. The a-Si:H samples were annealed at 450°C for one hour before ramping it to 600°C to remove the hydrogen from the sample .

After the SPC process, the samples received a rapid thermal anneal (RTA, CVD Equipment, USA) for 1 min at 1050°C in N_2 atmosphere , The samples were then hydrogenated in a PECVD system (AK800, Roth & Rau, Germany) at a glass temperature of 575°C. The electronic properties of the poly-Si film were determined after every processing step. The sheet resistance was determined using the four-point-probe technique and the majority carrier mobility was determined by means of the Hall effect. The Hall effect measurements were conducted at a magnetic field of 0.32 T and temperature of 300 K and the system was calibrated using a c-Si reference sample at the identical magnetic field. UV-VIS-NIR spectrophotometry (PerkinElmer, Lambda 950, UV/VIS Spectrometer) was used to determine the thickness of the poly-Si thin film.

RESULTS & DISCUSSION

n^+ poly-Si thin film

Figure 2 shows the Hall measurement results for the n^+ poly-Si films. Hall mobility and resistivity as a function of the majority carrier concentration are shown in Fig. 2(a) and 2(b), respectively. The Hall mobility and resistivity of n-type poly silicon prepared by the SPC method decreases for an increasing majority carrier concentration and follows a trend similar to that of single crystal Si. A Hall mobility of 40cm^2/Vs was obtained for n^+ poly-Si at the carrier concentration of 2.19×10^{19} cm^{-3}. The mobility increased to 71 cm^2/Vs (which is approximately 70% of the hall mobility of c-Si) after the RTA process. This is significantly higher than the results reported by T. Matsuyama *et al.* [7]. The significant increase in mobility by 77% after the RTA process can be understood from the fact that the RTA process removes a substantial fraction of crystallographic defects and activates the dopants in the SPC poly-Si film. The improvement in the electronic property of poly-Si thin film after the RTA suggest that the electronic properties of the poly- Si film are not determined by the a-Si:H deposition process, however are determined during the subsequent thermal processing steps in agreement with results published by Basore *et al.* [8]. The resistivity of n^+ poly-Si films decreases with increasing majority carrier concentration. There is a significant decrease in the resistivity at a given concentration after RTA. The resistivity obtained from Hall measurement system was in good agreement with the results obtained by four-point probe measurements. It was observed that hydrogenation steps had negligible impact on the resistivity. A slight decrease in carrier mobility was observed with the majority carrier concentration being nearly unchanged after hydrogenation. This can be explained from the fact that excess hydrogen has the ability to counteract the electrical activity of dopants as shown by Kitahara *et al.* [10]. Thus it is very important to understand and optimize the hydrogenation process to achieve the best electronic properties for a given poly-Si thin film. The carrier mobility and resistivity data for n^+ poly-Si thin film plotted in Figure 2 being so close to the single-crystal Si curves indicates the high quality of poly-Si thin film obtained by the SPC method.

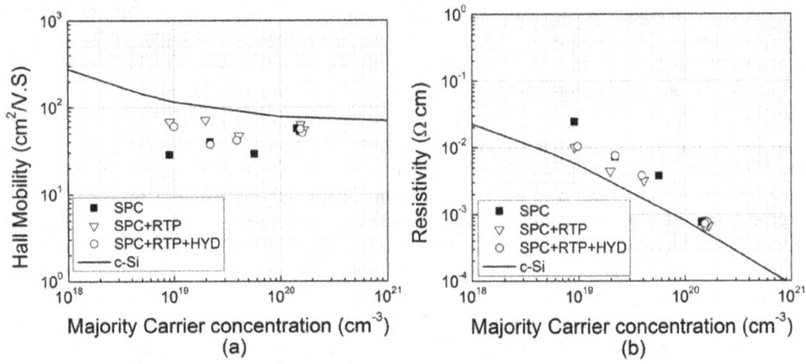

Figure 2. (a) Hall mobility of n^+ poly-Si thin film as a function of majority carrier concentration, (b) Resistivity of n^+ poly-Si thin film as a function of majority carrier concentration. The solid line indicates single crystal n-type Si [9].

p^+ poly-Si Thin film

Figure 3 shows the results of Hall measurement for the p^+ poly-Si films. Hall mobility and resistivity as a function of the majority carrier concentration are plotted as shown in Fig. 3(a) and 3(b), respectively.

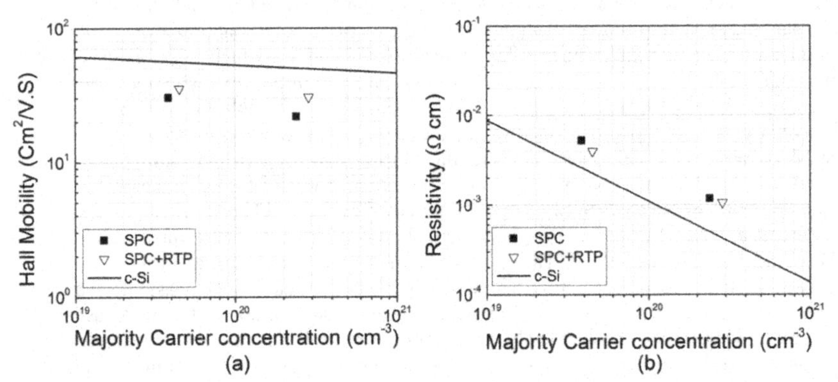

Figure 3. (a) Hall mobility of p^+ poly-Si thin film as a function of the majority carrier concentration, (b) Resistivity of p^+ poly-Si thin film as a function of majority carrier concentration. The solid line indicates single crystal p-type Si [9].

After the SPC anneal, a Hall mobility of 30 cm^2/Vs was obtained for p^+ poly-Si thin film at a carrier concentration of 3.8×10^{19} cm^{-3}. The Hall mobility increased to 35 cm^2/Vs at the majority carrier concentration of 4.5×10^{19} cm^{-3} after the RTA process. Both the Hall mobility and carrier concentration increased by 17% after the RTA process which is in good agreement with results reported for boron implanted poly-Si films by Almaggoussi A., *et al.* [11]. This impact of RTA on dopant concentration and mobility in the boron doped films has also been observed by Jeanjean, P., *et al.* [12]. The resistivity results obtained from Hall measurements were in good agreement with the results obtained from four-point probe measurements. The measurement data plotted in the figure 3, being so close to the c-Si curve, are a good indication of the high crystallinity of the poly-Si thin films.

CONCLUSIONS

High quality n^+ and p^+ poly-Si thin films were fabricated using SPC. The electronic properties of the poly-Si films were studied directly after SPC, after a subsequent RTA, and after a RTA plus hydrogenation step. The experimental results show that the RTA process significantly improves the electronic properties of the n^+ and p^+ poly-Si thin film. A high majority carrier mobility of 71 cm^2/Vs and 35 cm^2/Vs was subsequently obtained for n^+ and p^+ poly-Si thin films after the RTA process, at the carrier concentration of 2×10^{19} cm^{-3} and 4.5×10^{19} cm^{-3} respectively. The high carrier mobility demonstrates the good quality of the poly-Si film. However, it is important to optimize both the RTA and hydrogenation steps as they lead not only to the passivation of defects but also can lead to the formation of many complex defects.

ACKNOWLEDGEMENT

The Solar Energy Research Institute of Singapore (SERIS) is sponsored by the National University of Singapore (NUS) and the National Research Foundation (NRF) of Singapore through the Singapore Economic Development Board. This work was sponsored by a grant (Clean Energy Research Program) from the NRF. A. K. acknowledges a Clean Energy PhD scholarship from the NRF.

REFERENCES

1. W. P. Hirshman, G. Hering and M. Schmela, Survey on cell and module production 2006, *Photon International*, March 2007**,** 136 (2007).
2. Matsuyama, T., et al., *High-quality polycrystalline silicon thin film prepared by a solid phase crystallization method.* Journal of Non-Crystalline Solids, 1996. **198**: p. 940-944.
3. M.J. Keevers et al., Proc 22nd European Photovoltaic Solar Energy Conf., Milan, 2007 (WIP, Munich, 2007, p. 1783).
4. Baliga, B.J., *Morphology of silicon epitaxial layers grown by under cooling of a saturated tin melt. Journal of Crystal Growth*, 1977. **41**(2): p. 199-204.
5. Haberecht, R.R. and E.L. Kern, *Semiconductor silicon.* 1969: Electronics Division, Electrochemical Society.
6. Gat, L. Gerzlieyg, J.F. Gibbons, T.J. Magee, J. Peng and J.D. Hong, Appl. Phys. Lett. **33** (1978) 775.
7. Matsuyama, T., et al., *Preparation of High-Quality n-Type Poly-Si Films by the Solid Phase Crystallization (SPC) Method.* Japanese Journal of Applied Physics, 1990. **29**(part 1): p. 2327-2331.
8. P. A. Basore, *CSG-1 Manufacturing a new polycrystalline silicon PV technology , proc 4th world conference on Photovoltaic Energy Conversion*, Hawaii, 2006, pp. 874-876.

9. Hull, R. *Properties of crystalline silicon.* 1999: Institution of Electrical Engineers.
10. Kitahara, K., et al., *Correlation between electron mobility and silicon-hydrogen bonding configurations in plasma-hydrogenated polycrystalline silicon thin films.* Applied Physics Letters, 1998. **72**: p. 2436.
11. Almaggoussi, A., et al., *Electrical properties of highly boron implanted polycrystalline silicon after rapid or conventional thermal annealing.* Journal of applied physics, 1989. **66**(9): p. 4301-4304.
12. Jeanjean, P., et al., *Dopant activation and Hall mobility in B-and As-implanted polysilicon films after rapid or conventional thermal annealing.* Semiconductor science and technology, 1991. **6**: p. 1130.

Mater. Res. Soc. Symp. Proc. Vol. 1321 © 2011 Materials Research Society
DOI: 10.1557/opl.2011.807

Characterization of Green Laser Crystallized GeSi Thin Films

Balaji Rangarajan[1], Ihor Brunets[1], Peter Oesterlin[2], Alexey Y. Kovalgin[1] and Jurriaan Schmitz[1]
[1]MESA+ Institute for Nanotechnology, University of Twente, P.O. Box 217, 7500AE Enschede, The Netherlands.
[2]INNOVAVENT GmbH, Bertha-von-Suttner Str. 5, 37085 Gottingen, Germany.

ABSTRACT

Green laser crystallization of a-$Ge_{0.85}Si_{0.15}$ films deposited using Low Pressure Chemical Vapour Deposition is studied. Large grains of 8x2 μm^2 size were formed using a location-controlled approach. Characterization is done using Scanning Electron Microscopy, Atomic Force Microscopy, X-Ray Photoelectron Spectroscopy and X-Ray Diffraction.

INTRODUCTION

Germanium-silicon (GeSi) alloys deposited at lower temperatures (400-450 °C) can be incorporated in various 3D integration schemes (e.g. monolithic integration) [1, 2]. Making the temperature CMOS-back-end compatible has generated renewed interest in such films in recent years [3]. Electrical characteristics of the devices (e.g. transistors) fabricated on such films can however be affected by the randomly positioned grain boundaries as shown earlier for Si [4, 5].

In order to control the positioning of the grain boundaries, green laser crystallization was used in our study to crystallize a-$Ge_{0.85}Si_{0.15}$ films using a location-controlled approach involving pre-patterned a-GeSi lines. The context of the present work is in infrared detection and therefore the choice of this particular film composition.

EXPERIMENT

The a-$Ge_{0.85}Si_{0.15}$ and poly-$Ge_{0.85}Si_{0.15}$ films were deposited using Low Pressure Chemical Vapour Deposition (LPCVD). The GeSi films were deposited on top of 450-nm oxide which was grown on Si substrate using wet oxidation. The custom built LPCVD system has a horizontal furnace for batch deposition and uses resistive heating (to heat the furnace walls). A quartz wafer-boat is used in order to load the 4-inch wafers directly into the furnace. The system has a base pressure of 10^{-3} mbar. The precursor gases used for the deposition were SiH_4 and GeH_4. Deposition at 430 °C using 75 sccm of SiH_4 flow and 37 sccm of GeH_4 flow, with a total pressure of 6 mbar, resulted in the formation of a-$Ge_{0.85}Si_{0.15}$ films. On the other hand, deposition performed using the same set of process parameters except for a total pressure of 0.2 mbar resulted in poly-$Ge_{0.85}Si_{0.15}$ films. In order to form nucleation sites for GeSi film deposition on oxide, a few nanometers of a-Si were previously deposited using the same LPCVD system at 430 °C using 88 sccm of SiH_4 flow with a total pressure of 0.5 mbar. The variation in the composition of LPCVD poly-GeSi films with respect to different GeH_4 partial pressures has also been explored in this study. The as-deposited poly-$Ge_{0.85}Si_{0.15}$ films acted as the reference material while a-$Ge_{0.85}Si_{0.15}$ films were used for laser crystallization.

A green laser (λ = 515 nm) pulsed at a frequency of 10 kHz with a pulse duration of 285 ns was used for the laser crystallization of a-GeSi films. The laser beam had a Gaussian energy density profile with Full Width Half Maximum (FWHM) of 15 μm along the x-axis, whereas along the y-axis a uniform energy density profile was kept with FWHM of 2 mm (see Fig 1).

The a-$Ge_{0.85}Si_{0.15}$ films of a thickness of 50 nm were patterned to form 900 nm wide periodic lines with a pitch of 3μm. Further, these lines were covered with 100-nm-thick a-$Ge_{0.85}Si_{0.15}$ film deposited again using LPCVD. This was followed by the laser crystallization process as shown in Fig.1a. During the crystallization process this topography leads to a lateral, periodic temperature gradient, causing crystallization to start at predefined locations. As a result, the preformed lines dictate the location of the dominant grain boundaries. In order to confirm the effectiveness of the preformed lines over the crystallization process, a plain 100-nm-thick a-$Ge_{0.85}Si_{0.15}$ film was also laser crystallized under the same conditions as shown in Fig.1b.

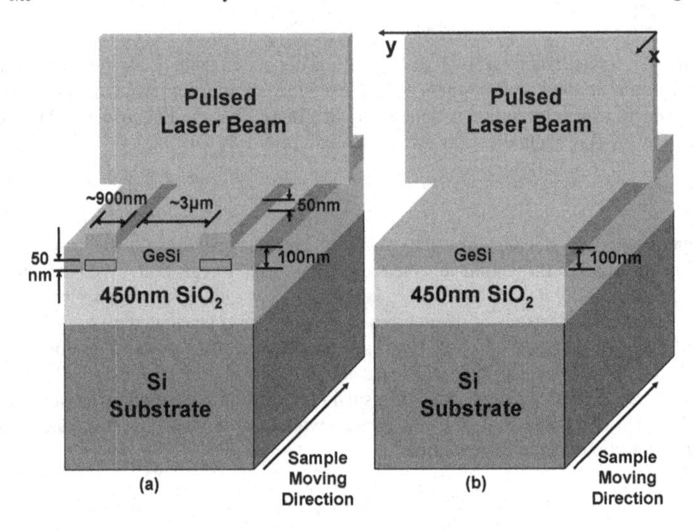

Figure 1. Overview of Laser Crystallization (a) with pre-patterned lines and (b) without lines.

Under the pulsed laser beam the samples were moved unidirectionally as shown in Fig.1a & 1b. Depending on the overlapping of consecutive laser pulses, as required for a lateral growth during the crystallization process, the scanning speed has to be adjusted accordingly. Laser pulse overlappings of 98% and 95% provided the optimal results (in terms of the formation of large grains). The samples were moved along the direction of the pre-patterned lines (as shown in Fig. 1a). Similar orientation was also adopted for the sample without the pre-patterned lines (Fig. 1b). Laser energy densities of 1.2 and 1.3 Jcm^{-2} were found to provide the best results of crystallization.

All the films were characterized using Scanning Electron Microscopy (SEM), Atomic Force Microscopy (AFM), X-Ray Photoelectron Spectroscopy (XPS) and X-Ray Diffraction (XRD).

DISCUSSION

LPCVD

Figure 2 shows the change in the composition of the deposited poly-GeSi films with respect to the relative GeH_4 partial pressure (P_{ger}). It can be noted that, even at the lowest P_{ger}, the obtained germanium content (measured using XPS) remained higher than 80%. This indicates the probable lower relative reactivity of SiH_4 at the above mentioned deposition temperature of 430 °C. A higher content of Si in the layers could be achieved by either using disilane or trisilane as one of the precursor gases replacing SiH_4 in these depositions [6]. This approach was however outside of the scope of this study and not further probed as we had already obtained our desired composition of 85% Ge in the poly-GeSi films.

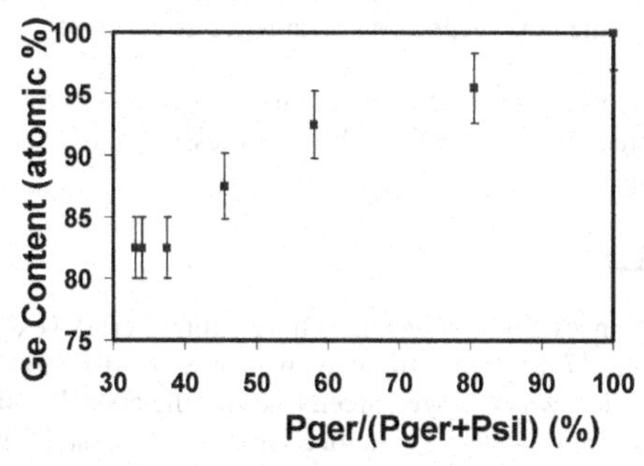

Figure 2. Variation in Ge content of the deposited poly-GeSi films versus relative (i.e. with respect to SiH_4 partial pressure - P_{sil}) GeH_4 partial pressure (P_{ger}).

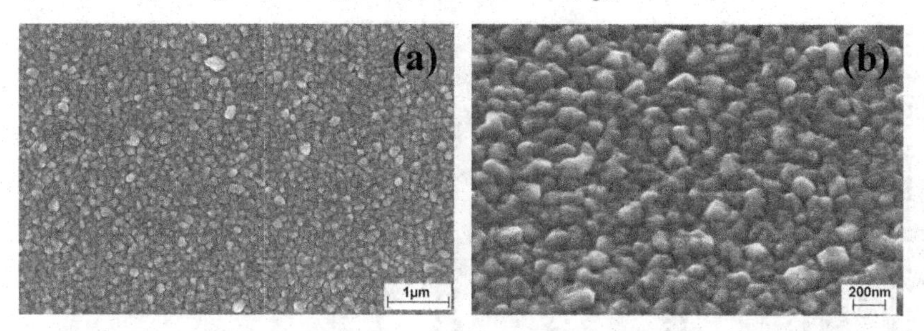

Figure 3. SEM images of the reference LPCVD poly-$Ge_{0.85}Si_{0.15}$ film: (a) top view; (b) tilted view at 50°.

The obtained reference poly-$Ge_{0.85}Si_{0.15}$ film was inspected in the SEM (Fig. 3) to reveal small grains with a typical size of 200 nm, and a surface roughness of 12.6 nm (RMS) as measured by AFM.

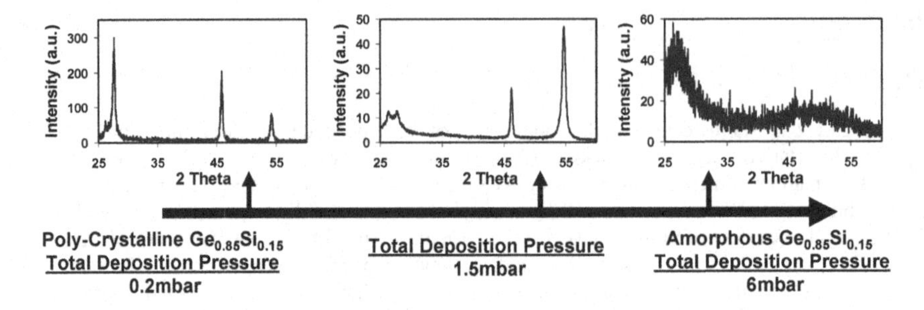

Figure 4. XRD graphs of the deposited poly-GeSi layers at three different total deposition pressures (while maintaining the other deposition parameters as constant).

The transition from polycrystalline to amorphous $Ge_{0.85}Si_{0.15}$ film was obtained using the same set of deposition parameters, by increasing the total pressure to 6 mbar, as depicted in Fig. 4 using the XRD graphs.

Green Laser Crystallization

There are earlier instances where researchers have explored (UV) excimer laser crystallization of Ge and GeSi [7, 8]. In our study, with an attempt to explore further ways to enhance material characteristics, we employed green laser for the crystallization of a-$Ge_{0.85}Si_{0.15}$ films. Fig. 5a shows the results of the laser crystallization of a-$Ge_{0.85}Si_{0.15}$ film with pre-patterned lines where large grains, typically 8x2 μm^2 in size, are formed and contained between the withered pre-patterned lines (which are represented by the dotted lines).

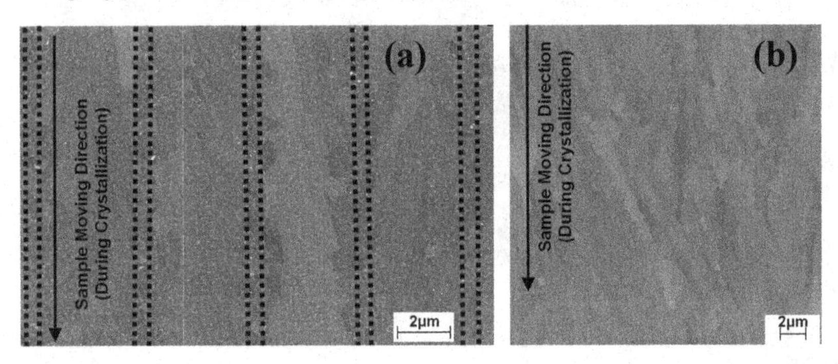

Figure 5. (a) SEM top view of laser crystallized film with pre-patterned lines; pre-patterning is indicated by the vertical dotted lines. (b) same, without pre-patterned lines.

A green laser was previously employed to crystallize a-Si films using the same approach, resulting in the dominant grain boundaries being located in between the pre-patterned lines [9].

The difference is that, in our case (see Fig. 5a), the grains originate anywhere in between the lines and eventually terminate at the pre-patterned lines. This difference in the crystallization behaviour between Si and GeSi films can be attributed to their respective thermal and optical properties. Fig. 5b shows the laser crystallization of a plain a-$Ge_{0.85}Si_{0.15}$ film resulting in the formation of elongated grains, typically 8x1 μm^2 in size, which are more randomly positioned. This shows the clear advantage of using pre-patterned lines during the crystallization process influencing the spatial orientation of the large grains. Such results have been observed for both 98% and 95% overlappings.

The surface roughness of both the films is around 2.5 nm (RMS) as shown by AFM. Both in terms of grain size and surface roughness, the laser crystallized $Ge_{0.85}Si_{0.15}$ films show a distinct advantage over LPCVD poly-$Ge_{0.85}Si_{0.15}$ films.

The XPS analysis shows a uniform composition profile along the depth for both as-deposited amorphous and poly $Ge_{0.85}Si_{0.15}$ films. The composition is measured along the depth by sputtering through the film using Ar ions. In contrast, the laser-crystallized $Ge_{0.85}Si_{0.15}$ films show that the top layers of the film are enriched with Ge compared to the bottom (Fig. 6). This Ge gradient did not exist in a-$Ge_{0.85}Si_{0.15}$ film prior to the laser treatment, indicating its occurrence is due to the laser crystallization process.

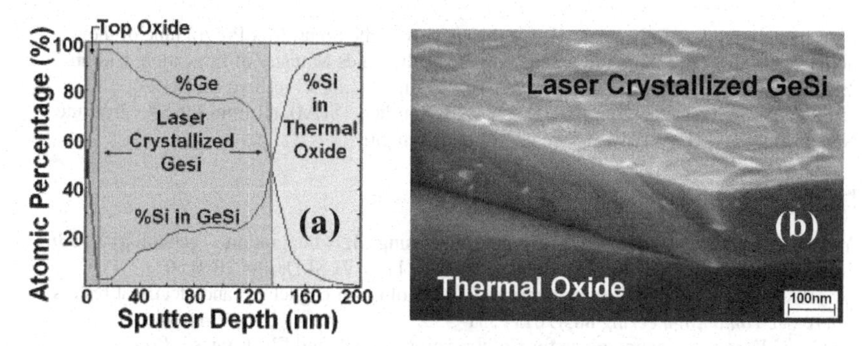

Figure 6. (a) XPS Depth profile for the laser crystallized $Ge_{0.85}Si_{0.15}$ film. (b) SEM cross-section of laser crystallized $Ge_{0.85}Si_{0.15}$ film.

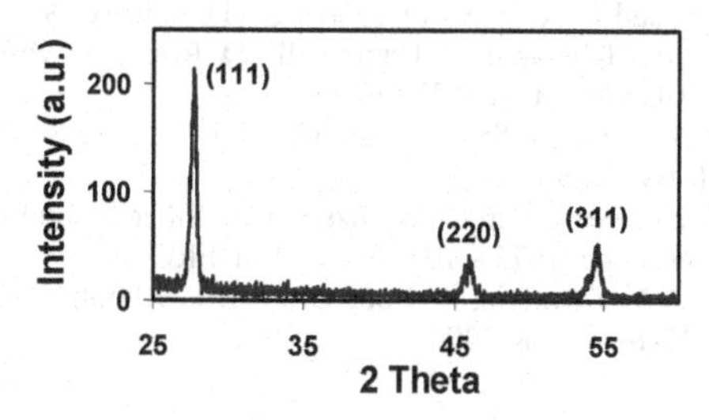

Figure 7. XRD graph of the laser crystallized $Ge_{0.85}Si_{0.15}$ film.

The XRD analysis (Fig. 7) shows the poly-crystalline nature of the laser crystallized $Ge_{0.85}Si_{0.15}$ films.

CONCLUSIONS

A custom-built LPCVD system was used to deposit poly- and amorphous-$Ge_{0.85}Si_{0.15}$ films at a temperature of 430 °C. The as-deposited poly-$Ge_{0.85}Si_{0.15}$ films showed small grains typically in the range of 200 nm.

The green laser crystallization of a-$Ge_{0.85}Si_{0.15}$ films, on the other hand, resulted in the formation of large-grained poly-$Ge_{0.85}Si_{0.15}$ films with a typical grain size of 8x2 μm^2. The pre-patterned lines influenced the positioning of grain boundaries during the crystallization process. Absence of the pre-patterned lines clearly led to more randomness in grain formation. The pre-determined-in-space grain locations are needed for the proper placement of the devices during further fabrication processes. This can likely enhance the characteristics of the device positioned properly on such films.

ACKNOWLEDGMENTS

The authors gratefully acknowledge the support of the Smart Mix Programme of the Netherlands Ministry of Economic Affairs and the Netherlands Ministry of Education, Culture and Science. The authors would like to thank Tom Aarnink (LPCVD), Jiwu Lu (XRD), Gerard Kip (XPS), Lan Anh Tran (AFM) and Mark Smithers (SEM), all from MESA+ Institute for Nanotechnology, University of Twente, for their support.

REFERENCES

1. Yuan Xie, "Processor Architecture Design Using 3D Integration Technology", *23rd International Conference on VLSI Design,* DOI 10.1109/VLSI.Design.2010.60.
2. M. Vinet *et al.*, "3D monolithic integration: Technological challenges and electrical results", *Microelectronic Engineering* **88**, (2011) 331–335.
3. Fedder G.K *et al.*, "Technologies for Cofabricating MEMS and Electronics", *Proc. Of IEEE,* **Vol.96**, No. 2, Feb. 2008, pp. 306-322.
4. J.G. Fossum and A. Ortiz-Conde, *IEEE Trans. Electron Devices,* **Vol. ED-30**, NO. 8, August 1983.
5. A. K. Ghosh, C. Fishman, and T. Feng, *J. Appl. Phys.* **51(1)**. January 1980.
6. A.Gouye, O. Kermarrec, A. Halimaoui, Y. Campidelli, D. Rouchon, M. Burdin, P. Holliger and D. Bensahel, *J. Cryst. Growth* **311,** 3522 (2009).
7. Tao Chen *et al.*, "High Performance Single-Grain Ge TFTs without Seed Substrate", *IEDM 2010*, pp. 496-499, IEEE International.
8. Ishihara. R *et al.*, M, "Excimer-Laser Crystallization of Silicon-Germanium", *Solid State Device Research Conference,* pp. 1075 – 107, 1996. ESSDERC '96.
9. I. Brunets, J. Holleman, A.Y. Kovalgin, A. Boogaard and J. Schmitz, *IEEE Trans. Electron Devices,* **Vol. 56**, pp. 1637-1644, Aug 2009.

Mater. Res. Soc. Symp. Proc. Vol. 1321 © 2011 Materials Research Society
DOI: 10.1557/opl.2011.1248

A Study of the Post-hydrogenation Passivation Mechanism of Crystallized Poly-Si Films

Chong Luo[1], Juan Li[1], He Li[1,2], Zhiguo Meng[1], Chunya Wu[1], Qian Huang[1], Xu Shengzhi[1], Hoi Sing Kwok[2] and Shaozhen Xiong[1♥]

1 Institute of Photo-Electronics, Nankai University, Tianjin 300071, P. R. China
2 Department of Electronic and Computer Engineering, The Hong Kong University of Science and Technology, Clear Water Bay, Kowloon, Hong Kong, P. R. China

ABSTRACT

The roles of hydrogen plasma radicals on passivation of several kinds of crystallized poly-Si thin films were investigated using optical emission spectroscopy (OES) combined with Hall mobility, Raman spectra, and absorption coefficient spectra. It was found that different kinds of hydrogen plasma radicals are responsible for passivation of dissimilar poly-Si crystallized by different method. Radicals H_α with lower energy are mainly responsible for passivating the poly-Si crystallized by solid phase crystallization (SPC) whose crystallization precursor was made by plasma enhanced chemical vapor deposition (PECVD). Higher energy radicals H^* are more effective in passivating defects left over by Ni in poly-Si crystallized by Metal Induced Crystallization (MIC). The highest energy radicals H_β and H_γ are needed to passivate the defects in poly-Si crystallized by SPC but whose precursor was made by low pressure CVD (LPCVD).

INTRODUCTION

Poly-crystalline silicon (poly-Si) thin films prepared on glass substrates at temperatures below 600℃ are used for large area electronic devices, because they cost less than mono-crystalline silicon, and have higher mobility and greater stability than amorphous silicon. They attract scientific interest because of their application in flat panel displays and solar cells. However, crystallized poly-Si films have more defects at grain boundaries [1] or intra-grains [2], which severely affect the performances and stabilities of the poly-Si devices. Hydrogen plasma treatment is one of the most effective methods to passivate the defects. Generally, hydrogen in silicon thin films is not only attracted to any strained regions but it also self-traps and removes dangling bonds. However, hydrogen plasmas etch Si atoms and bombard the surface of poly-Si, thereby producing some new defects [3]. As we know there are four radicals in H-plasma. Are all of the radicals playing the same role in passivation? In this work, we study the effect of hydrogen plasma radicals on passivation and analyze the relationship between H-plasma radicals and their specific role in different kinds of poly-Si films. We measured and compared the in-situ OES of the hydrogen plasma during passivation, and characterized the corresponding electrical and optical properties of passivated poly-Si thin films to check the passivation effect. Connecting the relation between each radical in OES and the passivation effect monitored by the electrical and optical properties of poly-Si, it was found that different hydrogen plasma radicals play different roles in passivating different kinds of poly-Si films.

♥ Corresponding author e-mail: xiongsz@nankai.edu.cn

EXPERIMENT

100nm thick a-Si thin films were deposited by LPCVD or PECVD, respectively, on the Eagle 2000 Corning glass and subsequently crystallized by SPC or MIC. The PECVD a-Si was deposited with hydrogen diluted silane at SiH_4:H_2=1:4. The gas pressure was 43 Pa at a substrate temperature of 270□. Radio Frequency (RF) active power of 10W was used. Dehydrogenation was performed before crystallization. The LPCVD a-Si was deposited with SiH_4 at 600 milli-Torr under temperature of 550□ and SPC crystallization was carried out for 24 hours at 600□ in a N_2 atmosphere. The samples crystallized by MIC had a Ni layer which was deposited by sputtering for 9 minutes with an RF power of 7W, followed by heating at 590℃ in N_2 for 6 hours. The Ni residue on the surface was etched. The sample whose crystallizing precursor of a-Si was made by PECVD and was crystallized by SPC will be named PE-SPC. The sample using LPCVD a-Si as precursor and crystallized by SPC will be named LP-SPC, but if it was crystallized by MIC it will be called LP-MIC. The power source for the hydrogen plasma treatment was a Radio Frequency (RF) power generator. The hydrogen plasma passivation was done normally with a H_2 flow rate of 30 SCCM, a reaction pressure of 800 milli-Torr, and an RF-power of 10-40W. The substrate temperature was changed from 200-550℃. The signal for the OES of the H-plasma was recorded by a PR650 spectrophotometer and focused at a point 2mm above the substrate. The passivated samples were characterized by Hall mobility using "Bio-Rad Model HL5500PC", and the Raman spectra were measured with "Renishaw in ViaRaman". The Raman TO peak at 520 cm^{-1} taken on a single crystal was used as a reference for all our measurements. In addition, the absorption coefficients were measured at low photon energies for determining the defect concentration.

RESULTS AND DISCUSSION

3.1 The passivation effect of the H-plasma

Fig.1 shows a comparison of the Raman spectrum before and after passivation at 350℃ with 40W for 15 minutes of three kinds of poly-Si films of PE-SPC, LP-SPC, and LP-MIC. Besides a shift of the TO peak from its crystalline Si value the H-plasma treatment narrows the Raman curve (specially for PE-SPC) which means the plasma promotes further crystallization to enlarge the grain size. But it caused some inner-stress according to the small shift of the TO peak to lower wave-numbers.

In order to see evidence of passivation by the H plasma, we analyzed the absorption coefficient spectra with special attention to the sub-band gap range, which results from absorption involving the defect states. Fig.2 shows the changes of sub-band gap absorption of LP-SPC poly-Si for several H-plasma treatment times. As we can see, at first, the sub-band gap absorption after passivation drops down by nearly one order of magnitude which indicates that the defect densities decreased dramatically due to the hydrogen plasma treatment. The sub-band gap absorption coefficient decreases from 0-20' treatment time. See Fig.2. After that the sub-band gap absorption coefficient increases again. This implies that, for LP-SPC poly-Si the treatment time of 20' is sufficient.

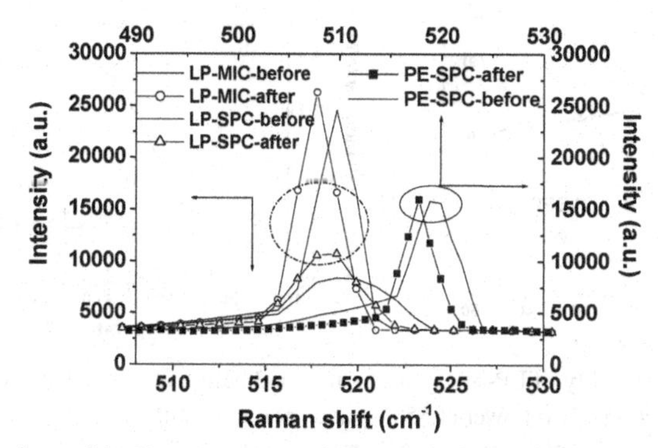

Fig.1 Comparison of the Raman spectrum before and after passivation by H-plasma at 350℃ and 40W and 15 minutes for three kinds of poly-Si samples

We also optimized the plasma conditions and selected a LP-SPC sample for the demonstration. The results are shown in Fig.3 (a), (b) respectively. The "Ratio of Hall mobility (a/b)" is the ratio of the mobility after to that before H-plasma treatment μ_{after}/μ_{before}. The mobility ratio declines at a plasma power higher than 10W, That means high power damages the quality of passivated poly-Si. Similarly lowering the temperature restricts the passivation effect. According to the data shown in Fig.3 (a) and (b), the best plasma condition is around 450℃, 10 W and choosing 20' for the treatment time of LP-SPC poly-Si films.

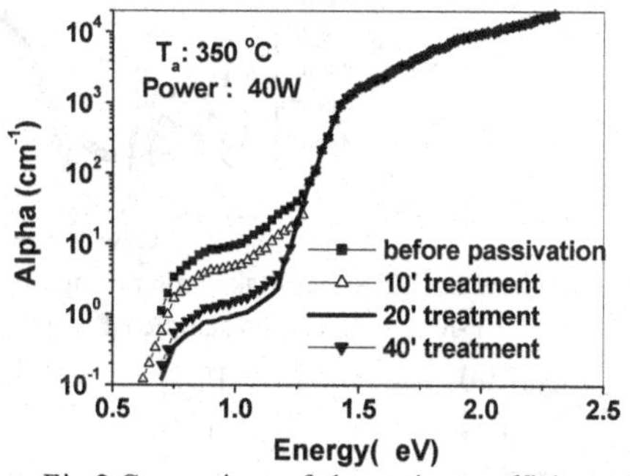

Fig.2 Comparison of absorption coefficient of LP-SPC before and after passivation treatment

3. 2. The role of H-plasma radicals on passivation

We used H-plasma to passivate the crystallized poly-Si, measured and recorded the in-situ OES of the hydrogen plasma during the passivation process. As we know there are four main kinds of H-radicals in the hydrogen plasma. These are radical H_α (peak at 656nm, corresponding to the n=3 excited state of H), H^* (602nm, excited state of H_2), H_β (486nm, corresponding to the n=4 excited state of H) and H_γ (434nm, n=5 excited state of H). Do these radicals play the same role in the passivation process?

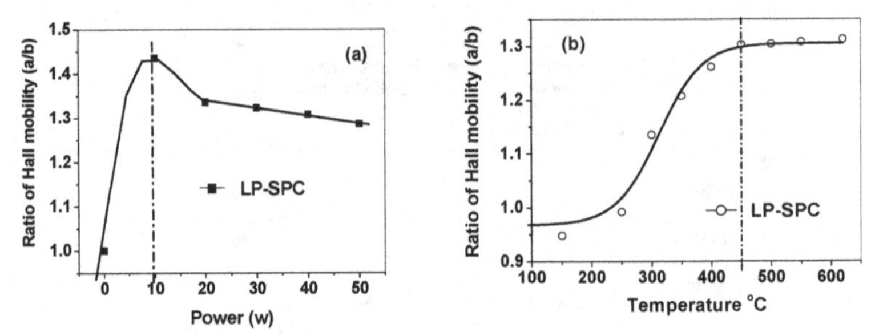

Fig.3 the ratio of Hall mobility of LP-SPC poly-Si after and before 20' treatment by H-plasma as function of (a) power at 550℃ and (b) temperature at 40W

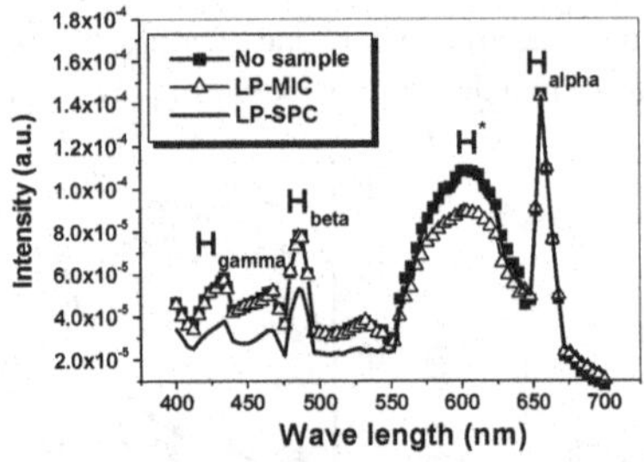

Fig.4 Comparison of OES signals taken after of hydrogen plasma treatment for 15 minutes of LP-SPC, LP-MIC (△) and no sample (■) in the chamber (note as "no sample")

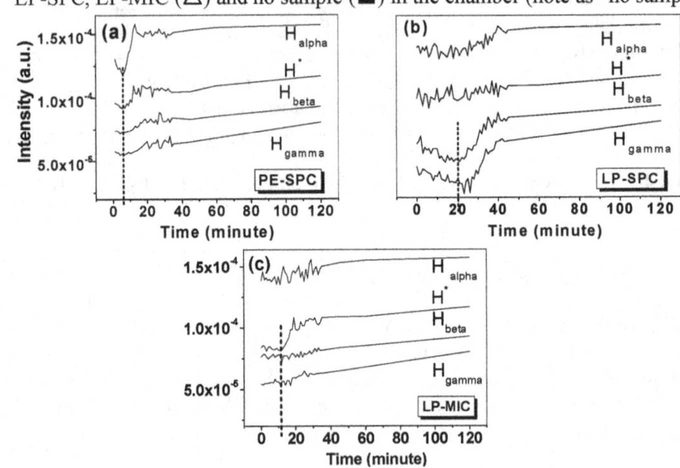

Fig.5 the intensity of four hydrogen radicals vs. time during H plasma treatment of (a) PE-SPC, (b) LP-SPC, (c) LP-MIC, respectively.

Fig.4 shows OES signals taken at 350℃ 15 minutes after hydrogen plasma treatment of LP-SPC, LP-MIC and no sample in the chamber (called "no sample"). The figure shows very clearly that the H_α signal is the same for the three cases. H* signals of LP-SPC and no-sample are the same but they are different to LP-MIC. H_β and H_γ of LP-MIC and no-sample are the same but they are different to LP-SPC. These results supply interesting information regarding the different effects on the dissimilar samples. We suppose that the decrease of some radical signal occurs because that radical was consumed during the passivation process.

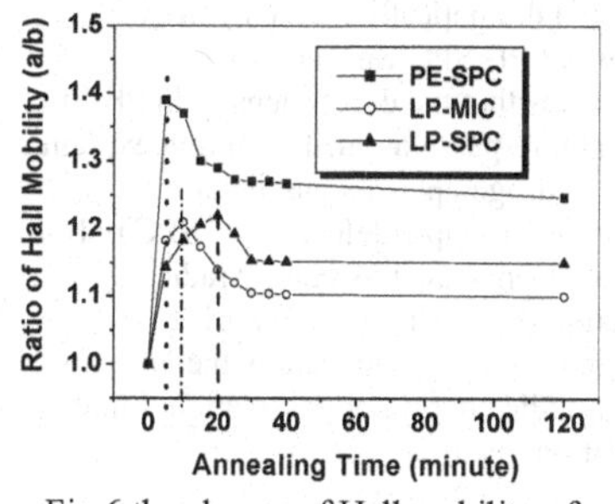

Fig.6 the change of Hall mobility of poly-Si with passivated time

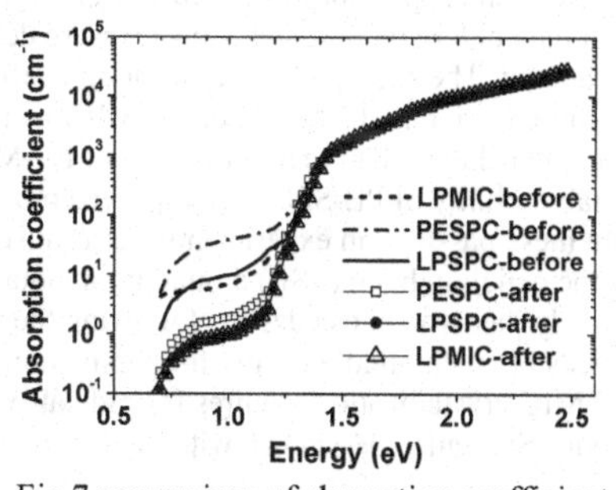

Fig.7 comparison of absorption coefficient spectra before and after passivation treatment

The changes in intensity of the four hydrogen radicals with time during H plasma treatment of three types of poly-Si are shown in Fig.5 (a), (b) and (c). For PE-SPC as Fig.5 (a) shows, the consumption of H_α is dominant during the first 5 minutes. After that time the intensity of H_α starts to increase. It is also clearly shown in Fig.5 (b), that the changes of H_β and H_γ are much greater than those of H_α and H^*, which implies that H_β and H_γ have a significant effect on the passivation of defects in LP-SPC poly-Si. The intensities of H_β and H_γ decreased during the beginning 20 minutes, and then increased after that time. We believe the decrease of the OES signal intensity is caused by the consumption of that hydrogen plasma radical for passivating the defects. After a certain time the radical signal starts to increase again which means the passivation process is gradually completed. For sample LP-MIC poly-Si as Fig.5(c) shows there is a similar situation but the responsible radical is H* and the increase of H* occurred after 10 minutes. These OES curves in Fig.5 indicate that each H-radical will just respond to the passivation requirement of its specific material. This means there are different defects in differently crystallized poly-Si.

We also analyzed the variations of the Hall mobility with the hydrogen plasma treatment time for three different poly-Si samples. The improvement status was monitored by the ratio of the Hall mobility after and before passivated treatment as shown in Fig.6. For the PE-SPC sample, with ~5 minute hydrogen plasma treatment, the Hall mobility increased by over 40% (from $11.5cm^2/V.s$ to $16.0\ cm^2/V.s$). However, longer treatments did not yield a better performance, instead, the Hall mobility decreased. Referring to the OES results shown in Fig.5 (a), we infer that with the consumption of H_α during the first ~ 5 minutes of hydrogen plasma treatment the PE-SPC poly-Si was passivated, so its Hall mobility increased. With longer plasma exposure, the consumption of H_α decreased (matching with H_α signal increase in OES), and the bombardment effect and etching effects become critical, which create new defects and degenerate the performance of poly-Si. Similarly, we can deduce from Fig.5 (b) and Fig.5(c) that LP-SPC was passivated by H_β and H_γ within ~20 minutes (its Hall mobility increases from 7.5

$cm^2/V.s$ to 9.0 $cm^2/V.s$) and LP-MIC was passivated by H* within 10 minutes (which mobility increases from 17.5 $cm^2/V.s$ to 21.1 $cm^2/V.s$). The OES and Hall mobility data agree with each other.

The results confirm that different poly-Si thin films crystallized by different methods need different hydrogen plasma radicals for passivation. We demonstrate the effectiveness of H plasma passivation by measuring the absorption coefficient spectra in the range of sub-band gap absorption, which is proportional to the defect concentration. As shown in Fig.7, all three types of poly-Si sub-band gap absorption coefficients were one order of magnitude lower after passivation than before their defect densities were decreased dramatically by the hydrogen plasma treatment. The sub-band gap absorption coefficient of PE-SPC was larger and it improved more than that of LP-SPC or LP-MIC, which means that the defect density in PE-SPC poly-Si sample is larger than that in LP-SPC or LP-MIC before passivation. This might explain why the Hall mobility of PE-SPC was improved more by hydrogen plasma passivation.

From these passivation experiments we surmise that the principal defect in PE-SPC poly-Si is mainly the dangling bond (≡Si-) around grain boundaries Hence the low energy radicals Hα can effectively passivate those defects following the reaction ≡Si- +Hα → ≡Si-H. LP-MIC poly-Si needs H* with middle energy for passivation. A kind of defect induced by the Ni leftover associated MIC crystallization requires H* radicals with middle energy for passivation. Finally, LP-MIC poly-Si requires H_β and H_γ with higher energy for passivation.

CONCLUSIONS

OES spectra showed that the peak intensity of hydrogen plasma radicals in hydrogen passivation of poly-Si thin films depends on the crystallization methods and the a-Si precursors of these films. Based on OES, absorption spectra, Raman spectra and Hall mobility results, we found that the reaction between hydrogen radicals in plasma and defects in poly-Si depends on the defect type in the crystallized poly-Si. For PE-SPC poly-Si, in which the major defects are the dangling bonds at grain boundaries, low energy radicals H_α perform the passivation. For LP-SPC poly-Si, which contain intra-grain defects, higher energy radicals H_β and H_γ are needed to perform the passivation. For LP-MIC poly-Si, which contains many defects related to Ni impurities, H^* radicals with middle energy are best to passivate this kind of defect. The optimum effective passivation is realized when the relation between H-plasma radicals and the plasma process conditions, such as pressure, power, and temperature have been optimized

ACKNOWLEDGEMENTS

This work was supported by the Flat-Panel Display Special Project of National 863 Plan (No.2008AA03A335). We would like to give our greatly appreciate to Prof. Helmut Fritzsche for his revision of this paper.

REFERENCES

1. N. H. Nickel, N. M. Johnson, and W. B. Jackson, Appl. Phys. Lett. **62**, 3285 (1993)
2. D. Van Gestel, M. J. Romero, I. Gordon, L. Carnel, J. D'Haen, G. Beaucarne, M. Al-Jassim, and J. Poortmans, Appl. Phys. Lett. **90**, 092103 (2007)
3. S. Darwichea, M. Nikravecha, D. Morvana, J. Amourouxa, D. Ballutaud, Solar Energy Materials & Solar Cells **91**, 195–200 (2007)

Mater. Res. Soc. Symp. Proc. Vol. 1321 © 2011 Materials Research Society
DOI: 10.1557/opl.2011.1249

The Role of H-Plasma in Aluminum Induced Crystallization of Amorphous Silicon

Chong Luo[1], Juan Li[1], He Li[1,2], Zhiguo Meng[1], Qian Huang[1], Shengzhi Xu[1], Hoi Sing Kwok[2], Shaozhen Xiong[1♣]

1.Institute of Photo-Electronics, Nankai University, The Tianjin Key Laboratory for Photo-Electronic Thin Film Devices and Technology, Tianjin 300071, PR China

2.Department of Electronic and Computer Engineering, The Hong Kong University of Science and Technology, Clear Water Bay, Kowloon, Hong Kong, P. R. China

ABSTRACT

A technique to improve and accelerate aluminum induced crystallization (AIC) by hydrogen plasma is proposed in this paper. Raman spectroscopy and Secondary Ion Mass Spectrometry of crystallized poly-Si thin films show that hydrogen plasma radicals reduce the crystallization time of AIC. This technique shortens the annealing time from 10 hours to 4 hours and increases the Hall mobility from 22.1 $cm^2/V{\cdot}s$ to 42.5 $cm^2/V{\cdot}s$. The possible mechanism of AIC assisted by hydrogen radicals will also be discussed.

1. INTRODUCTION

Poly-Si thin films prepared on glass substrate at temperatures lower than 600℃ are of scientific interest because of their application in flat panel displays [1], solar cells [2].

However, each preparation technique for poly-Si thin films, such as Solid Phase Crystallization (SPC) and Rapid Thermal Annealing（RTA), Laser Crystallization （LA） and Metal-induced Crystallization (MIC), has restrictive factors. For example, SPC and RTA high temperatures which sometimes exceed the glass substrate limit and Laser Crystallization needs expensive equipment. In contrast, MIC using Al as the inducing metal (called AIC) is a promising method because of its low cost and relatively short annealing time [3]. Furthermore, AIC can yield large grain size [4] and the high preferential (100) orientation [5] p-doped [6] material useful as high quality seed layer for solar cells.

However, AIC still would profit from shortening its relatively long annealing time of about ten or more hours at low temperatures (around 450℃) [7]. Furthermore, poly-Si thin films crystallized by any crystallization method always contain many defects resulted from grain boundaries [7] or intra-grains [8], which severely affect the performances and stabilities of the devices made by such poly-Si thin films. Hydrogen Plasma Treatment is one of the most common methods to deal with these problems because of its low cost and simple technique [9]. In this paper, AIC in hydrogen plasma surroundings is being proposed. This technique combines the crystallization and passivation into one process, which not only reduces the annealing time but also passivates the defects in the material

2. EXPERIMENT

At first, 1000 Å a-Si was deposited by low pressure chemical vapor deposition (LPCVD) on

♣ Corresponding author e-mail: xiongsz@nankai.edu.cn

Eagle 2000 Corning glass. This was followed by a less than 10-nm SiO_2 separation layer and a 1000Å thick evaporated Al film. The glass/a-Si/SiO_2/Al stack was put into a chamber which is normally used for plasma enhanced chemical vapor deposition (PECVD) and annealed in a hydrogen plasma with H_2 of 30sccm at 800 Milli-Torr and Ts=450℃ for several hours. This process of AIC under H-plasma environment is called HAIC. After completed crystallization, the residual Al on the top of surface was etched off using a standard Al etching solution. We did not check if the thickness of AIC or HAIC poly-Si thin films has been varied because the crystallized poly-Si completed by AIC through exchange of layer to layer and an etching process to wipe off the residual Al. For optimizing performances of Poly-Si we use a suitable lower power to prevent the bombardment effect. Because the sputtering yield of Al is lower so the used power could not result in Al contamination to the chamber during the annealing process.

In order to evaluate the effect of the H-plasma on crystallization we also carried out traditional AIC by just heating the samples in vacuum at 450℃ for crystallization. We compared the passivation effect of HAIC and AIC.

The poly-Silicon films were characterized by Raman spectroscopy (type of Renishaw in Via), Secondary Ion Mass Spectrometry (SIMS) and Hall mobility measurements (type of Bio-Rad Model HL5500PC). The results were compared and analyzed.

3. RESULTS AND DISCUSSION

3.1. The effect of promotion crystallization using Al as inducing source in H plasma [10]

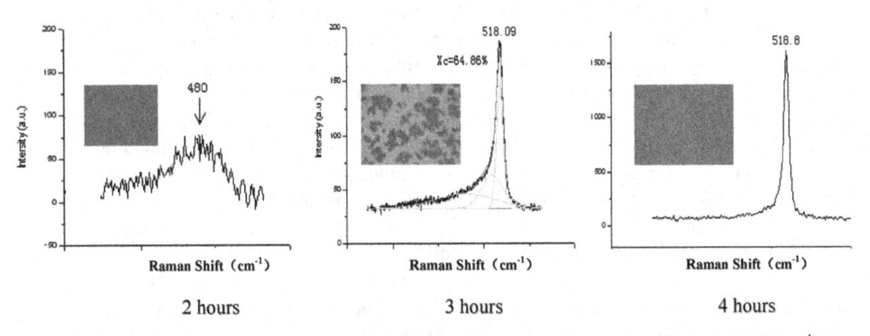

Fig.1 Raman spectrum of the sample after different time annealing Raman Shift (cm^{-1})

Fig.1 shows the Raman spectra of the samples annealed for 2-4 hours in hydrogen plasma. One finds the typical TO mode of a-Si at 480cm^{-1} after 2hs in the hydrogen plasma, which indicated that the sample did not crystallize. After 3 or 4 hours annealing in the same condition, we can see a t peak at about 520 cm^{-1} in both figures. Using three Gaussian curves with peaks at 480 cm^{-1}, 510 cm^{-1}, and 520 cm^{-1} to fit the Raman spectra, the crystalline volume ratio (Xc) was calculated. The peak at 510 cm^{-1} is commonly attributed to the presence of tensile strained Si-Si bonds at grain boundaries [11]. The Xc is given by the ratio of the integrated intensities of the Gaussian curves at 510 cm^{-1} and 520 cm^{-1} to the total integrated intensity. The peaks at 480cm^{-1}

and 520cm^{-1} are observed in the films after 3 hours which indicates that the samples are partly crystallized with Xc =64.8% obtained by a 3-peaks fitting results as shown in the curve labeled "3 hours". After treated 4 hour treatment, no signal around 480cm^{-1} can be observed and the peak is very sharp. This indicates that the samples are totally crystallized. The insert in each Raman figure shows a photo of the surface taken with the microscope of the micro-Raman apparatus. They match up with the crystallization situation shown by the Raman curves.

In order to compare the time needed for completely crystallization of HAIC with normal AIC, we measured Xc vs. annealing time of each sample as shown in Fig. 2. After 6 hours AIC started to crystallize and after 12 hours the crystallization was fully completed. For the curve named "HAIC", the nucleation time is just 2 hours. After 4 hours, the film are crystallized completely while annealing in the H-plasma.. Another feature of HAIC is fast crystal growth as reflected by the rise gradient. That means one obtains short and rapid crystallization by HAIC.

Would the short and rapid crystallization under H-plasma damage the performances of the HAIC poly-Si? We compare the Hall mobility of HAIC and AIC poly-Si materials as shown in Fig. 3. The mobility (42.5 $cm^2/V.s$) of samples crystallized with H-plasma is much higher (nearly in double) than that (22.1 $cm^2/V.s$) when no H-plasma accompanied crystallization.

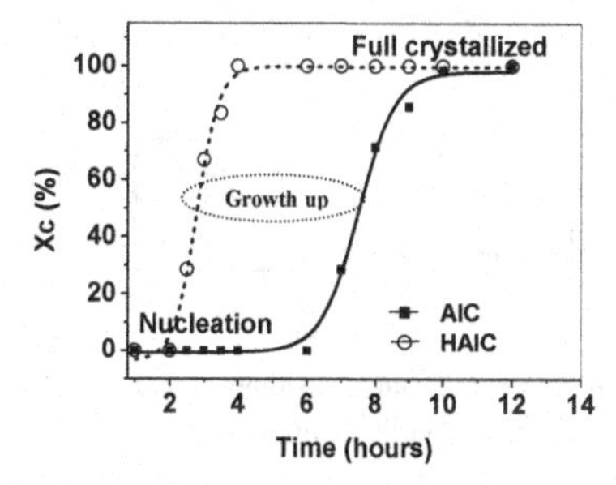

Fig.2 Crystalline volume factor (Xc) as a function of annealing time for samples exposed to hydrogen plasma (HAIC) or not (AIC)

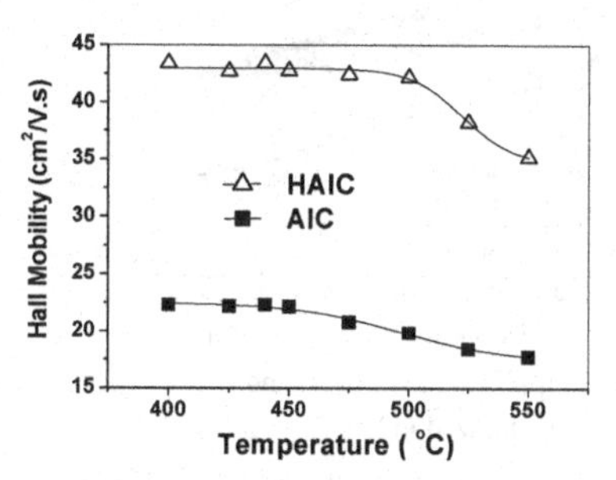

Fig.3 comparison of Hall mobility of the poly-Si crystallized by AIC with (HAIC) or not (AIC)

3. 2. What is the reason for accelerated crystallization in H-plasma?

As mentioned in reference [12], Al-Si alloy is a lower melting point system so Al induced-crystallization of Si involves five important steps. They are the following: Si atoms migrate (step ①) and dissolve into Aluminum (step ②). When the concentration of Si dissolved in Al is higher than its critical concentration then Si will segregate (step ③) and crystallize (step ④) from Al. The Al will also migrate into Si shown as step ⑤ in the opposite direction. This is a layer-exchange process. In this case the rapid migration and dissolution of Si atoms in the Al film will be an important factor for promoting the nucleation and crystallization.

Al atom distributions in fully crystallized poly-Si samples by HAIC and AIC are shown in Fig.4. The Al distribution (red square) in HAIC is steeper than that (blue square) in AIC. Because the Al and Si atoms move in opposite directions, the Al distribution in Si film should be similar to that of Si in Al films. The steeper gradient of Al in HAIC poly-Si implies a steeper gradient of Si in the Al part of HAIC poly-Si compared to that in AIC. The gradient is steeper when the diffusion is faster, so it is favorable to rapidly reach and exceed the critical concentration for nucleation which then shortens the crystallization time. There are abundant H ions (as black Solid Square shown) in HAIC poly-Si (see Fig.4). No H signal H could be found in AIC samples, which indicates that hydrogen plasma radicals could penetrate the Al film accompanied by vacancies formed at the surface [13] and could migrate through the very thin oxide [14-15]. The hydrogen could reach the silicon layer, interact with Si atoms and then passivate the defects in the material during the crystallizing process. Finally Fig.4 shows that the Al content in HAIC poly-Si annealed in H plasma is half an order of magnitude lower than that in AIC poly-Si annealing in vacuum. This means that the H plasma is also active to decrease the Al leftover in poly-Si films Therefore the electron mobility in HAIC poly-Si is much higher than that in AIC samples as shown in Fig.3.

The hydrogen plasma reduces the crystallization time from a-Si to poly-Si thin films. The H-plasma radicals provide an additional thermal budget for crystallization to aid and shorten the crystallization time and to improve the performance of the crystallized poly-Si at the same time.

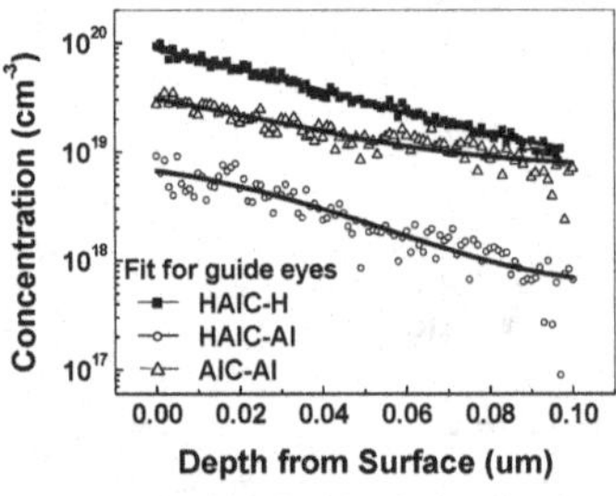

Fig.4 comparison SIMS of Al and H in poly-Si crystallized in H plasma or not

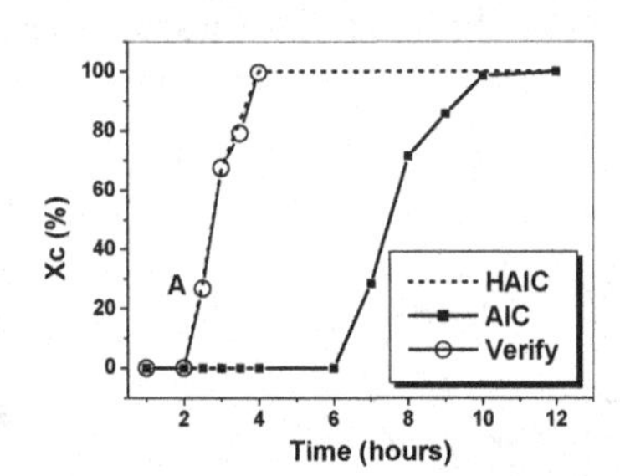

Fig.5 The verification experiment (here curve AIC just for comparison)

3.3. The mechanism of AIC enhanced by H plasma

We further explored a hypothesis to explain the mechanism of the enhancement effect of H-plasma. We first crystallized the sample using Al as an inducing source in hydrogen plasma just for 2.5 hours (point "A" shown in Fig.5), then turned off the plasma power and continued annealing at the same temperature in vacuum for several hrs to complete the crystallization. We measured the Raman spectrum and calculated Xc. The result is shown in Fig.6 which includes HAIC and normal AIC results for comparison. We selected the time of 2.5 hours because it just passed the nucleation process and starts the crystalline growth, as we know from the earlier results.

As shown by Fig. 5, the curves labeled as "HAIC" and "Verify (by open circle)" coincide. After 4 hours, the sample is totally crystallized after annealing for the last 1.5 hours just in vacuum without hydrogen plasma. That means the H plasma has provided enough Si concentration and a gradient of the Si atom distribution in the Al film to maintain subsequent crystalline growth under these experimental conditions. This result supports the view that the hydrogen plasma mainly contributes to Si transport thereby being beneficial to nucleation and grain grow.

The hydrogen radicals with higher energy in plasma are able to move through interstice around strained bonds of Al or Si, and react with Si to form Si-H bonds. The Si–H bond is subsequently easier to break and release single Si atoms because the Si–H bond enthalpy is smaller than that of Si–Si bonds [16]. Single atoms of Si neighboring Al then dissolve in the Al film and to create there a steeper gradient of Si atoms. The steeper slope of Si atoms in the Al film will promote the diffusion of Si and dissolution in Al. The concentration of Si dissolved in the Al film will reach and exceed the critical value of nucleation to form the Si grain nuclei separated out from the Al film thereby shortening the time of forming crystalline-nuclei. The steeper distribution of Al in HAIC shown by SIMS in Fig.5 implies that its steeper slope will also accelerate the layer-exchange process resulting in a higher growth rate. The rapid layer-exchange of Si and Al films will contribute to Al migration out of Si film. At the same time, the H in HAIC will passivate the crystallized poly-Si during the crystallization process; thereby improving the performance of crystallizing poly-Si.

4. CONCLUSION

We propose that aluminum induced crystallization in a hydrogen plasma combines crystallization and passivation into one process. Raman spectroscopy, SIMS and Hall mobility data reveal that it can not only reduce the annealing time of AIC but also enhance the performance of crystallized poly-Si. The mechanism of this technique is mainly that hydrogen plasma radicals penetrate the aluminum and the thick oxide layer, resulting in that more Si atom diffuse into the Al, and reach a critical concentration for nucleation, accelerating thereby crystallization. The growth time is also shortened by hydrogen plasma radicals possibly because of a steeper gradient of Si in the Al layer of HAIC sample than that in AIC.

ACKNOWLEDGEMENTS

This work was supported by the Flat-Panel Display Special Project of National 863 Plan (No.2008AA03A335). We would like to give our greatly appreciate to Prof. Helmut Fritzsche for his revision of this paper.

REFERENCES

1. U. Kroll, C. Bucher, S. Benagli, I. Schönbächler, J. Meier, A. Shah, J. Ballutaud, A. Howling, C. Hollenstein, A. Büchel, M. Poppeller, Thin Solid Films 451, 525 (2004)
2. Y. Nasuno, M. Kondo, A. Matsuda, Sol. Energy Mater. Sol. Cells 74, 497 (2000)
3. F.Demichelis, C.F. Pirri, E.Tresso, J. Appl. Phys 72(4), 1327 (1992)
4. T. Roschek, nanocrystalline silicon solar cells prepared by 13.56MHz, Ph.D. Thesis, IPV, Berichte des Forschungszentrums Jülich, 2003, pp.41-43.
5. S. A. Filonovich, H.Águas, I. Bernacka-Wojcik, C.Gaspar, M.Vilarigues, L.B.Silva, E.Forunato, R.Martins, Vacuum 83, 1253 (2009)
6. T. Matsui, M. Kondo, A. Matsuda, J. Non-crystal. Solid 338-340, 646 (2004)
7. Y. Zhao, X. D. Zhang, F. Zhu, Y.T. Gao, C.C. Wei, J.M. Xue, H.Z. Ren,. D.K. Zhang, G.F. Hou, J. Sun, X. H. Geng, in: 15th International Photovoltaic Science & Engineering Conference (PVSEC-15), Shanghai, PR China, 2005, pp. 65.
8. K. Adhikary, S. Ray, J. Non-Crystal. Solids 353, 2289 (2007)

9. Debajyoti Das, Madhusudan Jana, Mater. Lett. 58, 980 (2004)
10. T. Fujibayashi, M. Kondo, J. Appl. Phys 99, 043703 (2006)
11. T. Matsui, A. Matsuda, M. Kondo, Sol. Energy Mater. Sol. Cells 90, 3199 (2006)
12. X.D. Zhang, F.H. Sun, G.H. Wang, S.Z. Xu, C.C. Wei, G.F. Hou, J. Sun, S.Z. Xiong, X.H. Geng, and Y. Zhao, Phys. Status Solidi C 7(3-4), 1073(2010)
13. G.H. Wang, X.D. Zhang, S.Z. Xu, C.C. Wei, J. Sun, S.Z. Xiong, X.H. Geng, Y. Zhao, Phys. Status Solidi C 7(3-4), 1116(2010)

Mater. Res. Soc. Symp. Proc. Vol. 1321 © 2011 Materials Research Society
DOI: 10.1557/opl.2011.933

Excimer-Laser-Induced Melting and Solidification of PECVD a-Si films under Partial-Melting Conditions

Q. Hu[1], Catherine S. Lee[1, 2], T. Li[1], Y. Deng[1], U.J. Chung[1], A. B. Limanov[1], A. M. Chitu[1], M.O. Thompson[3] and James S. Im[1, 2]

[1]Program in Materials Science and Engineering, Department of Applied Physics and Applied Mathematics, Columbia University, New York, NY, USA

[2]Department of Materials Science and Engineering, College of Engineering, Korea Advanced Institute of Science and Technology, Daejeon, Korea

[3]Department of Materials Science and Engineering, College of Engineering, Cornell University, Ithaca, NY, USA

ABSTRACT

This paper reports on new experimental findings and conclusions regarding the pulsed-laser-induced melting-and-solidification behavior of PECVD a-Si films. The experimental findings reveal that, within the partial-melting regime, these a-Si films can melt and solidify in ways that are distinct from, and more complex than, those encountered in microcrystalline-cluster-rich LPCVD a-Si films. Specifically (1) spatially dispersed and temporally stochastic nucleation of crystalline solids occurring relatively effectively at the moving liquid-amorphous interface, (2) very defective crystal growth that leads to the formation of fine-grained Si proceeding, at least initially after the nucleation, at a sufficiently rapidly moving crystal solidification front, and (3) the propensity for local preferential remelting of the defective regions and grain boundaries (while the beam is still on) are identified as being some of the fundamental factors that can participate and affect how these PECVD films melt and solidify.

INTRODUCTION

Fundamentally, melting of a-Si corresponds to one of the simplest examples of phase transitions that can transpire between two metastable phases in condensed systems. The melting of a metastable solid phase inevitably requires rapid heating (as can, for instance, be accomplished through excimer laser irradiation), since slow heating of the material would kinetically lead to solid-phase transformation of the metastable solid phase into a more thermodynamically stable solid phase.

In general, irradiating amorphous or polycrystalline Si films on SiO_2 using a short-duration laser pulse can lead to several distinct melting and solidification scenarios. It is well established that the process can be characterized in terms of two major regimes (i.e., low-energy-density/partial-melting regime and high-energy-density/complete-melting regime) [1] and a sub-regime (i.e., near-complete-melting/super-lateral-growth regime) [2]. In this paper, we focus our attention on investigating the details associated with melting and solidification of dehydrogenated PECVD a-Si films irradiated within the low-energy-density/partial-melting regime. Whereas the behavior of microcrystalline-cluster-containing LPCVD a-Si films within the partial-melting regime was characterized and explained (in terms of the early and rapid microcrystal-triggered explosive crystallization of the films followed sequentially by uniform remelting and solidification of the explosively crystallized Si layer [3, 4]), a more intrinsic and technologically significant situation involving microcrystalline-cluster-deficient PECVD a-Si films in the partial melting regime has yet to be resolved. Addressing this situation constitutes the main aim of the present investigation. Technologically, the understanding of pulsed-laser-induced melting and solidification of PECVD a-Si films is important in that the usage of the method is rapidly expanding for fabricating thin-film transistors (TFT) utilized in manufacturing advanced flat-panel displays [5].

EXPERIMENTS

Single shot irradiation experiments using a XeCl excimer-laser-based system (308 nm wavelength) were conducted at various energy densities over the entire partial-melting window. A brief schematic diagram of our experimental setup is shown in Figure 1. The temporal profile of the laser pulse, which is shown in Figure 4, consists essentially of a simple Gaussian-like profile (and weak and insignificant residual fluctuations). This constitutes an important experimental element for a couple of reasons: (1) the condition is similar to those which have been used in previous partial melting investigations [6-11], which, in turn, permits more direct comparisons to be made to a number of previously attained experimental results, and (2) such a simple temporal profile leads to simpler transformation scenarios and more rigorous interpretations. *In situ* transformation analysis was performed using front-side as well as back-side transient reflectance measurements, and microstructural characterization of the irradiated films was conducted using TEM and AFM.

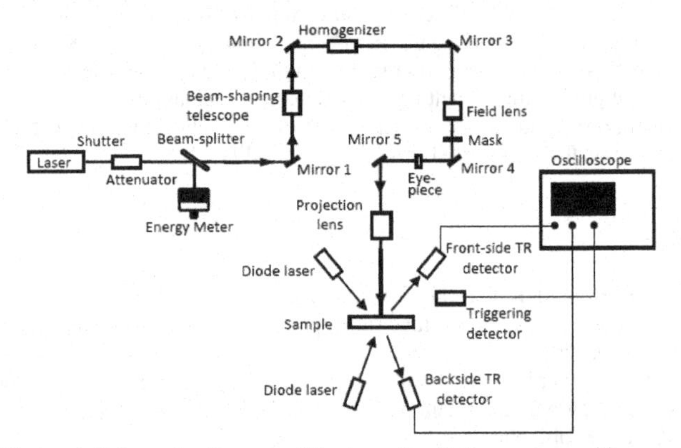

Figure 1. Schematic diagram of the laser irradiation system with *in situ* frontside and backside transient reflectance measurement setup.

RESULTS

The TEM micrographs in Figure 2 indicate that, at a low energy density just above the surface melting threshold (2(a)), the film is converted into a fine-grained (~10nm-sized) polycrystalline microstructure. As the energy density increases, the surface area is covered by radial disk shaped regions (DSRs) (2(b)), which become more prominent with increasing diameter at higher energy densities (2(c),(d)). At even higher energy densities (2(e),(f)), we can notice that the center regions of the DSRs become less defective, the diameter of the DSRs actually decreases, and the apparent density of the discernable DSRs decreases.

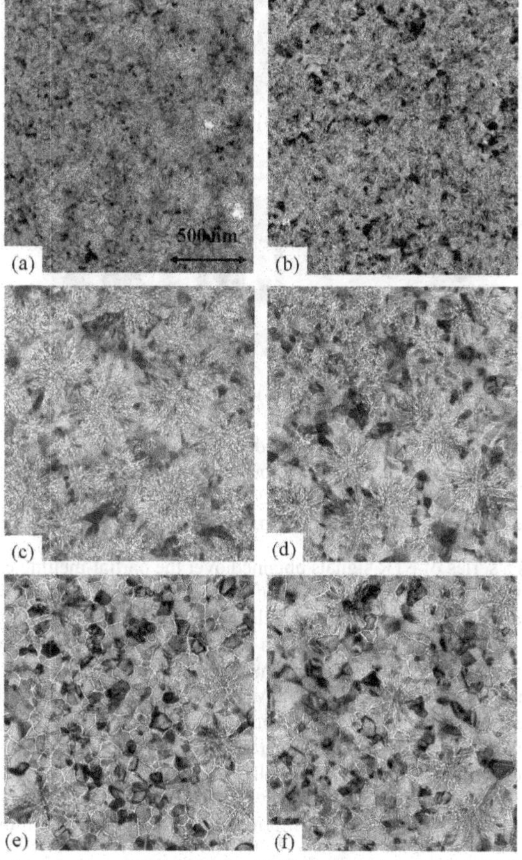

Figure 2. Planar view TEM images of 100nm dehydrogenated PECVD a-Si films single-shot irradiated at various energy densities: (a) 90 mJ/cm^2, (b) 109 mJ/cm^2, (c) 134 mJ/cm^2, (d) 171 mJ/cm^2, (e) 222 mJ/cm^2, (f) 238 mJ/cm^2. Complete melting threshold is 270 mJ/cm^2.

Experimental results obtained from AFM are also consistent with the above TEM-based observations. Figure 3 reveals that at low energy densities, DSRs cover most of the surface area. At higher energy densities, small protrusions start to appear in the center of the DSRs (resulting presumably from localized remelting of defective core regions) as discussed in the next section.

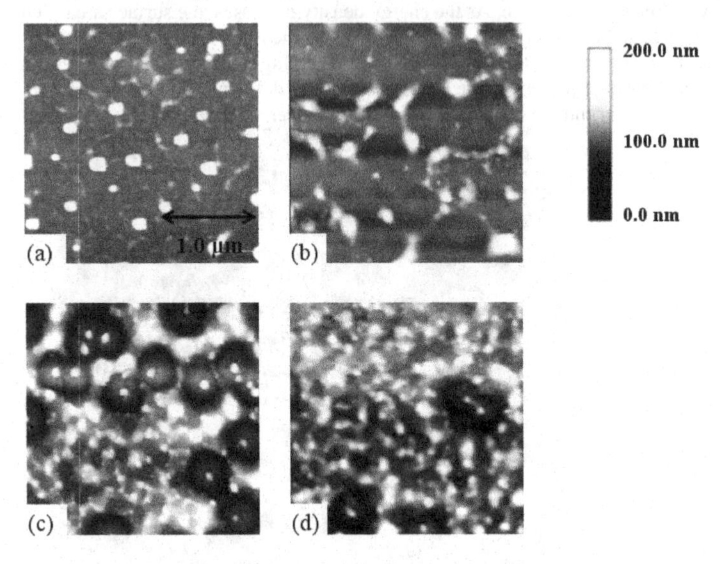

Figure 3. AFM images of 100nm dehydrogenated PECVD a-Si films single-shot irradiated at various energy densities: (a) 108 mJ/cm^2, (b) 150 mJ/cm^2, (c) 222 mJ/cm^2, (d) 238 mJ/cm^2. Complete melting threshold is 270 mJ/cm^2.

Figure 4 shows a series of FTR signals obtained over the partial-melting energy-density window. There are two observations we can identify as being noteworthy : (1) the signals simply rise and decay without exhibiting a prompt and clear oscillations previously observed in melting of LPCVD a-Si films [3, 12]. This difference can be identified as resulting from these PECVD films being devoid of microcrystalline clusters that were identified as being responsible for triggering early explosive crystallization in LPCVD films (which led to the early signal oscillations); (2) at sufficiently higher energy densities, a distinctive shoulder can be identified in the decaying portion of the signals. This observation can be viewed as being supportive of the idea that some and concurrent remelting of initially solidified materials is taking place, even as the films are being "overall" solidified. At energy densities above 247 mJ/cm^2, we observe a plateau in the FTR signal, which indicates uniform and sustained melting of the surface layer.

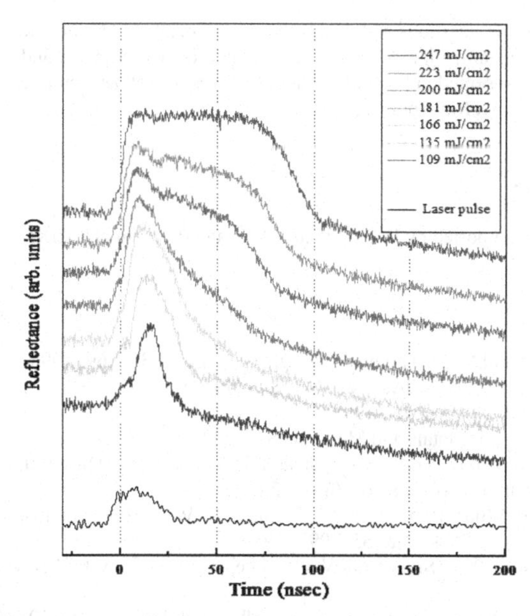

Figure 4. Front-side time-resolved reflectance (in arbitrary units) obtained using 650 nm CW laser. As shown at the bottom of the figure, the laser pulse starts at −10 nsec point in the time scale and the top of the laser pulse occurs at 8.4 nsec point. Complete melting threshold is 270 mJ/cm^2.

DISCUSSION

Based on the above experimental findings and recent progress involving the direct growth model of microcrystalline silicon [13, 14], we identify the following melting and solidification sequence for the experimental conditions employed in the present experiment. (1) Solidification is initiated (in these microcrystalline-cluster-free a-Si films) by temporarily stochastic and spatially dispersed heterogeneous nucleation of crystal Si occurring at the moving interface between liquid and amorphous (during the amorphous Si melting phase), (2) subsequent solidification following nucleation proceeds, at least initially, to generate extremely defective fine-grained Si via the defective mode of crystal growth, and (3) spatially localized preferential remelting and resolidification of defective regions and grain boundaries can take place, surprisingly, while the beam is still on.

We note that the above model can be recognized as being distinct from a number of arguments and models that were previously suggested by other investigators in dealing with the research topics that are related to the present work [3, 15, 16]. Our model appears to be singularly consistent with essentially all experimental results that have been obtained, and, in the process, addresses some of the questions and details that are associated with the initiation and propagation of explosive crystallization. We have carried out more experiments and obtained more findings than can be presented in this paper; these findings as well as a more extended discussion will be provided in a forthcoming paper.

ACKNOWLEDGMENTS

This work was in part supported by WCU Flexible Signage Program at KAIST.

REFERENCES

1. J. S. Im, H. J. Kim, and M. O. Thompson, Appl. Phys. Lett. **63**, 1969 (1993).
2. J. S. Im, and H. J. Kim, Appl. Phys. Lett. **64**, 2303 (1994).
3. J. H. Yoon, (Columbia University, 1998).
4. J. H. Yoon, and J. S. Im, Metals and Materials **5**, 525 (1999).
5. J. W. Hamer, A. Yamamoto, G. Rajeswaran, and S. A. Van Slyke, Digest of Technical papers - SID International Symposium **36**, 1902 (2005).
6. A. Polman, D. J. W. Mous, P. A. Stolk, W. C. Sinke, C. W. T. Bulle-Lieuwma, and D. E. W. Vandenhoudt, Appl. Phys. Lett. **55**, 1097 (1989).
7. A. Polman, S. Roorda, P. A. Stolk, and W. C. Sinke, Journal of Crystal Growth **108**, 114 (1991).
8. W. C. Sinke, A. Polman, S. Roorda, and P. A. Stolk, Appl. Surf. Sci. **43**, 128 (1989).
9. L. Mariucci, A. Pecora, G. Fortunato, C. Spinella, and C. Bongiorno, Thin Solid Films **427**, 91 (2003).
10. M. O. Thompson, G. J. Galvin, M. J. W., P. S. Peercy, J. M. Poate, D. C. Jacobson, A. G. Cullis, and N. G. Chew, Phys. Rev. Lett. **52**, 2360 (1984).
11. G. E. Jellison, and D. H. Lowndes, Appl. Phys. Lett. **51**, 352 (1987).
12. J. S. Im, H. J. Kim, and M. O. Thompson, Appl. Phys. Lett. **63**, 1969 (1993).
13. S. Hazair, P. C. Van Der Wilt, Y. Deng, U. J. Chung, A. B. Limanov, and J. S. Im, Mat. Res. Soc. Symp. Proc. **979**, HH11 (2006).
14. H. S. Cho, D. Kim, A. B. Limanov, M. A. Crowder, and J. S. Im, Mat. Res. Soc. Symp. Proc. **621**, Q9.9.1 (2001).
15. D. H. Lowndes, S. J. Pennycook, G. E. Jellison, Jr., S. P. Withrow, and D. N. Mashburn, Journal of Materials Research **2**, 648 (1987).
16. L. Mariucci, A. Pecora, G. Fortunato, C. Spinella, C. Bongiorno, Thin Solid Films **427**, 91 (2003).

Mater. Res. Soc. Symp. Proc. Vol. 1321 © 2011 Materials Research Society
DOI: 10.1557/opl.2011.1190

Growth of Large Grain Polycrystalline Silicon Thin Film on Soda-lime Glass at Low Temperature for Solar Cell Applications

K. Wang[1] and K. H. Wong
Department of Applied Physics and Materials Research Centre, The Hong Kong Polytechnic University, Hung Hom, Kowloon, Hong Kong, China

ABSTRACT

High quality polycrystalline silicon (poly-Si) thin film solar cell was successfully fabricated on soda-lime glass substrates by electron beam (Ebeam) evaporation at low processing temperature. The initial poly-Si seed layer (p^+-type 0.5 μm thick) was grown via the aluminum induced crystallization (AIC) method at 450 °C. Prominent interdiffusion and Si crystallization have been observed. X-ray diffraction (XRD) shows that (111) is the dominating crystalline orientation. Post annealing at 450 °C for six hours has produced densely packed Si grains with dimension of more than 10 μm in the plane of the film. Non-destructive Raman spectroscopy reveals the remarkable crystalline improvement for samples after thermal treatment. After removing the top diffused Al by chemical means, an absorber layer (p-type) of 0.9 μm thick was subsequently deposited onto the seed layer by Ebeam evaporation at 500 °C. Transmission electron microscopy (TEM) confirmed good homo-epitaxial growth. Without breaking the high vacuum, an n-type amorphous Si (a-Si) layer (0.7 μm thick) was coated onto the absorber layer to form p-n junction. The corresponding I-V characteristics suggest that our low temperature processing technique is applicable for production of poly-Si thin film solar cell on low cost substrates.

INTRODUCTION

Although crystalline silicon (c-Si) wafer is the most widely used photovoltaic materials in commercialized solar cell to date, its expensive purification processes have raised big question on its role for future economic green energy generation [1]. However, Si which is the second most abundant element in Earth's crust and has proven records of high photovoltaic efficiency and long term stability will remain as the most reliable solar cell materials for many years to come [2]. In order to reduce the cost of c-Si based solar cells, the use of polycrystalline Si (poly-Si) thin films in lieu of Si wafer is favored [3, 4]. Conventionally, the phase transition from amorphous to crystalline phase for Si requires temperature as high as 1000 °C. Such high temperature will definitely eliminate the degree of freedom on choosing inexpensive substrates, such as low temperature glass. In this work we aim at developing method to fabricate poly-Si thin films with large grain size (>10 μm) on soda-lime glass substrate at reduced temperature. The technique that we have adopted relies on a combined use of electron beam (Ebeam) evaporation and of metal-induced crystallization (MIC) of Si. Previous research has demonstrated that many metals, such as gold (Au) [5], nickel (Ni) [6], and aluminum (Al) [7], are effective with this method. The conventional MIC method requires long processing time and the poly-Si grain size achieved is relatively small. In our experiment, we chose Al because of its availability and being a group III element. Our scheme and the physical mechanism behind the Al-induced crystallization (AIC) of poly-Si are briefly explained in figure 1. A soda-lime glass

[1] Corresponding author: E-mail: wangkai369@hotmail.com. (K. Wang).

was used as a substrate (figure 1(a)). The evaporated Al forms a layer on the glass substrate at room temperature (figure 1(b)). Si was then deposited onto the Al layer at 450 °C. Meanwhile, the solid solution of Al and Si forms at substrate temperature below their eutectic temperature [8-10]. The formation of the solid solution is due to ineterdiffusion of Al and Si at 450 °C. Despite the formation of poly-Si on the glass (figure 1(c)), the co-existence as-coated a-Si suggests post-annealing at 450 °C is necessary for complete interchange of Al and Si (figure 1(d)).

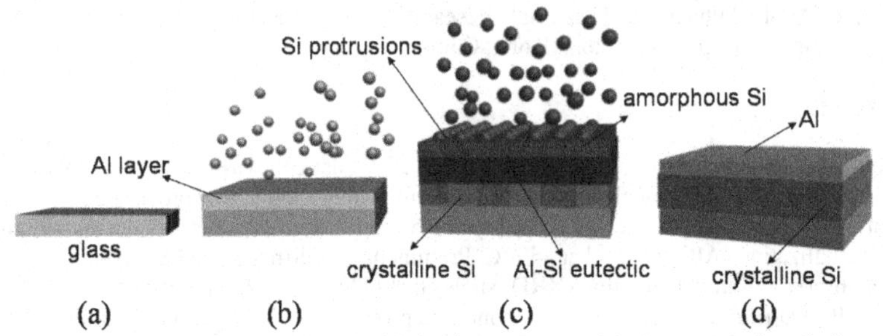

Figure 1. Schematic diagrams of AIC fabrication process. (a) As-prepared soda-lime glass substrate. (b) The deposition of Al layer at room temperature. (c) The deposition of Si at 450 °C. (d) The final stage of AIC process.

EXPERIMENTAL DETAILS

In the experiment, the soda-lime glass substrates were ultrasonically cleaned with acetone, ethanol and de-ionized water. Film fabrication was carried out in the Ebeam system under high vacuum (1×10^{-6} Torr). Both Al pellets (99.999% in purity) and intrinsic, p-type and n-type Si flakes (99.999% in purity) were bought commercially. Al with thickness of 200 nm was initially coated onto glass as room temperature. Without breaking the high vacuum, the intrinsic Si with thickness of 0.5 μm was deposited onto Al thin film at the substrate temperature of 450 °C. The in-situ post annealing commenced after completing the deposition of Si. A mixture of acids, 70 ml H_3PO_4, 5 ml HNO_3, 10 ml CH_3COOH and 15 ml DI-H_2O, was used to remove the top diffused Al layer. X-ray diffraction (XRD CuKα, $\lambda = 1.541$ Å) and non-destructive micro-Raman Spectroscopy ($\lambda = 488$ nm) were used to investigate the crystalline phase and crystallinity of poly-Si. The sample surface and the layer cross section were studies by Field Emission-Scanning Electron Microscopy (FE-SEM) and Transmission Electron Microscopy (TEM). The Hall measurement of the 0.5 μm poly-Si thin film was performed after the out-diffused Al was removed. Based the seed layer growth, a relatively thick p-type Si layer of about 900 nm in thickness was deposited at 500 °C. Then an n-type a-Si layer with 700 nm in thickness was subsequently coated at room temperature. The corresponding epitaxial film quality and current-voltage (I-V) characteristic were examined.

RESULTS AND DISCUSSION

Figure 2 shows the FE-SEM images of our samples grown on soda-lime glass substrates. On the Al coated glass, the as-deposited Si atoms experience interface migration at 450 °C. The Si atom is screen by the free electrons in Al which leads to the reduction of binding energy [11]. The excess energy is utilized for prompting the dissolution of Si atoms. Once the supersaturation of Al and Si eutectic is satisfied the precipitated Si atoms coalescence on the glass substrate. The accumulation of Si atoms leads to nucleation. As we can see from figure 2(a), c-Si appears close to the substrate. Due to fast deposition rate of Si, a thin a-Si layer is present on the top of the sample. The formation of Si protrusions is due to the surface tension. Between the c-Si and the a-Si, a broad eutectic phase of Al and Si can be observed. With 2 hours post annealing at 450 °C, the formation of large Si grains can be seen from figure 2(b). The average grain size is greater than 1 μm. Figure 2(c) shows the top view after etching away some partially out-diffused Al. The features of Si grains are made visible with residual Al at the grain boundaries. At this stage, the diffusion of Al along the surface or the grain boundaries of the poly-Si is called short-circuit diffusion since it is much faster than the diffusion of Al by passing through in poly-Si grains (lattice diffusion). Therefore, the diffusion mechanism here includes both surface and grain boundary diffusions; and, the correlated diffusion coefficient depends on the dimension of Si grain [12]. Prolonged annealing for 3 hours, as revealed in figure 2(d), can further enlarge the overall Si grain size. For sample with 6 hours annealing, the entire catalytic Al has diffused to the top surface. The corresponding FE-SEM image is displayed in figure 2(e). In figure 2(f), a smooth poly-Si thin film forms a sharp interface with soda-lime glass. There is no residual Al at the grain boundaries. The average poly-Si grain size exceeds 10 μm [13].

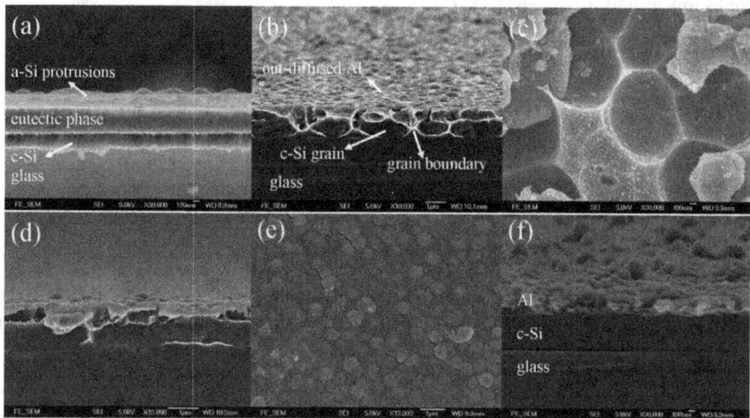

Figure 2. FE-SEM images of AIC fabrication process. (a) 500 nm Si fabricated onto 200 nm Al coated glass at 450 °C. (b) After 2 hours post annealing. (c) Top-view of poly-Si grains. (d) After 3 hours post annealing. (e) Top-view for the Al protrusions. (f) After 6 hours post annealing.

Further evaluation of the poly-Si thin film crystalline quality involves TEM characterization. The TEM image of the layer cross section was captured and displayed in figure 3(a). Selected area diffraction pattern (SADP) for each layer was taken and displayed in the insets. The white arrows connect the layers with their corresponding SADP. For the catalytic Al layer, three diffraction spots, (220), (002) and (111), have been indexed.

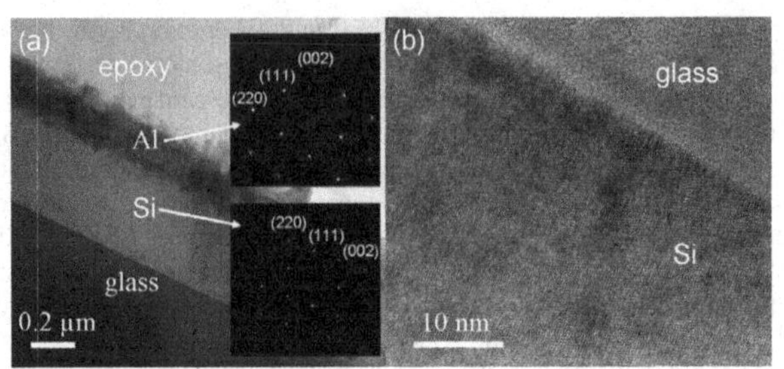

Figure 3. TEM images of AIC fabricated poly-Si thin film. (a) Crossing sectional view of glass/poly-Si (500 nm)/ Al (200 nm), the insets show the selected area diffraction patterns for both poly-Si and Al layers. (b) High resolution-TEM image for as-grown poly-Si thin film.

Similar calculation was made for the poly-Si layer. Three representative diffraction spots for face central cubic (f.c.c) structure of poly-Si are labeled. Figure 3(b) displays the high resolution-TEM (HR-TEM) image. The lattice arrangement of Si has been clearly revealed. There is no impurity phase at the poly-Si and soda-lime glass interface. Both SADP and HR-TEM results suggest that our fabrication technique can produce high quality poly-Si on soda-lime glass at temperature as low as 450 °C.

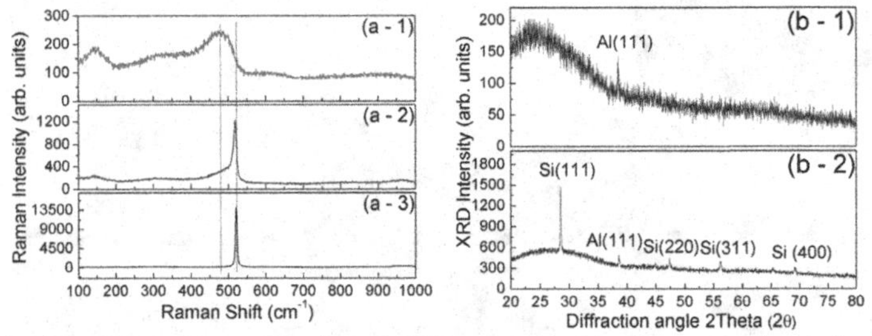

Figure 4. Micro-Raman spectroscopic and XRD spectroscopic characterizations of poly-Si. (a) Micro-Raman spectroscopic studies for (a-1) a-Si, (a-2) AIC fabrication without post annealing, and (2-3) AIC fabrication with 6 hours annealing. (b) XRD studies for (b-1) sample fabricated at room temperature, (b – 2) sample fabricated by AIC with 6 hours post annealing at 450 °C.

In addition to the microscopic studies, we also use non-destructive Raman spectroscopy to evaluate the Si thin films prepared under different conditions. Figure 4(a) shows the micro-Raman spectra for samples fabricated at room temperature and at 450 °C with and without post annealing. In figure 4(a-1), of which the sample was fabricated at room temperature, a dominating band centered at around 480 cm^{-1} with a full width at half maximum (FWHM) of 113.88 cm^{-1} is the a-Si phase. It is ascribed to the transverse optical (TO) mode. Another distinct peak for the a-Si at 150 cm^{-1} is associated with transverse acoustic (TA) vibrational mode. There is no discernable Si crystalline phase. Figure 4(a-2) shows the Raman spectrum for sample fabricated at 450 °C without post-thermal treatment. A sharp but asymmetric Raman peak occurs at 518 cm^{-1}. It represents the zone-center longitudinal optical (LO) and transverse optical (TO) phonon mode in c-Si. The Al gives rise to p-type Si formation. The Raman shift is smaller than the one of undoped single c-Si. The FWHM of this band is 10.44 cm^{-1}. Owing to the coexistence of amorphous phase, the calculated crystalline volume based on the intensity ratio of the peaks at 480 cm^{-1} and 518 cm^{-1} is approximately equal to 0.58. For the sample that has been post annealed at 450 °C for 6 hours we found that the Si was fully crystallized. As shown in figure 4(a-3), there is no observable amorphous phase. The FWHM is decreased to 5.00 cm^{-1} and the dominating Raman shift is exactly at 520 cm^{-1} of an undoped single c-Si. Figure 4(b) shows the XRD spectra for samples fabricated at room temperature and at 450 °C with 6 hours post annealing. In figure 4(b-1), the glass substrate results in a broad diffracted band extends from 20° to 35°. The X-ray diffraction at $2\theta = 38.47$ originates from Al (111) crystalline planes. There is no trace of c-Si peak. By contrast, figure 4(b-2) reveals out-of-plane crystalline orientations for the Si fabricated by AIC method. The dominating crystalline orientation is (111).

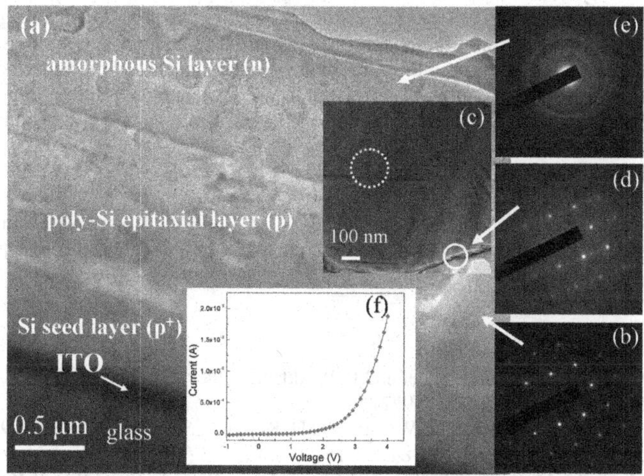

Figure 5. TEM image of a prototype poly-Si thin film solar cell. (a) Crossing sectional view of the solar cell. (b) SADP for the expitaxial layer. (c) LM-TEM image captured from poly-Si grain boundaries. (d) SADP taken from poly-Si grain boundaries. (e) SADP for the n-type a-Si layer. (f) IV-characteristic curve.

Owing to the use of catalytic Al during the poly-Si seed layer growth, this gives rise to the heavily doped p-type (p^+-type) Si formation [14]. Hall measurement of this 0.5 μm poly-Si layer revealed a resistivity (ρ), hole concentration (n_h) and mobility (μ_h) to be 3.28×10^{-2} Ω·cm, 5.81 $\times10^{18}$ cm^{-3} and 34.11 $cm^2 \cdot V^{-1} \cdot s^{-1}$ respectively [14, 15]. In order to demonstrate the feasibility of our poly-Si thin film fabrication technique for future solar cell application, a p-n junction structure partially based on poly-Si thin film was fabricated on an ITO coated soda-lime glass substrate. The TEM image of the films cross section is shown in figure 5 (a). The Si seed layer growth is based on the AIC technique. The fabrication temperature for such seed layer was 450 °C and the post-annealing was carried out at the same temperature. Then, a p-type poly-Si absorber layer was epitaxial grown on the p^+-type poly-Si seed layer at 500 °C via Ebeam evaporation. Excellent crystalline quality of the poly-Si epitaxial layer was confirmed by SADP (figure 5 (b)). Figure 5 (c) shows the low magnification-TEM (LM-TEM) image taken around the grain boundaries (solid white circle). The corresponding SADP is displayed in figure 5 (d). Occasionally, we observed the appearance of dislocations which has been highlighted by white dotted circle. Subsequently, an n-type a-Si layer was coated onto the p-type expitaxial layer at room temperature by Ebeam evaporation in order to form a p-n junction. Figure 5 (e) displays the SADP of this n-type layer. In figure 5 (f), good rectifying profile has been clearly demonstrated.

CONCLUSIONS

In summary, we have shown that high quality poly-Si thin film can be fabricated on soda-lime glass at 450 °C by Ebeam evaporation. Excellent crystalline quality of the as-prepared poly-Si thin film has been verified by TEM and micro-Raman spectroscopy. The dominating grain orientation is Si (111). We have shown that grain size of more than 10 μm can be achieved. Such layer can be utilized as an epitaxial seed layer. From the structural quality of the as-fabricated solar cell and its corresponding IV-characteristics, our low temperature poly-Si thin film fabrication technique can be used for cost-reduction and large-scale solar cell production.

ACKNOWLEDGEMENTS

This work was partially supported by a research grant of The Hong Kong Polytechnic University (PolyU J-BB9Q). Kai Wang is grateful for the award of a research studentship by The Hong Kong Polytechnic University.

REFERENCES

1. A. Müller, M. Ghosh, R. Sonnenschein and P. Woditsch, *Mater. Sci. Eng. B* **134**, 257 (2006).
2. A. L. Hammond, *Science* **178**, 732 (1972).
3. A. Shah, P. Torres, R. Tscharner, N. Wyrsch and H. Keppner, *Science* **285**, 692 (1999).
4. M. Green, *Appl. Phys. A: Mater. Sci. & Process.* **96**, 153 (2009).
5. Z. Tan, S. M. Heald, M. Rapposch, C. E. Bouldin and J. C. Woicik, *Phys. Rev. B* **46**, 9505 (1992).
6. Z. Jin, G. A. Bhat, M. Yeung, H. S. Kwok and M. Wong, *J. Appl. Phys.* **84**, 194 (1998).
7. O. Nast and A. J. Hartmann, *J. Appl. Phys.* **88**, 716 (2000).
8. E. A. Guliants and W. A. Anderson, *J. Appl. Phys.* **89**, 4648 (2001).
9. D. He, J. Y. Wang and E. J. Mittemeijer, *J. Appl. Phys.* **97**, 093524 (2005).
10. Z. M. Wang, J. Y. Wang, L. P. H. Jeurgens and E. J. Mittemeijer, *Phys. Rev. B* **77**, 045424 (2008).
11. A. Hiraki, *Surf. Sci. Rep.* **3**, 357 (1983).
12. H. Gleiter and B. Chalmer, *Prog. Mater. Sci.* **16**, 77 (1972).
13. H. Kim, D. Kim, G. Lee, D. Kim and S. H. Lee, *Sol. Energy Mater. Sol. Cells* **74**, 323 (2002).
14. S. F. Gong, H. T. G. Hentzell, A. E. Robertsson, L. Hultman, S.-E. Hornstrom and G. Radnoczi, *J. Appl. Phys.* **62**, 3726 (1987).
15. S. Huang, S. Xu, Q. Cheng, J. Long and K. Ostrikov, *Appl. Phys. A: Mater. Sci. & Process.* **97**, 375 (2009).

Mater. Res. Soc. Symp. Proc. Vol. 1321 © 2011 Materials Research Society
DOI: 10.1557/opl.2011.808

Non-melt Laser Thermal Annealing of Shallow Boron Implantation for Back Surface Passivation of Backside-Illuminated CMOS Image Sensors

Zahra AIT FQIR ALI-GUERRY[1,2], Karim HUET[3], Didier DUTARTRE[1], Rémi BENEYTON[1], Daniel BENSAHEL[1], Philippe NORMANDON[1] and Guo-Neng LU[2]
[1]STMicroelectronics, 850 rue Jean Monnet, 38920 Crolles, France
[2]Institut des Nanotechnologies Lyon (INL – site UCB), UMR 5270, Bât. Léon Brillouin, Université Lyon1, 43 bd du 11 novembre 1918, 69622 Villeurbanne Cedex, France
[3]Excico, 13-21 Quai des Gresillons, 92230 Gennevilliers, France

ABSTRACT

Back surface passivation is one of the major challenges in the backside illuminated sensor technology. Ion implantation followed by non-melt pulsed Laser Thermal Annealing (LTA) has been identified as a promising candidate to address this issue. In this work, a shallow B-doped layer is implanted at the backside, further activated using LTA in the non-melt regime. LTA process effectiveness in terms of crystal damage recovery as well as dopant diffusion and activation is studied through room-temperature photoluminescence, Secondary Ion Mass Spectroscopy and four-point probe sheet resistance. These studies demonstrate that non-melt LTA with multiple pulses induces high activation without visible diffusion with an effective curing of the implantation-induced crystalline defects. This is made possible thanks to a sub-microsecond process timescale coupled to a reasonable number of shots as shown by thermal simulations and simple diffusion estimations.

INTRODUCTION

Advanced CMOS Image Sensors (CIS) manufacturers are seeking new architectures in order to decrease pixel size while maintaining or enhancing electro-optical performances. For this purpose, Backside Illumination (BSI) technology is one of the most promising candidates. The major challenges for this technology are pixel crosstalk and high dark current level due to the backside surface generation/recombination centers. Therefore, the creation of a junction at the backside reducing dark current and further crosstalk by drifting the photo-generated carriers to the photodiode region appears as a key process step for introducing this technology [1-2]. However, this junction should be thin enough to improve blue wavelength sensitivity.

Several studies have shown that pulsed LTA in the nanosecond regime is particularly well-suited for back-surface process developments of BSI-CIS, which require very low thermal budgets to avoid any damage of the buried device and metal layers [2-3]. Indeed, the back-surface passivation can be realized by low-energy implantation followed by very short and localized heating provided by the LTA process. In the melt regime, box shaped profiles with activation rates close to 100% and excellent surface uniformity were obtained, which were suitable for backside passivation. In the non-melt regime, preliminary results also showed some potential, especially for multiple pulse conditions. Indeed, high level of electrical activation with practically no visible boron diffusion can be achieved thanks to the globally low thermal budget in the sub-microsecond time scale.

In this work, we investigate the implementation of highly doped, ultra shallow Boron implanted layers using the multi-pulsed non-melt LTA. The characteristics of the formed junctions were studied using room-temperature photoluminescence (RT-PL) mapping, sheet resistance (R_s) measurements and secondary ion mass spectroscopy (SIMS) analysis. The results were qualitatively interpreted thanks to thermal simulations and dopant diffusion calculations.

EXPERIMENT

The analyzed samples are 4 µm-thick p-type SOI wafers implanted with 2 keV B+ ions with the doses of $1x10^{13}$, $3x10^{13}$, $1x10^{14}$ and $3x10^{14}$ at/cm². Thermal treatment was realized with Excico LTA equipment, based on UV (λ = 308 nm, pulse duration < 200 ns) laser head, with a maximum beam size of 15x15 mm^2. Energy densities (ED) are given relatively to the Si melting threshold condition. Each spot was treated using single or multi-pulsed irradiation, with laser ED ranging from 80% to 95% of Si melting threshold. The multiple pulse conditions were carried out with a repetition rate below 10 Hz. PL measurements were carried out using SiPHER® (Silicon Photo Enhanced Recombination) tool which provides an integrated PL intensity over an energy range of [0.6 - 1.2 eV]. PL measurements were performed by scanning the sample surface with a focused green laser beam of 1 µm spot size (λ ~ 532 nm, P ~ 15,47 mW) using the micro-scan method. The micro-scans were done inside each annealed zone on 100x100 µm² surface areas with 0.5 µm scan step. Dopant profiles were analyzed using CAMECA tool with 1keV oxygen beam. Doping activation was estimated by measuring sheet resistance (Rs) with a four-point probe method. The measured resistance for boron-doped layers is given by:

$$Rs = \frac{\rho}{e} = \frac{1}{eq\mu p}$$

where ρ is the Resistivity (Ohm.cm) and e the doped layer thickness that can be extracted from SIMS analysis. The Rs measurements provide the activated dose of the annealed layer. The dopant activation rate is estimated as the ratio of the activated dopant dose to the implanted dose.

RESULTS AND DISCUSSION

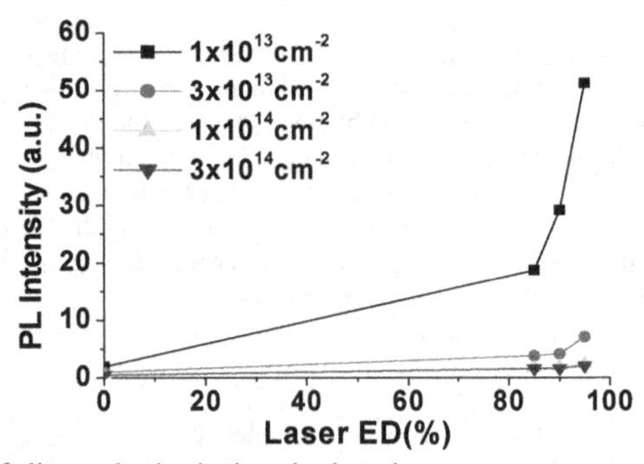

Figure 1. PL intensity of all samples in single pulsed mode.

Fig. 1 shows the PL intensity in the single pulse case as a function of ED in the non-melt regime. Important PL enhancements were observed, especially for the lowest implanted dose. Indeed, higher doses induce much more damage, which therefore requires higher thermal budget to be significantly cured.

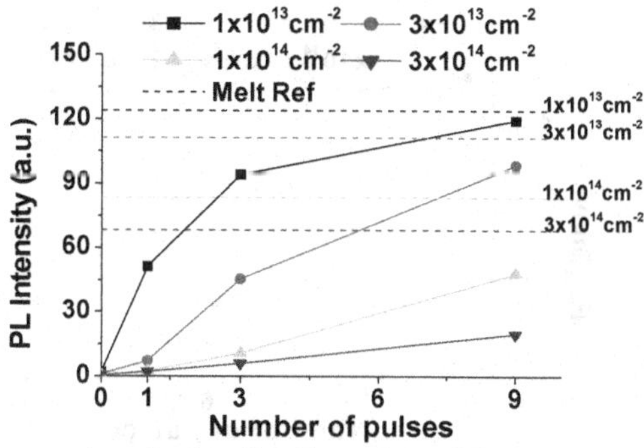

Figure 2. PL intensity vs. number of pulses for all samples at 95% LTA condition. Dashed lines: PL intensities obtained in the melt regime for each dose.

Fig. 2 shows PL intensity results as a function of the number of pulses in non-melt regime for all samples. Melt regime references are reported for each case for comparison. As it can be seen, the PL intensity increases as the number of laser pulses increases. Results close to the melt reference are observed, especially in the lowest dose case. This behavior shows that the multi-pulsed annealing is more efficient than the single-pulsed one in terms of damage-curing [4]. Thus, by allowing a higher overall thermal budget while keeping the buried structures at low temperatures, the multi-pulsed mode in non-melt regime would be an answer for curing implantation-induced crystalline damage.

Sheet resistance measurements are shown in Fig. 3 and Fig. 4 as a function of ED and number of pulses. Melt regime references are reported in Fig. 4. As shown in Fig. 3, Rs decreases with increasing the ED. Furthermore, as shown in Fig. 4, for a given ED, Rs exhibits much better electrical activation in the multi-pulse case compared to single-pulse one [5]. Indeed, the activation rate reached for the 3e13 cm^{-2} dose undergoing 90% ED is about 45% with 9 pulses compared to some 27% for 1 pulse. These activation rates are still low when compared to the melt regime. However, as shown by the PL study, most of the crystalline defects are cured, which makes this kind of process suitable for backside junction formation.

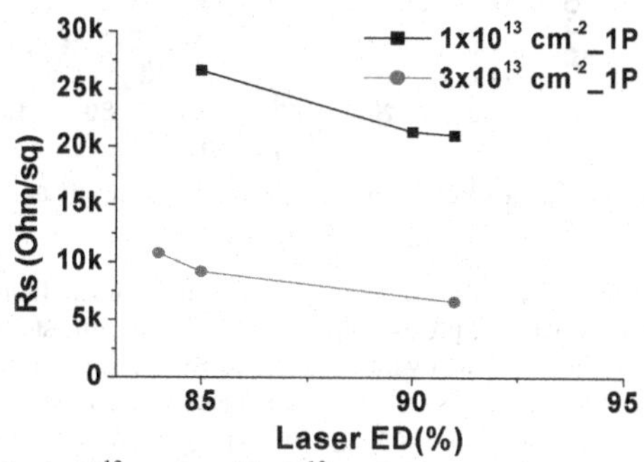

Figure 3. Rs vs. laser ED for $1x10^{13}$ cm^{-2} and $3x10^{13}$ cm^{-2} doses for single-pulse LTA.

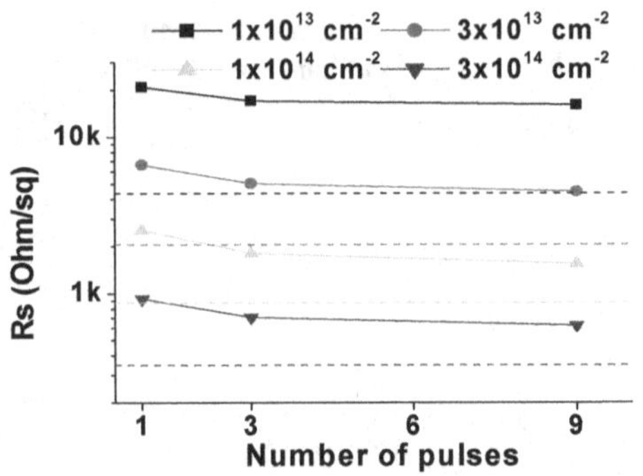

Figure 4. Rs vs. number of pulses at 90% LTA condition for (from top to bottom) $1x10^{13}$, $3x10^{13}$, $1x10^{14}$ and $3x10^{14}$ cm^{-2} doses. The dashed lines represent the Rs values obtained in the melt regime for each dose.

Boron SIMS profile in the $3x10^{13}$ cm^{-2} case are shown in Fig. 5 for various non-melt LTA conditions. It can be seen that non-melt annealing does not alter the implanted profile even in the multi-pulsed case, thus demonstrating the diffusion-less character of the LTA process. A shallow junction depth of about 35 nm (extracted at 10^{18} cm^{-3} concentration) has been achieved. As shown previously [6], in the melt regime, the junction is defined by the melting depth, which in turn is strongly related to the laser energy density. Here, in the non-melt case, the junction depth is mainly limited by the implantation conditions.

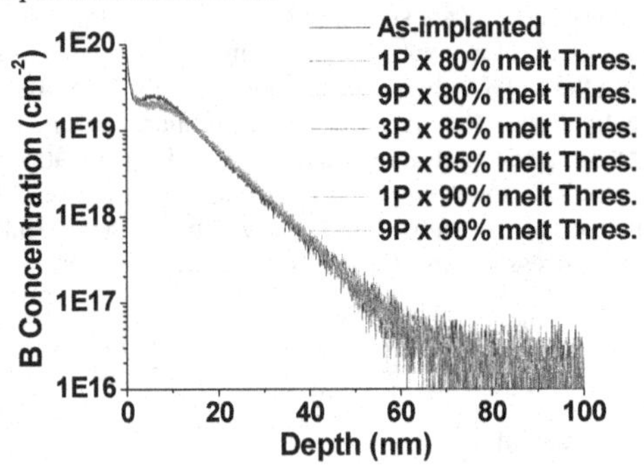

Figure 5. SIMS profiles on the 3e13 cm^{-2} implanted sample for single and multi-pulsed non-melt process.

Fig. 6 shows the thermal simulation results near the melt threshold in the single pulse case. As it can be seen on the temperature profiles shown for different time steps, the high temperature region is limited to the surface and the overall process timescale is sub-microsecond. Typically, as shown in the insert of Fig. 6, the first 100 nm are subjected to temperatures above 1300°C for about 100 ns. This kind of process is fast enough to keep the thermal budget low and leave the buried structures undamaged. In the multiple pulse process cases, the repetition rate is typically below 10 Hz (>100 ms). Given the great difference in timescales, the heat generated by one pulse is already dissipated when the next pulse begins and there is therefore no preheating effect. However, the Rs and PL results indicate that there is a cumulative effect which leads to an increase in activation and defect curing with the number of pulses.

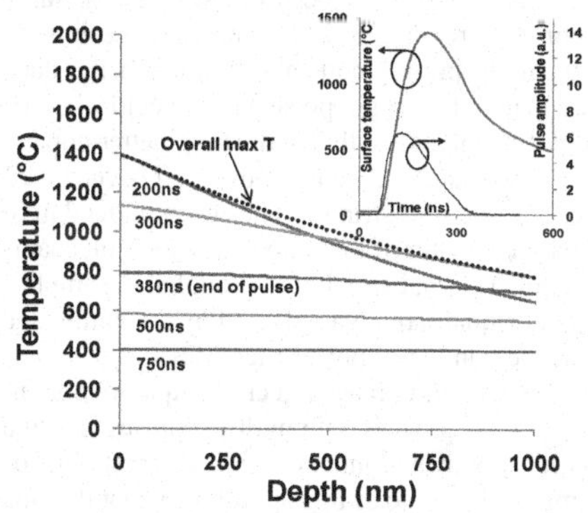

Figure 6. Simulated temperature profiles at different times of the process (t=200, to 750 ns). Dashed line: maximum temperature reached. Insert: Surface temperature vs. time superposed to a typical laser pulse shape.

B diffusion length results are shown in Fig. 7, obtained with simple calculations taking into account Boron diffusion in a limited dopant source with a Gaussian-like profile [7]. The cumulative effect can be qualitatively understood thanks to the diffusion induced by the thermal budget applied by each pulse. Indeed, the single pulse process enables the B atoms to diffuse up to 2 Å, which is in the same order of magnitude as the Si lattice constant. This may lead to some limited activation. In the multiple pulse case, considering that the diffusion times cumulate for each pulse, the diffusion length may reach values greater than the interatomic distance, as shown here for the 6 and 9 pulse cases. This may explain the cumulative effect observed. It should be noted that this behavior is favored by the use of longer pulse duration than other Excimer laser sources, which are typically closer to 30 ns.

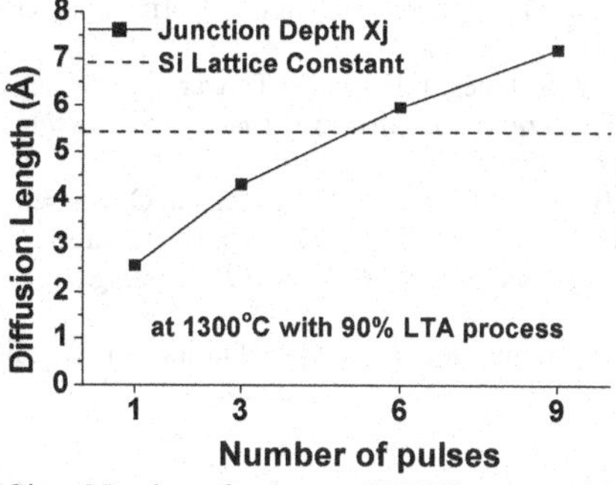

Figure 7. B - Diff. length in Si vs. Number of pulses at 1300°C generated by 90% LTA process.

CONCLUSIONS

Shallow junction formation using dopant implantation and pulsed LTA in the non-melt regime for the backside passivation of BSI-CIS application has been investigated. The effect of non-melt LTA on boron implanted layers has been investigated for various implantations and anneal conditions. In the non-melt regime, photoluminescence results close those obtained in the melt regime indicate that the multi-pulsed mode allows much better damage curing and crystalline quality recovery than in the single pulse case. Sheet Resistance has been shown to decrease as the number of pulses increases, thus indicating better dopant activation. An activation rate of about 45% was achieved using 9 pulses LTA versus 27% in the single pulse case. Furthermore, no visible boron re-distribution has been noticed in SIMS analysis, demonstrating the solid-phase character of the recrystallization mechanism and the diffusion-less dopant activation achievement. Thermal simulations coupled to sample diffusion calculations confirm that such results can be qualitatively explained by the cumulated thermal budget applied to the implanted junction as the number of pulses increases.

To sum up, although the activation rates and crystal quality obtained are not as good as in the melt regime, the non-melt LTA process with multiple pulses is a suitable candidate for the formation of shallow and highly activated junctions, with a crystalline quality suitable for good device performance. The main advantage of this approach is that the final dopant profile remains similar to the as-implanted profile. Therefore, it is a promising solution to obtain a good back-surface passivation, while maintaining electrical performances of back-illuminated sensors in terms of dark current and short wavelengths sensitivity.

REFERENCES

1. R.C. Westhoff, B.E. Burke, H.R. Clark, A.H. Loomis, D.J. Young, J.A. Gregory, R.K. Reich, *Proc. SPIE*, edited by E. Bodegom and V. Nguyen, San Jose, CA (2009), Vol. 7249, p. 1-11.
2. H. Bourdon, A. Halimaoui, J. Venturini, F. Gonzatti, D. Dutartre, *15th International Conference on Advanced Thermal Processing of Semiconductors*, Catania (2007), p. 275.
3. Z. Ait Fqir Ali-Guerry, M. Marty, R. Beneyton, N. Moussy, J. Venturini, K. Huet, G.N, Lu, D. Dutartre, *21th IEEE International Conference on Microelectronics,* Marrakech (2009), p. 386-389.
4. Z. Ait fqir ali-Guerry, D.Dutartre, R. Beneyton, P. Normandon, G.-N. Lu, *Sens Lett, (to be published)*
5. S. Earles, M.K. Law, K.S. Jones, J. Fraser, S. Talwar, D. Downey and E. Arevalo, *12th IEEE international Conference on Advanced Thermal Processing of Semiconducteurs*, Oregon (2004), p. 143-147.
6. K. Huet, R. Lin, C. Boniface, F. Desse, D.H. Petersen, O. Hansen, N. Variam, A. La Magna, M. Schuhmacher, A. Jensen, P.F. Nielsen, H. Besaucele, J. Venturing, *17th IEEE International Conference on Advanced Thermal Processing of Semiconductors*, Gainesville (2009).
7. S. Fransilla, Diffusion, in Introduction to Microfabrication, 2nd edition, edited by Wiley & Sons, 2010, p. 165-172.

Thin Film Silicon Alloys

Mater. Res. Soc. Symp. Proc. Vol. 1321 © 2011 Materials Research Society
DOI: 10.1557/opl.2011.1191

Effect of Substrate Temperature on Hardness and Transparency of SiOC(–H) Thin Films Synthesized by Atmospheric Pressure Plasma Enhanced CVD Method

Mayui Noborisaka[1], So Nagashima[1], Hidetaka Hayashi[2], Naoharu Ueda[2], Kyoko Kumagai[2], Akira Shirakura[3], and Tetsuya Suzuki[1]
[1]Center for Science of Environment, Resources and Energy, Graduate School of Science and Technology, Keio University, 3-14-1 Hiyoshi, Kohoku-ku, Yokohama 223-8522, Japan
[2]Research and Development Center, Toyota industries Co., Japan, 8, Chaya, Kyowa-cho, Obu-shi, Aichi 474-8601, Japan
[3]Kanagawa Academy of Science and Technology, 3-2-1 Sakado, Takatsu-ku, Kawasaki 213-0012, Japan

ABSTRACT

Silicon-based films have gained much interest as protective coatings for transparent polymeric materials. In this study, SiOC(–H) thin films were deposited on polycarbonate (PC) or Si substrates from trimethylsilane (TrMS) gas diluted with He gas by atmospheric pressure plasma enhanced CVD (AP-PECVD) method with varying substrate temperature, and transparency and hardness of the films were investigated. The films exhibited a good optical transparency with an optical transmittance of about 90% irrespective of the substrate temperature, and the hardness increased from 0.6 to 1.3 GPa as the substrate temperature increased from 60 to 140°C. The results are discussed in terms of chemical structural changes in the films according to the substrate temperature.

INTRODUCTION

Transparent polymeric materials such as polycarbonate (PC) and polyethylene terephthalate (PET) have attracted much attention and been employed in industry as substitutes for glass owing to their excellent properties (e.g. lightweight, highly transparent). For example, polymer substrates are currently used for engineering materials such as eyewear lenses and liquid crystal displays for cell-phones [1,2]. PC is one of the leading candidates for a lightweight material because of their thermal resistance [3]; however, since PC is inferior to glass in hardness and scratch resistance and can be easily damaged [4], it should be improved in mechanical properties to be used as a substitute for glass.

Coating PC with protective films is an effective approach to improving the properties. Silicon-based films such as SiC, SiN and SiO_2 as protective coatings for polymers have been studied actively along with an increase of industrial application of polymeric materials [5-7]. Rats *et al.* and Damansceno *et al.* investigated the mechanical properties of silicon-based films deposited on PC by PECVD method. PECVD is a commonly used technique with simple but powerful control of the deposition and generally operated at low pressure. However, the vacuum technique has proven to be costly because of the expensive pumping equipment and the time-consuming vacuum process. An atmospheric pressure glow discharge (APG) system is cost-effective and employing the system in the PECVD process thus serves as one approach to overcoming the problems. Okazaki *et al.* reported on the mechanism of stabilization of glow plasma under atmospheric pressure and succeeded in preventing a transition from glow to arc discharge [8-10]. Highly functionalized films are much desired and considerable efforts have been made to synthesize functional films by the APG system; however, films synthesized under atmospheric pressure are generally inferior to those synthesized at low pressure in hardness [11]. In a previous paper, we reported that hardness of amorphous carbon films synthesized under atmospheric pressure was dependent on the substrate temperature [12].

In this study, SiOC(–H) films were synthesized by the AP-PECVD method with varying substrate temperature and we investigated the effect of the substrate temperature on the hardness and optical transparency of the films. The aim of this study was to synthesize hard and transparent SiOC(–H) films.

EXPERIMENT

Figure 1 shows a schematic view of the AP plasma enhanced CVD equipment we built. Trimethylsilane ($(CH_3)_3SiH$, TrMS) gas diluted with He gas was introduced between the parallel-plate electrodes with a distance of 0.5 mm and the chamber was purged with the process gas. The plasma was generated and sustained between the electrodes. The frequency of the power source and the voltage applied were 10 kHz and 20 kV, respectively. The upper electrode of 50 mm square was covered with dielectric material and chilled to 15°C and the lower electrode of 70 mm square was heated. The difference in size between the upper and lower electrodes was to avoid arc discharge occurring at the edges of electrodes. The deposition of the films was carried out at substrate temperatures of 60, 100 and 140°C. PC sheet with a thickness of 0.1 mm (FE2000, MGC Filsheet Co., Ltd., Japan) was used as the substrate for ultraviolet-visible spectroscopy measurement and single crystal Si (100) wafers were used for all the other analyses.

The film thickness was measured by a contact-type surface profiler (Dektak3030, Veeco, USA). The measurements were performed at 5 different surface locations and the results are expressed as the mean of 5 replicates and the corresponding standard deviation. Chemical bonding structure of the films was analyzed by Fourier transform infrared (FT-IR) spectroscopy (ALPHA-T, Bruker., Germany) in absorbance mode and bare silicon substrates were scanned to obtain reference spectra. Hardness of the films was measured using a triboindenter (TI-900, Hysitron, USA) with a Berkovich diamond indenter. The penetration depth was 50 nm and the load-displacement curve was analyzed using the Oliver and Pharr method [13]. When analyzing the chemical bonding structure and the mechanical properties of the films, the deposition time was adjusted to obtain 1 μm thick films. The measurements of film thickness and hardness were performed at 5 different surface locations and the results are expressed as the mean of 5 replicates and the corresponding standard deviation.

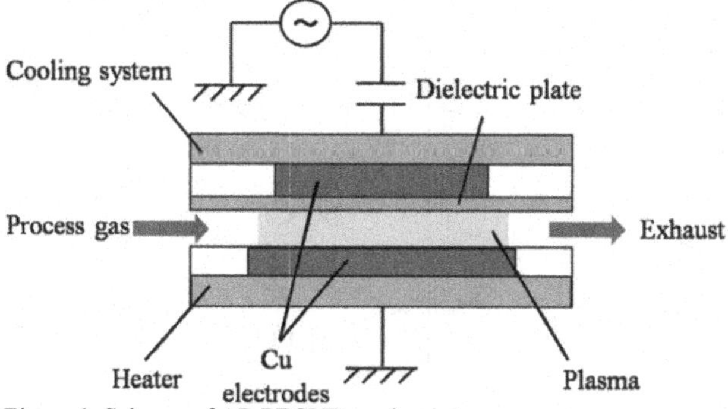

Figure 1. Scheme of AP-PECVD equipment.

DISCUSSION

Deposition rate, Hardness and Transparency

Figure 2 shows the deposition rate and hardness of SiOC(–H) films as a function of the substrate temperature. The deposition rate ranged approximately from 4 to 9 nm/sec and the deposition rate decreased with increasing the substrate temperature. Kuo *et al.* reported a similar relationship for amorphous silicon carbonitride [Si(C,N)] synthesized under low pressure [14]. As the substrate temperature increased, the hardness increased from 0.6 to 1.3 GPa, which was more than four times harder than PC substrates.

Figure 3 shows the optical transmission spectra of SiOC(–H) films on the PC substrates. All the films exhibited a good optical transparency with an optical transmittance of about 90% throughout the visible range. Such a result is of great importance for the protection of transparent polymers without altering their aspect.

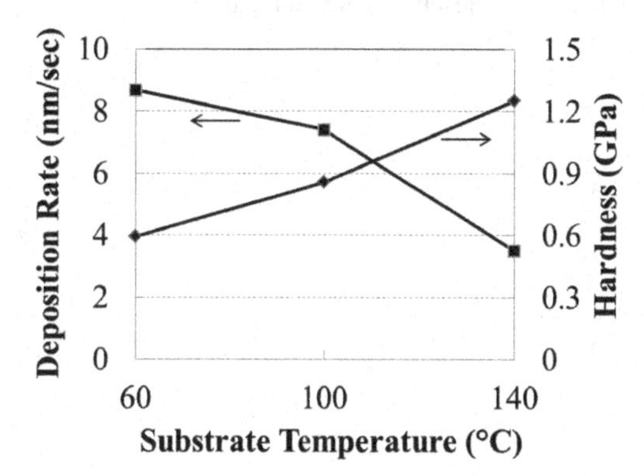

Figure 2. Deposition rate and hardness of SiOC(–H) films as a function of the substrate temperature.

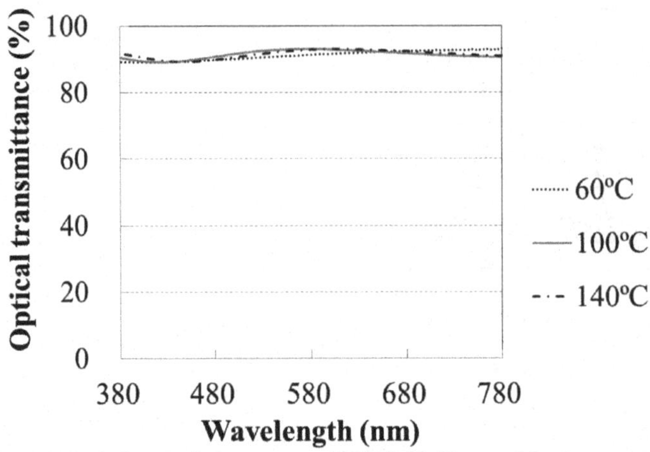

Figure 3. Optical transmission spectra of SiOC(–H) films on PC substrates using various substrate temperatures.

Effect of chemical bonding structure on hardness

Figure 4 presents the FT-IR spectra of SiOC(–H) films. The strong broad band at 1300–600 cm^{-1} observed for all the films is most likely composed of overlapping vibrations from Si–O–Si, H–Si–O and Si–$(CH_3)_x$ (x=1, 2, 3) [15-19]. Absorption peaks were observed at approximately 2900 cm^{-1} and 2100 cm^{-1}, which could be assigned to the asymmetric and symmetric stretching mode of C–H_x (x=2, 3) and the symmetric stretching mode of Si–H [17], respectively. To confirm the variation of the integrated absorption area of Si–O–Si, H–Si–O and Si–(CH_3) bonds according to the substrate temperature, the spectra in the range from 1300 to 600 cm^{-1} were deconvoluted by Gaussian peak fitting and the result is shown in Figure 5 (A).

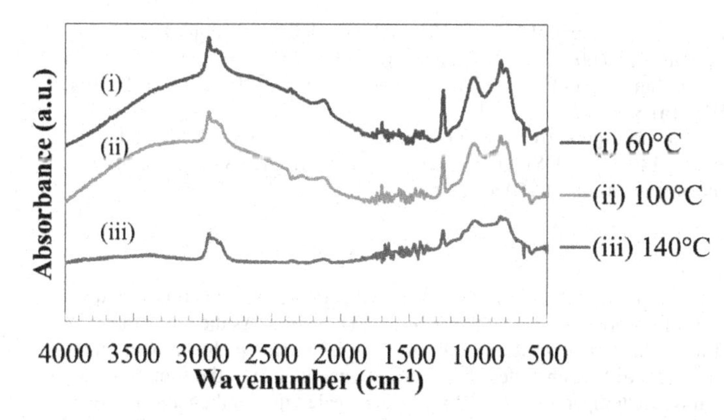

Figure 4. Full FT-IR spectra of SiOC(–H) films deposited at substrate temperatures of (i) 60°C, (ii) 100°C and (iii) 140°C.

Figure 5 (B) shows the peak area ratio of FT-IR absorption of each peak to the area of the Si–(CH_3) absorption (1275 cm^{-1}) as a function of the substrate temperature. As the substrate temperature increased, the peak area ratio of Si–O–Si-related bonds decreased, while those of Si–$(CH_3)_x$-related bonds increased. These results indicate that a decrease of Si–O–Si-related bonds and an increase of Si–$(CH_3)_x$-related bonds lead to a rise of hardness.

Further investigations are ongoing that focus on the relationship between the hardness and chemical bonding structure of the films or other properties (e.g. adhesion and scratch resistance) of the films.

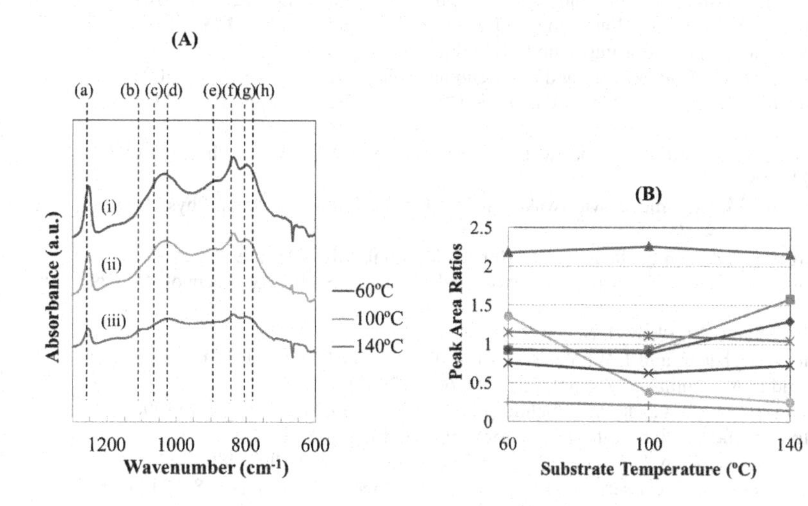

Figure 5. (A) FT-IR spectra in the range of 600 to 1300 cm^{-1} recorded from samples with different substrate temperatures (i) 60°C, (ii) 100°C and (iii) 140°C. (a) Si– (CH_3), (b) Si–O–Si (angle ~ 150°), (c) Si–O–Si (angle ~ 144°), (d) Si–O–Si (angle < 140°), (e) H–Si–O, (f) H–Si–O / Si–(CH_3)$_3$, (g) Si–(CH_3)$_2$, (h) Si–(CH_3) / Si–(CH_3)$_3$.
(B) Peak area ratios of FTIR absorption of (b) Si–O–Si (angle ~ 150°), (c) Si–O–Si (angle ~ 144°), (d) Si–O–Si (angle < 140°), (e) H–Si–O, (f) H–Si–O / Si–(CH_3)$_3$, (g) Si–(CH_3)$_2$, (h) Si–(CH_3) / Si–(CH_3)$_3$ to the area of (a) Si– (CH_3) absorption.

CONCLUSIONS

In this work, we investigated the effect of substrate temperature on the hardness and transparency of SiOC(–H) films synthesized by the AP-PECVD method. As the temperature increased from 60 to 140°C, the hardness increased from 0.6 to 1.3 GPa, which was about four times harder than the hardness of PC substrates. The optical transmittance of the films was about 90% regardless of the substrate temperature. FT-IR analysis revealed that as the temperature increased, Si–O–Si-related bonds decreased and Si–(CH_3)$_x$-related bonds increased. These results indicate that the substrate temperature changes the chemical bonding structure of SiOC(–H) films, which eventually control the hardness.

REFERENCES

1. J.-J. Ho, C.-Y. Chen, C.-M. Huang, W.J. Lee, W.-R. Liou, and C.-C. Chang, Appl. Opt. **44**, 6176 (2005).
2. C. Charitidis, A. Laskarakis, S. Kassavetis, C. Gravalidis, and S. Logothetidis, Superlattices Microstruct. **36**, 171 (2004).
3. K. Kirwan and G. Smith, Plast. Rubbers Compos. **33**, 452 (2004).
4. Y. Nojima, M. Okoshi, H. Nojiri, and N. Inoue, Jpn. J. Appl. Phys. **49**, 072703 (2010).
5. H. Anma, Y. Yoshimoto, M. Warashina, and Y. Hatanaka, Appl. Sur. Sci. **175**, 484 (2001).
6. D. Rats, V. Hajek, and L. Martinu, Thin Solid Films **340**, 33 (1999).
7. J.C. Damasceno, S.S. Camargo Jr., and M. Cremona, Thin Solid Films **433**, 199 (2003).
8. S. Kanazawa, M. Kogoma, T. Moriwaki, and S. Okazaki, J. Phys. D. Appl. Phys. **21**, 838 (1988).
9. T. Yokoyama, M. Kogoma, S. Kanazawa, T. Moriwaki, and S. Okazaki, J. Phys. D. Appl. Phys. **23**, 374 (1990).
10. T. Yokoyama, M. Kogoma, T. Moriwaki, and S. Okazaki, J. Phys. D. Appl. Phys. **23**, 1125 (1990).
11. T. Suzuki and H. Kodama, Diamond and Related Materials **18**, 990 (2009).
12. T. Sakata, H. Kodama, H. Hayashi, T. Shimo, and T. Suzuki, Surf. Coat. Technol. **183**, 295 (2004).
13. W.C. Oliver and G.M. Pharr, J. Mater. Res. 7, **6**, 1564 (1992).
14. K.-T. Rie, A. Gebauer, and J. Whole, Surf. Coat. Technol. **74-75**, 362 (1995).
15. A. Grill and D.A. Neumayer, J. Appl. Phys. **94**, 6697 (2003).
16. M.J. Loboda, C.M. Grove, and R.F. Schneider, J. Electrochem. Soc. **145**, 2861 (1998).
17. M.G. Albrecht and C. Blanchette, J. Electrochem. Soc. **145**, 4019 (1998).
18. Y.-H. Kim, M.S. Hwang, H.J. Kim, J.Y. Kim, and Y. Lee, J. Appl. Phys. **90**, 3367 (2001).
19. H.G. Pryce Lewis, T.B. Casserly, and K.K. Gleason, J. Electrochem. Soc. **148**, F212 (2001).

Mater. Res. Soc. Symp. Proc. Vol. 1321 © 2011 Materials Research Society
DOI: 10.1557/opl.2011.1253

Use of a-SiC:H multilayer transducers for detection of fluorescence signals from reactive cyan and yellow fluorophores

P. Louro[1,2], M. Vieira[1,2,3], M. A. Vieira[1,2], J. Costa[1,2], M. Fernandes[1,2], A. Karmali[4]
[1] Electronics Telecommunications and Computer Dept, ISEL, Lisbon, Portugal.
[2] CTS-UNINOVA, Lisbon, Portugal.
[3] DEE-FCT-UNL, Quinta da Torre, Monte da Caparica, 2829-516, Caparica, Portugal
[4] CIEB-ISEL, Lisbon, Portugal

ABSTRACT

The transducer consists of a p-i'(a-SiC:H)-n/p-i(a-Si:H)-n heterostructures produced by PECVD and optimized for the detection of the fluorescence resonance energy transfer between fluorophores with excitation in the violet(400 nm) and emissions in the cyan (470 nm) and yellow (588 nm) range of the spectrum. The thickness and the absorption coefficient of the i'- and i- layers were tailored for cyan and yellow optical confinement, respectively in the front and back photodiodes acting both as optical filters. The devices were characterized through transmittance and spectral response measurements and under different electrical.

To simulate the FRET pairs and the excitation light a chromatic time dependent combination of violet, cyan and yellow wavelengths was applied to the device. The generated photocurrent was measured under negative and positive bias to readout the combined spectra. The independent test signals were chosen in order to sample all the possible chromatic. Different wavelength backgrounds were also superimposed.

Results show that under negative bias the phorocurrent signal presents eight separate levels each one assigned to the different polychromatic mixtures. If a blue background is superimposed the yellow channel is enhanced and the cyan suppressed while under red irradiation the opposite behavior occurs. So under appropriated steady state optical bias the sensor will detect separately the cyan and yellow fluorescence pairs. An electrical model, supported by a numerical simulation, gives insight into the transduction mechanism.

INTRODUCTION

There is great interest in developing semiconductor devices able of sensing the distance between two molecular species in real time for medical and biological applications. One possible approach is to take advantage of the mechanism of Fluorescence Resonance Energy Transfer (FRET), by which the fluorescence wavelength of two labeled molecules is shifted when they are within close range. This technique has grown in popularity due to the emergence of various fluorescent mutant proteins with shifted spectral properties [1, 2].

In the past, optical color sensors based on multilayered a-SiC:H heterostructures have been used as voltage controlled optical filters in the visible range. In this paper we extend the application of these devices to detect violet, cyan and yellow fluorophores. The optical transducer has to accomplish the detection of the transient fluorescent signals coming from the different fluorescent proteins without losing any information about wavelength and intensity. The advantage of this type of sensor is that it does not rely on mechanical parts; it is compact and cost effective.

In alternative to FRET measurements, the sensor could be also interesting to measure fluorescent signals that show peak values in the cyan and yellow regions. In this paper we have used a purified preparation of glucose oxidase (EC 1.1.3.4) from *Aspergillus niger* [3]. The fluorescent signals obtained by excitation at 450 nm were measured with a spectrofluorimeter. Changes in the peak values of fluorescence in the cyan and yellow region can be used to detect the presence of glucose. The sensor was studied also having this application in consideration.

DEVICE CONFIGURATION

The device described herein works from 400 to 700 nm which makes it suitable for FRET applications using phlurophores operating in the visible spectrum. The device is a multilayered heterostructure based on a-Si:H and a-SiC:H. The configuration of the device includes two stacked p-i-n structures between two electrical and transparent contacts (Fig. 1). Both front (pin1) and back (pin2) structures act as optical filters confining, respectively, the short and the long optical carriers, while the intermediate wavelengths are absorbed across both [4, 5].

The active device consists of a p-i'(a-SiC:H)-n / p-i(a-Si:H)-n heterostructure with low conductivity doped layers. The thicknesses and optical gap of the thin i'- (200nm; 2.1 eV) and thick i- (1000nm; 1.8 eV) layers are optimized for light absorption in the short and long visible ranges, respectively.

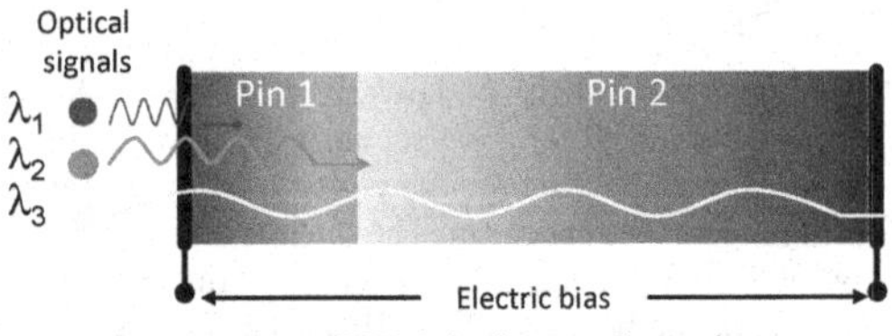

Figure 1 WDM device configuration.

The device was operated within the visible range using as optical signals to simulate the excitation light and the emitted fluorescent signals (yellow and cyan) of glucose oxidase from *Aspergillus niger* [1], the modulated light (external regulation of frequency and intensity) supplied by a violet (λ_1), a cyan (λ_2) and a yellow (λ_3) LED with wavelengths of 400 nm, 470 nm and 588 nm, respectively.

FLUORESCENCE SPECTRUM

In Fig. 2 it is displayed the fluorescence emission spectrum of glucose oxidase from Aspergillus niger prepared with purified enzyme (specific activity of 125 U/mg protein) dissolved in 20 mM phosphate buffer pH 7.0 (5 mg/l). A wavelength of 450 nm was used as excitation source. The emission spectrum shows fluorescence in the range 480-600 nm with a peak located at 520 nm. The deconvolution of the experimental data Gaussian fit originates two peaks located at 518 nm and 562 nm, which corresponds respectively to the cyan and yellow emissions.

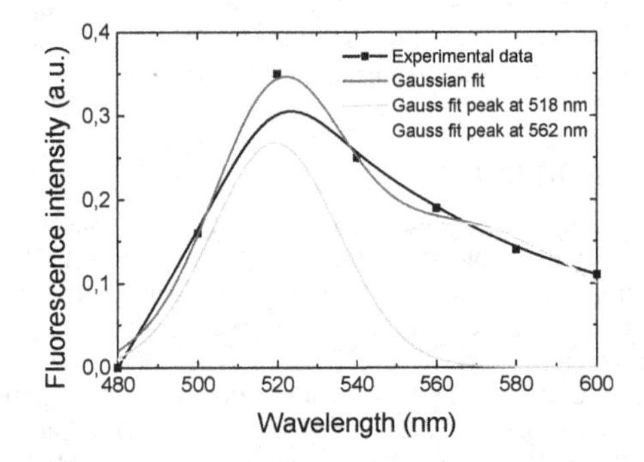

Figure 2 Fluorescence emission of glucose oxidase.

DEVICE OPTOELECTRONIC PROPERTIES

Spectral sensitivity

Figure 3 displays the spectral photocurrent, measured along the visible spectrum, under reverse and forward bias without and with optical light bias of different wavelengths.

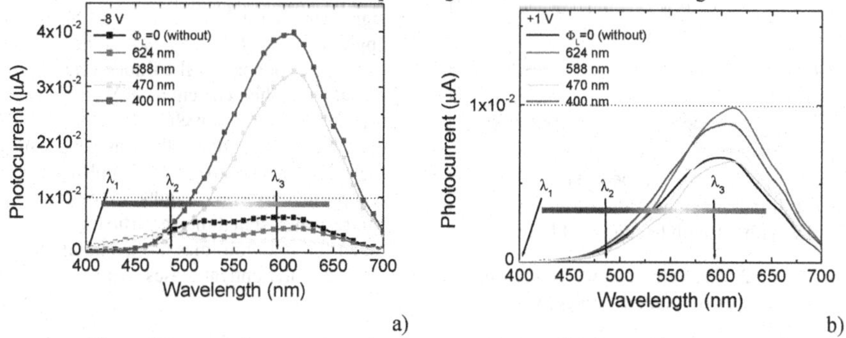

Figure 3 Spectral photocurrent under: a) reverse (-8 V) and b) forward bias (+1 V).

Results show that for wavelengths longer than 500 nm the use of short wavelengths (400 nm and 470 nm) for optical biasing the device enhances the photocurrent when compared to the absence of optical bias (Φ_L=0), while longer wavelengths of the optical bias (588 nm and 624 nm) cause a decrease of the signal. On the remaining part of the spectrum and opposite behavior is observed as the amplification effect occurs for longer light bias wavelengths and the reduction of the signal for shorter wavelengths. Thus, at reverse bias, the use of a 400 nm light source as optical bias amplifies both fluorescent signals (λ_2 and λ_3) and reduces the excitation signal (λ_1).

This result is an effective method for tuning the device sensitivity in order to filter undesirable light prone to overlap with the fluorescent signals. From the comparison of data from Fig. 3) and fig. 3b) obtained results show that, without optical bias, in the long wavelengths range (> 600 nm) the spectral response is independent on the applied bias while in the short wavelength the collection strongly increases with the reverse bias. This means that under forward bias the violet excitation optical signal at 400 nm (λ_1) can be fully suppressed without changing the intensity of the fluorescent signal at 588 nm (λ_3). Under optical bias the variation of the electrical bias induces always a change of the photocurrent along the whole spectrum.

Transient photocurrent signals

The photocurrent signal obtained at reverse (-8 V, solid lines) and forward (+1 V, dash lines) electric bias with (dark lines) and without background light (400nm, 470 nm, 624 nm, color lines)) is displayed in Fig. 4 and 5, respectively, for the excitation and fluorescent optical signals.(624 nm) optical bias and at reverse bias the photocurrent waveform brought on by the excitation signal is highly amplified. The magnitude of the amplification factor is around 12. The other background lights (470 nm and 400 nm) do not change significantly the signal, inducing a slight decrease.

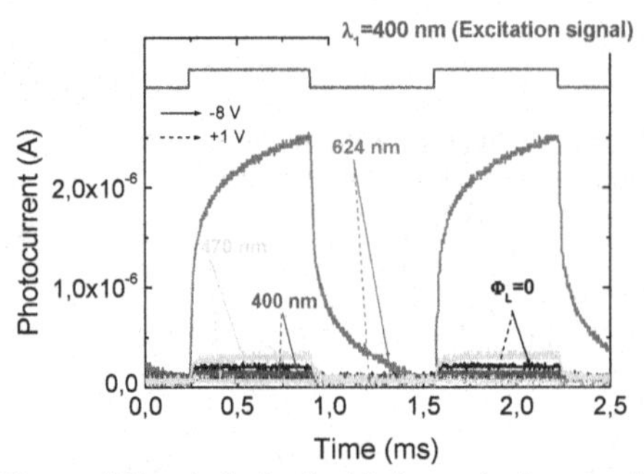

Figure 4 Signal obtained with the excitation signal at -8 V/+1V without/with background lights.

As already seen in Fig. 3, this shrinkage of the output signal can be used to filter the effect of the excitation signal and improve the signal to noise ratio.

The amplification of the excitation signal can be used in the scope of this application, to check the state of the excitation source. For the fluorescent signals, the photocurrent at -8 V is reduced when the wavelength of the optical bias is similar to the wavelength of the signal to be detected and amplified for complimentary wavelengths. In the cyan fluorescence signal the amplification under red (624 nm) bias has a magnitude factor around 2. The yellow signal is amplified under cyan and violet optical biases, with magnitude factors of 3 and 4, respectively.

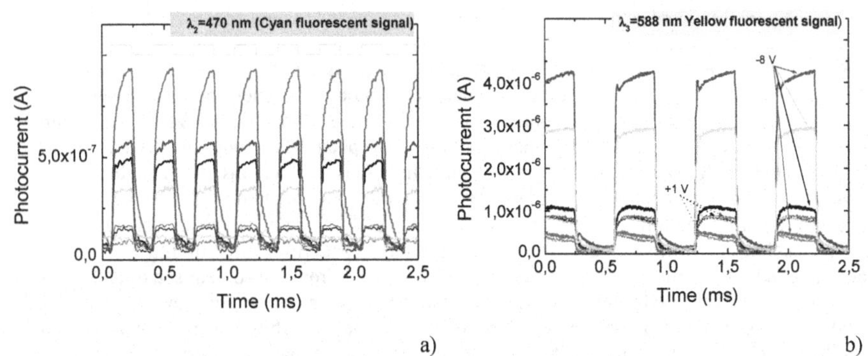

Figure 5 Photocurrent signal obtained at reverse (-8 V, solid lines) and forward (+1 V, dash lines) bias without (dark lines) and with under different background lights (color lines) with the: a) cyan (470 nm) and b) yellow (588 nm) fluorescent signals.

A chromatic time dependent wavelength combination (3000Hz) of λ_1 (400 nm), λ_2 (470 nm) and λ_3 (588 nm) pulsed input optical signals, was used to generate a multiplexed signal in the device. The output photocurrents, with (color lines) and without (dark lines) optical background light are displayed in Figure 6 a) and b) under forward (dash arrows) and reverse (solid arrows) voltages, respectively. The reference level was assumed to be the signal when all the input optical signals channels OFF. At the top of the figure, the individual optical signals are displayed to guide the eyes in relation to the different ON-OFF states. The independent test signals were chosen in order to sample all the possible chromatic mixtures for a pulse rate of 3kHz (6000bps). Under forward bias (Fig. 6a) the photocurrent signals with or without optical bias are similar and their waveform follows the yellow fluorescence signal, due to the lower sensitivity of the device to the cyan and violet (Figures 2 and 3) signals.

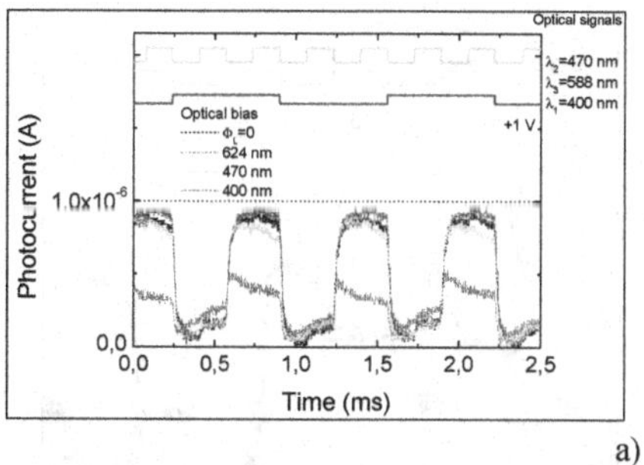

a)

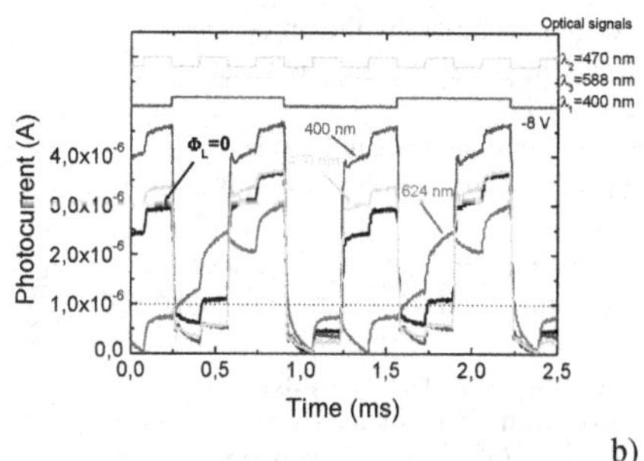

b)

Figure 6 Output photocurrent signals without (Φ_L=0) and with optical bias (624 nm, 470 nm, 400 nm) at: a) forward bias (+1 V) and b) reverse bias (-8 V). The optical signals waveforms are shown at the top of the figure.

This feature allows immediate decoding of the yellow fluorescence signal. Under reverse bias (Fig. 6b) the photocurrent signal is more complex. Each waveform is 8-level encoding (2^3) due to the different combinations of the input optical signals.

Once the yellow signal is decoded by the use of forward bias, the other fluorescence signal can be obtained from the photocurrent signal at -8V taking into account the amplitude dependence on the applied bias (Figure 5). To recover the CFP and YFP emission intensities, red optical bias was used. Under red irradiation the yellow channel is quenched, the blue enhanced and the violet strongly amplified (Figure 3b and Figure 6a). So the highest four levels in the photocurrent signal under 624 nm (Figure 6b) corresponds to the presence of the excitation light (violet) and the lowest four levels to its absence. The yellow channel is decoded under forward bias (2-level encoding, Fig. 6a). By subtracting the yellow coding to the signal under 624 nm (Fig. 6b) a 4 level encoding is obtained. Here the highest amplitude corresponds to the presence of both violet and cyan ´signals ON, the lowest to their absence and the two intermediated levels respectively to the presence of the violet or of the cyan signal. By using this simple algorithm the emission spectra of the CFP and YFP can be recovered, in real time, and its ratio can be correlated with the distance between the fluorophores.

ELECTRICAL SIMULATION

Based on the experimental results and device configuration an optoelectronic model was developed [6]. The device was modeled by a two single-tuned stages circuit with two variable capacitors and interconnected phototransistors through a resistor. Two optical gate connections ascribed to the different light penetration depths across the front and back phototransistors (Figure 1) were considered to allow independent Violet (I_V), yellow (I_Y) and cyan (I_C) optical signals transmission.

The operation is based upon the following principle: the flow of current through the resistor connecting the two front and back transistor bases is proportional to the difference in the voltages across both capacitors (charge storage buckets). So, it uses a changing capacitance to control the power delivered to the load acting as a state variable filter circuit.

In Figure 7 the simulated current without and under red backgrounds is displayed (symbols) using the same test signal of Fig.6. The input channels (I_V, I_C , I_Y, Figure 4 and 5) are also displayed (lines). To simulate the red background, current sources intensities (input channels) were multiplied by the on/off ratio between the input channels with and without red optical bias (α_C , α_Y, α_V, Figure 4 and 5).Good agreement between experimental and simulated data was observed.

The eight expected levels, under reversed bias, and their reduction under red irradiation, are clearly seen.Under red background the expected optical amplification of the cyan channel and the quenching of the yellow one were observed due to the effect of the active multiple-feedback filter when the back diode is light triggered.

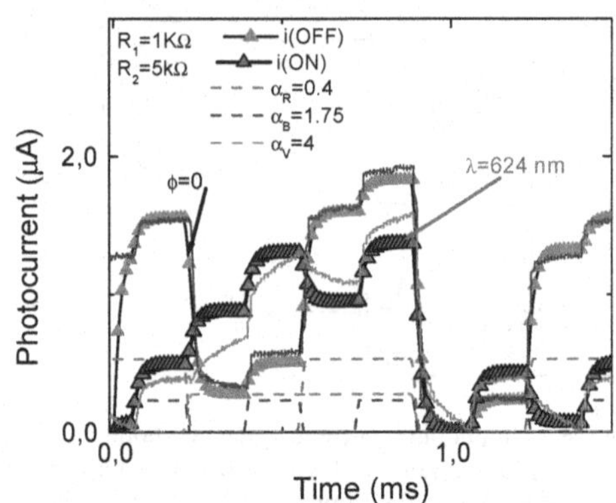

Figure 7 Multiplexed simulated (symbols), input channels (dash lines) and experimental (solid lines): under negative dc bias and red background.

CONCLUSIONS

In this paper a novel device for the detection of FRET signals was analyzed. The active optical filtering properties of the device were optimized for the detection of the fluorescent signals with simultaneous suppression of the excitation signal in order to minimize spectral overlap. Future work comprises linearity and intensity detection limits of the transducer, as well as tests with fluorescence signals from different samples of glucose oxidase.

ACKNOWLEDGEMENTS

This work was supported by FCT (CTS multi annual funding) through the PIDDAC Program funds and PTDC/EEA-ELC/111854/2009.

REFERENCES

1 D.A LaVan, Terry McGuire, Robert Langer, Nat. Biotechnol. 21 (10), 2003, 1184–1191.

2 D. Grace, Medical Product Manufacturing News, 12, 2008, 22–23.

3 K. Karmali, A. Karmali, A. Teixeira, M. J. Marcelo Curto "Assay of glucose oxidase from *Penicillium amagasakiense and Aspergillus niger* by Fourier Transform Infrared Spectroscopy". (2004) Analytical Biochemistry 333, 320-327.

4 P. Louro, M. Vieira, M.A. Vieira, M. Fernandes, A. Fantoni, C. Francisco, M. Barata, Physica E: Low-dimensional Systems and Nanostructures, 41 (2009) 1082-1085.

5 P. Louro, M. Vieira, M. Fernandes, J. Costa, M. A. Vieira, J. Caeiro, N. Neves, M. Barata, Phys. Status Solidi C 7, No. 3–4, 1188– 1191 (2010).

6 M. A. Vieira, M. Vieira, J. Costa, P. Louro, M. Fernandes, A. Fantoni, in Sensors & Transducers Journal Vol. 9, Special Issue, December 2010, pp.96-120.

Mater. Res. Soc. Symp. Proc. Vol. 1321 © 2011 Materials Research Society
DOI: 10.1557/opl.2011.1094

Properties of A-(Si,Ge) Materials and Devices grown using Chemical Annealing

Ashutosh Shyam, Daniel Congreve, Max Noack and Vikram Dalal
Iowa State University, Dept. of Electrical and Computer Engr. and Microelectronics Research Center, Ames, Iowa 50011

ABSTRACT

Chemical annealing is a powerful technique for controlling H bonding and optical absorption in amorphous semiconductors. We have shown previously that the use of careful chemical annealing by Argon can lower the bandgap of a-Si:H while maintaining electronic properties in both films and devices. In this work, we describe new work on chemical annealing of A-(Si,Ge):H films and devices. The technique consists in growing very thin layers (1-3 nm) of A-(Si,Ge) from mixtures of hydrogen, Silane and Germane, and then subjecting this thin layer to ion bombardment by Ar. The cycle is repeated many times to achieve the desired thickness of the intrinsic layer. The resulting film and device were measured for their composition using energy dispersive spectroscopy (EDS) analysis. We discovered that the composition itself, namely the Ge:Si ratio in the film, could be varied by changing the ion bombardment conditions. Lower energy bombardment led to a higher Ge:Si ratio for the same germane/Silane ratio in the gas phase. By controlling ion bombardment during the Ar annealing cycle, we were able to reduce the H content of the film and achieve good electronic properties. It will be shown that by appropriate control over ion energies, one can obtain films and devices which are of good quality and low bandgap as well.

INTRODUCTION

A-SiGe is an attractive candidate for low bandgap material. However owing to its poor electronic properties it presents considerable challenge in fabricating high quality solar cell. This study is dedicated to the fabrication of low bandgap, but high quality solar cells using A-SiGe.

In our previous work[1] we had shown that it is possible to fabricate high quality (FF ~ 65%), low bandgap (Eg~ 1.62 Ev) device using a layer by layer approach, which is also known as chemical annealing[2,4] in which a thin layer of A-Si was grown (~10Å) followed by Ar Ion bombardment. The objective of this study is to try and apply the same technique to A-SiGe. We have on one hand, focused on effect of ion bombardment of Ar on material property and bandgap, and on the other hand discussed the effect of RF power on the Ge content of devices. It has also been shown that it is vital to compare bandgaps of continuously grown or non annealed (NON CA) and chemical annealed (CA) devices with identical Ge-Si composition. Mainly p-i-n devices and films were studied for the purpose of this study.

EXPERIMENTAL DETAILS

The growth of devices was done using PECVD growth technique, in which frequency used was in the VHF range, ~ 45 MHz. The growth temperature was about 275C. The growth was done using a mixture of Silane, Germane, Hydrogen and Argon, and the annealing was done using just Argon. For comparison we used devices with the same SiGe composition as chemically annealed devices - this was ensured by doing extensive EDS studies of CA and NON CA devices.

The growth and anneal cycle is represented schematically in Figure 1. The duration of growth time (T1) was varied which allowed us to control the thickness of the layer grown per individual cycle. This was done because Argon cannot penetrate too deeply into the surface unless its ion energy is too high, which is undesirable as it leads to distortion in the lattice and damage during growth [5]. So we have to keep Argon energies small, which necessitates that the layer thickness/cycle should be kept small so that the ions can effectively influence the structure of the entire thickness.

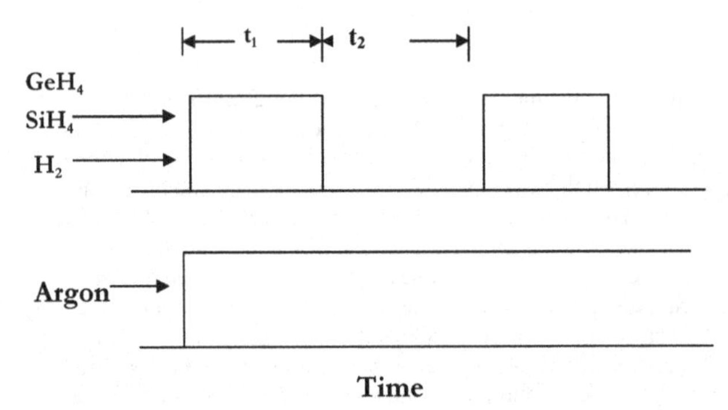

Figure 1 Schematic representation of chemical annealing

The devices were of the standard p-i-n type (with ITO top contact), with the middle i layer being deposited using chemical annealing. The outer p+ and n +layers were deposited using continuous deposition conditions; in other words they were not grown by layer by layer growth method. An ITO contact was used on the top of the p layer. All devices were deposited on stainless steel substrates without any back reflector. The device characterization included light I-V curves, quantum efficiency (QE) vs. photon energy under both zero bias and forward bias conditions to study hole transport properties and finally QE vs. photon energy in the subgap region so as to determine Urbach energy of valence band tails [5]. A comparison of such subgap QE data for devices made using annealed and unannealed i layers allows one to estimate the Tauc bandgap of the i layers.

The alpha values (α) needed for plotting such absorption curves can be determined from QE equation under strong reverse bias:-

$$QE = (1-R) \exp(-\alpha t_1)[1- \exp(-\alpha t_2)] \qquad (1)$$

Where R is the reflection from the cell, and t_1 and t_2 are the thicknesses of the p and i layers respectively.

RESULTS AND DISCUSSION

Initial problems with CA film and devices - trend of reverse band gap

In the first CA film fabricated under the same conditions as continuous grown film (4.5 W RF power, 50 mtorr pressure) it was found that the CA film had higher bandgap than non CA film. This was observed for devices as well. This was completely contrary to what had been seen for a-Si. Absorption vs. energy curves for films and devices showing the anomalous characteristics are shown below in Figure 2.

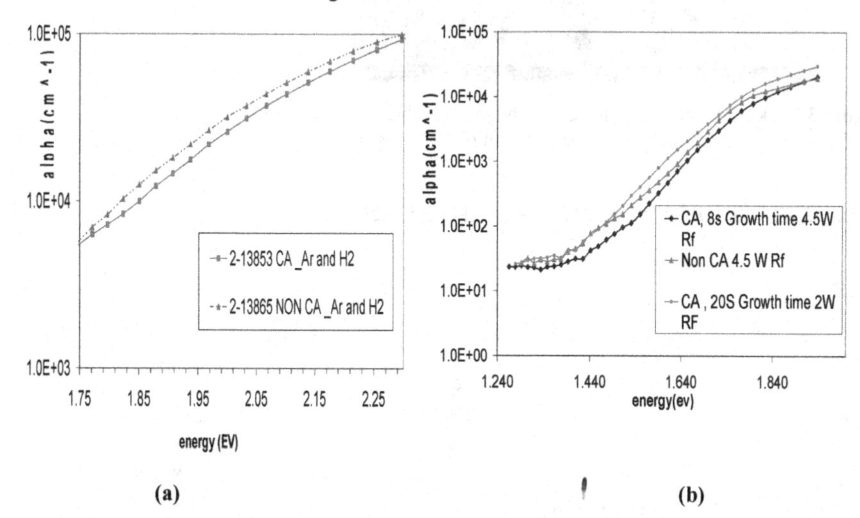

Figure 2 Alpha vs. energy curves for (a) Ca and non CA films (b) Effect of power on CA device (alpha energy plot for 4.5 W vs 2 W RF)

There could be two reasons for such a trend; either chemical annealing was not successful or else SiGe composition was not identical in CA and non CA. An EDS experiment was done to check the composition of the various films. The EDS data showed that the Ge content in non CA was higher than CA. The back scattered image shown in Figure 3 verified this fact- the non CA was lighter than the CA , which can happen only if its Ge content is more than CA (as only then it would reflect more). The problem was solved when we made a CA device at 2 W RF with 20 s growth time. The reason for choosing 20 s as growth time was to maintain same growth per cycle as for the case when using the 4.5 W RF power CA cycle.

The data presented in Fig 2 shows that at 4.5 W RF power, Ar during the annealing stage was etching away Ge. We discovered from EDS studies that when RF power was reduced to 2W, the preferential etching of Ge was reduced significantly, so that we could now compare samples prepared using CA and Non-CA with similar Ge contents, approximately 14%. Upon chemical annealing at low power, the absorption curve shifted to lower energies, as shown in Fig. 2, a result similar to our earlier results on the bandgap of a-Si, where CA reduced the bandgap.

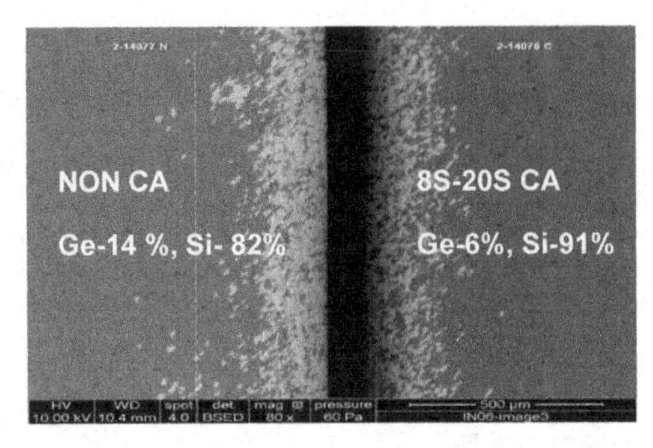

Figure 3 Back scattered image for CA vs NON CA fabricated under identical conditions. The numbers inset indicate the Ge, Si composition of two samples

Effect of pressure and growth per cycle- CA experiments

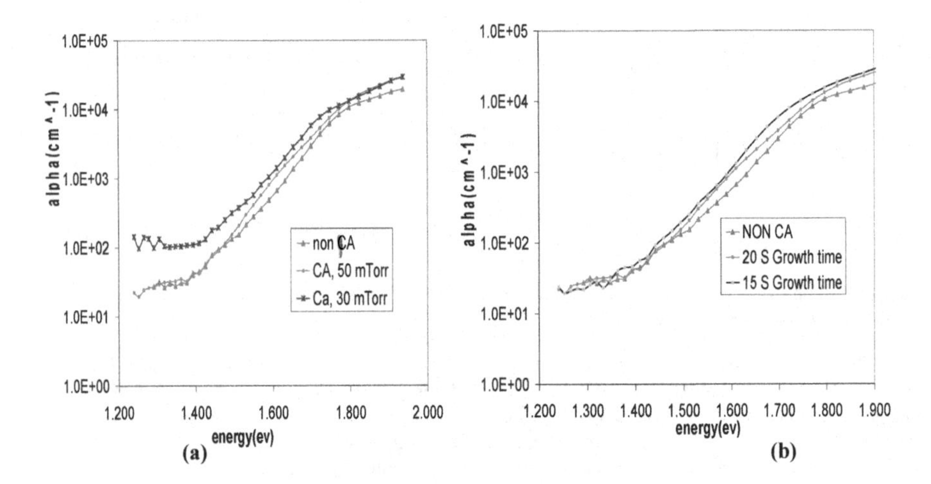

Figure 4 Effect of (a) pressure and (b) growth time on absorption curve.

As mentioned before and shown in our previous work [1] it has been seen that the most important parameter that allows us to lower bandgap is growth per cycle. When we decrease the pressure, the Ar Ion energy increases leading to heavier bombardment of the film, thus facilitating reduction in bandgap by reducing H content. However, if the pressure is made too low (30 mtorr in this case), then excessive ion bombardment will lead to high midgap defect densities. This is indicated, in Fig 4(a) by higher absorption shoulder for 30 mtorr CA as compared to 50 mtorr CA. Hence optimum pressure should be selected for chemical annealing. Lower growth per cycle implies lower thickness of film per cycle, which will facilitate more complete penetration of the film by Ar, thereby leading to lowering of H content throughout the film and hence to a smaller bandgap. This fact is shown in the figure 4 (b) where the bandgap is seen to reduce as the growth time per cycle reduces.

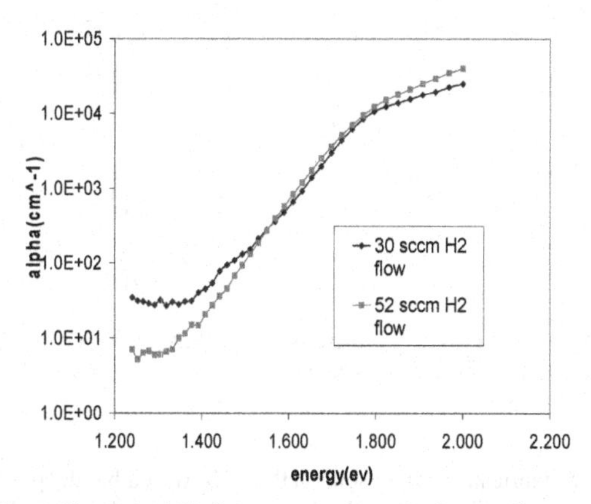

Figure 5 Effect of increased Hydrogen content on device quality

Effect of High Hydrogen content – improvement in device quality

Till now the experiments had been done on CA and NON CA devices with 30 sccm H_2 flow, Typical FF for the best CA device with this hydrogen flow was~ 60%. When absorption curves of two NON CA devices fabricated under higher and lower hydrogen dilution were compared (fig 5), the former showed a lower absorption shoulder indicating high quality, which is in agreement with similar studies reported by Ganguly et al [6]. High hydrogen dilution may lead to a more ordered film structure and thus improves device quality.

To see whether a higher hydrogen dilution led to better devices, we fabricated devices at a pressure of 50 mtorr with a two times increase in hydrogen dilution. The i- layer thickness was about .3 micrometer. The I-V curve shown in Fig. 6(a) has better fill factor, ~65%, indicating that the CA device using a higher hydrogen dilution during growth is of better quality. The efficiency of the solar cell is about 6.7% and its corresponding growth per cycle about 12 A/cycle. The corresponding NON CA device has a higher Voc of .82, which indicates that its bandgap is higher than that of Ca devices. From alpha energy plot, CA device has a Tauc gap of about 1.56 eV as against 1.62 eV for non CA. CA device has Urbach energy of about 48 meV. This clearly indicates the success of chemical annealing; on one hand bandgap has been reduced, while on the other hand device quality has not been affected.

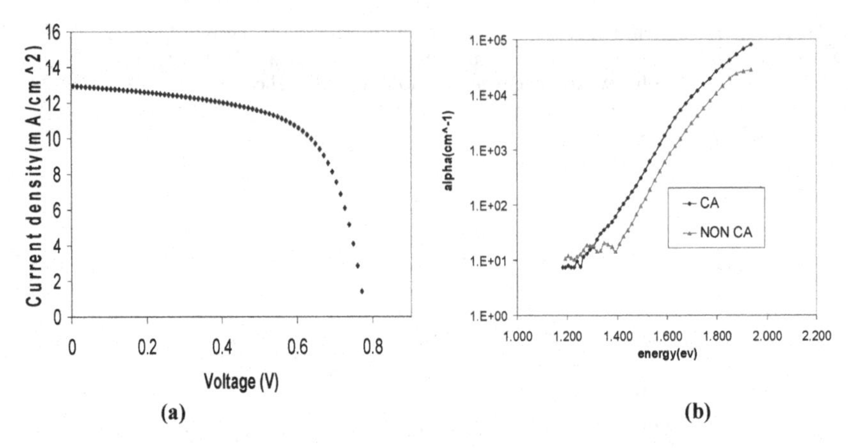

Figure 6 (a) Better quality CA device - I layer =.3 microns, Jsc=13mA/cm^2, Voc=.78V, FF=65.4% (b) comparison of absorption curve for CA vs non CA

CONCLUSIONS

We were successfully able to fabricate A-SiGe solar cell device with a bandgap of 1.56 eV, Urbach energy of about 48 meV and a FF ~ 65% . We have shown that CA, when done under high ion bombardment conditions for alloy systems, leads to changes in Si:Ge ratio. We have showed that CA will reduce the bandgap when done under the right conditions. As part of future work, we would like to compare the stability of the CA with NON CA devices upon light soaking.

ACKNOWLEDGEMENTS

This work was supported in part by a subcontract from Powerfund and a grant from NSF. We thank our colleagues at Iowa State, particularly J Bhattacharya, K Han, P Webster, S. Konduri and S Kajjam for their experimental help in this project.

REFERENCES

1. V. Dalal, A. Shyam and D. Congreve, *Proc. Of MRS* **1245,** 77-79 (2010)
2. H. Sato, K. Fukutani, W. Futako and I. Shimizu, *Solar Energy Mater.and Solar Cells* **66**, 321 (2001).
3. W. Futako, T. Kamiya, C. Fortmann and I. Shimizu, *J. Non-Cryst. Solids* **266**, 630 (2000).
4. M. Kambe, Y. Yamamoto, K. Fukutani, C. Fortmann and I. Shimizu, *Proc. Of MRS* **507**, 205 (1999)
5. A. Shah, J. Meier and E. Vallat-Sauvain, *Solar Energy Mater. and Solar Cells* **78**, 469-491 (2003)
6. G. Ganguly and A. Matsuda, *J. Non-Crys. Solids* **198**, 559-562 (1996)

Mater. Res. Soc. Symp. Proc. Vol. 1321 © 2011 Materials Research Society
DOI: 10.1557/opl.2011.810

Effect of Dynamic Bias Stress (AC) In Short-Channel (L=1.5μm) p-Type polycrystalline Silicon (poly-Si) Thin Film Transistors (TFTs) on the glass substrate

Sung-Hwan Choi[1], Yeon-Gon Mo[2] and Min-Koo Han[1]

[1]School of Electrical Engineering and Computer Science, Seoul National University, 599 Gwanangno, Gwanak-gu, Seoul, 151-744, Korea.
[2]Corporate R&D Center, Samsung Mobile Display Co. Ltd., Yongin-City, Kyeonggi-do, Korea.

ABSTRACT

We have investigated the stability of short channel (1.5μm) p-Type polycrystalline silicon (poly-Si) Thin Film Transistors (TFTs) on the glass substrate under AC bias stress. The variation of threshold voltage in short channel poly-Si TFT was considerably higher than that of long channel poly-Si TFT. Threshold voltage of the short channel TFT was considerably moved to the positive direction during AC bias stress, whereas the threshold voltage of a long channel was rarely moved. The variation of threshold voltage in the short channel p-type TFT under AC bias stess was more compared to that under DC bias stress. The threshold voltage of short channel (L=1.5μm) poly-Si TFT was increased about -7.44V from -0.305V to -7.745V when V_{GS} = 5 (base value) ~ -15V (peak value), V_{DS} = -15V was applied for 3,000 seconds. This positive shift of threshold voltage and significantly degraded s-swing value in the short channel TFT under dynamic stress (AC) may be due to the increase of the stress-induced trap state density at gate insulator / channel interface region.

INTRODUCTION

Recently, excimer laser annealed (ELA) poly-Si TFTs have gained a lot of attention as a pixel element of driving circuit in the high-resolution flat panel display (FDP) technology [1,2]. Short-channel ELA poly-Si TFTs are needed for high-resolution displays attribute to their high current driving ability as the resolution of displays increases. Short channel poly-Si TFT may be suitable for integrated circuit such as gate driver which require large driving current. However, the reliability of short-channel poly-Si TFTs has been a serious problem when the channel length is decreased [3]. The reliability of short-channel poly-Si TFTs under DC bias stress have been previously reported [4-6]. However, up to now, the reliability of short channel poly-Si TFTs under dynamic bias stress has been scarcely reported. Practically, the poly-Si TFTs used as switching elements for advanced displays, which are operated in an AC mode on the gate electrodes. Hence, it is significantly important to understand the instability mechanisms for the short-channel poly-Si TFTs under dynamic bias stress (AC).

In this work, we investigated the reliability of the ELA poly-Si TFT under AC bias stress with various channel length from 1.5μm to 7μm. The purpose of our work is to investigate the electrical characteristics of short channel p-type ELA poly-Si TFTs under dynamic bias stress (AC) (V_{GS} = 5 ~ -15V, V_{DS} = -15V) relative to static bias stress (DC). The influence of operating frequency, duty ratio, substrate temperatures into p-type short channel poly-Si TFT with AC stress is also discussed. It was already reported in our previous paper that the reliability of short channel a-Si:H TFT under AC bias stress could be far better than that of long channel a-Si:H

TFT [7]. In contrast, threshold voltage shift of the short channel poly-Si TFT was significantly increased about -7.74V during AC bias stress, whereas the threshold voltage shift of a long channel was little.

FABRICATION OF POLY-SI TFT

The p-type ELA poly-Si TFT used in the experiments was top gate low-temperature poly-Si TFTs on glass substrates with a buffer oxide layer. At first, amorphous silicon (a-Si) thin film (50nm) was deposited on the buffer oxide layer by plasma-enhanced chemical vapor deposition (PECVD). And it was crystallized by XeCl excimer laser annealing (ELA) (wavelength=308nm) for a low temperature process, and then patterning of a poly-Si film into the channel layer was followed. A gate oxide (SiO_2, 80nm thick) and an interposed silicon nitride (SiNx, 40nm thick) films were sequentially deposited.

A dielectric interlayer was deposited by PECVD. Afterwards, dopant activation was carried out thermally. Cross-sectional views of poly-Si TFTs are shown in Fig. 1 for SiO_2 / SiNx double gate insulators. In order to investigate the stability of poly-Si TFTs under high gate and drain voltage stress, transistor with threshold voltage about V_{TH} = -1V were subjected to electrical bias stress for 3,000 seconds.

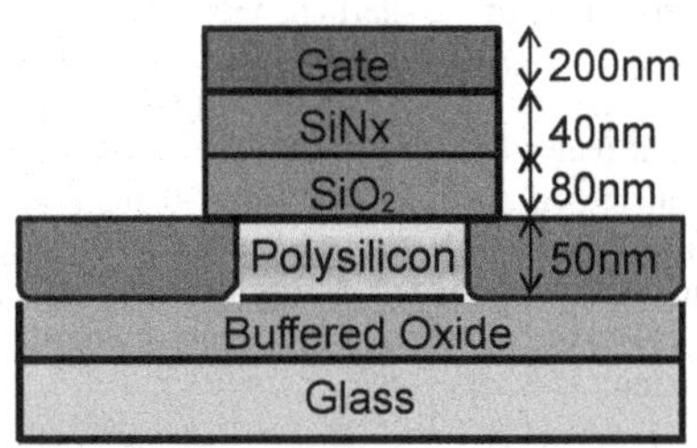

Figure 1. Top-gate p-channel Poly-Si TFT structure fabricated on SiO_2 buffer layer on glass substrate.

RESULTS AND DISCUSSION

1. Reliability of Poly-Si TFT for various channel lengths (DC bias stress)
1-1. Off-state condition (V_{GS}=5V, V_{DS}=-15V)

The transfer characteristics of Poly-Si TFT with various channel length. As the channel length of Poly-Si TFT was decreased, the on current of Poly-Si TFT was increased. In first, we have measured transfer characteristics of excimer laser annealing (ELA) poly-Si TFT with off state bias stress according to stress times. We have applied a fixed gate bias V_{GS} = 5V and drain bias V_{DS} = -15V to ELA poly-Si TFT for 3,000 seconds in order to investigate an effect of the off state bias stress on the transfer characteristic of the ELA poly-Si TFT. After the off state bias annealing, the threshold voltage of ELA poly-Si TFT with L = 1.5μm was decreased by 0.38V (from -1.14V to -0.76V, ΔV_{TH} = 0.38V at V_{DS} = -0.1V) as shown in Fig. 2, whereas that of device with L = 7μm was reduced by only 0.02V (from -1.98V to -1.96V, ΔV_{TH} = 0.02V).

This result was occurred due to electron traps were generated in the gate oxide and hole charges induced in channel near drain junction by the off state bias stress. Due to the existence of trapped electron charges, negative gate voltage could be effectively applied. Corresponding with the trapped electrons, hole charges were induced in the channel region near drain junction. Because an electric field in the TFT does not depend on the channel length, the damaged region (electron trapping occurred) is almost same for the each gate lengths. So the effective channel length seemed to be shortened and mobility appeared to be slightly increased in the short channel device. For these reasons, the threshold voltage shift of short channel TFT showed a relatively larger value than that of long channel TFT.

Threshold voltage of short channel poly-Si TFT (L = 1.5 μm) was significantly decreased as a stress time passed on. So it was decreased below the initial threshold voltage for a whole 3,000s stress times consistently. It is mainly due to the generation of the trap states near the drain junction.

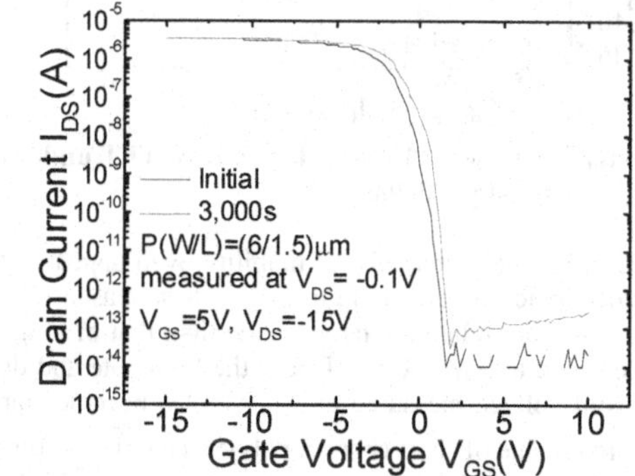

Figure 2. Transfer characteristic of ELA poly-Si TFT (L=1.5μm) under off state bias stress (V_{GS} = 5V, V_{DS}= -15V, 3,000s).

1-2. On-state condition (V_{GS}=-15V, V_{DS}=-15V)

The transfer characteristics of excimer laser annealing (ELA) poly-Si TFT with high gate and drain bias stress (V_{GS} = V_{DS} = -15V) with stress time, as shown in Fig. 2. We have applied a fixed gate and drain bias V_{GS} = V_{DS} = -15V to ELA poly-Si TFT for 3,000 seconds in order to investigate an effect of the self-heating stress on the transfer characteristic of the ELA poly-Si TFT. After the self-heating stress, the threshold voltage of ELA poly-Si TFT with L = 1.5μm was increased by 2.08V (from -0.99 V to -3.07 V at V_{DS} = -0.1V), whereas that of device with L = 7μm was increased by only 0.10V (from -2.09 V to -2.19 V at V_{DS} = -0.1V).

And Fig. 3 also shows that the minimum leakage current of the short channel ELA poly-Si TFT with L = 1.5μm was relatively increased with a constant bias stress for 3,000 seconds. Even on-current was dramatically decreased compared with that of unstressed devices after the high gate and drain bias stress, as shown in Fig. 3. This result was occurred due to the injection of hole charges was generated between the gate oxide and the poly-Si channel layer by impact ionization during the high gate and drain bias stress time (3,000 seconds). The interface state generation was also simultaneously occurred after the release of hydrogen atom from the poly-Si layer due to the existence of hot-hole charges. It can explain both the degradation of the subthreshold swing (s-swing) and the reduction of mobility. Due to the existence of interface trap states, negative gate voltage should be more applied to obtain a similar current compared to unstressed devices.

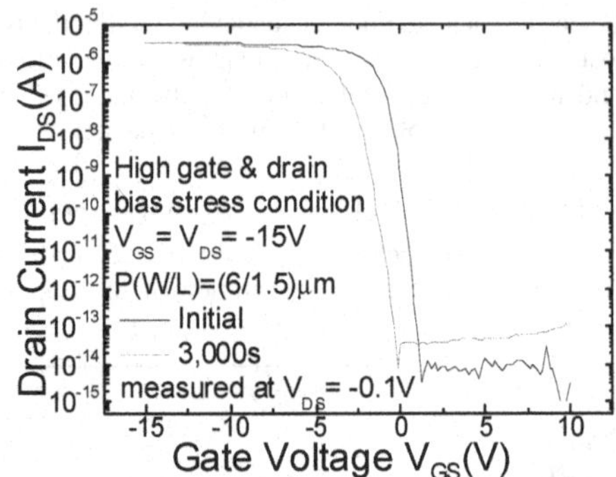

Figure 3. Transfer characteristic (V_{DS}=-0.1V) of ELA poly-Si TFT under the high-gate and drain bias stress time ($V_{GS} = V_{DS}$= -15V, 3,000s).

The threshold voltage seemed to be increased and mobility even appeared to be significantly decreased in the short channel device as shown in Fig. 3. For these reasons, the threshold voltage shift of short channel TFT showed a relatively larger value than that of long channel TFT. The threshold voltages were measured at V_{TH}= -0.1V. Under the high gate and drain voltage stress, the negative shift of threshold voltage measured at V_{TH}= -0.1V becomes large as the channel length was shorter. Threshold voltage of short channel poly-Si TFT (L=1.5 μm) was significantly increased as a stress time passed on.

In the low temperature poly-Si TFT with channel length 1.5μm, the threshold voltage is substantially increased so that it moves above the initial threshold voltage value (V_{TH} = -0.99 at V_{DS} = -0.1V) for a whole 3,000s stress times consistently. It was happened because the trapping of hot hole charges and interface state generation mainly occurs at front interface between poly-Si film and gate oxide layer.

As the stress time passed on, opposite phenomenon was observed. Generation of deep trap state between gate oxide and poly-Si active layer followed after the negative charge trapping in the gate oxide near the drain junction due to the existence of grain boundaries which enlarge the vertical electric field. This leads to decrease of mobility (reflecting the transconductance g_m), which is consistent with the generation of hot-hole induced interface states [8-10]. It leads to a reduced drain current by grain boundary defect generation.

2. Reliability of Poly-Si TFT for various channel lengths (AC bias stress)

We applied AC-bias stress on the short channel (L=1.5μm) ELA poly-Si TFTs with constant drain-source bias of -15V for 3,000 seconds. The diagram of applied AC bias is shown in Figure 4(a). The period of AC pulse was 16.67ms which was compatible to 60Hz frequency. The base value of AC bias was 5V and the peak value of it was -15V. The peak value and base value was applied equally within one period. When the base value of AC pulse was applied, the V_{GS}=-15V, V_{DS}=-15V were applied on the Poly-Si TFT. It was identical to constant bias stress (on-state condition).

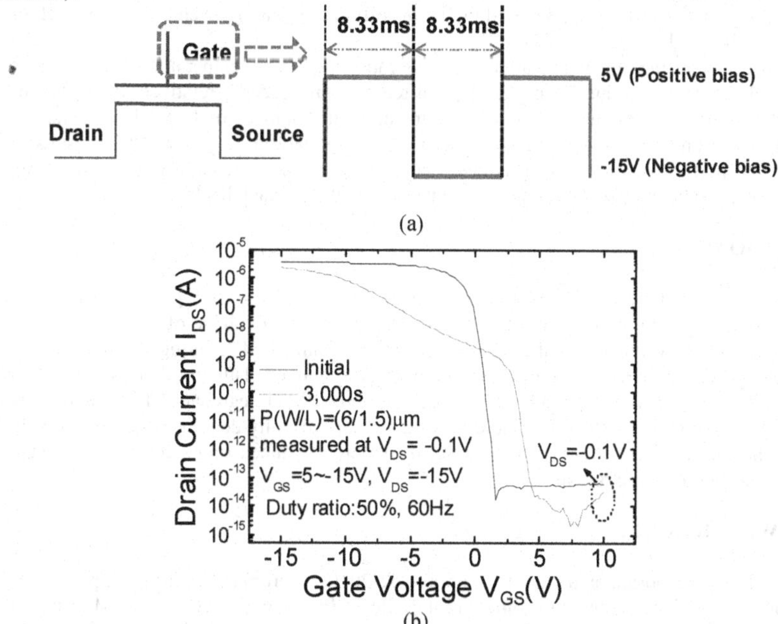

(b)

Figure 4. (a) The pulse diagram of applied AC-bias and (b) transfer characteristic of short channel (L=1.5μm) poly-Si TFT after AC bias stress (3,000s).

When the period of base value AC bias was applied, the V_{GS}=5V, V_{DS}=-15V were applied on the Poly-Si TFT. The electron carriers could be trapped in the gate insulator under this bias condition. The transfer characteristic of short channel (L=1.5μm) poly-Si TFT after AC bias stress was shown in Figure 4(b). We have found that the threshold voltage of short channel TFT (1.5μm) was significantly shifted to negative direction due to AC bias stress (from -0.305V to -7.745V, ΔV_{TH} = -7.44V), whereas that of long channel TFT (7μm) was rarely moved to negative direction (ΔV_{TH} = -0.02V). This negative shift of threshold voltage in the short channel TFT may be attributed to creation of interface states during AC bias stress. Also, s-swing value was severely degraded.

Because the time duration of applied AC-bias stress time on the short channel Poly-Si TFT was 3,000 seconds, and the ratio of peak value period was 50%, the total DC-bias stress time (on-state condition) was about 1,500 seconds during the whole AC-bias stress time. As shown in Figure 3, the threshold voltage of short channel Poly-Si TFT was increased about -1.5V for 1500 seconds under DC bias stress (on-state condition). The variation of threshold voltage during AC-bias stress was increased dramatically relative to DC bias stress (on-state condition) during AC-bias stress.

The effects of stress frequency (1Hz / 60Hz / 1MHz) on AC bias stress were also investigated. As we decrease the applied frequencies from 1MHz to 1Hz, we confirmed that reduction of on-current in linear regime was occurred. It also indicates that the stress-induced trap state density at gate insulator / channel interface increased by the distortion of the transfer characteristic in short channel poly-Si TFT [11,12].

The electrical characteristic with variety of duty ratio and stress temperature after AC-bias stress was also studied. By increasing the applied temperature (125°C), relatively large shift of transfer characteristics was occurred due to augmenting the number of trapped hole charges compared to that under room temperature. Because the relatively large V_{TH} variation results in malfunctioned performance of driving circuits, poor bias instability of poly-Si TFT under AC bias stress needs to be improved to achieve a high-resolution flat panel display.

CONCLUSIONS

We have investigated the stability of short channel (L=1.5μm) p-type ELA poly-Si TFTs under AC bias stress compared with that under DC bias stress. Threshold voltage of the short channel TFT was considerably moved to the positive direction about -7.74V during AC bias stress, whereas the threshold voltage of a long channel was rarely moved. This threshold voltage shift means that dynamic stress (AC) results in significantly more NBTI degradation relative to static stress (DC). This positive shift of threshold voltage and significantly degraded s-swing value in the short channel TFT may be due to increase of the stress-induced trap state density at gate insulator / channel interface region.

ACKNOWLEDGMENT

This research was supported by a grant (No. F0004062-2010-33) from Information Display R&D Center, one of the Knowledge Economy Frontier R&D Program funded by the Ministry of Knowledge Economy of Korean government.

REFERENCES

1. M. Stewart, IEEE Trans. on Electron Devices, **48**, 845 (2001).
2. A. Kumar K. P. and J. K. O. Shin, *in IEDM Tech. Dig.*, pp. 515 (1997).
3. J. W. Lee, N. I. Lee, and C. H. Han, IEEE Electron Devices Lett., **19**, 458 (1998).
4. J.-W. Lee , N.-I. Lee and C.-H. Han, Jpn. J. Appl. Phys., **37**, 1047 (1998).
5. S.-H. Choi, S.-J. Kim, Y.-G. Mo, H.-D. Kim, and M.-K. Han, Jpn. J. Appl. Phys., **49**, 03CA04-1 (2010).
6. S.-H. Choi, H.-S. Shin, Y.-G. Mo, H.-D. Kim, and M.-K. Han, Jpn. J. Appl. Phys., **48**, 03B011-1 (2009).
7. S.-G. Park, S.-Y. Lee, J.-S. Woo, J.-S. Yoo and M.-K. Han, ECS Trans., **33**, 2010.
8. S.Inoue, H.Ohshima, and T.Shimoda, *in IEDM Tech. Dig.*, pp.527-530 (1997).
9. T. Tsuchiya, J. Frey, IEEE Electron Device Lett., **6**, 8 (1985).
10. M. Rodder, IEEE Electron Devices Lett., **11**, 346 (1990).
11. M. Hack, A. G. Lewis, and I. Wu, IEEE Trans. on Electron Devices, **40**, 890, (1993).
12. C.-Y. Huang, T.-H. Teng, J.-W. Tsai and H.-C. Cheng, Jpn. J. Appl. Phys., **39**, 3867 (2000).

Mater. Res. Soc. Symp. Proc. Vol. 1321 © 2011 Materials Research Society
DOI: 10.1557/opl.2011.1374

The suppression of leakage current in the solid phase crystallized silicon (SPC-Si) TFT employing off-state bias annealing under light illumination.

Sang-Geun Park, Seung-Hee Kuk, Jong-Seok Woo and Min-Koo Han
mkh@snu.ac.kr
Seoul National University

ABSTRACT

We fabricated PMOS SPC-Si TFTs which show better current uniformity than ELA poly-Si TFTs and superior stability compare to a-Si:H TFT on a glass substrate employing alternating magnetic field crystallization. However the leakage current of SPC-Si TFT was rather high for circuit element of AMOLED display due to many grain boundaries which could be electron hole generation centers. We applied off-state bias annealing of V_{GS}=5V, V_{DS}=-20V in order to suppress the leakage current of SPC-Si TFT. When the off-state bias annealing was applied on the SPC-Si TFT, the electron carriers were trapped in the gate insulator by high gate-drain voltage (25V). The trapped electron carriers could reduce the gate-drain field, so that the leakage current of SPC-Si TFT was reduced after off-state bias annealing. . We also applied same off state bias annealing at SPC-Si TFT with 20,000 lx light illumination in order to verify the reduction of leakage current of SPC-Si TFT under light illumination. The leakage current of SPC-Si TFT was reduced successfully even under light illumination during off-state bias annealing. The off-state bias annealed SPC-Si TFT could be used as pixel element of high quality AMOLED display.

INTRODUCTION

Hydrogenated amorphous silicon (a-Si:H) TFTs are considered as the pixel element of active matrix organic light emitting diode (AMOLED) due to excellent uniformity in large areas. However the threshold voltage of a-Si:H TFT is increased easily under electrical bias stress [1-2]. Poly-Si TFTs employing excimer laser annealing (ELA) show very good electric characteristics such as high mobility and good stability. But the non-uniformity of poly-Si TFT caused by inherent fluctuation of excimer laser should be improved. Recently solid phase crystallized silicon TFT (SPC-Si TFT) on the glass substrate has gained a considerable attention for pixel element of AMOLED due to a better current stability than a-Si:H TFTs, and an improved current uniformity compare to ELA poly-Si TFTs. The leakage current of SPC-Si TFT is rather high due to fairly many grain boundaries in the channel which behaves generation and recombination centers[3].

The purpose of our work is to report the suppression of the leakage current in SPC-Si TFT employing the off sate bias annealing under light illumination. The SPC-Si TFT could be exposed by backlight or OLED when it is applied at actually display pixel element, the suppression of the leakage current in SPC-Si TFT under light illumination is important.

EXPERIMENT

Fabrication of device

We fabricated PMOS SPC-Si TFTs of coplanar structures on the glass substrate. Amorphous Si was deposited by 50nm. It was crystallized at 700°C for 15 minutes employing alternating magnetic field crystallization (AMFC). The glass substrate was not damaged during the crystallization. The grain size of crystallized Si was around 300Å. The hydrogen plasma treatment was applied in order to improve the s-slope and threshold voltage of SPC-Si TFT [4]. SiO_2 was deposited by 59nm as a gate insulator after active layer pattering. Then the gate electrode was patterned and source, drain were defined by ion implantation (self-aligned structure). The channel width of fabricated SPC-Si TFT was 10μm and the channel length was 7 μm. Figure 1 shows the cross section view of the fabricated SPC-Si TFT.

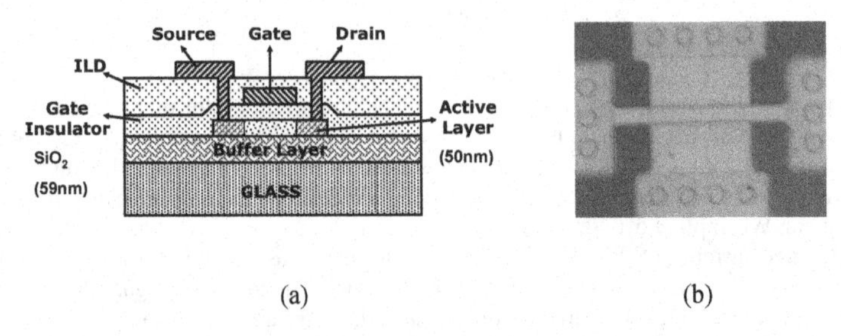

Figure 1. (a) The schematic of the fabricated SPC-Si TFT. (b) The microscopic picture of the fabricated SPC-Si TFT

Off-sate bias annealing

Figure 2 shows transfer characteristics of SPC-Si TFT. When the leakage current of SPC-Si TFT was measured at V_{GS} = 5V, V_{DS} = -5V, it was about 1.49 pA. It was rather high for pixel circuit of AMOLED display. We applied off-state bias annealing with V_{GS} = 5V, V_{DS} = -20V for 1,000 seconds in order to suppress the leakage current of SPC-Si TFT. The leakage current of the SPC-Si TFT was reduced from 1.49 pA to 0.47 pA at V_{GS} = 5V, V_{DS} = -5V due to gate filed reduction caused by trapped electron carriers in the gate insulator. The leakage current was not increased even at the high gate voltage while it was increased as the gate voltage increased at the initial SPC-Si TFT.

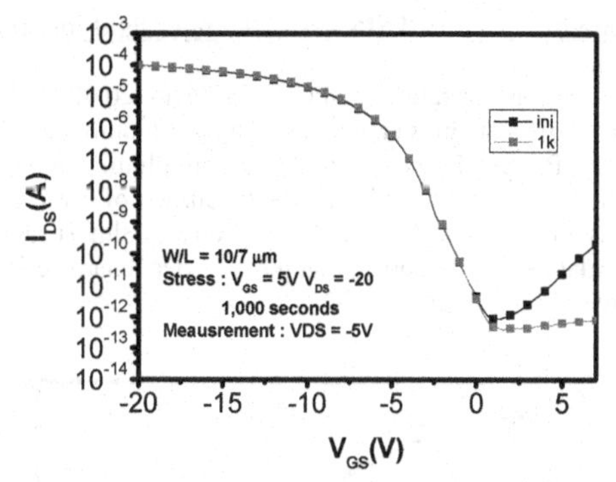

Figure 2. The transfer characteristics of SPC-Si TFT at initial and after off-state bias annealing. The leakage current was reduced significantly after off-state bias annealing.

Thermal annealing of off-state bias annealed SPC-Si TFT

We applied thermal annealing at the off-state bias annealed SPC-Si TFT to verify that the trapped electron carriers could be de-trapped by thermal energy under practical AMOLED display application. We applied the off-state bias annealing of V_{GS}=10V, V_{DS} = -20V for 1,000seconds in order to suppress the leakage current of SPC-Si TFT. The gate-drain voltage(30V) during off-bias annealing, was sufficient for reducing the leakage current of SPC-Si TFT. Then we annealed it at 60 °C, 90°C and 120°C for 1 hour respectively. As shown in figure 3, the leakage current of the SPC-Si TFT was not altered after thermal annealing. We can deduce that the trapped electron carriers could not be released by thermal energy under 120°C. Since the practical operation temperature of AMOLED display is under 60°C, the reduced leakage current could be remained low at the practical application.

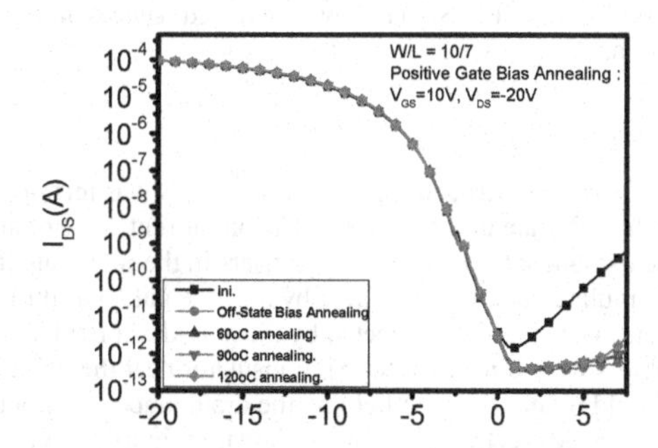

Figure 3. Thermal annealing of off-state bias annealed SPC-Si TFT. The leakage current of SPC-Si TFT was remained low after the thermal annealing.

Bias stress at the off-state bias annealed SPC-Si TFT under light illumination

We applied off-state bias annealing at SPC-Si TFT with 20,000 lx light illumination. Figure 4 shows the result of off-state bias annealing with light illumination. The 20 V of V_{GS}, -20V of V_{DS} was applied at SPC-Si TFT under 20,000 lx light illumination for 1,000 seconds. The leakage current of SPC-Si TFT was reduced while the on current of it was not altered almost. The result of off-state bias annealing with light illumination was almost same with that of off sate bias annealing under dark state. It showed that the electrical characteristic of SPC-Si TFT was not altered almost by light illumination.

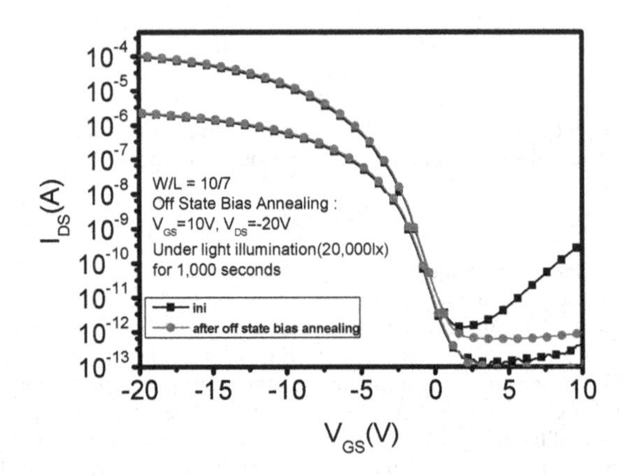

Figure 4. The off-state-bias annealing result on SPC-Si TFT with light illumination. The 20 V of V_{GS}, -20V of V_{DS} was applied at SPC-Si TFT under 20,000 lx light illumination for 1,000 seconds. The leakage current of SPC-Si TFT was reduced successfully by off-state bias annealing with light illumination.

DISCUSSION

The mobility of SPC-Si TFT was about 28.41 V/cm^2 sec. It was increased slightly to 29.38 V/cm^2 sec (3.4%) after off-state bias annealing. The on-current was not altered by off-sate bias annealing. It could be explained trapped electron carriers in the SiO_2 gate insulator. The electron carriers were accumulated at the active layer by positive gate bias during the positive bias annealing. These electron carriers were affected by the gate-drain field near the drain junction. The electron carriers were trapped at the SiO_2 insulator near the drain junction. This trapped electron carriers could reduce the gate field on the drain junction. As a result the leakage current was reduced. This trapped electron carriers are shown in figure 5. The leakage current at the high gate voltage could be explained by band to band tunneling mechanism [5]. The energy band of valence band and conduction band was bended by high gate voltage. The energy gap of valence band and conduction band became narrow so that many electron tunneling could occur.

However after positive gate bias annealing the trapped electron charges in the gate insulator reduced gate-drain field, the energy band could not be banded so much. The leakage current induced by band to band tunneling was suppressed successfully employing positive gate bias annealing.

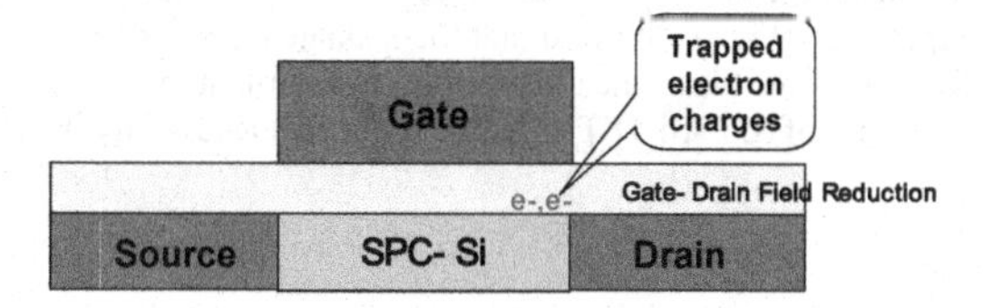

Figure 5. The trapped electron carriers at the SiO_2 gate insulator. It could reduce gate-drain field so the leakage current of SPC-Si TFT could be reduced.

In order to identify the trapped electron carriers in a gate insulator by the off-state bias annealing, we also measured C-V characteristics of SPC-Si TFT before and after off-state bias annealing. The result is shown in Figure 6. The normalized value of gate-drain capacitance was shifted positively after the off-state bias annealing due to the trapped electron charges in a gate insulator near the drain junction [6].

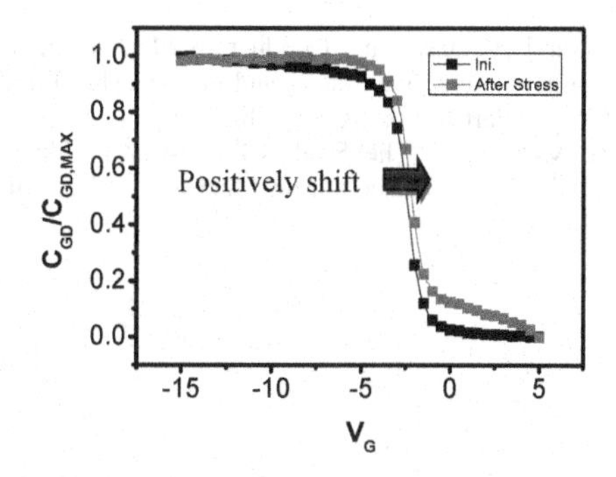

Figure 6. The normalized value of gate-drain capacitance in SPC-Si TFT before and after the off-sate bias annealing. The normalized value of capacitance was shifted positively after the off-state bias annealing due to the trapped charges in a gate insulator

CONCLUSIONS

We fabricated PMOS SPC-Si TFT on a glass substrate employing alternating magnetic field crystallization. The SPC-Si TFT showed better current uniformity than excimer laser annealing poly-Si TFT and better stability than a-Si:H TFT. However the leakage current of SPC-Si TFT was rather high for circuit element of AMOLED display. In this work we applied off-state bias annealing of V_{GS}=5V, V_{DS}=-20V in order to suppress the leakage current of SPC-Si TFT. When the off-state bias annealing was applied in the SPC-Si TFT, the electron carriers were trapped in the gate insulator by high gate-drain field. The trapped electron carriers could reduce the gate-drain field, so that the leakage current of SPC-Si TFT was decreased even at the high gate voltage. The off-state bias annealing was applied at SPC-Si TFT with 20,000 lx light illumination. The experimental result showed that the leakage current of SPC-Si TFT was also reduced successfully by off-sate bias annealing with light illumination.

The leakage current of SPC-Si TFT could be reduced successfully by off-state bias annealing.

REFERENCES

1. P. G. Lecomber, W. E. Spear, and A. G. Ghaith, "Amorphous Silicon Field Effect Device and Possible Application", *IEEE Electron Lett.*, Vol. 15, p.179, 1979.
2. M.J.Powell, "Charge trapping instabilities in amorphous silicon-silicon nitride thin-film transistors", *Appl. Phys. Lett. 43(6)*, pp.597-599, 1983.
3. F. V. Farmakis, J. Brini, G. Kamarinos, C. T. Angelis, C. A. Dimitriadis, and M. Miyasaka, "On-Current Modeling of Large-Grain Polycrystalline Silicon Thin-Film Transistors", *IEEE TRANSACTIONS ON ELECTRON DEVICES,* VOL. 48, No. 4, Apr. 2001
4. F. Kail, A. Hadjadj, P. Roca i Cabarrocas, "Hydrogen diffusion and induced-crystallization in intrinsic and doped hydrogenated amorphous silicon films", *Thin Solid Films*, vol. 487, Issues 1-2, 1 pp. 126-131, Sep, 2005
5. M. Yazakis, S. Takenaka1 and H. Ohshima, "Conduction Mechanism of Leakage Current Observed in Metal-Oxide-Semiconductor Transistors and Poly-Si Thin-Film Transistors", *Jpn. J. Appl. Phys.* Vol. 31 pp.206-209 Part 1, No. 2A, 15 Feb., 1992
6. K-C Moon, J-H Lee, and M-K Han, "The Study of Hot-Carrier Stress on Poly-Si TFT Employing C-V Measurement", IEEE Transaction on Electron Device, vol.52, No.4, pp.512-517, April, 2005

Mater. Res. Soc. Symp. Proc. Vol. 1321 © 2011 Materials Research Society
DOI: 10.1557/opl.2011.1091

Investigation of Amorphous IGZO TFT Employing Ti/Cu Source/Drain and SiNx Passivation

Young Wook Lee[1], Sung-Hwan Choi[1], Jeong-Soo Lee[1], Jang-Yeon Kwon[2] and Min-Koo Han[1]

[1]School of Electrical Engineering, Seoul National University, Gwanak-gu, Seoul 151-742, Korea
[2] Department of Material Science and Engineering, Seoul National University, Gwanak-gu, Seoul 151-742, Korea

ABSTRACT

We successfully fabricated a-IGZO TFTs employing a Ti/Cu source/drain (S/D) and SiNx passivation in order to reduce the line-resistance, as compared to most oxide TFTs that use Mo (or TCO) and SiO_2 for their S/D and passivation, respectively. Although passivated with SiNx, the TFT exhibits good transfer characteristics without a negative shift. However, the TFT employing a Mo S/D exhibited conductor-like characteristics when passivated with SiNx. Our investigation suggests that the IGZO oxygen vacancies found in the Ti/Cu S/D are controlled, resulting in low concentrations, and so prevent the SiNx-passivated TFT from having a negative shift.

INTRODUCTION

Amorphous Indium-Gallium-Zinc-Oxide Thin Film Transistors (a-IGZO TFT) have attracted considerable attention because they exhibit excellent electrical properties, such as a high field-effect mobility and a large on-current along with a small leakage current [1], [2] Its low sub-threshold swing enables fast on/off switching; the uniformity of a-IGZO TFTs is also good [3]. TFT-LCDs and AMOLEDs employing a-IGZO TFTs suggest that an oxide semiconductor is a good candidate for switching devices in displays [3-6].

High resolution and large display sizes require a low data-line resistivity along with high mobility TFTs. When display devices increase in size, resolution, and refresh rate, the required short charging time makes it difficult to charge the pixel electrodes up to the necessary data voltage. High mobility TFTs can overcome this charging problem, however, signal distortions due to the RC-delay of the data-line are still a liability. The RC-delay, therefore, needs to be decreased in order to avoid image distortion. Since the capacitance is usually fixed by the data-line width, the resistivity of the data-line needs to be reduced in order to decrease the RC-delay. Al is widely used for the data-line in displays. Cu, which is less resistive than Al by about 30%, is being considered for the reduction of the data-line resistance [7]. Because the data line is formed simultaneously with the source/drain (S/D) electrodes, employing Cu for a TFT S/D is very important.

Oxide semiconductor TFTs are vulnerable to moisture and oxygen[8]. For that reason, they need a passivation layer, such as SiO_2, Al_2O_3 or PMMA, in order to block ambient gases [8-10]. Among them, SiO_2 is widely used because it can be deposited using the conventional PECVD methods and it exhibits superior stability to the environment[8]. However, SiNx passivation, which is widely used in a-Si TFTs, is difficult to apply to an oxide semiconductor TFT. Because oxide semiconductors are easily damaged by hydrogen plasma, SiNx passivation, which is formed from precursors such as SiH_4, NH_3 and H_2, is likely to damage oxide semiconductors. In actuality, oxide semiconductor TFT with SiNx passivation shifts negatively or becomes a conductor, due to the increased carriers[10, 11].

Unlike other metals, Cu has difficulties when combined with SiO_2 passivation, because Cu-SiO_2-diffusion occurs and so increases the resistance of the Cu data-line[12]. Therefore, it is desirable to passivate with SiNx when employing Cu S/Ds in oxide semiconductor TFTs.

The purpose of this paper is to report on a-IGZO TFTs using Ti/Cu S/Ds, which exhibit good transfer characteristics even though they employ SiNx passivation. In our experiment, we fabricated a-IGZO TFTs employing Ti/Cu for the S/D, and passivated the TFTs with either SiNx or SiO_2 in order to investigate the influence the passivation had on them. Identical experiments were conducted employing Mo S/Ds in order to compare the results.

EXPERIMENT

We fabricated the TFTs with an inverted-staggered etch-stopper structure. The gate metal (Mo) was deposited using DC sputtering on a glass substrate. After the Mo deposition, the gate was patterned by photo-lithography and wet-etching. 250nm-thick-SiO2 was deposited by PECVD for the gate insulator. 40nm-thick-IGZO (In_2O_3 : Ga_2O_3 : ZnO =1:1:1 atomic %) was deposited by DC sputtering and patterned for the active layer. In order to prevent active layer damage from the post processes, such as S/D etching, 50nm-thick-SiO2 was deposited for an etch-stopper (E/S) layer. We deposited a Ti/Cu bi-layer (30nm/300nm), which was patterned by wet-etching for the S/D electrodes. In order to investigate the passivation influences, we deposited two types of passivation, either a 200nm-thick-SiNx or a 200nm-thick-SiO_2 layer. For comparison, we also fabricated TFTs employing a Mo S/D instead of a Ti/Cu S/D. Transfer Length Method (TLM) patterns were fabricated in order to measure the resistivity of the IGZO layer using same process as for the TFTs, except that the gate pattern was removed. Besides the S/D and passivation processes, all of the other processes were performed on all of the devices at the same time. After the fabrication was completed, all of the TFTs were annealed at 300℃ in the air.

RESULTS AND DISCUSSION

Figure 1 shows the transfer curves of the four types of TFTs, which employ different S/Ds and passivations. We can verify the obvious differences in the transfer characteristics according to the S/D. In case of the Ti/Cu electrodes (Figure 1 (a)), although the TFT employed SiNx passivation, it exhibited good transfer characteristics except for a slight negative turn-on compared to the TFT using SiO_2 passivation.

However, in the case of the Mo S/Ds (Figure 1 (b)), the TFT with SiO_2 passivation shows good characteristics, whereas the transfer curve of the TFT with SiNx passivation is flat, resembling a conductor. This is attributed to the hydrogen injected into the IGZO layer bonds with the oxygen and therefore increases the number of oxygen vacancies in the IGZO channel layer [10, 11], i.e. the hydrogen behaves like a dopant. Because more hydrogen is injected into the IGZO layer during the SiNx deposition compared to the SiO_2 deposition, the TFT with the SiNx passivation tends to shift negatively; i.e. it becomes like a conductor [10]. The different results with the different S/D metals suggest that Ti/Cu can suppress the carrier increase in spite of the SiNx passivation.

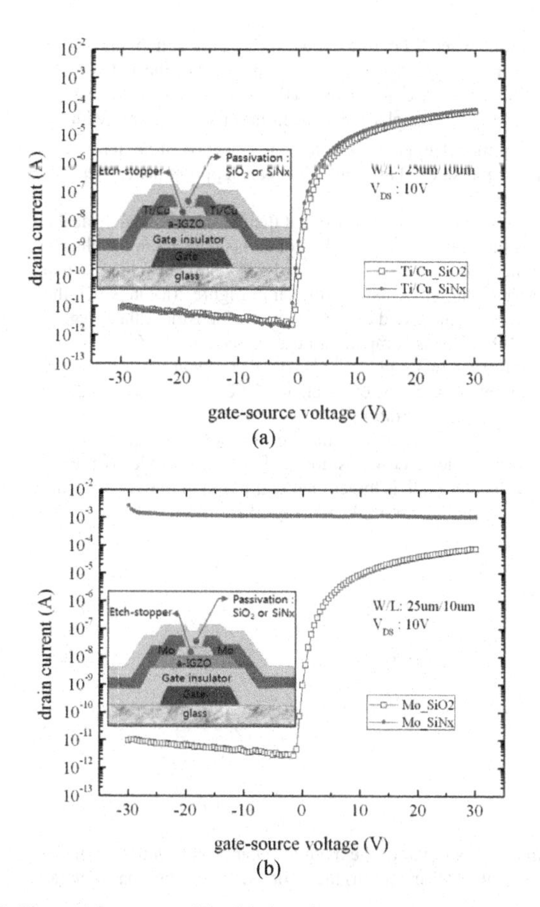

Figure 1. The transfer curves of the fabricated TFTs using the different passivations: (a) the Ti/Cu source/drain, (b) the Mo source/drain (the insets show the structure of the TFTs)

Many studies on the characteristic of TFTs which employ various metals for S/D electrodes have been reported [4, 13-15], however, most of them have focused on the contact behavior; there has been no satisfactory information about the influence of the S/D metal on the oxide semiconductor channel, especially regarding the carrier concentration.

In order to compare the influence of the S/D on the IGZO channel layer, we measured the resistance of the various channel lengths with the TLM-pattern, and determined the IGZO film resistivity [Ω/um], which is the slope value of the plot of the resistivity versus the channel length. For the SiO_2 passivation there is little difference in the IGZO film resistivity between the Mo ($5*10^6$Ω/um) and the Ti/Cu ($4*10^6$Ω/um), whereas, in the SiNx case, the resistivity of the IGZO with the Ti/Cu ($5*10^5$Ω/um) is about ten-times larger than that of the

Mo ($4*10^4\Omega$/um). In other words, when SiNx was deposited for passivation, the IGZO conductivity for the Mo S/D is ten times larger than that for the case of the Ti/Cu S/D. Assuming that the change of the carrier mobility was small, we might estimate, using the conductivity equation ($\sigma = q\mu_n n$), that the electron concentration of the IGZO with the Mo is about ten times higher than that of Ti/Cu, i.e. the carrier concentration of the IGZO channel layer depends on the S/D metal and so the Ti/Cu electrode can suppress the carrier increase in the channel even under SiNx passivation.

Through the further investigation, we surmise that the increased oxygen vacancy donor concentration, which is thought to equal the electron concentration in the channel, during SiNx passivation can be reduced by the Ti-contact-IGZO, similar to the mechanism illustrated in Figure 2. We assume an initial state, like seen in Figure 2(a), in which the oxygen vacancies of the channel are increased during the SiNx deposition; the oxygen vacancy of the Ti-contact-IGZO is low as compared to the Mo-contact-IGZO. It is well known that oxygen vacancies diffuse easily [4, 16-18]. This means that when the oxygen vacancy concentration is different between the channel and the contact region, like seen in Figure 2(a), it induces oxygen vacancy migration.

After migration, the oxygen vacancies in channel of the Ti/Cu S/D decrease, so that the carrier increase incurred during the SiNx deposition will be reduced. Meanwhile, in the case Mo, the oxygen vacancy decrease rate is lower than that found in the Ti/Cu, as described in Figure 2(b), so that the suppressing effect might be insignificant.

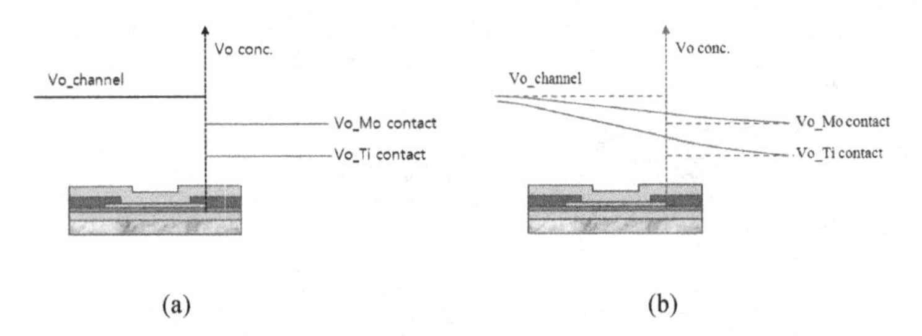

Figure 2. The schematic illustration of the oxygen vacancy concentration difference in the channel region according to the S/D metal: (a) the initial state, (b) the final state (after the oxygen vacancy diffusion)

In order to support the above mechanism, there is the need for proof, i.e. exhibiting the difference in oxygen vacancy concentrations in the Ti-contact-IGZO and the Mo-contact-IGZO. However, it is rather difficult to measure oxygen vacancy directly, so we measured and compared the oxygen intensity instead.

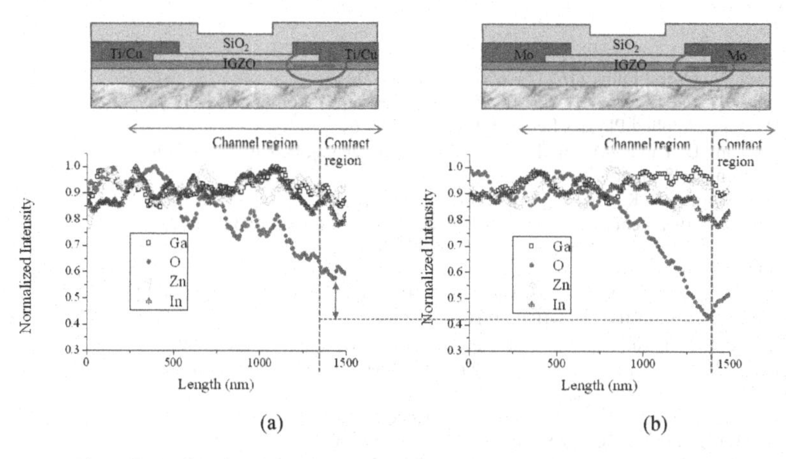

Figure 3. Analyzed regions (depicted by the red line in the cross-section schematics) and the normalized intensities of the IGZO elements: (a) the Ti/Cu S/D, (b) the Mo S/D

We analyzed the intensity of each element in the IGZOs by employing Energy Dispersive Spectrometry (EDS) taken from the channel region to the contact region. The samples taken to analyze are the TFTs passivated with SiO_2. Figure 3 shows the analyzed region (depicted by the red line in the cross-section schematics) and the results of the normalized intensity, which was evaluated from the intensity divided by the maximum intensity of each element. We notice that there is a difference in the oxygen intensity between the channel and the contact region, whereas there is little difference in the In, Ga and Zn intensities. In addition, the oxygen intensity of the Mo-contact-IGZO is lower than that found for the Ti-contact-IGZO. If the oxygen intensity is inversely proportional to the oxygen vacancy, we might regard that the oxygen vacancy concentration of the Mo-contact-IGZO is higher than that of the Ti-contact-IGZO.

In this manner, the Ti-contact-IGZO ends up with a lower oxygen vacancy concentration than the Mo-contact-IGZO, and the diffusion of the oxygen vacancies between the channel and the contact region is the cause for the difference in the oxygen vacancies in the channel according to the S/D metal as shown in Figure 2.

CONCLUSIONS

When Cu is employed for the S/D, SiO_2 passivation cannot be used due to Cu-diffusion, and so SiNx passivation is desirable. However, oxide semiconductors are easily damaged by hydrogen or hydrogen plasma; this makes it difficult to use SiNx passivation because it has much hydrogen sourced during deposition. In order to solve this conflict, we investigated and implemented a Ti/Cu structure for the S/D electrode, which could control the control carrier concentration in the channel by 1/10 times below that of a Mo S/D. The origin of low carrier concentration in the channel is that Ti-contact-IGZO has lower oxygen vacancy concentration than Mo-contact-IGZO. This can suppress the carrier increase in the channel caused by oxygen vacancy diffusion.

REFERENCES

1. H. Hosono, Thin Solid Films **515** (15), 6000-6014 (2007).
2. H. Yabuta, M. Sano, K. Abe, T. Aiba, T. Den, H. Kumomi, K. Nomura, T. Kamiya and H. Hosono, Applied Physics Letters **89**, 112123 (2006).
3. J. Lee, D. Kim, D. Yang, S. Hong, K. Yoon, P. Hong, C. Jeong, H. Park, S. Y. Kim and S. K. Lim, 2008 (unpublished).
4. T. Arai, N. Morosawa, K. Tokunaga, Y. Terai, E. Fukumoto, T. Fujimori, T. Nakayama, T. Yamaguchi and T. Sasaoka, Proceeding of SID, 1033-1036 (2010).
5. H.-H. Lu, H.-C. Ting, T.-H. Shih, C.-Y. Chen, C.-S. Chuang and Y. Lin, Proceeding of SID, 1136-1138 (2010).
6. Y. G. Mo, M. Kim, C. K. Kang, J. H. Jeong, Y. S. Park, C. G. Choi, H. D. Kim and S. S. Kim, Proceeding of SID, 1037-1040 (2010).
7. S. Takasawa, S. Ishibashi and T. Masuda, Proceeding of SID, 1313-1316 (2009).
8. J. K. Jeong, H. Won Yang, J. H. Jeong, Y.-G. Mo and H. D. Kim, Applied Physics Letters **93** (12), 123508 (2008).
9. D. Hong and J. F. Wager, Journal of Vacuum Science & Technology B: Microelectronics and Nanometer Structures **23** (6), L25 (2005).
10. K.-S. Son, T.-S. Kim, J.-S. Jung, M.-K. Ryu, K.-B. Park, B.-W. Yoo, K. Park, J.-Y. Kwon, S.-Y. Lee and J.-M. Kim, Electrochemical and Solid-State Letters **12** (1), H26 (2009).
11. S. Narushima, H. Hosono, J. Jisun, T. Yoko and K. Shimakawa, Journal of Non-Crystalline Solids **274**, 313-318 (2000).
12. F. Braud, J. Torres, J. Palleau, J. L. Mermet, C. Marcadal and E. Richard, Microelectronic Engineering **33**, 293-300 (1997).
13. P. Barquinha, A. M. Vilà, G. Gonçalves, L. Pereira, R. Martins, J. R. Morante and E. Fortunato, IEEE TRANSACTIONS ON ELECTRON DEVICES **55** (4), 954-960 (2008).
14. H.-K. Kim, S.-H. Han, T.-Y. Seong and W. K. Choi, Journal of The Electrochemical Society **148** (3), G114 (2001).
15. Y. Shimura, K. Nomura, H. Yanagi, T. Kamiya, M. Hirano and H. Hosono, Thin Solid Films **516** (17), 5899-5902 (2008).
16. G.-Y. Huang, C.-Y. Wang and J.-T. Wang, Journal of Physics: Condensed Matter **21** (19), 195403 (2009).
17. K. Jug, N. N. Nair and T. Bredow, Phys . Chem. Chem. Phys **7**, 2616-2621 (2005).
18. R. Schaub, E. Wahlstro¨m, A. Rønnau, E. Lægsgaard, I. Stensgaard and F. Besenbacher, SCIENCE **299**, 377-379 (2003).

Mater. Res. Soc. Symp. Proc. Vol. 1321 © 2011 Materials Research Society
DOI: 10.1557/opl.2011.950

Reliability of Oxide Thin Film Transistors under the Gate Bias Stress with 400 nm Wavelength Light Illumination

Soo-Yeon Lee[1], Sun-Jae Kim[1], Yongwook Lee[1,2], Woo-Geun Lee[2], Kap-Soo Yoon[2], Jang-Yeon Kwon[3], and Min-Koo Han[1]

[1]School of Electrical Engineering and Computer Science, Seoul National University, Seoul, Republic of Korea
[2]Samsung Electronics, Yongin-Si, Republic of Korea
[3]Department of Materials Science and Engineering, Seoul National University, Seoul, Republic of Korea

ABSTRACT

We have investigated the reliability of the inverted-staggered etch stopper structure oxide-based TFTs under negative gate bias stress combined with 400 nm wavelength light illumination and the relationship between the carrier concentration at the channel and the extent of V_{th} shift. It was found that the photo-induced holes cause the severe V_{th} degradation at the beginning of stress and the hole trapping rate of a single hole is not altered with the increase of the hole concentration. In oxide-based TFTs, the hole concentration at the channel is the determinant factor of the reliability.

INTRODUCTION

Recently, amorphous oxide-based thin-film transistors (TFTs) have attracted considerable attention due to their higher field-effect mobility and good uniformity. Oxide-based TFTs such as IGZO TFTs can be fabricated at low temperature(less than 350 C°) by simple process compatible to the commercially available Si:H TFTs fabrication process[1-7]. Therefore, amorphous oxide-based TFTs are considered as the pixel element for active-matrix liquid crystal displays and active-matrix organic light emitting diode displays.

However, oxide-based TFTs are very sensitive to electrical stress and ambient condition such as light, temperature and moisture [8-12]. The changes in the characteristics of the TFT, such as the threshold voltage (V_{th}) shift, are caused by not only electrical and optical stress but also the atmospheric conditions. Especially, the electrical stress of oxide TFTs has been researched intensively and it is considered that the main reason of the V_{th} shift is charge trapping [10-11]. As for the light illumination, degradation under negative gate bias stress combined with light is reported because the TFTs are exposed to light and negative gate bias stress in display applications [8-9]. When light illumination is combined with negative gate bias stress, V_{th} is degraded more than that in the dark state and it is reported that hole carriers generated by light illumination contribute to charge trapping and accelerate the V_{th} shift [8-9]. However, the systematic analysis about the carrier concentration of active layer and the acceleration of V_{th} shift has not been accomplished yet. Suresh et al., have reported that as the carrier concentration of the active layer increases, the V_{th} shift increases under bias stress [10]. In this case, because the carrier concentration was controlled by the oxygen partial pressure for the deposition of the active layer, it is possible that not only the carrier concentration but also the other characteristic of the active layer is changed.

In this paper, we chose the 400 nm wavelength light, having larger photon energy than the optical band gap (E_{opt}) of the IGZO semiconductor layer, to induce the band to band carrier generation and examined negative gate bias stress. Because the photo-induced carrier concentration is proportional to the intensity of light, we could analyze the relationship between the carrier concentration at the channel and the extent of V_{th} shift systematically.

FABRICATION AND EXPERIMENT

We fabricated IGZO TFTs with an inverted-staggered etch stopper structure. The gate metal (Mo) was deposited by DC sputtering on a glass substrate. The gate insulator layer of SiN_x was deposited by plasma enhanced chemical vapor deposition (PECVD) and the active layer was deposited by sputtering. A SiO_2 layer was formed as an etch stopper. After the active island was patterned, the source and drain electrodes (Mo) were deposited by sputtering. The dimension of our device was 50 μm/25 μm (*W*/*L*). We used 400 nm wavelength light with a band-pass filter from a Xenon lamp light source. A semiconductor analyzer was used to measure the device characteristics.

RESULTS AND DISCUSSION

When we performed the negative gate bias stress in the dark state for 7000 sec, V_{th} was shifted negatively by less than 1 V. When 400 nm light was combined with the negative bias stress and the intensity of light was 0.025 mW/cm^2, V_{th} decreased by about 8.68 V. Figure 1 shows the time evolution of the transfer curve of the oxide TFTs. At the beginning of stress, V_{th} was shifted rapidly. Because the SS was not significantly changed, it is implied that the extent of V_{th} shift (ΔV_{th}) was caused by charge trapping as reported in the literature [8-9].

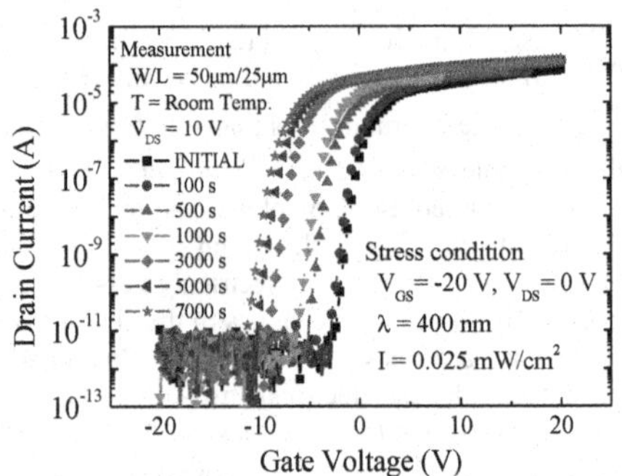

Figure 1. Transfer curves of oxide TFTs subjected to negative gate bias (-20 V) stress with 400 nm wavelength light illumination (0.025 mW/cm^2).

We examined negative bias stress under various intensities of light and the ΔV_{th} according to the stress time is shown in figure 2. The wavelength was 400 nm and the intensities of light was 0.025, 0.05, 0.1, 0.27, 0.54 and 0.84 mW/cm^2. As the intensity of light increased, the ΔV_{th} increased more drastically at the beginning of bias stress. The rate of increase of ΔV_{th} can be expressed numerically by $d\Delta V_{th}/dt$ at stress time (t) is close to 0 s. $d\Delta V_{th}/dt$ has a linear relationship with the intensity of light as shown in figure 2 (b). Because the carrier concentration at the channel is proportional to the intensity of light, it seemed that the photo-induced holes contributed to the charge trapping and made the change of ΔV_{th} rapid. As the intensity of light increases and the photo-induced holes get larger, the hole trapping rate might increase proportionally to the photo-induced holes so that the severe V_{th} degradation is observed.

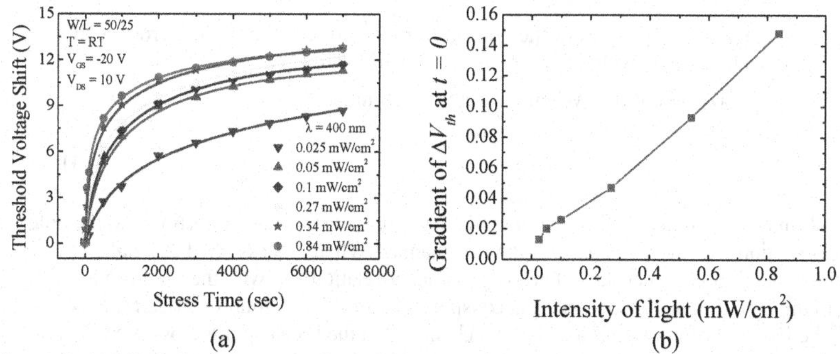

Figure 2. (a)ΔV_{th} versus the stress time of IGZO TFTs under negative gate bias stress (-20 V) with light illumination. The solid lines represent the fitted line with logarithmic function. (b)$d\Delta V_{th}/dt$ when the stress time is 0 s.

To investigate the quantified relationship between the carrier concentration and ΔV_{th}, we tried to fit the time evolution of ΔV_{th} with the following logarithmic function which expresses the charge trapping,

$$\Delta V_{th} \sim r_0\{\ln(t/t_0 + 1) - t/t_m\}, \tag{1}$$

where r_0 is proportional to the trap density of the gate insulator (N_0) and de Broglie wavelength (λ), $1/t_0$ is the tunneling rate and $1/t_m$ is introduced for obtaining realistic value [14]. It is assumed that the charge trapping is caused by the direct tunneling to the trap and all trapped charges contribute to ΔV_{th}. The time rate of change in the trap density of the gate insulator layer ($N_t(t)$) is proportional to $N_t(t)$ and the tunneling probability as follows,

$$dN_t(t)/dt = -t_0^{-1}\exp(-x/\lambda)N_t(t), \tag{2}$$

where $t_0^{-1}\exp(-x/\lambda)$ is the time independent tunneling probability and x is the location of the trap site in the gate insulator away from the interface between the active layer and the gate insulator layer. $d\Delta V_{th}/dt$ at t = 0 can be expressed approximately as r_0/t_0. Because r_0 is proportional to the trap density and the trap density may not be changed by the intensity of light, it is considered that r_0 is constant according to the intensity of light. Therefore, we found that $1/t_0$ is directly related to the ΔV_{th} at the beginning of stress.

When it is considered that the rate of hole trapping is determined by the product of the density of holes in the channel, p_{ch}, and the density of trap sites in the insulator, the relationship between carrier concentration at channel and $1/t_0$ can be found. $dN_t(t)/dt$ can be expressed using the concept of the recombination of electrons and holes in semiconductor. In semiconductor, the recombination rate of electrons and holes is expressed by product of the electron and hole concentration and the proportionality constant [15]. Proportionality constant includes the capture cross section and the carrier velocity. In the charge trapping, the eq. (2) can also be expressed as,

$$dN_t(t)/\,dt = -t_0^{-1}\exp(-x/\lambda)N_t(t) \approx -\alpha p_{ch}N_t(t), \tag{3}$$

where α should be defined by product of the hole carrier velocity and the capture cross section at the interface between the gate insulator and the active layer. From eq. (3),
$t_0^{-1}\exp(-x/\lambda) \approx -\alpha p_{ch}N_t(t)$ is deduced. When x variable is eliminated,

$$1/t_0 \propto \alpha p_{ch}, \tag{4}$$

is deduced ultimately. It means that $d\Delta V_{th}/dt$ at the beginning of stress is proportional to the hole capture cross section and the hole concentration in channel. When it is assumed that hole concentration is negligible under dark state, $1/t_0$ has linear relationship with the photo-induced hole carrier concentration at channel and α is constant. Therefore, it is found that $d\Delta V_{th}/dt$ is because of the increase of photo-induced hole at channel and the hole capture cross section at interface between gate insulator and active layer is not altered with the change of hole concentration. In other words, the increase of V_{th} degradation under light illumination is caused by the photo-induced holes and trapping probability corresponding to a single hole is constant.

From the result of figure 2 (b), the variation of the trapping rate of holes can be determined according to the incident photon flux. At first, using the $d\Delta V_{th}/dt$ value of the y axis of figure 2(b), the trapping rate of holes [#/cm^2-s] can be expressed. Then, the intensity of light of the x-axis of figure 2(b) can be converted to the photon flux [#/cm^2-s]. The result is also listed in Fig. 2(b). When the gradient is extracted, we can find the ratio of the incident photon to the trapped holes and it was 1.53×10^{-3} % as shown in figure 3. The ratio of the incident photon to the trapped holes will be dependent on the properties of the gate insulator and charge transfer between the active layer and gate insulator.

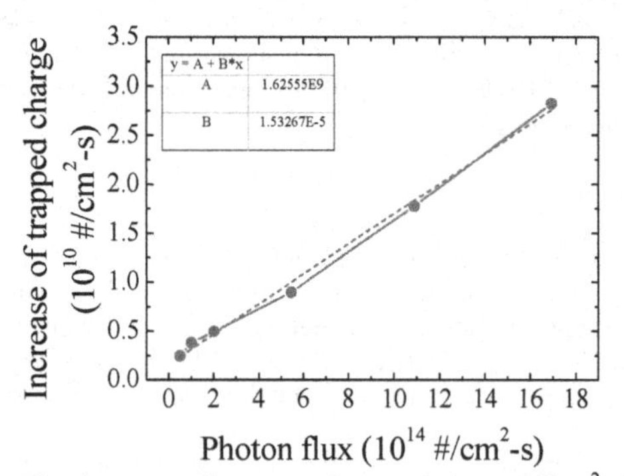

Figure 3. The time rate of increase of trapped charge [#/cm^2-s] according to illuminated photon flux.

CONCLUSIONS

We investigated the relationship between carrier concentration at the channel and ΔV_{th} with systematic analysis. In the oxide TFTs, the degradation of the V_{th} increased when the light illumination is combined with negative gate bias stress and V_{th} was shifted rapidly at the beginning of stress as the intensity of light increased. $d\Delta V_{th}/dt$ had a linear relationship with the intensity of light at the beginning of stress and this phenomenon was analyzed using the logarithmic charge trapping model. We found that the severe degradation is because of the photo-induced holes at the channel and he hole capture cross section at the interface between the gate insulator and the active layer was not changed even though the hole carrier increased.

REFERENCES

1. K. Nomura, H. Ohta, K. Ueda, T. Kamiya, M. Hirano, and H. Hosono, *Science*, 300, 1269 (2003).
2. K. Nomura, H. Ohta, A. Takagi, T. Kamiya, M. Hirano, and H. Hosono, Nature (London) 432, 488 (2004).
3. E. Fortunato, P. Barquinha, A. Pimentel, A. Goncalves, A. Margues, L. Pereira, and R. 4. Martins, Adv. Mater.(Weinheim, Ger.), **17,** 590 (2005).
4. H. Yabuta, M. Sano, K. Abe, T. Aiba, T. Den, H. Kumomi, K. Nomura, T. Kamiya, and H. Hosono, Appl. Phys. Lett., 89, 112123 (2006).
5. A. Suresh, P. Gollakota, P. Wellenius, A. Dhawan, and J. F. Muth, Thin Solid Films 516, 1326 (2008).
6. P. Barquinha, L. Pereira, G. Goncalves, R. Martins, and E. Fortunato, J. Electrochem. Soc., 156, H161 (2009).
7. J. Y. Kwon, K. S. Son, J. S. Jung, T. S. Kim, M. K. Ryu, K. B. Park, B. W. Yoo, J. W. Kim, Y. G. Lee, K. C. Park, S. Y. Lee, and J. M. Kim, IEEE Electron Device Lett. **29**, 1309 (2008).
8. J. S. Park, T. S. Kim, S. Son, J. S. Jung, K.-H. Lee, J.-Y. Kwon, B. Koo, and S. Lee, IEEE Electron Device Lett. 31, 440 (2010).
9. K.-H. Lee, J. S. Jung, K. S. Son, J. S. Park, T. S. Kim, R. Choi, J. K. Jeong, J.-Y. Kwon, B. Koo, and S. Lee, Appl. Phys. Lett. 95, 232106 (2009).
10. A. Suresh, and J. F. Muth, Appl. Phys. Lett. 92, 033502 (2008).
11. K. Hoshino, D. Hong, H. Q. Chiang, and J. F. Wager, IEEE Trans. Electron Devices, 56, 1365 (2009).
12. M. E. Lopes, H. L Gomes, M. C. R. Medeiros, P. Barquinha, L. Pereira, Appl. Phys. Lett. 95, 063502 (2009).
13. M. J. Powell, Appl. Phys. Lett. 43, 597 (1983).
14. A. V. Ferris-Prabhu, IEEE Trans. Electron Devices ED-24, 524 (1977).
15. A. S. Grove, Physics and Technology of Semiconductor Devices(John Wiley and Sons, 1967) p.128.

Mater. Res. Soc. Symp. Proc. Vol. 1321 © 2011 Materials Research Society
DOI: 10.1557/opl.2011.954

DC and AC Gate-Bias Stability of Nanocrystalline Silicon Thin-Film Transistors Made on Colorless Polyimide Foil Substrates

I-Chung Chiu[1], I-Chun Cheng[1,2,*], Jian Z. Chen[3], Jung-Jie Huang[4], Yung-Pei Chen[5]
[1] Graduate Institute of Photonics and Optoelectronics, National Taiwan University, Taipei, 10617 Taiwan
[2] Department of Electrical Engineering, National Taiwan University, Taipei, 10617 Taiwan
[3] Institute of Applied Mechanics, National Taiwan University, Taipei, 10617 Taiwan
[4] Department of Materials Science and Engineering, MingDao University, Changhua, 52345 Taiwan
[5] Display Technology Center, Industrial Technology Research Institute, Hsinchu, 31040 Taiwan

ABSTRACT

Staggered bottom-gate hydrogenated nanocrystalline silicon (nc-Si:H) thin-film transistors (TFTs) were demonstrated on flexible colorless polyimide substrates. The dc and ac bias-stress stability of these TFTs were investigated with and without mechanical tensile stress applied in parallel to the current flow direction. The findings indicate that the threshold voltage shift caused by an ac gate-bias stress was smaller compared to that caused by a dc gate-bias stress. Frequency dependence of threshold voltage shift was pronounced in the negative gate-bias stress experiments. Compared to TFTs under pure electrical gate-bias stressing, the stability of the nc-Si:H TFTs degrades further when the mechanical tensile strain is applied together with an electrical gate-bias stress.

INTRODUCTION

In recent years, flexible electronics have attracted considerable interest. They are the foundational technology for building electronics on foldable, rollable, and even stretchable surfaces. Their main application comprises electronic books, cell phones, roll-up displays, portable displays, and so on. Hydrogenated amorphous silicon (a-Si:H) thin-film transistors (TFTs) are the industrial technology standard for active matrix (AM) liquid crystal displays (LCDs). This leads to extensive studies on flexible a-Si:H TFTs [1-3]. Hydrogenated nanocrystalline silicon (nc-Si:H) has been considered an alternative active material for TFTs due to its higher electron mobility compared to a-Si:H [4-6], with CMOS-capability [7], and demonstrating better stability against light-induced degradation [8] and electrical bias stress [9]. Most significant is its fabrication process being fully compatible with current display mass-production facilities.

During the fabrication or operation of flexible electronics, the TFTs may suffer from mechanical stress. The on-plastic TFTs must sustain functionality either during or after mechanical deformation. This study demonstrated nc-Si:H TFTs on colorless polyimide foil substrates and investigated their electrical stability against a dc and an ac bias stress with and without mechanical bending.

EXPERIMENT

A glass plate was used as the supporting substrate. A layer of debonding material and a 20 μm experimental colorless polyimide were consecutively coated onto the supporting substrate. Prior to TFT fabrication, a 300 nm SiN_x buffer layer was deposited to prevent moisture and contaminants from the substrate and to improve adhesion to the subsequent deposited device layers. The nc-Si:H TFTs were fabricated in a staggered bottom-gate back-channel-passivated structure. Firstly, an approximate 200 nm-thick Ti/Al/Ti metallic thin film was deposited by rf-sputtering, and patterned to form the gate. The silicon tri-layer, comprising: (i) 300 nm SiN_x as a gate insulator; (ii) 200 nm intrinsic silicon nc-Si:H as an active layer; (iii) and 50 nm n^+ a-Si thin film as a source/drain layer, was deposited, followed by the deposition of a source/drain contact metal layer via rf-sputtering. After the island definition and source/drain patterning by wet and dry etching, the SiN_x passivation layer was deposited. Finally, contact holes were opened to reach the source, drain and gate electrodes. The thin films for substrate passivation, gate dielectric, active layer, and doped layer were obtained by plasma enhanced chemical vapor deposition (PECVD) at a substrate temperature of 200°C.

To investigate the mechanical strain effect on the electrical performance and stability of the on-plastic nc-Si:H TFTs, this study performed dc and ac electrical gate-bias stress experiments on these TFTs when they were kept flat or subjected to outward bending with a bending axis perpendicular to the current flow direction. A dc or ac gate-bias of +20 V or -20V was applied to the gate, with the source and drain being grounded for a certain time period. After electrical stressing, the transfer characteristics were evaluated using a Keithley 2636A Dual Channel Source Meter. Prior to the next gate-bias experiment, the TFT was annealed at 180°C for 30 min to restore their initial current-voltage characteristics. The saturation mobility was obtained by $\mu_{sat}=2(g_{ms})^2(L/W)/C_{ox}$ [10], with g_{ms} as the slope of the linear fit of $I_{DS}^{1/2}$ versus V_G curve at V_{DS}=10 V; L and W as the channel length and width; and C_{ox} as the capacitance per unit area. The threshold voltage was determined by the gate-voltage intercept of the linear fit. The dc or ac electrical stability was subsequently evaluated by the threshold voltage shift, ΔV_{th} after a dc or ac gate-bias stress was applied.

RESULTS AND DISCUSSION

Figure 1 shows the saturation mobilities of nc-Si:H TFTs with different channel length under various mechanical bending curvature κ, where κ = 1 / bending radius. The field-effect mobility increases in conjunction with the bending curvature, that is, the tensile strain increases. The rise in mobility under tensile strain is plausible due to the decrease of conduction band tail width, similar to that observed in a-Si:H TFTs [11,12].

This study consequently examined the effect of mechanical strain on the dc gate-bias stability of the nc-Si:H TFTs. Figure 2 shows the threshold voltage shifts of an nc-Si:H TFT (W/L= 5μm/40μm) under combined dc electrical gate-bias stressing and mechanical straining. Various outward bending curvatures were applied to the TFT. The results showed that ΔV_{th} increased with the stress time, and the amount of ΔV_{th} became greater for larger bending curvature, indicating that the mechanical tensile strain further deteriorated the stability of nc-Si:H TFTs, in addition to that caused by the dc gate-bias stress.

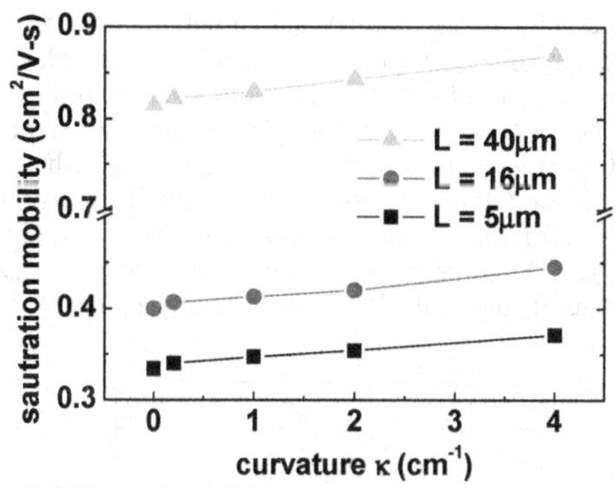

Figure 1. Saturation mobilities of nc-Si:H TFTs with various L = 5 μm, 16 μm and 40 μm, and fixed W = 5μm under various mechanical bending curvature κ. The lines shown in the figure are visual guideline.

Because nc-Si:H is a material comprising Si nanocrystals imbedded in an amorphous matrix, and the structure of TFTs used in this study is bottom-gated, we inferred the results from the studies of a-Si:H TFTs under mechanical flexing. The instability of nc-Si:H TFTs might have been caused by two mechanisms: charge trapping in the gate dielectric, and creation of meta-stable dangling bonds in the channel near the gate dielectric interface [13-15]. The fitting curves in Figure 2 show that the relationship between threshold voltage shift, ΔV_{th}, and gate-bias stressing time, t, is governed by power law, $\Delta V_{th} \propto t^{\beta}$ with $\beta \approx 0.32$, which suggests that the increase of threshold voltage shifts is due mainly to the state creation at the interface between the nc-Si:H channel and gate dielectric [12].

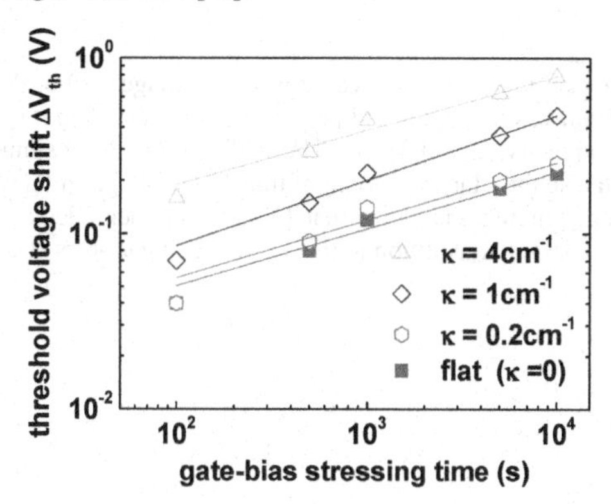

Figure 2. Threshold voltage shifts of on-polyimide nc-Si:H TFTs (W/L=5μm/40μm) versus dc gate-bias stressing time under various mechanical bending curvatures κ. The bias voltage is 20 V. The lines shown in the figure are fitting lines.

Figure 3 shows the threshold voltage shifts of the on-polyimide nc-Si:H TFTs caused by 10,000 sec of ac electrical stress at various frequencies, with a bias amplitude of +20V and -20V. Under a positive ac gate-bias stress condition, the threshold voltage shift generally decreased with an increase in ac frequency with weak frequency dependence. Different from positive ac gate-bias stress results, the threshold voltage shifts under negative gate-bias stress conditions showed an apparent dependence on the bias stress frequency, which was similar to the results observed in a-Si:H TFTs. A cut-off frequency of approximately 100 Hz was obtained, which was close to that of a-Si:H TFTs. The abrupt degradation in instability under a negative ac gate-bias stress at low frequency was plausible due to the RC time-delay effect [16-18].

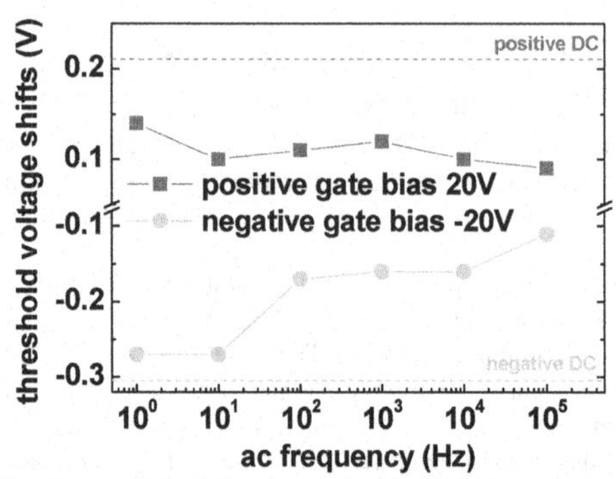

Figure 3. Threshold voltage shift of the on-polyimide nc-Si:H TFT (W/L=80μm/16μm) versus ac gate-bias stressing frequency. The stressing time for both the dc and ac gate-bias is 10,000 sec, and the duty cycle of the ac gate-bias stress is 50 %. The lines shown in the figure are a visual guideline.

Figure 4 shows the ac gate-bias stress-induced threshold voltage shift of the on-polyimide nc-Si:H TFT (W/L= 80μm/16μm) as a function of bias stress frequency when it is kept flat or subjected to an outward bending curvature of 4 cm^{-1}. The findings revealed a similar frequency-dependent trend, though with a slightly larger amount of threshold voltage shifts when the TFT was under mechanical outward bending, indicating that TFT stability degraded further when mechanical tensile strain was applied, in addition to the ac electrical gate-bias stress.

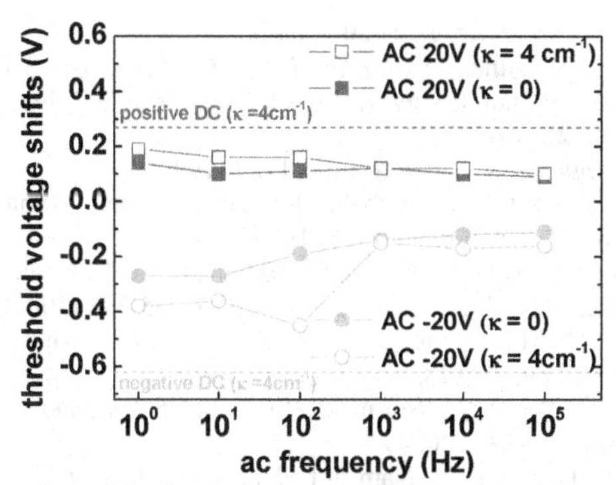

Figure 4. Threshold voltage shift of the on-polyimide nc-Si:H TFT (W/L=80μm/16μm) under a bending curvature of 0 (flat) and 4 cm^{-1} versus ac gate-bias stressing frequency. The stressing time for both the dc and ac gate-bias is 10,000 sec, and the duty cycle of the ac gate-bias is 50 %. The lines shown in the figure are a visual guideline.

CONCLUSION

This study investigated the electrical performance and the dc and ac gate-bias stability of on-polyimide nc-Si TFTs at various mechanical tensile strain levels. The electron field-effect mobilities of these TFTs increased in conjunction with the applied tensile strain, though their dc and ac electrical stability deteriorated under mechanical strain. The degree of deterioration was positively correlated with the amount of strain applied. The results revealed an apparent frequency dependence of the threshold voltage shifts under negative ac gate-bias stressing, which possibly resulted from an RC time-delay effect.

ACKNOWLEDGMENTS

I-Chun Cheng and Jian-Zhang Chen are grateful for and acknowledge the funding support from the National Science Council, Taiwan R.O.C. under grant nos. NSC 99-2628-E-002-203 and NSC 99-2627-M-002-008, respectively.

REFERENCES

1. H. Gleskova and S. Wagner, *IEEE Electron Device Lett.*, **20**, 473-475 (1999).
2. K. Long, A. Z. Kattamis, I-C. Cheng, H. Gleskova, S. Wagner, and J. C. Sturm, *IEEE Electron Device Lett.*, **27**, 111-113 (2006).
3. J. H. Cheon, W. G. Lee, T. H. Lim, and J. Jang, *Electrochem. and Solid-State Lett.*, **12**, 25-28 (2009).

4. I-C. Cheng and S. Wagner, *Appl. Phys. Lett.*, **80**, 440-442 (2002).
5. C. H. Lee, A. Sazonov, and A. Nathan, *Appl. Phys. Lett.*, **86**, 222106-1-3 (2005).
6. A.Z. Kattamis, R.J. Holmes, I-Chun Cheng; K. Long, J.C. Sturm, S.R. Forrest, S. Wagner, *IEEE Electron Device Lett.*, **27**, 49-51 (2006).
7. Y. Chen and S. Wagner, *Appl. Phys. Lett.*, **75**, 1125-1127 (1999).
8. N. Beck, J. Meier, J. Fric, Z. Remeš, A. Poruba, R. Flückiger, J. Pohl, A. Shah and M. Vaněček, *J. Non-Cryst. Solids*, **198-200**, 903-906 (1996).
9. A. Orpella, C. Voz, J. Puigdollers, D. Dosev, M. Fonrodona, D. Soler, J. Bertomeu, J. M Asensi, J. Andreu and R. Alcubilla, *Thin Solid Films*, **395**, 335-338 (2001).
10. J. Kanicki and S. Martin, "Thin Film Transistors," C. R. Kagan and P. Andry, Eds. (New York: Marccel Dekker, 2003) pp. 71-138.
11. H. Gleskova, P. I. Hsu, Z. Xi, J. C. Sturm, Z. Suo, and S. Wagner, *Journal of Non-Crystalline Solids*, **338-340**, 732-735 (2004).
12. S. H. Won, J. K. Chung, C. B. Lee, H. C. Nam, J. H. Hur, and J. Jang, *J. Electrochem. Soc.*, **151**, G167-G170 (2004).
13. Y. Kaneko, A. Sasano, and T. Tsukada, *J. Appl. Phys.*, **69**, 7301-7305 (1991).
14. B. Hekmatshoar, K. H. Cherenack, S. Wagner and, J. C. Sturm, *Digest of 2008 International Electron Devices Meeting*, 89-91 (2008).
15. J. Z. Chen, I-C. Cheng, S. Wagner, W. Jackson, C. Perlov, and C. Taussig, *Proc. Mater. Res. Soc.*, **989**, 0989-A09-04 (2007).
16. C. S. Chiang, J. Kanicki and K. Takechi, *Japanese Journal of Applied Physics,* **37**, 4704-4710 (1998).
17. C. Y. Huang, T. H. Teng, J. W. Tsai and H. C. Cheng, *Japanese Journal of Applied Physics,* **39**, 3867-3871 (2000).
18. R. Oritsuki, T. Horii, A. Sasano, K. Tsutsui, T. Koizumi,Y. Kaneko and T. Tsukada, *Japanese Journal of Applied Physics,* **30**, 3719-3723 (1991).

Simulation and Characterization

Mater. Res. Soc. Symp. Proc. Vol. 1321 © 2011 Materials Research Society
DOI: 10.1557/opl.2011.1115

Wide-spectral-range, Expanded-beam Spectroscopic Ellipsometer and its Application for Imaging/Mapping of Graded Nanocrystalline Si:H Films

A. Nemeth[1,2], D. Attygalle[1], L. R. Dahal[1], P. Aryal[1], Z. Huang[1], C. Salupo[1], P. Petrik[2], G. Juhasz[2], C. Major[2], O. Polgar[2], M. Fried[2], B. Pecz[2], R. W. Collins[1]
[1]Center for Photovoltaics Innovation & Commercialization
University of Toledo, Toledo, OH, USA
[2]Photonics, MFA, Budapest, Hungary

ABSTRACT

A prototype expanded-beam spectroscopic ellipsometer has been developed that uses uncollimated (non-parallel, diffuse) illumination with a detection system consisting of an angle-of-incidence-sensitive pinhole camera for high-speed, large-area imaging/mapping applications. The performance of this novel instrument is being tested for imaging/mapping of mixed-phase hydrogenated silicon films having graded amorphous (a-Si:H) and nanocrystalline (nc-Si:H) components throughout the film depth. The speed of the measurement system makes the instrument suitable for use on production lines. The precision enables detection of subnanometer thicknesses, and refractive index and extinction coefficient changes of 0.01. Angle-of-incidence and mirror calibrations are made via well-known sample structures. Alternative commercial instrumentation for mapping by spectroscopic ellipsometry must translate the sample or ellipsometer in two dimensions. For this instrumentation, even a 15 x 15 cm^2 sample with cm^2 resolution requires > 200 measurements and at least 15 min. By imaging along one dimension in parallel, the expanded-beam system can measure with similar resolution in < 2 min. The focus of recent instrumentation efforts is on improving the overall system spectral range and its performance.

INTRODUCTION

The highest efficiencies measured on laboratory-sized thin film solar cells are generally well ahead of the best production module efficiencies. Thus, in photovoltaic (PV) module production for commercialization, many key problems are related to scale-up. The difference is primarily caused by fluctuations in area uniformity of properties caused by each individual processing step. In order to learn more about uniformity, several mapping measurements and evaluation methods are available [e.g., single-spot reflectance and spectroscopic ellipsometry (SE), Kelvin probe, laser beam induced currents (LBIC), etc.]. The primary problem with many probes is their reduced utility due to the long measurement time that makes in-line mapping a challenge. With long measurement times, the ability to detect non-uniformities is limited, which in turn limits feedback in process monitoring and control for optimization. The aim of our study is to develop and demonstrate a high-speed and high-resolution measurement method to monitor the thin film PV process in-line over large areas.

EXPERIMENTAL DETAILS

Hydrogenated amorphous silicon (a-Si:H), nanocrystalline silicon (nc-Si:H), and mixed phase layers were deposited by very high frequency plasma enhanced chemical vapor deposition (vhf PECVD) onto ~ 15 x 15 cm^2 soda lime glass (SLG) and ~ 5 x 5 cm^2 Si wafer substrates. A thin Cr layer was sputtered onto the SLG prior the Si:H layer deposition to ensure adhesion and form a standard reflective interface for further SE measurement. The 5 x 30 cm^2 Cr target was located at a distance of 7 cm from the substrate. The sputtering in all cases was performed for 40 minutes with 5 mTorr Ar gas and at 40 W rf power. The parallel depositions on 5 cm Si wafer substrates were carried out with the same parameters in order to study the layers by cross sectional transmission electron microscopy (XTEM).

The deposition conditions for the Si:H films studied here are summarized in Table 1. The vhf PECVD electrode size is approximately 15 x 15 cm^2, and the applied vhf frequency was 70 MHz. Prior to deposition all substrates were preheated to a temperate of 200 °C. The measured samples were deposited as part of an ongoing program for tandem a-Si:H/nc-Si:H silicon solar cell development. In the case of these solar materials, analyses of the processes for phase evolution and uniformity are crucial. In this study, the focus was the deposition and optical characterization of the most critical layer, the intrinsic nc-Si:H thin film. XTEM images are shown for two such films in Fig. 1. The film prepared with R=20 (left) evolves from a-Si:H through mixed-phase Si:H to nc-Si:H; the film prepared with R=50 (right) nucleates immediately from the substrate as nc-Si:H.

Figure 1: XTEM images for a sample prepared at R=20 (left; MCS 25) and one prepared at R=50 (right; MCS 26). The Si substrate is seen in the upper-left corner. The a-Si:H, mixed-phase (a-Si:H + nc-Si:H), and nc-Si:H sub-layers can be seen in the R=20 sample from the upper to lower part of the image..

Table 1. Deposition parameters of the Si:H samples prepared onto Si wafer and Cr-coated SLG substrates.

Sample no.	Time [min]	Pressure [mTorr]	RF power [W]	SiH_4 [sccm]	H_2 [sccm]	*R*
MCS 23	10	300	4	10	5	0.5
MCS 25	20	1500	40	5	100	20
MCS 26	20	1500	40	5	250	50

The thin films were measured during deposition in-situ and in real-time using a rotating-compensator multichannel spectroscopic ellipsometer (J.A. Woollam Co., M-2000DI) at one point close to the center of the 15 x 15 cm^2 sample. The single-spot measurements of a series of ellipsometric spectra obtained versus time during deposition permit the development of an appropriate multilayer optical model, which provides starting parameters in least-squares regression for the analysis of the mapping data acquired by our newly developed prototype spectroscopic ellipsometer for imaging/mapping purposes.

This new device requires only one measurement cycle (one rotation period of a polarizer or analyzer) for the acquisition of a two-dimensional array of data points [1]. Our new measurement technique serves as a novel form of imaging spectroscopic ellipsometry, using a divergent (non-collimated, diffuse) source and a detection system consisting of an angle-of-incidence-sensitive pinhole camera [2]. By adding broad-band lamps on the source side and a concave grating on the detector side, the instrument provides continuous high-resolution spectra along a line image. With this capability, information on multilayer photovoltaics stacks can be obtained over large areas (several dm^2) at high speed [3]. The technique can be expanded to even larger areas by scaling-up the geometry. The lateral resolution is limited by the minimum resolved-angle determined by the detection system, mainly by the diameter of the pin-hole. The diameter of the pin-hole is a compromise between the light irradiance and the lateral resolution. Ray-tracing simulations show that a 0.1-0.2 mm diameter pin-hole is a suitable compomise. This diameter makes 30 x 30 point resolution possible in the case of 2D imaging on an area of any size; it also makes 30 point resolution possible in the case of 1D imaging along a line of any length, the latter in conjunction with continuous spectroscopy (presently over the 350-1000 nm wavelength range). Small-aperture polarizers (25 mm diameter) are incorporated into the instrument that allow for a wider variety of low cost options in the selection of the polarizers.

The prototype is designed to enable in situ imaging/mapping within one of the chambers of a cluster tool system at the Center for Photovoltaics Innovation and Commercialization (PVIC), University of Toledo (Ohio).

RESULTS AND DISCUSSION

Demonstration mapping measurements have been performed ex situ on mixed-phase (a-Si:H + nc-Si:H) silicon films having the structure: nc-Si/(nc-Si+a-Si)/a-Si/Cr/glass. Imaging/mapping for initial demonstration purposes, however, was performed on a single phase a-Si:H/Cr/glass sample (see Fig. 2) and compared with independent mapping measurements made by a commercially-available instrument (J.A. Woollam Co.; AccuMap) that is capable of point-by-point mapping measurements on larger size samples. With settings of the deposition

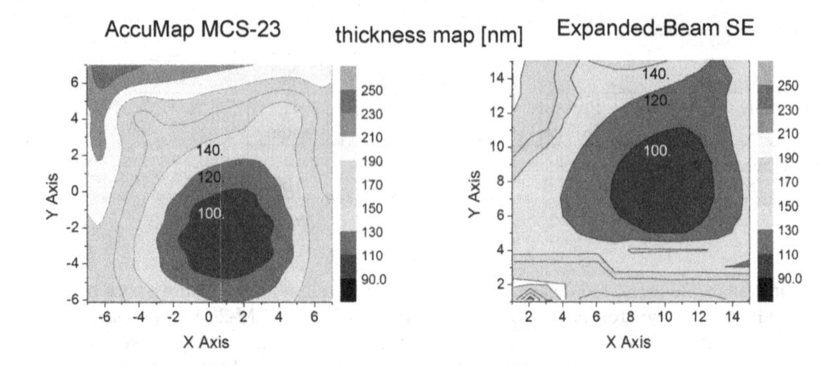

Figure 2: Mapping with a commercial instrument (left) and imaging/mapping with an expanded beam instrument (right) showing reasonable agreement; thickness maps in nm were obtained for a 15 x 15 cm² size sample of single-phase hydrogenated amorphous silicon (a-Si:H) layer on Cr. The map generated by the commercial instrument denotes points on a square grid at the sample surface; the map generated by the expanded-beam ellipsometer is described in terms of pixel-group coordinates.

parameters designed to achieve high spatial non-uniformity, a purely amorphous layer was prepared in order to demonstrate the utility of our method. A relatively simple optical model was used to analyze the mapping measurement: opaque Cr as a substrate, and a-Si:H with a Tauc-Lorentz oscillator dispersion model. The parameters of the Tauc-Lorentz oscillator were taken from the real time SE measurement.

We performed the second evaluation of the mapping measurement using the R=20 sample of Table 1 (see Fig. 3, right) and applying a 3-layer optical model consisting of the following stack: nc-Si:H/(a-Si:H+nc-Si:H)/a-Si:H/Cr, which was developed from single-spot, real-time spectroscopic ellipsometry (SE) analysis [3]. In addition to the appropriate model, this real time SE measurement also provided good starting parameters in least-squares regression for the analysis of the mapping data using the model. We also compared our mapping results with the results (using the same optical model) of independently performed ex-situ measurements with the commercial mapping instrument (see Fig. 3, left). We must note that this latter instrument obtains ellipsometric spectra at mm-size spots on a square grid, whereas the expanded-beam instrument obtains spectra for ~ 5 x 5 mm² areas. So, we cannot ensure that the individual ellipsometers are collecting spectra from precisely the same areas with the same resolution.

In spite of this, the basic agreement in Fig. 3 is very good, considering the complexity of the model which consists of three individual Si:H layers. The agreement in the spatial uniformity of the total thickness is best; a greater challenge is to extract information of the uniformity of the three component layers; these provide information on the spatially-dependent deposition phase diagram. The circular areas of reduced thickness on the maps arise from the use of a top cathode plate with two holes that permit passage of the beam for real time SE. When larger area devices are made, these specialized electrodes are replaced with standard ones.

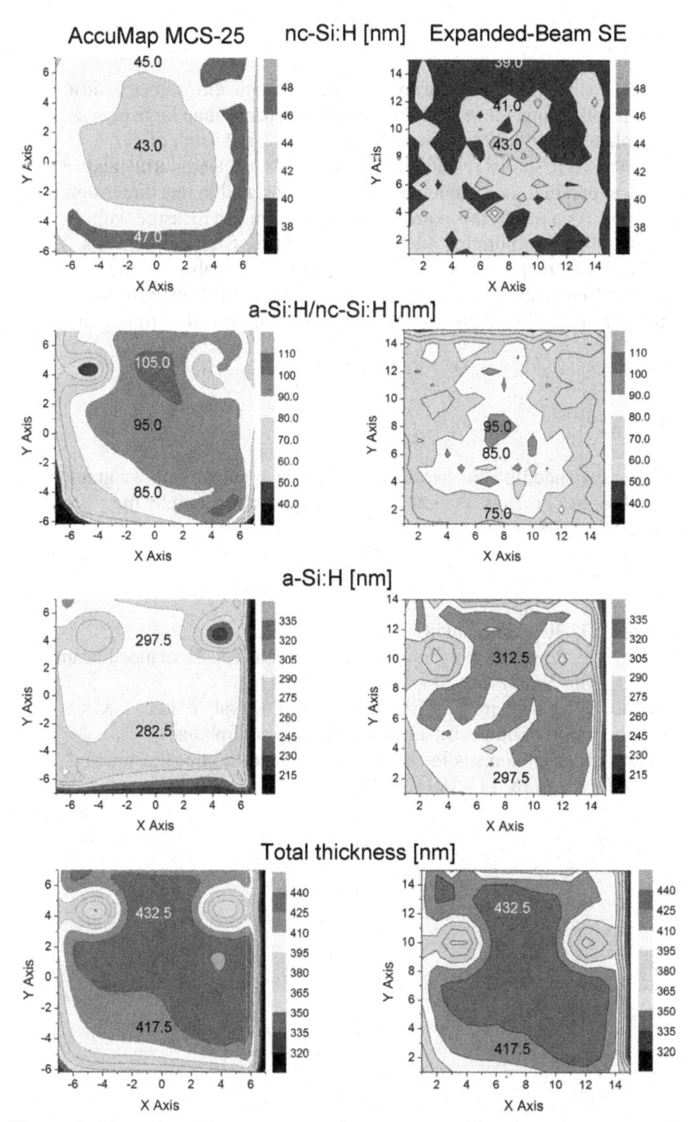

Figure 3: Mapping with a commercial instrument and imaging/mapping with the expanded-beam instrument for an thin Si:H film with R=20 using the 3-layer optical model given by nc-Si:H/(a-Si:H+nc-Si:H)/a-Si:H /Cr. The total thickness is the sum of the thicknesses of the three sub-layers for each spatial point.

CONCLUSIONS

Commercial instruments using 1D detector arrays for spectroscopic ellipsometry must translate either the sample or the ellipsometer in two dimensions in order to map large area samples. A 15 x 15 cm^2 sample requires > 200 measurements and at least 15 min of measurement time to achieve cm-scale spatial resolution. By using a 2D detector array and imaging along one dimension in parallel, the sample need only be translated in one dimension in order to map large area samples. As a result, the expanded-beam system can measure with similar spatial resolution in < 2 min. Incorporating the results of real time SE measurements, which provide good starting parameters in least-squares regression for the analysis of the imaging/mapping data, expanded-beam ellipsometry can be an effective in-line monitor of thickness and compositional uniformity in PV production lines using roll-to-roll or rigid plate substrates.

ACKNOWLEDGMENTS

The development of the expanded-beam instrument at PVIC was supported by a grant from the State of Ohio Third Frontier Projects, Wright Centers of Innovation, and by the Hungarian NKTH Office by PVMET08 project.

REFERENCES

1. Patent pending: P0700366, PCT/HU2008/000058.
2. M. Fried, G. Juhasz, C. Major, P. Petrik, O. Polgar, Z. Horvath, A. Nutsch, "Expanded beam (macro-imaging) ellipsometry", *Thin Solid Films* **519**, 2730-2736 (2011).
3. R. W. Collins, A. S. Ferlauto, G. M. Ferreira, C. Chen, J. Koh, R. J. Koval, Y. Lee, J. M. Pearce, and C. R. Wronski, "Evolution of microstructure and phase in amorphous, protocrystalline, and micro crystalline silicon studied by real time spectroscopic ellipsometry", *Solar Energy Materials and Solar Cells* **78**, 143-180 (2003).

Mater. Res. Soc. Symp. Proc. Vol. 1321 © 2011 Materials Research Society
DOI: 10.1557/opl.2011.934

Numerical 3D-Simulation of Micromorph Silicon Thin Film Solar Cells

Stefan Geißendörfer[1], Karsten von Maydell[1] and Carsten Agert[1]
[1]NEXT ENERGY · EWE-Forschungszentrum für Energietechnologie an der Carl von Ossietzky Universität , Carl-von-Ossietzky-Str. 15, 26129 Oldenburg, Germany

ABSTRACT

In this contribution 1, 2 and 3-dimensional simulations of micromorph silicon solar cells are presented. In order to simulate solar cells with rough interfaces, the surface topographies were measured via atomic force microscopy (AFM) and transferred into the commercial software Sentaurus TCAD (Synopsys). The model of the structure includes layer thicknesses and optoelectronic parameters like complex refractive index and defect structure. Results of the space resolved optical generation rates by using of the optical solver Raytracer are presented. The space resolved optical generation rate inside the semiconductor layers depends on the structure of the transparent conductive oxides (TCO) interface. In this contribution the influence of different optical generation rates on the electrical characteristics of the solar cell device are investigated. Furthermore, the optical and electrical results of the 1D, 2D and 3D structures, which have equal layer thicknesses and optoelectronic parameters, are compared.

INTRODUCTION

For silicon based thin film solar cells, TCO's like SnO_2:F or ZnO:Al can be used as front contact. The propagation of the incident light inside the solar cell depends on the TCO roughness and, therefore, the quantity of photo generated charge carriers. Many theoretical and experimental investigations of the short circuit current enhancement due to surface roughness are published e.g. [1,2] and the topic of several research groups is the improvement of the so-called light trapping effect.

One strategy to calculate the rate of photo generated charge carriers is to create virtual device structures in two or three dimensions. Therefore, a measured TCO surface topography has to be transferred to the simulation software. After construction of the virtual stack according to the data of experimentally deposited layers, different optical solvers can be used to calculate the local optical generation rate G(x,y,z). The finite-difference-time-domain method (FDTD) solves the Maxwell equations in time and space and considers interference effects of the incident light. Previews work using this method for silicon thin film solar cells was published by Lacombe et al [3]. However, the optical solver Raytracer, which calculates the light beam propagation depending on incidence angle to device interfaces and refractive indices of interface materials, can also be used in the case of rough interfaces. After modeling of the light beam propagation through the cell stack, the absorption will be calculated by the Lambert-Beer law. With that method it is not possible to regard interference effects. Though, in solar cell devices with good light trapping concepts and the corresponding TCO roughness the interference effects will become negligible. A similar behaviour of the lateral and vertical G(x,y,z)-distribution by using FDTD and Raytracer is observed. Hence, inside this work the optical solver Raytracer was used for the electrical investigations due to faster calculation time.

MODELING DETAILS

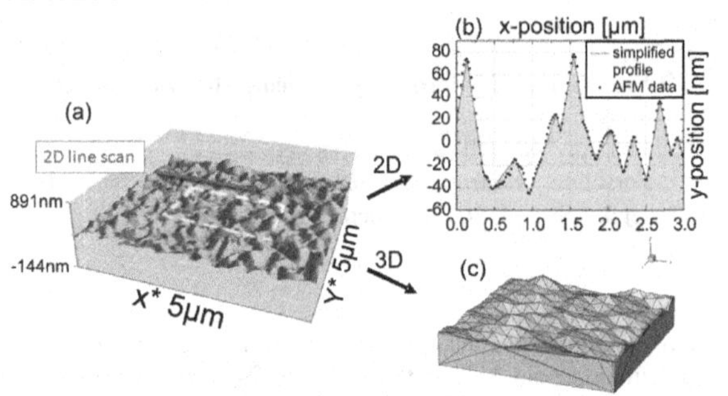

Figure 1. AFM surface scan (a) of a commercially available ASAHI-U (SnO_2:F) Front-TCO, the exported 2D line scan for modeling of rough interfaces is visualized. 2D line scan data of measured and simplified surface (b). Constructed virtual three dimensional TCO layer (c) appropriate the illustrated AFM. The edge length in x- and y-direction is 2µm.

Figure 1a shows the measured surface topography via AFM of a ASAHI-U (SnO_2:F) Front-TCO with a resolution of 256 pts x 256 pts over an area of 5µm x 5µm. For 2 dimensional modeling one typical line scan (solid line in Figure 1a) of the data was exported and reduced to a simplified profile as illustrated in Figure 1b. The cutback of the data points to characteristic maxima and minima accelerates the calculation time considerably because of the ability of a wider-meshed virtual structure. For 3 dimensional modeling the data points of the middle were taken (dashed lines in Figure 1a) and reduced to a resolution of 21pts x 21pts. Figure 1c shows the subsequent surface structure of a TCO layer with edges length of 2µm in x- and y-direction. It is obvious that the diminishing of data points modifies the pyramid-like constitution. But a higher resolution increases the calculation time significantly and a huge increase of the quantity of photo generated charge carrier at virtual devices with a surface resolution of 40pts x 40pts was not observed. The optical generation rate is calculated in different ways: i) 1-dimensional (Fig 2 left), ii) 2-dimensional (Fig 2 right), iii) 3-dimensional and iv) with a simplified constant generation rate (Fig 3). In the following, devices with flat interface will be identified with 1D because of their constant physical modeling settings and behaviors in lateral direction.

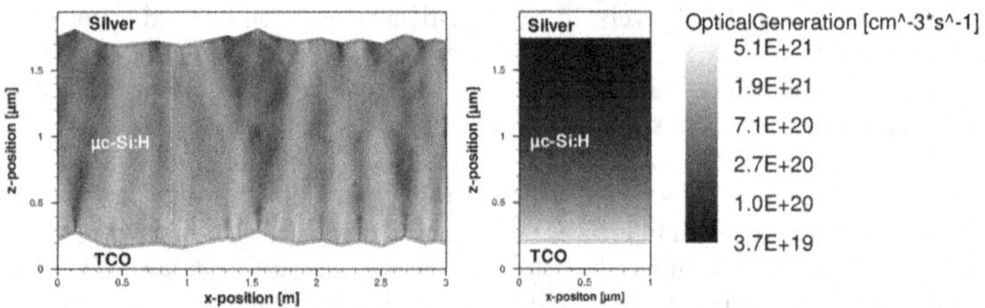

Figure 2. Profile of the calculated optical generation rates inside the semiconductor layers of a microcrystalline single solar cell. 1D device with flat interfaces (left). 2D device with rough interfaces (right). All semiconductor interfaces have the structure of the exported and simplified line scan as shown in Figure 1b.

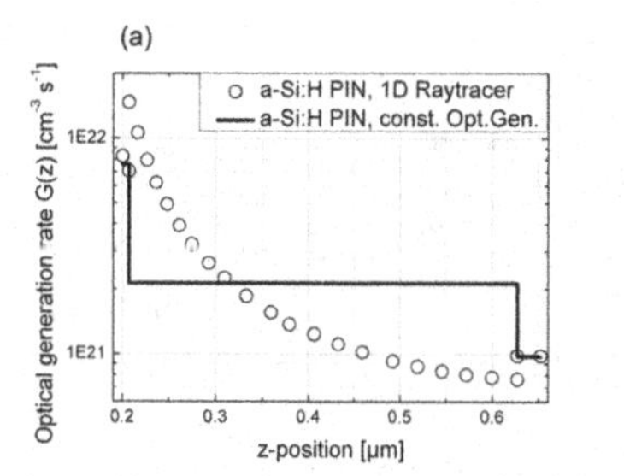

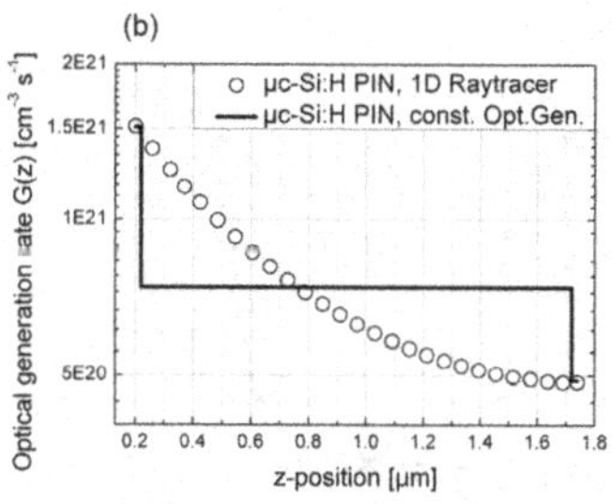

Figure 3. Vertical profile of the optical generation rates G(z) of 1D-devices, calculated by the optical solver Raytracer and specified as constant Rate (cons. Opt. Gen.) for the three different semiconductor layers (p,i,n) of an a-Si:H-PIN (a) and a μc-Si:H-PIN (b) single solar cell.

Figure 2 shows a microcrystalline (μc-Si:H) single solar cell and its PIN structure with flat (1D) and rough (2D) interfaces. In both cases, the thicknesses of the different semiconductor layers are equal (p: 20nm / i: 1500nm / n: 20nm). The light excitation is the standard spectrum AM1.5G with its light intensity of 100 mW/cm². The origin of the light source is the x-y-plane at z=0 and the propagation is the positive z-direction. The optical boundary conditions are set to 100% reflectance for edges and areas, which are perpendicular to the x-y-plane. Most of the used electrical parameters like trap distribution, charge carrier mobility, etc. are published in [4]. The calculated optical generation rates G inside the semiconductor layers are illustrated in Figure 2. In case of the 1D-simulation, only a decrease of G in z-direction can be observed compared with the 2D-simulation that has also lateral differences.

Especially semiconductor regions on top of pronounced TCO hills show a considerable decrease of G compared with regions on TCO valleys. The latter forms columns with higher generation rates through the whole semiconductor stack and therefore, regions with a higher charge carrier density. Hence, an increase of the current density in the electrical simulation can be observed and main transport paths through the intrinsic layer are determined in the columns. Note that the simulations assume only a TCO layer with a thickness of 200 nm is assumed. Also the glass substrate and its junctions to air and TCO are not considered for our model. However, the influences of different G(x,y,z)-distributions to the electrical characteristic parameters of solar cells should be demonstrated in this paper. Therefore, these simplifications are sufficient. Figure 3 shows 1D optical generation rates inside the semiconductor layers of a-Si:H and μc-Si:H single solar cells, which are calculated with the Raytracer method (dotted line). The assumed constant rates (solid line) deliver the same photo generated current density for doped and intrinsic layers respectively referred to equation 1.

The photo generated current density j_{ph} is calculated as follows:

$$j_{ph} = q/A * \int G(x,y,z)*dV \qquad (1)$$

It is the integration of the optical generation rate G(x,y,z) of a volume e.g. all semiconductor layers with q as the elementary charge and A the area of the device to obtain the current density. For comparisons of some devices, j_{ph} can also be calculated layer by layer.

DISCUSSION

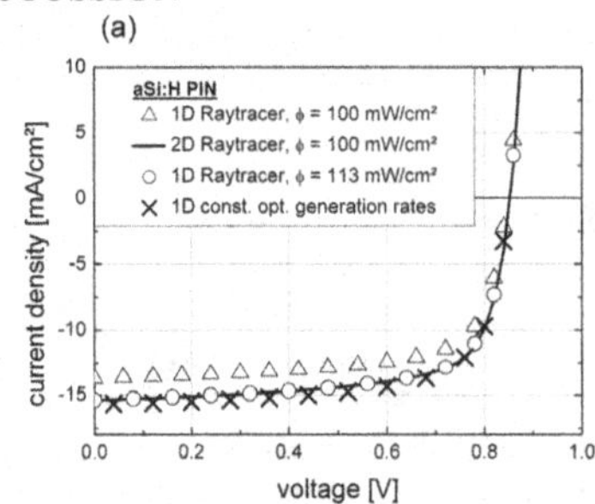

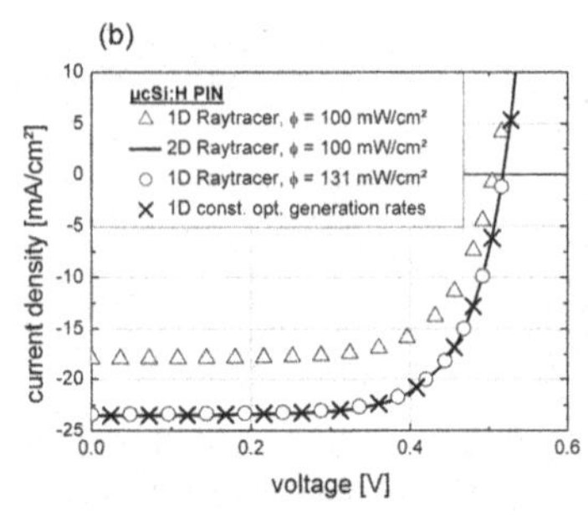

Figure 4. IV-characteristics of a-Si:H (a) and µc-Si:H (b) single solar cells, calculated with different optical models (cf. Table I & II). The legends depict the device structure with flat (1D) or rough (2D) interfaces and the used optical solver Raytracer with its light intensity Φ or assumed constant optical generation rates respectively (Figure 3).

Figure 4 shows the simulated IV characteristics of single solar cells in one and two dimensions, whose optical generation rates G are determined by the optical solver Raytracer with different light intensities Φ_{light} and constant optical generation rates. In both cases for amorphous (left) and microcrystalline (right) cells, an increase of the short circuit current density j_{sc} can be observed by comparison of devices with flat (1D) and rough (2D) interfaces and the same light intensity of 100 mW/cm². This behavior can be explained by the light-trapping-effect and the associated ray path extension through the absorber layers. Furthermore, the 2D values of j_{sc} are in good consistence with published short circuit currents with structured TCOs [5,6]. Tables I & II show the characteristic parameters after electrical modeling. But fill factors FF and open circuit voltages V_{oc} are too complex for comparison due to the different photo generated charge carriers inside the semiconductor layers. The light intensities of the Raytracer were increased so that photo generated current densities j_{ph} in each layer compared with the 2D Raytracer model and Φ_{light} = 100mW/cm² are conform. This increase leads to a perfect matching of the IV characteristics (circles and solid line in Figure 4) and the electrical parameters j_{sc}, V_{oc} and FF are nearly identical. The same behavior can be also observed for calculations with constant optical generation rates (crosses, Fig.4) if the constant values inside the different doped and intrinsic layers equivalent to j_{ph} of the 2D Raytracer model.

In Table I & II the calculated j_{ph} with the 3D Raytracer model are also listed and the corresponded interface structure is shown in Figure 1c. A further increase of j_{ph} compared to the 2D Raytracer model is observed. These behaviors can also explained by the possibility of ray path extension in the additional dimension (y-direction). Otherwise, the huge increase of V_{oc} and FF is not realistic and the discrepancy compared to the 1D model with respective constant optical generation rates can be explained by the wide-meshed 3D-device. In that case the error of electrical 3D-modeling is huge but the approach of downscaling to 1D leads to plausible electrical parameters and was verified just now. Attempts to calculate the 3D-device with smaller sized meshes were aborted due to the long calculation time. So, the calculation time of electrical 3D-modeling strongly depends on the mesh size and is a disadvantage compared to 2D-modeling, which is much faster.

Table III shows the calculated IV-characteristics of micromorph tandem solar cells and the used thicknesses of the intrinsic layers (a-Si:H(i) and μc-Si:H(i)). Due to the light trapping effect in the case of the 2D- and 3D-models, more charge carriers are photo generated in the μc-Si:H(i) layer and a thicker a-Si:H(i) layer is necessary for current matching of the two subcells regarding to the 1D model with flat interfaces. The current matching is demonstrated by equal values of $j_{ph,\ a\text{-}Si:H(i)}$ and $j_{ph,\ \mu c\text{-}Si:H(i)}$. It is obvious that different layer thicknesses of a-Si:H(i) in one, two and three dimensions lead to different electrical parameters. However, the layer thicknesses of the 3D model are more realistic in comparison to tandem solar cells, which were prepared in our lab. However, the comparability of j_{sc}, V_{oc} and FF between 1D, 2D and 3D is very complex because of three different layer thicknesses.

Table I. Different optical models of a-Si:H PIN single cells and their calculated photo generated current density j_{ph} and the electrical parameters of the IV curves as shown in Figure 4 (a).

Optical model (a-Si:H PIN)	Φ_{light} (mW/cm²)	j_{ph} (mA/cm²)	j_{sc} (mA/cm²)	V_{oc} (V)	FF (%)	η (%)
1D Raytracer	100	15,6	13,7	0,847	71	8,2
1D Raytracer	113	17,7	15,6	0,851	71	9,4
1D const. opt. generation rates	---	17,6	15,7	0,850	71	9,5
2D Raytracer	100	17,7	15,4	0,852	71	9,3
3D Raytracer	100	18,8	16,5	0,888	74	10,9
1D const. opt. generation rates	---	18,8	16,6	0,851	71	10,0

Table II. Optical models and parameters of calculated IV-characteristics of μc-Si:H PIN single cells as shown in Figure 4 (b).

Optical model (μc-Si:H PIN)	Φ_{light} (mW/cm²)	j_{ph} (mA/cm²)	j_{sc} (mA/cm²)	V_{oc} (V)	FF (%)	η (%)
1D Raytracer	100	18,4	17,9	0,506	69	6,3
1D Raytracer	131	24,1	23,5	0,518	70	8,5
1D const. opt. generation rates	---	24,1	23,5	0,518	70	8,5
2D Raytracer	100	24,1	23,5	0,518	70	8,5
3D Raytracer	100	25,7	25,2	0,545	81	11,1
1D const. opt. generation rates	---	25,7	25,1	0,521	70	9,1

Table III. Layer thicknesses of intrinsic layers, their calculated photogenerated current densities j_{ph} and the characteristical parameters of micromorph tandem solar cells under illumination. Values in brackets are calculated with a 1D-model and constant optical generation rates.

Parameter	1D simulation	2D simulation	3D simulation
Thickness a-Si:H(i) (nm)	120	210	250
Thickness μc-Si:H(i) (nm)	1500	1500	1500
$j_{ph,\ a\text{-}Si:H(i)}$ (mA/cm²)	9.1	11.4	12.2
$j_{ph,\ \mu c\text{-}Si:H(i)}$ (mA/cm²)	9.1	11.4	12.1
j_{sc} (mA/cm²)	9.1	11.4	(12,2)
V_{OC} (V)	1.296	1.327	(1,333)
FF (%)	76	73	(73)
η (%)	9.0	11.1	(11,9)

CONCLUSIONS

For electrical simulation, the demonstrated results of single and tandem solar cells reflect the importance of realistically calculated generation rates G in each semiconductor layer. However, the distribution of optical generation rates G(x,y,z) in homogenous active layers without any local differences in morphology or trap distributions has only negligible influences to the electrical parameters j_{sc}, V_{oc}, and FF. The reason of this behavior could be the sufficient mobilities of the majority and minority charge carriers of the intrinsic layer. Therefore, the most of the photo generated charge carrier reach the respective doped layer and can be extracted from the solar cell without recombination. Our used electrical parameter set [4] is comparable to other published parameter sets [7,8] and therefore, unrealistic parameter settings could not be the reason for the almost generation profile independent current voltage characteristics under illumination. Furthermore, it was demonstrated that the light trapping effect due to the rough interfaces results in an increase of j_{sc}. Especially for the modeling of tandem solar cells the light trapping effect has to be well simulated because of the resulting differences in layer thicknesses. The resulting electrical parameters are dependent on layer thickness variations and the comparability is more complex because of different photo generated current densities j_{ph}.

Finally, modeling in more dimensions leads to realistic j_{sc}, however, electrical 1D modeling is sufficient for layers with homogenous morphology in relation to the other electrical parameters V_{oc} and FF.

REFERENCES

1. A. Čampa, O. Isabella, R. van Erven, P. Peeters, H. Borg, J. Krč, M. Topič and M. Zeman, *Progress in Photovoltaics: Research and Applications* **18**, 160–167 (2010).
2. C. Pflaum et al., in *Proceedings of the 24th European Photovoltaic Solar Energy Conference*, 2310-2312, Hamburg (2009).
3. J. Lacombe et al., in *Proceedings of the 35th IEEE Photovoltaic Specialist Conference*, 1535-1539, Hawaii (2010).
4. S. Geißendörfer et al., in *Proceedings of the 25th European Photovoltaic Solar Energy Conference*, 3133, Valencia (2010).
5. T. Kilper et al., in *Proceedings of the 20th European Photovoltaic Solar Energy Conference*, 1544-1547, Barcelona (2005).
6. B. Rech, *PhD. Thesis*, Forschungszentrum Jülich, Jül-3427, 1997.
7. J.A. Willemen, *PhD. Thesis*, Delft University of Technology, 1998.
8. A. Sturiale, Hongbo T. Li, J. K. Rath, R. E. I. Schropp, and F. A. Rubinelli, *J. Appl. Phys.* **106**, 14502 (2009).

Mater. Res. Soc. Symp. Proc. Vol. 1321 © 2011 Materials Research Society
DOI: 10.1557/opl.2011.1192

Correlated photoluminescence spectroscopy investigation of grain boundaries and diffusion processes in nanocrystalline and amorphous silicon (nc-Si:H) mixtures

Jeremy D. Fields[1], K. G. Kiriluk[1], D. C. Bobela[2], L. Gedvilas[2], P. C. Taylor[1]

1. Department of Physics, Colorado School of Mines, Golden, CO, United States.
2. National Renewable Energy Laboratory, Golden, CO, United States.

ABSTRACT

Photoluminescence (PL) spectra obtained with correlated set of experiments investigating grain boundary characteristics and diffusion processes in nanocrystalline silicon alloys (nc-Si:H), provide insight regarding formation and passivation of electronic defects in these regions. Based upon current results and previous works we believe thermally driven processes induce a PL band centered at 0.7 eV upon thermal annealing, and most likely involve diffusion of hydrogen and oxygen near interfaces. A nc-Si:H sample set with varied crystal volume fraction, X_c, was subject to thermal annealing treatments at different temperatures – each exceeding the deposition temperature. Fourier-transform photoluminescence (FTPL) and Fourier-transform infrared absorption spectroscopy (FTIR), were employed to correlate the relative 0.7 eV defect band emergence with compositional changes indicative of Si–H_x and Si–O species, for each sample, at each temperature, respectively. Hydrogen effusion data provide additional perspective.

We find the X_c to strongly affect susceptibility of nc-Si:H to oxygen related effects. The higher the X_c, the more readily oxygen penetrates the nc-Si:H network. We attribute this relationship to elevated diffusivity of oxygen in highly crystalline nc-Si:H materials, owing to their abundance of gain boundaries and interfaces, which serve as pathways for impurity migration. These findings corroborate the expectation that oxygen impurities and diffusion processes contribute to development of microstructural features giving rise to radiative recombination through deep defects in nc-Si:H.

INTRODUCTION

Photoluminescence (PL) spectra obtained with correlated set of experiments investigating grain boundary characteristics and diffusion processes in nanocrystalline silicon alloys (nc-Si:H), provide insight regarding formation and passivation of electronic defects in these regions. Based upon current results and previous works we believe thermally driven processes induce a PL band centered at 0.7 eV upon thermal annealing, and most likely involve diffusion of hydrogen and oxygen near interfaces.

Reports from other authors attribute a 0.7 eV PL band in crystalline-Si (c-Si), polycrystalline-Si (poly-Si), and microcrystalline-Si (μc-Si:H) systems, emerging in response to thermal exposure, to oxygen related deep electronic defects [1-3]. Our previous investigations found the 0.7 eV PL band in nc-Si:H, labeled here a *defect band*, to originate from the nanocrystalline phase, and to involve defects of an analogous nature [4,5]. Observing correlations between oxygen contamination, thermal exposure, and defect band onset in all cases leads to speculation about the roles of oxygen agglomeration and other thermally activated phenomena that may contribute to the defect formation mechanism. Building our understanding of the mechanism(s) responsible for this behavior in nc-Si:H will enable technologies built upon this promising material.

A nc-Si:H sample set with varied crystal volume fraction, X_c, was subject to thermal annealing treatments at different temperatures – each exceeding the deposition temperature. Fourier-transform photoluminescence (FTPL) and Fourier-transform infrared absorption spectroscopy (FTIR), were employed to correlate the relative 0.7 eV defect band emergence with compositional changes indicative of Si–H_x and Si–O species, for each sample, at each temperature, respectively. These findings corroborate the expectation that oxygen impurities and diffusion processes contribute to development of microstructural features giving rise to radiative recombination through deep defects in nc-Si:H.

EXPERIMENTAL DETAILS

Samples for this study were PECVD grown at ~ 200 °C, in a vacuum system, under hydrogen dilution. The amount of hydrogen dilution was varied to produce different X_c values. The thickness of each nc-Si:H layer was about 1 μm. A WiTech confocal Raman system with doubled YAG (532 nm) incident light, and another Raman system with a SPEX 1877 spectrometer and Kr 647 nm excitation, each in the backscattering configuration, were used to perform Raman measurements. Comparing spectra obtained with 647 nm and 532 nm light showed essentially uniform crystallinity with depth in the regions probed by PL. Crystalline volume fractions were assigned to each film according to the standard 3 Gaussian peak fit method as reported by reference [6], with bands centered at 520 cm^{-1} and 480 cm^{-1} attributed to the crystalline phase and hydrogenated amorphous silicon (a-Si:H) regions, respectively, and with a third band representing grain boundary contributions, centered at ~ 510 cm^{-1}.

To investigate PL from each nc-Si:H sample, a Nicolet Magna 560 interferometer, equipped with InGaAs and HgCdTe photodiode detectors, was used. A liquid He cryostat provided a measurement temperature of 18 K, and an Ar laser on the 514.5 nm emission line, at ~ 50 mW, excited the samples. FTIR measurements were performed at the National Renewable Energy Laboratory, in Golden, CO.

RESULTS

From the Raman data we assign approximately 5% crystallinity to Sample A, 60% to Sample B, 80% to Sample C, and 90% to Sample D (Fig. 1 (a)). Hydrogen effusion spectra (Fig. 1 (b)) appear bi-modal in nature, with a low temperature maximum (LTM) and a high temperature maximum (HTM) observed for each of these samples. We attribute the LTM to hydrogen evolving from surfaces at voids and grain boundary regions, and the HTM to hydrogen from a-Si:H bulk. Sample D differs from the others due to its relatively weak LTM, and an HTM occurring at lower temperature than observed for Samples A, B and C. In general, hydrogen desorption onset, and the LTM, occurs at lower temperature the higher the X_c. For Samples B, C, and D hydrogen effusion begins at ~ 375 °C, with a LTM at approximately 400 °C, whereas Sample A does not lose hydrogen at an appreciable rate until ~ 400 °C, with a LTM at 425 °C.

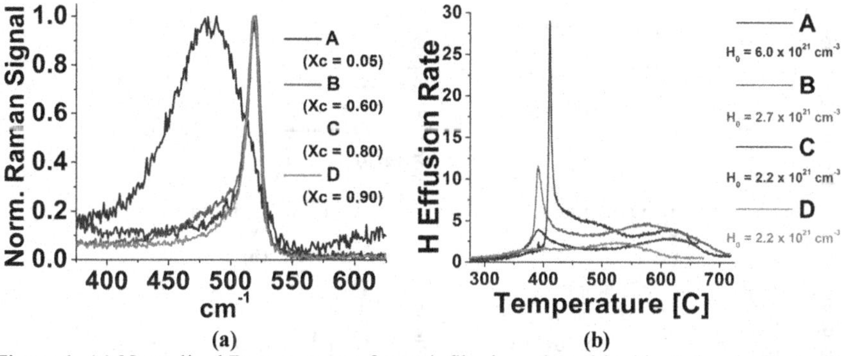

Figure 1: (a) Normalized Raman spectra for each film investigated in this study, obtained with the doubled YAG, 532 nm, excitation, and **(b)** corresponding normalized hydrogen effusion data.

Relative crystalline volume fractions apparent in the PL spectra (FIG. 2) agree with the Raman and effusion data sets, by comparative band tail-to-band tail transition contributions from nanocrystalline grain boundary regions and a-Si:H bulk regions, at ~ 0.95 eV and 1.20 – 1.30 eV, respectively [5]. Sample A shows emission originating from a-Si:H regions, with a maximum at approximately 1.20 eV initially, and which red-shifts about 100 meV in response to annealing. Samples possessing higher X_c (Samples B, C, and D) possess a PL band centered at about 0.95 eV, and also some emission extending below 0.75 eV.

Insufficient emission intensity for FTPL measurement employing an HgCdTe detector with wider spectral operating range than that obtained with the InGaAs detector, restricted PL measurements to monitoring emission with energy > 0.75 eV. Therefore, contributions from defect states producing emission below this threshold are cut-off on the low energy side of the spectra, and only shoulders indicative of the 0.7 eV PL band were observed in specimens possessing the deep oxygen related states (Fig. 2 (b), (c), and (d)). Defect PL was subtly existent prior to annealing, and became much more prevalent accompanying thermal exposure in samples containing a substantial crystalline volume fraction. The defect band onset occurred most strongly for Sample D, which had the highest crystalline volume fraction.

IR absorption data for each sample show a systematic decrease in absorption at around 1900 cm^{-1} to 2100 cm^{-1}, attributed to Si–H_x (x = 1,2,3) stretch modes, after each thermal annealing treatment (Fig. 3). Absorption in this range can be deconvoluted into contributions from so-called low stretching modes (LSM) said to arise from Si-monohydride stretch in a-Si:H bulk, at ~ 2000 cm^{-1}, and high stretching modes (HSM) from Si–H stretching modes on crystallite and void surfaces, at (2090 cm^{-1} – 2110 cm^{-1}) [7]. Generally, as X_c increases, absorption of nc-Si:H film becomes more dominated by HSM. This trend was previously observed (see Ref. [8]), and our spectra show the same relationship. For Samples A and B (low X_c) LSM absorption dominates; for the higher X_c samples, Samples C and D, HSM absorption dominates (Fig. 3 (a)).

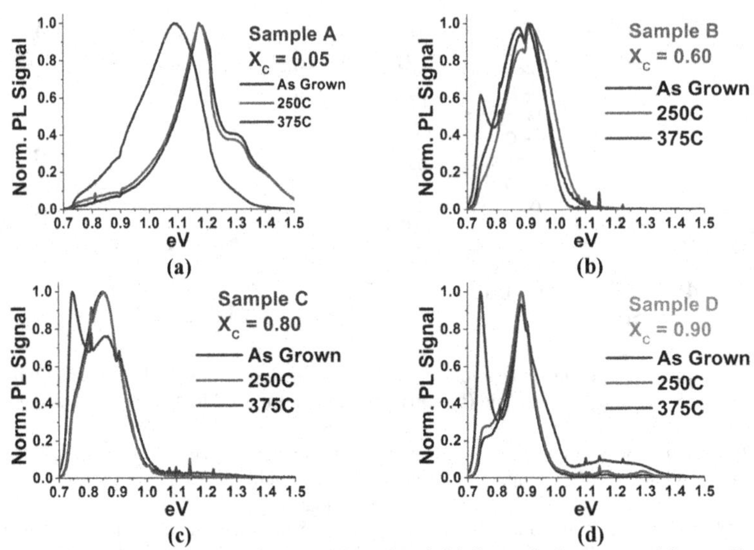

Figure 2: PL spectra from samples A, B, C, and D, initially, and after sequential annealing.

Absorption at around 1000 – 1200 cm^{-1} is well known to arise from asymmetric Si–O–Si bonds [7,9]. Oxidation of crystallite surfaces has been reported to produce this type of absorption in aged nc-Si:H films, especially for highly crystalline mixtures [7]. A stronger onset of oxygen related absorption was observed in high X_c Samples C and D, than with Samples A and B. Furthermore, subtle absorption at ~ 2250 cm^{-1} seen for Sample D, after annealing to 400 °C, signifies oxygen back-bonded in Si–H bonds at crystallite surfaces [7,10]. From these results, we expect oxygen to diffuse and react along crystalline grain boundary regions during thermal annealing.

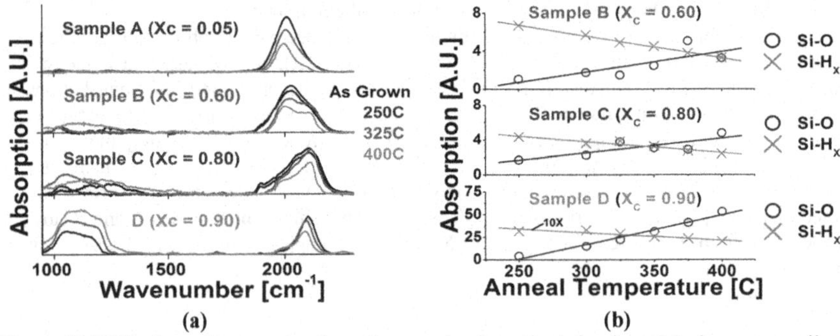

Figure 3: FTIR absorption spectra from the samples investigated, where **(a)** shows normalized absorption profiles, and **(b)** shows integrated intensity of Si–O–Si and Si–H_x species absorption.

DISCUSSION

We find the crystalline volume fraction to strongly affect susceptibility of nc-Si:H to oxygen related effects. First, the higher the crystalline volume fraction, the more readily oxygen penetrates the nc-Si:H network, as has been observed by other authors [7]. The FTIR data show this relationship the most clearly. On the left side of the spectra (Fig. 3 (a)), the strongest increase in absorption in the 950 cm^{-1} to ~ 1300 cm^{-1} range appears in sample D, samples B and C show Si–O–Si related absorption increasing after annealing, and sample A shows almost no such onset. Comparing integrated absorption intensities (Fig. 3 (b)) we see about a factor of 10 times more Si–O–Si absorption after annealing Sample D than was observed for the other samples.

The PL data show the same trend (Fig. 2). On the low energy side of these spectra the shoulder of the defect band emerges much more predominantly in samples D and C, than for B, and for A not at all. This trend has been observed previously [5]. Thus, if the defect band does involve oxygen contamination, as all evidence suggests, then more crystalline nc-Si:H films are more likely to develop the electronic centers producing the radiative transitions.

Elevated diffusivity of oxygen in highly crystalline nc-Si:H materials, owing to their abundance of gain boundaries and interfaces, which serve as pathways for impurity migration, explains these trends. Hydrogen desorption beginning at lower temperatures in high X_c samples (Fig. 1 (b)) also strengthens this interpretation. All of these findings corroborate the expectation that oxygen plays a central role in the 0.7 eV defect formation process.

Oxygen accumulates and diffuses along grain boundary regions in nc-Si:H [11], and the annealing temperature ranges shown to instigate radiative recombination via deep defects and that corresponding to the onset of hydrogen desorption from the same regions have been repeatedly demonstrated to coincide. Therefore, we expect the absence of hydrogen passivation at crystallite/crystallite and crystallite/amorphous-silicon interfaces to enable oxygen impurity coordination changes, and thermal annealing to provide the energy needed to overcome the barrier for reconfiguration. The FTIR data showing a trade-off between Si–H and Si–O–Si bonding, and also the H effusion data, agree with this proposed mechanism. The nature of the resultant microstructural environment after this process likely involves oxygen agglomerates and extended defects, as has been reported previously [5]. This study elucidates some of the consequences of hydrogen and oxygen diffusion in nc-Si:H, promoted by thermal annealing, with numerous implications for considering oxygen related deep defect formation mechanism(s) in nc-Si:H.

ACKNOWLEDGEMENTS

This research was partially supported by a DOE grant through United Solar Ovonics, Inc., under the Solar America Initiative Program Contract, No. DE-FC36-07 GO 17053, and by an NSF cooperative agreement through the Renewable Energy MRSEC at Colorado School of Mines, under contract DMR 0820518.

REFERENCE

1. Michio Tajima, Journal Of Crystal Growth **103**, 1-7 (1990).
2. S. S. Ostapenko, A. U. Savchuk, G. Nowak, J. Lagowski, and L. Jastzebski, Materials Science Forum **196 - 201**, 1897-1902 (1995).
3. T. Merdzhanova, R. Carius, S. Klein, F. Finger, and D. Dimova-malinovska, Thin Solid Films **512**, 394 - 398 (2006).
4. J. D. Fields, P. C. Taylor, J. G. Radziszewski, D. A. Baker, G. Yue, and B. Yan, Mater. Res. Soc. Symp. Proc. **1153**, 5-10 (2009).
5. J. D. Fields, P. C. Taylor, D. C. Bobela, B. Yan, and G. Yue, Mater. Res. Soc. Symp. Proc. **1245**, 1-6 (2010).
6. C Droz, Solar Energy Materials and Solar Cells **81**, 61-71 (2004).
7. A.C. Bronneberg, A.H.M. Smets, M. Creatore, and M.C.M. van De Sanden, Journal Of Non-Crystalline Solids **357**, 884-887 (2011).
8. W. Beyer, P. Hapke, and U. Zastrow, Mat. Res. Soc. Symp. Proc. **467**, 343 - 348 (1997).
9. C.T. Kirk, Phys. Rev. B **38**, 1255 - 1273 (1988).
10. P. C. P. Bronsveld, H. J. Van Der Wagt, J. K. Rath, R. E. I. Schropp, and W. Beyer, Thin Solid Films **515**, 7495 - 7498 (2007).
11. Toshihiro Kamei, Takehito Wada, and Akihisa Matsuda, IEEE Conf. Proc. 784-787 (2000).

Mater. Res. Soc. Symp. Proc. Vol. 1321 © 2011 Materials Research Society
DOI: 10.1557/opl.2011.809

Hopping transport in doped co-deposited mixed-phase hydrogenated amorphous/nanocrystalline silicon thin films

L. R. Wienkes, C. Blackwell, and J. Kakalios
School of Physics and Astronomy, University of Minnesota, Minneapolis, MN 55455

ABSTRACT

Studies of the electronic transport properties of n-type doped hydrogenated amorphous/nanocrystalline silicon (a/nc-Si:H) films deposited in a dual-plasma co-deposition reactor are described. For these doped a/nc-Si:H, the conductivity increases monotonically for increasing crystal fractions up to 60% and displays marked deviations from a simple thermally activated temperature dependence. Analysis of the temperature dependence of the activation energy for these films finds that the dark conductivity is best described by a power-law temperature dependence, $\sigma = \sigma_o (T/T_o)^n$ where n = 1 – 4, suggesting multiphonon hopping as the main transport mechanism. These results suggest that electronic transport in mixed-phase films occurs through the a-Si:H matrix at lower nanocrystal concentrations and shifts to hopping conduction between clusters of nanocrystals at higher nanocrystal densities.

INTRODUCTION

Photovoltaic devices employing mixed-phase silicon thin films, consisting of silicon nanocrystals embedded in hydrogenated amorphous silicon (a-Si:H) as the photovoltaic material, have recently attracted considerable attention due to their high solar conversion efficiencies, high deposition rates and improved resistance to light-induced defect creation (the Staebler-Wronski effect) without a degradation of the nominal opto-electronic properties [1-3]. These mixed-phase films are typically synthesized in either a capacitively-coupled single-chamber plasma system, using high gas pressures and a heavily hydrogen-diluted silane precursor [4-8], or in a dual-chamber co-deposition system [9,10], which enables the separate optimization of the deposition conditions for nanocrystallite formation and growth of the surrounding a-Si:H matrix. While intrinsic mixed-phase silicon plays an important role as the absorber layer in photovoltaic devices, an understanding of the transport in doped mixed-phase films is also crucial for device optimization. In this report we present detailed electronic transport measurements on n-type doped hydrogenated amorphous/nanocrystalline silicon (a/nc-Si:H) synthesized in a dual-chamber reactor.

EXPERIMENTAL DETAILS

The films used in this study were grown in a two-chamber co-deposition system [10], in which the nanoparticles are synthesized in a plasma reactor using silane (SiH_4), argon and phosphine (PH_3), with the phosphine-silane ratio set at $[PH_3]/[SiH_4] = 6 \times 10^{-4}$ for all of the films studied here. The nanocrystallites are grown in the particle reactor chamber, consisting of a 3/8 inch quartz tube with ring electrodes, where the pressure and power of the particle reactor (1.7 Torr and 70W) are chosen to promote the growth of crystalline nanoparticles. The nanocrystallites are then entrained by the argon and injected into a second PECVD chamber (a traditional, capacitively coupled reactor, with an electrode area of 53.5 cm^2) where the a-Si:H film is deposited using the residual silane/dopant gas mixture from the nanoparticle chamber. Conditions for the a-Si:H matrix in the second chamber are closer to conditions for optimum amorphous films, with a total pressure of 700 mTorr and an RF power of 4W at the standard frequency of 13.56 MHz. The total gas flow was held at 90 sccm for all the films described here, but the ratio of silane/dopant mixture to argon was adjusted in order to vary the crystalline content in the mixed-phase films. The substrate electrode is heated to 250°C as is typical for high-quality a-Si:H films, while the RF electrode is unheated. The films were deposited on Corning 1737 glass substrates and vary in thickness from 250 nm to 4 microns. A sketch of the deposition system can be seen in Ref [10].

The films reported here have crystalline contents ranging from < 1% to 60%, as determined by Raman spectroscopy measurements using a Witec Alpha 300R confocal Raman microscope, equipped with an UHTS 200 spectrometer. The excitation source is an argon ion laser of wavelength 514.5 nm at a power of 5 - 7 mW focused to an area of ~ 2-3 μm diameter. The crystal fraction is quantified using the typical formula:

$$X_C = \frac{I_{520} + I_{500}}{I_{520} + I_{500} + I_{480}} \qquad (1)$$

Here, I_{480} represents the area under the amorphous (480 cm^{-1}) TO peak, while the nanocrystalline TO mode includes both the 520 cm^{-1} (I_{520}) and the 500 cm^{-1} (I_{500}) mode. For simplicity, the Raman scattering cross section, often set to be between 0.8 and 0.9 [11-14], is taken to be unity; any error in a determination of the crystalline content by neglecting this factor will affect the absolute crystal fraction, but not the relative values for the series of films studied here. Raman measurements for a few of the films in this study can be found in Fig. 1; for clarity, only the most crystalline films are shown ($X_C \geq 20\%$). Based on x-ray diffraction analysis of the films using the Scherrer formula, the grain sizes for the films are 8.5 nm in the (220) and (311) directions and 20 nm in (111) direction, with the (111) peak being the most intense.

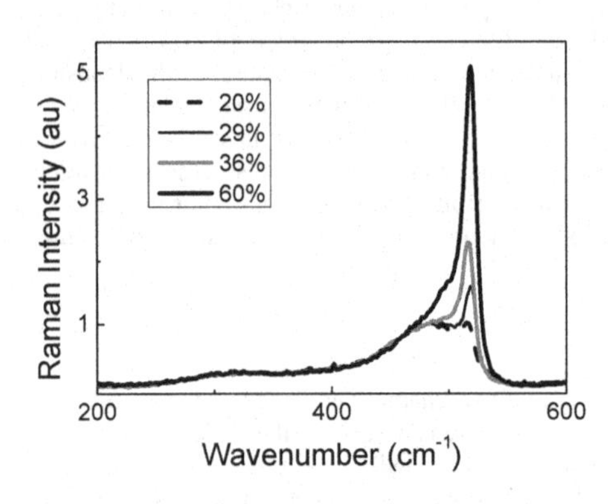

Figure 1: Raman measurements used to determine the relative crystallinity of the films. Only films with $X_C \geq$ 20% are shown for readability. All the plots have been scaled to the same background levels to allow for a visual comparison of the nanocrystalline silicon TO peak.

For the conductivity data, the sample is first annealed at 470K to eliminate any effects from light-induced defects or surface absorbates. The film is then cooled slowly at ~1 K/min to at least 350K to avoid differences due to thermal equilibration effects [15,16] before being rapidly cooled to 80K. The data are taken by applying a constant voltage and measuring the current upon warming at 1 K/min. All films were first confirmed to have Ohmic current-voltage characteristics down to the lowest measurement temperature.

RESULTS AND DISCUSSION

The results of the conductivity measurements are displayed in a conventional Arrhenius plot in Fig. 2. For the films with low crystalline content (<1%, 7%), the expected activated behavior is observed. For films with higher crystalline content, starting with X_C = 15%, clear non-Arrhenius behavior is found, with a concurrent drastic increase in magnitude of the low temperature conductivity. From 29% to 36%, we see another large jump in conductivity, while still maintaining the non-activated behavior. Such non-activated behavior is often ascribed to hopping conduction, such as $T^{-1/4}$ Mott variable range hopping (M-VRH) [17], $T^{-1/2}$ Efros-Shklovskii variable range hopping (ES-VRH) [18], or multiphonon hopping (MPH) [19-21].

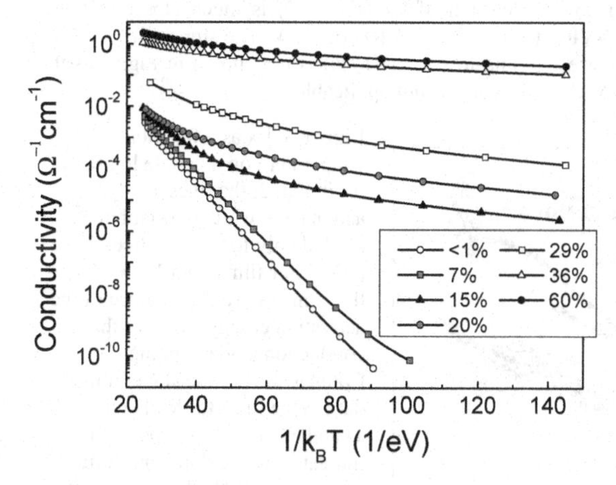

Figure 2: The conductivity for all the films in this study, showing the non-Arrhenius behavior for the highly crystalline films starting with X_C = 15%. The lines represent the actual data, while, for clarity, we have plotted the symbol only for every fifteenth point.

Using simply a "quality of fit" criteria for plots with different abscissa (T^{-1}, $T^{-1/4}$, $T^{-1/2}$, log(T), etc) in order to distinguish the type of hopping exhibited is extremely difficult owing to the limited temperature range over which the non-Arrhenius behavior is observed. In order to avoid this pitfall, an analysis similar to Zabrodskii [22,23] and Hill [24] was employed to directly extract information about the functional form of the data and hence the type of hopping observed. This is done by calculating the activation energy (E_A), defined in eqns. 2, as a function of temperature and fitting it to extract the temperature dependence. Assuming a conductivity of the form of eqn. 2a, where we allow E_A to be temperature dependent, a logarithmic derivative of the conductivity with respect to $\beta = 1/k_BT$, results in:

$$\sigma = \sigma_0 \exp(-\beta E_A) \tag{2a}$$

$$-\frac{d\ln\sigma}{d\beta} = E_A \tag{2b}$$

Where T is the temperature and k_B is Boltzmann's constant. Using eqn. 2b, we fit a double logarithmic plot of E_A against temperature to determine the activation energy's temperature dependence. For Mott VRH, we expect an exponent of ¾, meaning $E_A \approx T^{3/4}$, while for Efros-Shklovskii VRH we expect $E_A \approx T^{1/2}$ [25]. In contrast, multiphonon hopping MPH, which does not have an exponential dependence, but rather a power-law temperature dependence, would display an unphysical exponent, that is $E_A \approx T^n$, where n is equal to or greater than unity. This plot can be seen in Fig. 3 along with straight lines to represent M-VRH (n = ¾) and MPH ($n \geq 1$). For the low X_C films (1, 7%), we see a temperature independent E_A at low temperatures followed by the well-known shift in E_A due to thermal equilibration effects [15,16]. For the 15% film, an activation energy characteristic of hopping at low temperatures is found, followed by a shift to a constant E_A and finally the shift in E_A due to thermal equilibration. For films with $X_C > 20\%$, hopping conduction is indicated by the temperature dependent activation energy. Furthermore, for all cases of hopping, the power-law exponent n of E_A ($E_A \approx T^n$) is much closer to that expected for MPH ($n \geq 1$) than for M-VRH (n = ¾) or ES-VRH (n = ½). The slope of the MPH line in Fig. 3 is set to 1. A slope corresponding to ES-VRH (½) is not shown in Fig. 3 as it would be even shallower than the M-VRH and clearly is not applicable.

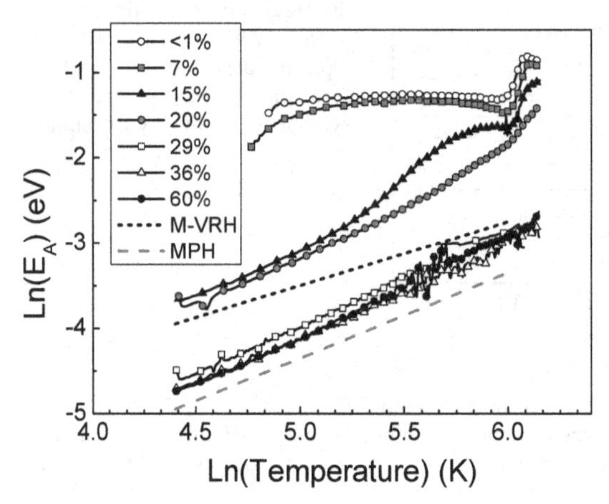

Figure 3: E_A as a function of temperature on a double log scale. As in Fig. 2, the lines are the actual data, while the symbol is plotted only for every fifteenth point. For films with $X_C > 15\%$, the temperature dependence of the activation energy suggests that conduction is via hopping. Comparisons of the slope of the data to the lines for MPH (slope = 1) and Mott-VRH (slope = ¾), indicate that the dominant form of hopping is MPH. The jump in the 29% film is where two data sets covering different temperature ranges were appended together.

Fig. 4 shows fits of the data to a power-law temperature dependence, along with the corresponding power-law exponents. The exponent decreases with increasing crystal fraction, with a large jump between the X_C = 29 and 36% films. We hypothesize that for highly nanocrystalline films ($X_C \geq 15\%$), the transport, particularly at low temperatures, is dominated by multiphonon hopping between clusters of nanocrystals. Within each nanocrystalline cluster, the electron wavefunction is relatively delocalized, allowing the electrons to move with ease. The rate limiting step is the transition between the nanocrystalline clusters, which requires phonon-assisted hopping. As the wavefunction is partially delocalized within the nanocrystalline clusters, the electrons would couple most strongly to phonons with a wavelength comparable to the radius of the electron wavefunction [26], that is, long wavelength, low energy phonons, thus making it necessary for more than one phonon to contribute to each hop. The power-law exponent represents the average number of phonons required for each hop; a distribution of hop energies would result in a non-integer number. Such a mode of transport has been proposed to exist in transition metal oxide glasses, where the hops are between metallic clusters [27], as well as between defect clusters in amorphous germanium [28] and localized π bonds in amorphous carbon [29]. The deviations at high temperature are attributed to competition between activated conduction in the surrounding a-Si:H matrix and hopping through the particles.

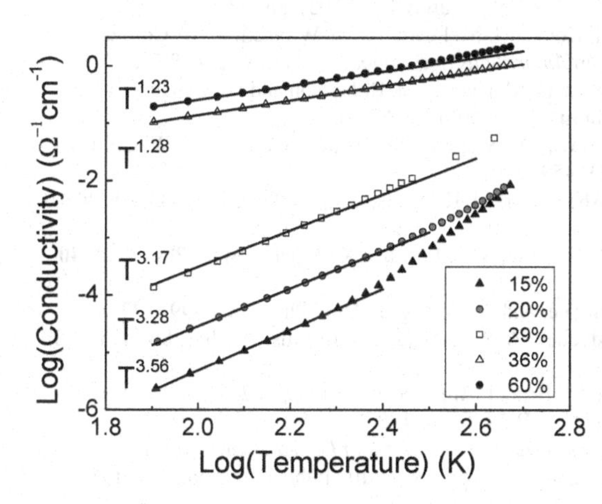

Figure 4: Double logarithmic plot of the conductivity against temperature. For multiphonon hopping (MPH), we expect to see a straight line on such a plot. Fits, along with corresponding power-law exponents are displayed. Only the films that exhibit MPH are shown. Deviations at high temperature are likely due to the competition between MPH and activated conduction. For clarity, only every fifteenth data point is shown.

CONCLUSIONS

We have shown for the first time, that in n-type doped mixed-phase silicon, multiphonon hopping (MPH) is the dominate transport mechanism. Analysis of the temperature dependence of the activation energy provides a definitive determination that MPH provides a superior fit to the conductivity data than either ES-VRH or M-VRH. As the current temperature range extends to liquid nitrogen temperatures from 470K, we cannot rule out the possibility that a transition to a different form of hopping does not occur at lower temperatures. Experiments at liquid helium temperatures are underway in order to determine whether or not this is the case.

This work was partially supported by NSF grant DMR-0705675, the NINN Characterization Facility, the Xcel Energy grant under RDF contract #RD3-25, and the University of Minnesota.

REFERENCES

1. R. Butte, M. Meaudre, R. Meaudre, S. Vignoli, C. Longeaud, J. P. Kleider, and P. R. Cabarrocas, Phil. Mag. B **79**, 1079 (1999).
2. Y. Lubianiker, J. D. Cohen, H. -. Jin, and J. R. Abelson, Phys. Rev. B **60**, 4434 (1999).
3. T. Kamei, P. Stradins, and A. Matsuda, Appl. Phys. Lett. **74**, 1707 (1999).
4. A. Fontcuberta i Morral, H. Hofmeister, and P. Roca i Cabarrocas, J. Non-Crystal. Solids **299-302**, 284 (2002).
5. J. Yang, K. Lord, S. Guha, and S. R. Ovshinski, in *Amorphous and Heterogeneous Silicon Thin Films*, edited by R. W. Collins, H. M. Branz, M. Stutzmann, S. Guha and H. Okamoto (Mater. Res. Soc. Symp. Proc. **609**, Pittsburgh, PA, 2000), pp.A15.4.1.
6. R. W. Collins, A. S. Ferlauto, G. M. Ferreira, J. Koh, C. Chen, R. J. Koval, J. M. Pearce, C. R. Wronski, M. M. Al-Jassim, and K. M. Jones, in *Amorphous and Nanocrystalline Silicon-Based Films*, edited by J. R. Abelson, G. Ganguly, H. Matsumura, J. Robertson and E. A. Schiff (Mater. Res. Soc. Symp. Proc. **762**, Pittsburgh, PA, 2003), pp.A.10.1.
7. A. S. Ferlauto, G. M. Ferreira, R. J. Koveal, J. M. Pearce, C. R. Wronski, R. W. Collins, M. M. Al-Jassima, and K. M. Jones, in *Amorphous and Nanocrystalline Silicon-Based Films*, edited by J. R. Abelson, G. Ganguly, H. Matsumura, J. Robertson and E. A. Schiff (Mater. Res. Soc. Symp. Proc. **762**, Pittsburgh, PA, 2003), pp.A5.10.
8. C. R. Wronski, J. M. Pearce, R. J. Koval, X. Niu, A. S. Ferlauto, J. Koh, and R. W. Collins, Mater. Res. Soc. Symp. Proc. **715**, 459 (2002).
9. Y. Adjallah, C. Anderson, U. Kortshagen, and J. Kakalios, J. Appl. Phys. **107**, 043704 (2010).
10. J. Kakalios, in this Proceedings.
11. R. Tsu, J. Gonzalez-Hernandez, S. S. Chao, S. C. Lee, and K. Tanaka, Appl. Phys. Lett. **40**, 534 (1982).
12. A. T. Voutsas, M. K. Hatalis, J. Boyce, and A. Chiang, J. Appl. Phys. **78**, 6999 (1995).
13. V. Golubev, V. Davydov, A. Medvedev, A. Pevtsov, and N. Feoktistov, Phys. Solid State **39**, 1197 (1997).
14. E. Bustarret, M. A. Hachicha, and M. Brunel, Appl. Phys. Lett. **52**, 1675 (1988).
15. J. Kakalios and R. A. Street, Phys. Rev. B **34**, 6014 (1986).
16. J. Kakalios and R. A. Street, *Thermal Equilibrium Effects in Doped Hydrogenated Amorphous Silicon*, edited by H. Fritzsche (World Scientific Publishing Company, 1988), p. 165.
17. N. F. Mott, J. Non-Crystal. Solids **1**, 1 (1968).
18. A. L. Efros and B. I. Shklovskii, J. Phys. C **8**, L49 (1975).
19. N. F. Mott and E. A. Davis, *Electronic processes in non-crystalline materials,* (Oxford, Clarendon Press, Oxford, 1979).
20. N. Robertson and L. Friedman, Phil. Mag. **36**, 1013 (1977).
21. N. Robertson and L. Friedman, Phil. Mag. **33**, 753 (1976).
22. A. G. Zabrodskii and I. S. Shlimak, Sov. Phys. Semicond. **9**, 391 (1975).
23. A. G. Zabrodskii, Sov. Phys. Semicond. **11**, 345 (1977).
24. R. M. Hill, Phys. Stat. Sol. (a) **35**, K29 (1976).
25. B. I. Shklovskii and A. L. Efros, *Electronic properties of doped semiconductors* (Springer-Verlag, Berlin, Heidelberg, New York, Tokyo, 1984).
26. D. Emin, Phys. Rev. Lett. **32**, 303 (1974).
27. K. Shimakawa, Phil. Mag. B **60**, 377 (1989).
28. K. Shimakawa, Phys. Rev. B **39**, 12933 (1989).
29. K. Shimakawa and K. Miyake, Phys. Rev. B **39**, 7578 (1989).

Mater. Res. Soc. Symp. Proc. Vol. 1321 © 2011 Materials Research Society
DOI: 10.1557/opl.2011.1150

Photocarrier Excitation and Transport in Hyperdoped Planar Silicon Devices

Peter D. Persans[1], Nathaniel E. Berry[1], Daniel Recht[2], David Hutchinson[1], Aurore J. Said[2], Jeffrey M. Warrender[3], Hannah Peterson[1,3], Anthony DiFranzo[1], Christina McGahan[1], Jessica Clark[1], Will Cunningham[1], and Michael J. Aziz[2]

[1]Rensselaer Polytechnic Institute, 110 8th Street, Troy NY 12180
[2] Harvard, School of Engineering and Applied Science, 29 Oxford Street, Cambridge, MA 02138
[3]US Army – ARDEC, Benet Laboratories, Watervliet, NY 12189

ABSTRACT

We report an experimental study of photocarrier lifetime, transport, and excitation spectra in silicon-on-insulator doped with sulfur far above thermodynamic saturation. The spectral dependence of photocurrent in coplanar structures is consistent with photocarrier generation throughout the hyperdoped and undoped sub-layers, limited by collection of holes transported along the undoped layer. Holes photoexcited in the hyperdoped layer are able to diffuse to the undoped layer, implying $(\mu\tau)_h \sim 5\times10^{-9}$ cm^2/V. Although high absorptance of hyperdoped silicon is observed from 1200 to 2000 nm in transmission experiments, the number of collected electrons per absorbed photon is 10^{-4} of the above-bandgap response of the device, consistent with $(\mu\tau)_e < 1\times10^{-7}$ cm^2/V.

INTRODUCTION

Pulsed laser irradiation enables the production semiconductors that have compositions not readily attainable with other methods and that exhibit interesting and potentially useful optical properties[1, 2]. Recently, optically smooth sulfur-supersaturated crystalline silicon has been fabricated in thin film form using ion implantation followed by nanosecond pulsed laser melting [3]. This new material exhibits an unexpectedly high sub-bandgap absorption coefficient [2, 3]. The nature of the transitions that give rise to this absorption has not yet been determined, however free carrier absorption has been ruled out. The present work is an attempt to elucidate the excitation and photoconduction mechanisms in hyperdoped silicon devices.

EXPERIMENT

Hyperdoped layers were prepared by implantation of sulfur into a thin crystalline silicon-on-insulator (SOI) layer, followed by pulsed laser melting to rapidly recrystallize the implanted layer while retaining a high dopant concentration [3]. For the current study, the SOI layer was 260 nm thick with a 1000 nm oxide. The peak implant density of ~10^{20} cm^{-3} is about 110 nm deep. After laser melting of the top 200 nm, the peak S density is about 2.5×10^{19} cm^{-3}. The back 50 nm of the SOI layer is not melted in order to act as a seed layer for crystalline regrowth. Au surface contacts for coplanar measurements were evaporated with a 0.8 mm gap.

The optical absorption coefficient for a hyperdoped layer prepared by implantation of 10^{15} S ions per cm^2 followed by pulsed laser melting is plotted against wavelength in Fig. 1 [4, 5]. Dilute sulfur doping leads to donor impurity levels 0.3 eV below the conduction band [6, 7]. It has been proposed that hyper-doping with S leads to an impurity band [3]. About 10% of S impurities lead to carriers in the conduction band [8].

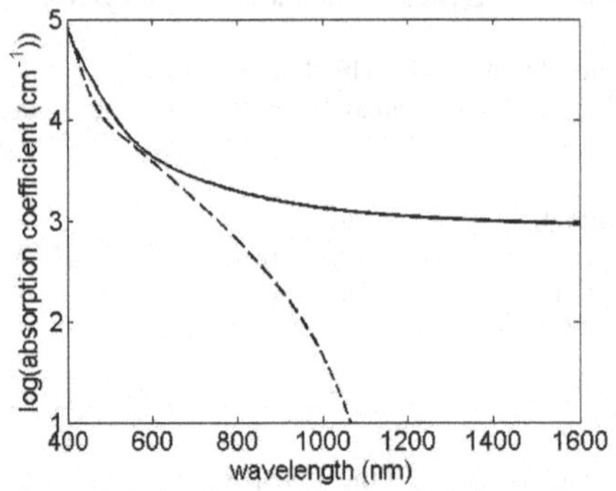

Figure 1. Optical absorption coefficient plotted as a function of wavelength for undoped Si (dashed line) and chalcogen hyperdoped Si (solid line). Sulfur implantation dose for the hyperdoped sample is $1x10^{15}cm^{-2}$. (from Pan et al. [5])

Room temperature dark current-voltage curves were ohmic and n-type, yielding a sample resistance of 1440 ohms and layer conductivity of 20 $\Omega^{-1}cm^{-1}$, consistent with carrier density of ~10^{19} cm^{-3} [9, 10]. Electron Hall mobility of ~ 50 cm^2/Vs has been reported in similar materials [7]. Photocurrent under typical illumination levels is orders of magnitude smaller than dark current, so steady state photocurrent was detected using chopping and lock-in techniques [11]. Transient photocurrent was measured using a 5 nanosecond nitrogen laser pumped dye laser at 650 nm and oscilloscope.

RESULTS AND DISCUSSION

Photocurrent magnitude was nearly linear in voltage for applied voltages less than 0.5 V and displayed saturation for voltage greater than ~1.2V (Fig. 2). For a given applied voltage, cw photocurrent was linear with intensity. Transient photocurrent measurements, extrapolated to zero applied field, yielded exponential current decay with decay lifetime of ~40 μs. We note that this is a surprisingly large value for heavily doped material – typical free carrier lifetimes for heavily doped material are less than 10 ns [12]. Saturation of photocurrent with increasing voltage occurs in metal-semiconductor-metal devices either when primary photocarriers are not re-injected or when the minority carrier drift length $L_{drift} = \mu\tau E$ exceeds the length of the sample [11]. Reinjection of electrons is likely, due to the ohmic contacts, so photocurrent is likely limited by minority carrier extraction. Our observed saturation field of ~20 V/cm for a sample width of 0.8 mm yields $\mu\tau = 3\times10^{-3}$ cm^2/V for minority carriers. Using the decay lifetime of 40 μs, we find a drift mobility of ~75 cm^2/Vs.

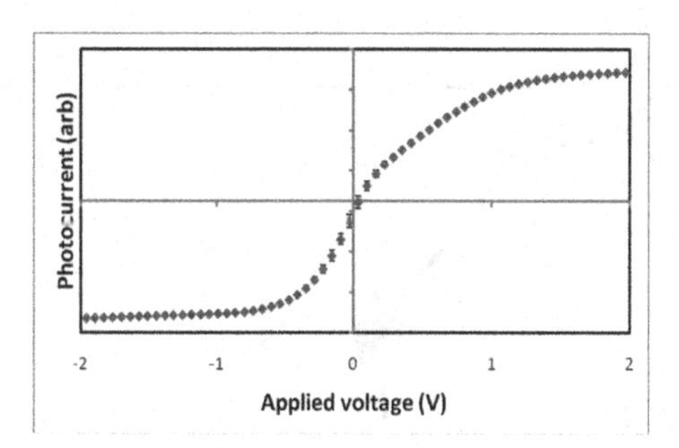

Figure 2. Photocurrent as a function of voltage applied across the sample. Excitation wavelength was 660 nm.

The combination of high drift mobility and long decay lifetime is not expected for photocarrier transport in hyperdoped Si – long decay lifetime implies carrier trapping and trapping decreases drift mobility [13]. We therefore propose a photocarrier model in which electron and hole photocarriers are quickly separated into different layers of the sample. The structure of the sample, sketched in Figure 3, provides a mechanism. The doping profile is abrupt, leaving the bottom ~50 nm seed layer with much less active sulfur [3, 14]. This leads to a profile in which photoexcited holes can diffuse or drift a short distance to the undoped seed layer and electrons remain in the hyperdoped layer, as shown in case A in the sketch. If the $(\mu\tau)$ product for holes in the seed layer exceeds that of electrons in the hyperdoped layer, then photocurrent saturation occurs when holes are swept from the bottom layer, recombining with electrons in the external circuit.

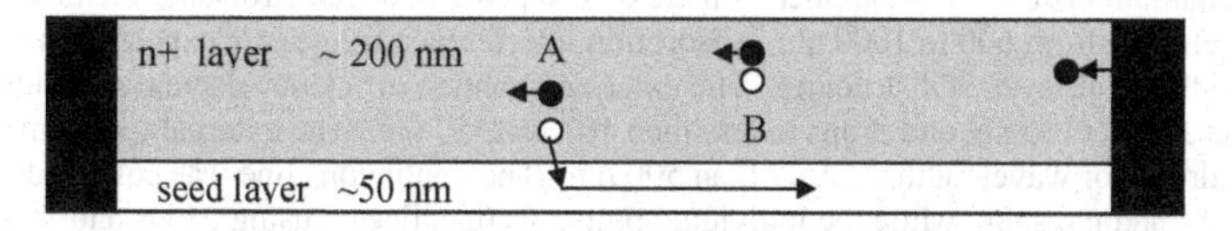

Figure 3. A schematic side-view of the silicon-on-insulator layer showing the proposed motion of photogenerated electrons (black) and holes (white). The doped layer is about 200 nm thick and the undoped seed layer is about 50 nm thick. Free holes generated in the doped layer leave the doped layer and travel laterally in the seed layer (case A). Electrons and trapped holes remain in the doped layer (case B). The sketch is not to scale. The aspect ratio of thickness to length is ~$3x10^{-3}$.

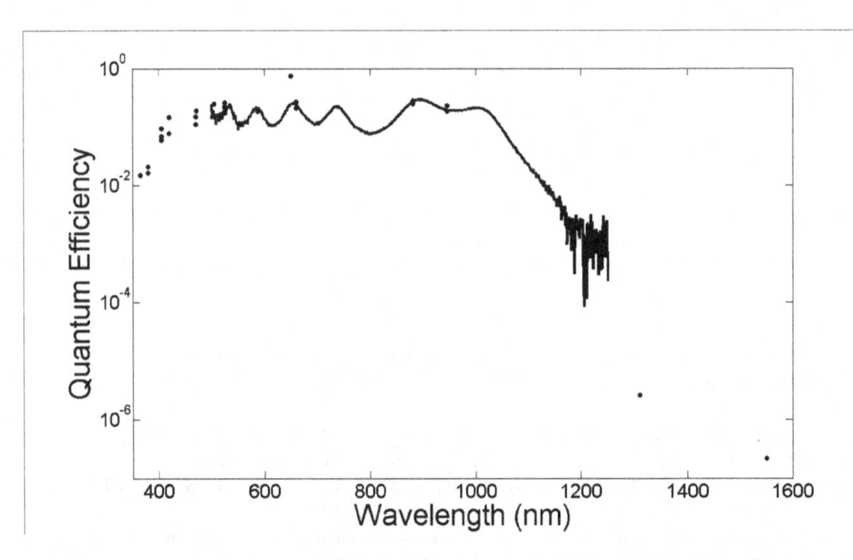

Figure 4. Spectral response as a function of wavelength. Solid line is collected with a scanned monochromator with a tungsten-halide lamp source. Points are collected with LED and laser sources. The two data points at 1310 and 1550 nm are upper bounds on the signal.

The photocurrent spectrum, normalized to display electronic charge collected per incident photon (external quantum efficiency), is plotted in Fig. 4 for wavelength from 365 nm to 1550 nm. External quantum efficiency for applied voltage of 1 V is 0.1 to 0.3 electrons per incident photon for wavelength from 600 to 1000 nm. Absorption interference fringes are expected and observed due to the multilayer SOI structure. The external quantum efficiency decreases rapidly for wavelengths above 1100 nm and drops to less than 10^{-7} at 1550 nm. The external quantum efficiency also drops for wavelengths shorter than 500 nm. The continuous line was collected using a monochromator system, while the individual points were collected using LEDs and lasers.

The absence of response beyond 1200 nm further constrains the photoconduction models. Collection efficiency for sub-bandgap excitation is at least 10^4 times smaller than would be expected if the photocarrier were the same for above and below gap excitation. Sub-bandgap absorption should result in a free electron and a trapped hole in an impurity state in the hyperdoped layer (illustrated as case B in the sketch in Fig. 3). The electron collection efficiency would then be $L_{drift} / L = \mu_n \tau_n V / L^2$, where L is the distance between contacts. Absorptance in the hyperdoped layer at 1550 nm is about $A=10^{-2}$, so the observed collection efficiency of $< 3\text{x}10^{-7}$ implies $\mu_n \tau_n < 2\text{x}10^{-7}$ cm^2/V in the hyperdoped layer.

The free electron lifetime in the hyperdoped layer is expected to be shorter than 1 ns because the equilibrium density of holes (in impurity states) is high (~10^{18} cm^{-3}). Assuming reasonable values from the literature of free carrier mobility of 50 cm^2/Vs [7] and lifetime of $\tau < 10^{-9}$ s [10], yields an expected upper bound of $(\mu\tau)_e < 10^{-7}$ cm^2/V.

Photocurrent response at 365 nm, where absorption is entirely in the hyperdoped material, is 10^5 times greater than in the infrared. This is explained if holes generated at the top surface can diffuse or drift about 150 nm to the undoped layer. Assuming diffusion-limited transport, this yields an estimate of $(\mu\tau)_h \cong 5\times10^{-9}$ cm^2/V for holes in the hyperdoped layer.

CONCLUSIONS

Experimental measurement of coplanar photoresponse in a 260 nm thick silicon on insulator layer sets limits on photocarrier excitation, recombination, and transport. We propose that:

i) Electron-hole pairs excited with above bandgap radiation separate into hyperdoped and seed layers, leading to longer than expected lifetimes.
ii) The n+ contact-metal band structure allows holes to escape easily, but they cannot be re-injected. (This leads to saturation.)
iii) Electron-hole pairs excited with sub-bandgap radiation leave holes trapped on the impurity states in the hyperdoped layer, which provide sites for rapid recombination.
iv) The fall off of response for wavelengths shorter than 400 nm is consistent with the short lifetime and diffusion length proposed for holes and electrons in the hyperdoped layer.

These model assumptions lead to limits on the free carrier mobility-lifetime product for carriers in S-hyperdoped material of $\mu_n\tau_n < 2x10^{-7}$ cm^2/V for electrons and $\mu_p\tau_p \approx 5x10^{-9}$ cm^2/V for holes.

ACKNOWLEDGMENTS

This work is supported in part by the Army Research Office under contract No. W911NF0910470, US Army ARDEC under contract No. W15QKN-07-P-0092, and by the NSF REU programs at Harvard and at Rensselaer. Prepublication optical data were provided by Athena Pan (Brandeis University). Prepublication carrier density and majority carrier mobility data were supplied by Mark Winkler (Harvard University and MIT). We have benefitted from discussions with James Carey of SiOnyx Inc., Eric Schiff (Syracuse University), and Tonio Buonassisi (MIT).

REFERENCES

1. J.E. Carey, C.H. Crouch, M.Y. Shen, and E. Mazur, *Visible and near-infrared responsivity of femtosecond-laser microstructured silicon photodiodes.* Optics Letters, 2005. **30**(14): p. 1773-1775.
2. C.H. Crouch, J.E. Carey, J.M. Warrender, M.J. Aziz, E. Mazur, and F.Y. Genin, *Comparison of structure and properties of femtosecond and nanosecond laser structured silicon.* Appl. Phys. Lett., 2004. **84**: p. 1850.
3. T.G. Kim, J.M. Warrender, and M.J. Aziz, *Strong sub-band-gap infrared absorption in silicon supersaturated with sulfur.* Appl. Phys. Lett., 2006(241902).

4. S.H.A. Pan, *Enhanced Visible Absorption of Ion Implanted and Pulse Laser Melted Si Supersaturated with Chalcogens*. 2010, Brandeis University.
5. S.H. Pan, D. Recht, S. Charnvanichborikarn, J. Williams, S., and M.J. Aziz, *Enhanced visible and near-infrared optical absorption in silicon supersaturated with chalcogens.* Appl. Phys. Lett., 2011. **98**: p. 121913.
6. H.G. Grimmeiss and E. Janzen, in *Handbook on Semiconductors*, T.S. Moss and S. Mahajan, Editors. 1994, Elsevier: Amsterdam.
7. M. Winkler, *Non-Equilibrium Chalcogen Concentrations in Silicon: Physical Structure, Electronic Transport, and Photovoltaic Potential*, in *Physics*. 2010, Harvard: Cambridge, MA.
8. M. Winkler, D. Recht, M.-J. Sher, A.J. Said, E. Mazur, and M.J. Aziz, *Insulator to Metal Transition in Sulfur Doped Silicon.* Phys. Rev. Lett., 2011. **106**: p. 178701.
9. H.F. Wolf, *Semiconductors*. 1971, New York: Wiley- Interscience.
10. S.M. Sze, *Physics of Semiconductor Devices*. 1981, New York: Wiley.
11. S.M. Ryvkin, *Photoelectric effects in semiconductors*. 1964, New York: Consultants Bureau.
12. M.S. Tyagi and R. VanOverstraeten, *Minority carrier recombination in heavily doped silicon.* Solid State Electronics, 1983. **26**: p. 577-598.
13. E.A. Schiff, *Transit-time measurements of charge carriers in disordered silicons: Amorphous, microcrystalline and porous.* Phil. Mag., 2009. **89**: p. 2505-2518.
14. M. Tabbal, T.G. Kim, J.M. Warrender, M.J. Aziz, B.L. Cardozo, and R.S. Goldman, *Formation of single crystal sulfur supersaturated silicon based junctions by pulsed laser annealing.* J. Vac. Sci. Technol. B, 2007. **25**: p. 1847.

Mater. Res. Soc. Symp. Proc. Vol. 1321 © 2011 Materials Research Society
DOI: 10.1557/opl.2011.1095

Theoretical Studies of Structure and Doping of Hydrogenated Amorphous Silicon

Bin Cai and D. A. Drabold
Department of Physics and Astronomy, Ohio University
Athens, OH 45701, U.S.A.

ABSTRACT

In a-Si:H, large concentrations of B or P (of order 1%) are required to dope the material, suggesting that doping mechanisms are very different than for the crystal for which much smaller concentrations are required. In this paper, we report simulations on B and P introduced into realistic models of a-Si:H and a-Si, with concentrations ranging from 1.6% to 12.5% of B or P in the amorphous host. The results indicate that tetrahedral B and P are effective doping configurations in a-Si, but high impurity concentrations introduce many defect states. For a-Si:H, we report that both B(3,1) and P(3,1) (B or P atom bonded with three Si atoms and one H atom) are effective doping configurations. We investigate H passivation in both cases. For both B and P, there exists a "hydrogen poison range" of order 6 Å for which H in a bond-center site can suppress doping. For B doping, nearby H prefers to stay at the bond-center of Si-Si, leaves B four-fold and neutralizes the doping configuration; for P doping, nearby H spoils the doping by inducing a reconstruction rendering initially tetrahedral P three-fold.

INTRODUCTION

By introducing B or P, a-Si:H may be doped either n-type or p-type [1], a point of profound technological importance. In c-Si, doping has been extensively studied. Because of translational invariance, impurities are compelled to have the same local tetrahedral environment as Si. According to the *8-N* rule, B atoms create a hole when they have T_d symmetry, and P similarly donate an electron. The doping efficiency is almost 100%. In c-Si:H, H atoms passivate doping by relaxing strain, and rendering B or P doping-inactive by enabling the impurities to become three-fold [1]. In this short paper, we make no pretense of properly reviewing the substantial literature on the subject and recommend the book of Street[1] as a suitable introduction.

In a-Si or a-Si:H, the absence of a unique atomic environment leads to site-dependent doping, as seen in studies with low concentration of Boron [2]. However, theoretical studies on P doping and high concentration of B are still needed. Early experiments show a very low doping efficiency in a-Si:H, with a doping efficiency rollover observed in experiment when impurity density is around 1% [1,3]. NMR [4] shows that, for B doping, 40% of the B has a nearby H at 1.6 Å; for P doping, 50% of P has a H at 2.6 Å. Boyce and Ready have conjectured that the sluggish doping may be due to H passivation [5]. But the atomistic mechanism of H passivation in doped a-Si:H is unclear.

In this paper, we report molecular dynamic simulations on B and P doped a-Si and a-Si:H, focusing on the impurity geometry and associated electronic structure. For a-Si, we report the electronic density of states (EDOS) for various impurity concentrations based on substitutional doping. We seek to determine the effective doping and non-doping configurations. For a-Si:H, we focus primarily on describing the H passivation mechanisms for the system with 1.6%

impurities. By manually placing H in the models, and allowing the system to relax, we are able to investigate energetically preferred impurity positions and explore consequent electronic structure. We should point out that this work leaves many stones unturned; the discussion of doping efficiency versus dopant density in a-Si:H, the thermal stability of effective doping configuration, and dopant diffusion in a-Si:H are not included here. We will report more details of those important topics in a future paper.

THEORY

All calculations were performed using the plane wave LDA code VASP with ultrasoft pseudopotentials and the local density approximation [6]. A previously generated defect-free 64-atom a-Si model was used as the initial configuration. B or P atoms were introduced into the network by substituting for Si atoms. Conjugate-gradient relaxations were performed at constant volume. We obtained relaxed 1.6%, 3.1%, 7.8% and 12.5% B or P doped a-Si models. To study H passivation in B- or P-doped a-Si:H, we introduced H atoms at particular sites of the 1.6% B- or P-doped a-Si models with various distances from impurities.

DISCUSSION

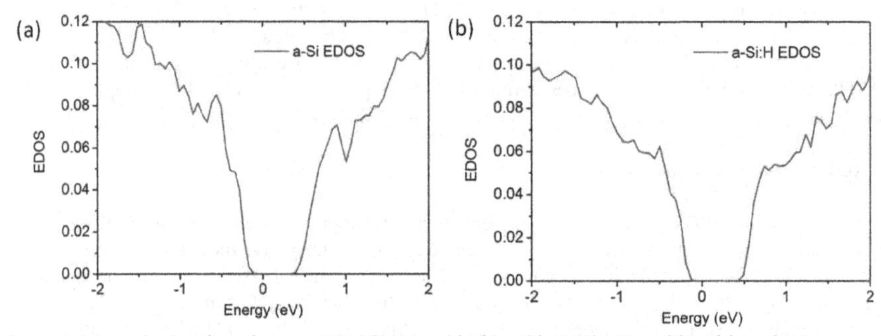

Fig. 1 Electronic density of states (EDOS) (a) a-Si. (b) a-Si:H. The Fermi level is at 0eV.

We first present the electronic density of states of a 64-atom a-Si model and a 70-atom a-Si:H model with 8.5% H in Fig. 1. Both a-Si and a-Si:H models exhibit gaps unsullied by defect states. In the following, we first discuss B- and P-doped a-Si with impurity concentrations from 1.6% to 12.5%. Then, we investigate the mechanism of H passivation in both systems.

Boron doped a-Si

We plot the EDOS of seven B-doped models [configuration (1)-(7)] in Fig. 2. Overall, when B atoms are introduced substitutionally into the network, the Fermi level shifts toward the valence edge, signaling doping. In configuration (1), (2), (4) and (7), all B atoms are four-fold and prefer to form three shorter bonds and one longer bond with Si. As the concentration increases, more valence tail states are formed, and states move into the gap. Where B dimers or clusters are concerned, it seems that B-B bonds won't impact the doping as long as B atoms are four-fold as shown in configuration (3). However, when more B dimers or clusters are formed, additional defect states appear near the conduction tail, and they clutter the gap, as shown in configuration (5) and (6). Those defect states are associated with under- and over-coordinated Si atoms.

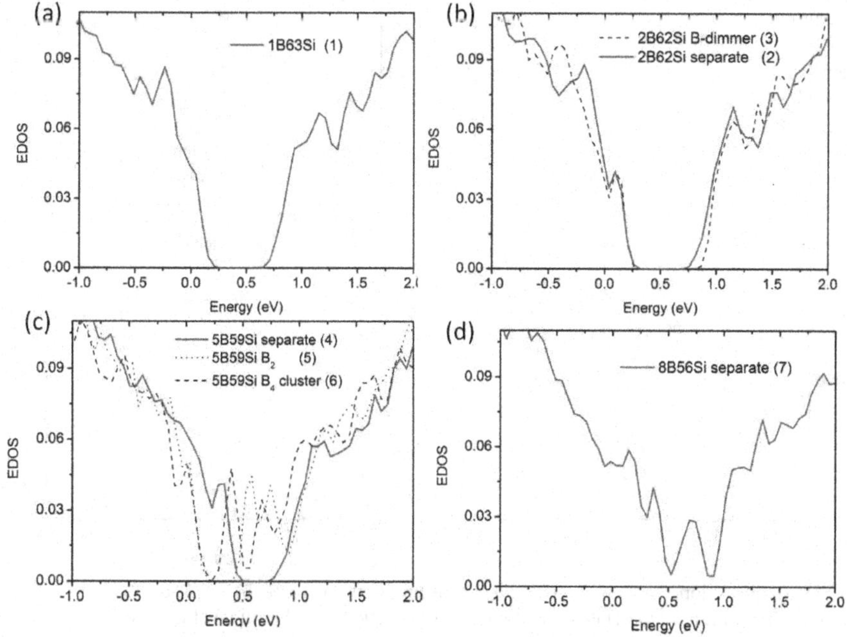

Fig. 2 Boron doped a-Si with different impurity concentrations. (a) 1.6% (b) 3.1% (c) 7.8% (d) 12.5%. In configuration (1), (2), (4) and (7), all B are bonded with four Si atoms. One B dimer is formed in configuration (3). Two B dimmers exist in configuration (5) and one B_4 cluster formed in configuration (6). The Fermi level is at 0 eV.

We conclude that tetrahedral B makes the Fermi level shift from mid-gap into the valence band tail and effectively dopes the system. However, as B concentration increase, more valence tail states are formed. B clusters may introduce tail and mid-gap states, which impacts the doping and surely transport.

Phosphorus doped a-Si

We plot the EDOS of eight models [(1)-(8)] of P-doped a-Si in Fig. 3. We found that tetrahedral P forms deep donor states and the Fermi level shifts toward the conduction band tail. As P concentration increases, more defect states appear, and the gap closes (configurations (1), (2), (5), and (8)). P dimers and clusters dope the system so long as all P are four-fold

(Configuration (3),(5)), but they also lead to defects which give rise to tail states. If P atoms are three-fold (configuration (4)), the configuration is non-doping.

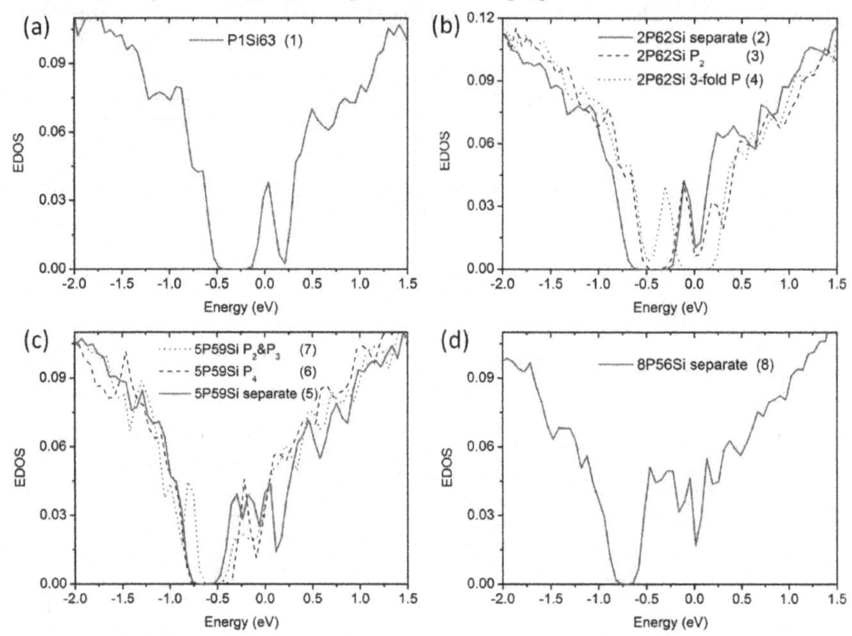

Fig. 3 Phosphorus doped a-Si with different P concentration. (a) 1.6% (b) 3.1% (c) 7.8% (d) 12.5%. In (b) configuration (4) is a non-doping configuration. The Fermi level is at 0eV.

Thus, for P doping, tetrahedral P dopes the system by shifting the Fermi level into the conduction tail. As for B doping, as concentration increases, more defect states move into gap. Three-fold P is a non-doping configuration.

H passivation in B doped a-Si:H

We next report that B (3,1) (B atoms bonded with three Si atoms and one H atom) is an effective doping configuration. The structural evolution and corresponding EDOS is plotted in Fig. 4 and Fig. 5.

H is initially attached to a B atom and makes B form a metastable unit B(4,1) as in Fig. 4 (a). After relaxation, H breaks one Si-B bond forming a B(3,1) structure and leaves one Si DB, as in Fig. 4 (b). The EDOS of this configuration reveals that the Fermi level shifts back into gap. One mid-gap state forms due to the Si DB. However, if another H passivates the Si DB, as in Fig.4 (c), the Fermi level shifts into valence band tail once again and the mid-gap state disappears. Thus, we conclude that the B(3,1) without a Si DB is an effective doping conformation. This confirms the simulation results reported in Ref [5].

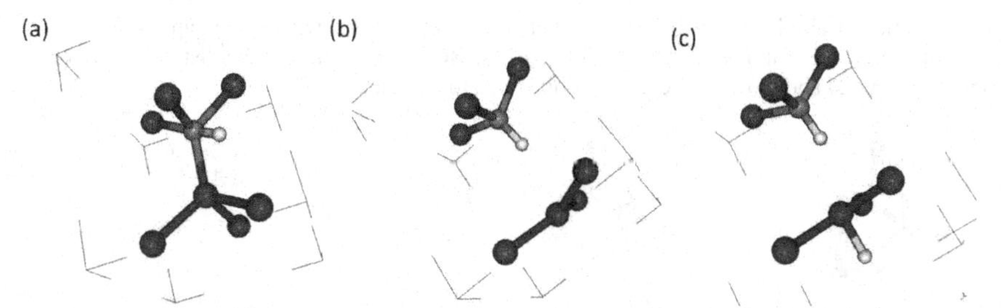

Fig. 4 H passivation at B site. (a) H initially bonds with B and B forms B(4,1) structure. (b) After relaxation, B becomes B(3,1) with a vestigial Si DB. (c) Another H passivates the Si DB. The cyan (light, big) atom is B, blue(dark, big) atoms are Si and white(light, small) atom is H.

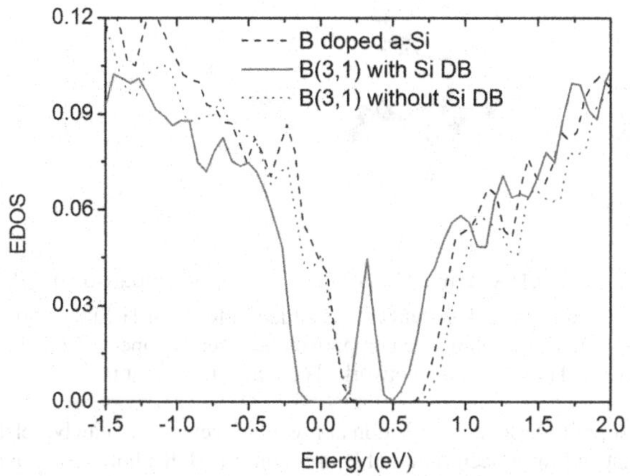

Fig. 5 EDOS comparison of configurations during H passivation. The black (dashed) line shows the original B doped a-Si. The red (solid) line shows the EDOS of B(3,1) and one Si DB. The blue (dotted) line shows the EDOS when Si DB is passivated by another H. The Fermi level is at 0eV.

Next, we show that H prefers to stay at the bond center (BC) when near a B atom and this BC H suppresses the doping. We show two cases of H passivation with different H-B distances in Fig. 6. The top panel (a) of Fig. 6 shows the situation in which H is initially bonded to a Si neighbor of the B atom. After relaxation, H breaks the Si-Si bond and stays at the BC forming a B-Si-H-Si structure, the top panel [6(b)]. The EDOS shows that the Fermi level shifts back from the valence tail into the gap, indicating that the BC H kills the doping. The bottom panel of Fig. 6 shows similar passivation with H bonded to a second neighbor Si of B atom and finally forming a B-Si-Si-H-Si structure. The EDOS of the B-Si-Si-H-Si structure shows that it is also a non-doping configuration. These results indicate that BC H sufficiently near a B atom will neutralize the doping. Notice that, in all cases, there is no reconstruction of B atoms -- which are still 4-fold after relaxation. There is no Si DB left in the network and no defect states in the gap. Further calculations indicate that there is an "H kill range" for BC H passivation. If the distance between H and B is beyond about 6.0 Å, the passivation seems to not occur.

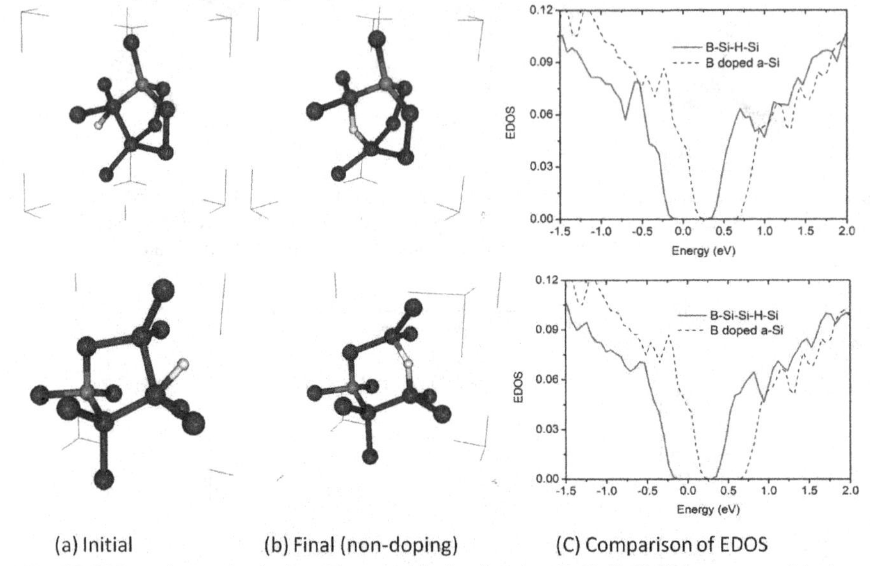

(a) Initial (b) Final (non-doping) (C) Comparison of EDOS

Fig. 6 BC H passivates the doping. Top panel is the situation for B-Si-H-Si; bottom panel is the case for B-Si-Si-H-Si. The EDOS show the comparison between B doped a-Si and the final relaxation result when BC H exists in the network. The Fermi level is at 0eV.

The mechanism of this BC H passivation at low B concentration may be related to charge interactions. In Ref [2], at low concentration of B, it is confirmed that holes could be trapped at strained Si-Si bond centers. Thus, the H may compensate by staying at the BC and killing the doping structure. Another calculation in a-Si:H also confirm this. In a-Si:H, the Si (4,1) structure (Si bonded with four Si and one H) is stable. However, if one electron is removed from the system, H tends to break a Si-Si bond and occupy the BC position. Considering the "H poisoning range", it may be related to the exciton radius [7], which is about 5.9 Å in a-Si.

H passivation in P doped a-Si:H

In analogy with H passivation in B doped a-Si:H, we investigate H passivation in P-doped a-Si:H. We first report that P(3,1) is an effective doping configuration. The simulation is shown in Fig.7 and Fig. 8. H is originally bonded to P forming a P(4,1) metastable structure. After relaxation, a P-Si bond breaks. H sticks to the P, forming a P(3,1) structure with a Si DB. The corresponding EDOS indicates that the deep donor state disappears, the Fermi level shifts back to the gap, and there is one mid-gap state formed (due to the Si DB). The configuration becomes non-doping. However, if another H passivates the Si DB as shown in Fig.7 (c), the

Fermi level again shifts to the conduction tail. Thus, we conclude that P(3,1) is an effective doping structure, but the Si DB in the network may eliminate the doping.

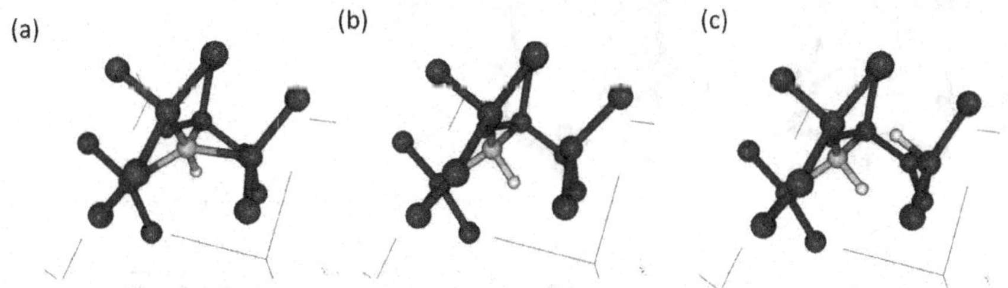

Fig.7 H passivation at P site. (a) H bond with P forming a P(4,1) structure. (b) H breaks one P-Si bond and makes P form P(3,1) with one Si DB. (c) another H passivates the Si DB. Green (light, big) atom is P. Blue(dark, big) atom is Si. White (small) atoms is H.

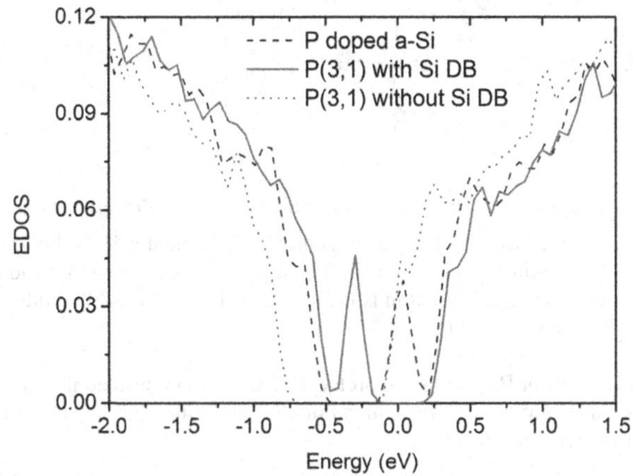

Fig. 8 EDOS: comparison of configurations during H passivation. Black (dashed) line is P doped a-Si. Red (solid) line is P(3,1) with one Si DB. Blue (dotted) line is the P(3,1) with another H passivate the Si DB. The Fermi level is at 0eV.

We then show two cases of H passivation in P doped a-Si:H in Fig.9. Unlike H passivation in B-doped a-Si:H, H in this network does not prefer BC position. In the top panel (a) of Fig.9, H is originally at BC of a P-Si bond and makes a P(3,1) structure. However, after relaxation, the H-P bond ruptures, the H bonds with Si, and P becomes three-fold. The EDOS becomes non-doping as the deep donor state disappears and the Fermi level is now in the gap. The bottom panel of Fig. 9 shows another case. H is initially at a bond center of Si-Si and forms a P-Si-H-Si structure. After relaxation, the network reconstructs as P becomes three-fold and H sticks to a Si DB. The doping is rendered inactive. Notice that there are no defect states in the EDOS of final configuration. Further calculations show that again there exists an "H kill range" of ~ 6.0 Å. When H is sufficiently near a P site in a BC configuration, the network will reconstruct so that P becomes three-fold, H sticks to Si DB, and neutralizes the doping structure.

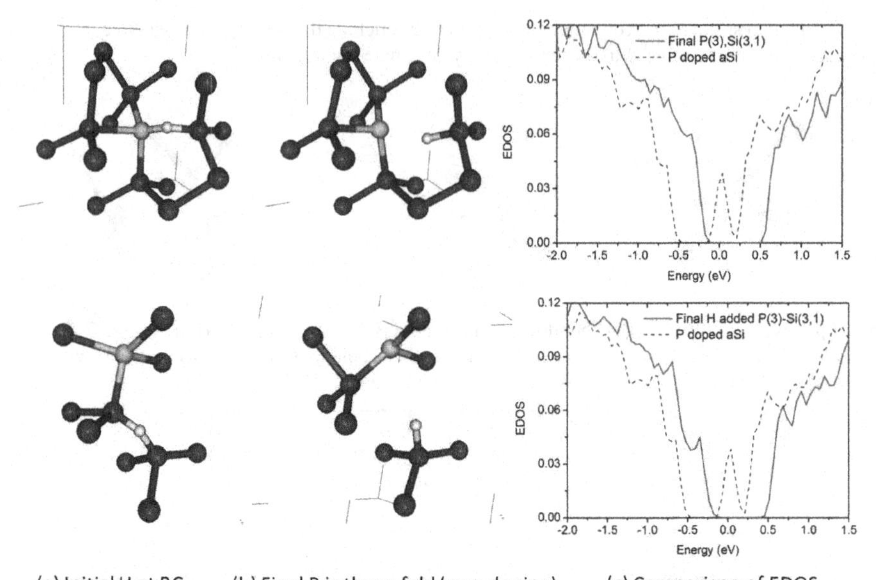

(a) Initial H at BC (b) Final P is three-fold (non-doping). (c) Comparison of EDOS.

Fig. 9 H passivation in P doped a-Si:H. Top panel, H originally formed P-H-Si; bottom panel, H originally formed p-Si-H-Si. After relaxation, in both cases, P becomes three fold and non-doping configurations. Green (light, big) atom is P. Blue(dark, big) atom is Si. White (light, small) atom is H. The Fermi level is at 0eV.

Unlike H passivation for B, H does not prefer the BC position, instead it prefers to bond with Si. This result is consistent with NMR, which implies that P often has an H neighbor around 2.6 Å away (*not* the first neighbor distance).

CONCLUSIONS

Tetrahedral B and P dope the system, but uniformly distributed high concentrations of impurities greatly reduce the gap. Clusters also create defect states in the gap. Strongly supporting experimental conjectures[4], H passivation is a key to understanding doping efficiency. We also make it clear that the basic mechanism of doping is quite different from the crystal and we can gauge success not by the appearance of a shallow level, but my monitoring the position of the Fermi level. The configurations B(3,1), P(3,1), Si(3,1) are effective doping states. There exists an "H kill range": for B doping: when BC H sufficiently is near B, H breaks the bond and stays at the bond center; for P doping, H atoms bond to Si and induce a reconstruction leaving P three-fold and thus electronically inert.

Most calculations in this paper are based on direct relaxations. We are currently performing long MD runs to study the thermal stability of B, P tetrahedral structure in a-Si, the interaction between H and impurities at high temperatures in a-Si:H, and H passivation in a-Si:H with high concentrations of impurities distributed in various ways. Another topic for future study would be to decipher the importance of charge states on the doping and defect equilibrium. The key finding of this paper is that BC H has a powerful and negative effect on doping a-Si:H both for B and P impurities, and in the case of P tends to convert tetrahedral P into trigonal P, resulting in suppression of doping. For B too, H suppresses doping, though it seems less able to induce a trigonal reconstruction at the B site.

ACKNOWLEDGMENTS

We thank Dr. W. Windl and Dr. S. Estreicher for helpful suggestions. We thank the ARO under MURI W91NF-06-2-0026 and NSF under DMR 09-03225 for supporting this work.

REFERENCES

1. R. A. Street, Hydrogenated Amorphous Silicon, Cambridge University Press, Cambridge, UK (2002).
2. I. Santos, P. Castrillo, W. Windl, D. A. Drabold, L. Pelaz and L.A. Marques, Phys. Rev. B **81**, 033203 (2010).
3. M. Stutzmann, Mat. Res. Soc. Symp. Proc. Vol. 70, (1986). M. Stutzmann, D. K. Biegelsen, R. A. Street, Phys. Rev. B **35**, 5666-5701 (1987).
4. J. B. Boyce and S. E. Ready, Phys. Rev. B **38**, 11008 (1998).
5. P. A. Fedders and D. A. Drabold, Phys. Rev. B **56**, 1864 (1997).
6. G. Kresse and J. Furthmuller, Phys. Rev. B **54**, 11169. http://cmp.univie.ac.at/vasp/.
7. D. A. Drabold, European Physical Journal B **68**, 1(2009). See esp. pp. 18-19.

Mater. Res. Soc. Symp. Proc. Vol. 1321 © 2011 Materials Research Society
DOI: 10.1557/opl.2011.811

Ab Initio Structure Characterization for the Amorphous Assembly of Si Clusters Encapsulating Transition Metal

Takehide Miyazaki[1], Noriyuki Uchida[2], and Toshihiko Kanayama[3]
[1]Nanosystem Research Institute, National Institute for Advanced Industrial Science and Technology,
1-1-1 Umezono, Tsukuba, Ibaraki 305-8568, Japan
[2]Nanoelectronics Research Institute, National Institute for Advanced Industrial Science and Technology,
1-1-1 Higashi, Tsukuba, Ibaraki 305-8562, Japan
[3]National Institute for Advanced Industrial Science and Technology,
1-1-1 Umezono, Tsukuba, Ibaraki 305-8568, Japan

ABSTRACT

We present a first-principles lattice dynamics for the assembly of the transition-metal (*M*)-encapsulated Si_n clusters in amorphous phase (a-MSi_n), which has been proposed as a potential candidate for the channel material of the next-generation thin-film transistors (TFTs) [N. Uchida *et al., Appl. Phys. Express* **1**, 121502 (2008)]. The shape of calculated vibrational density of states (VDOS) curve of a-$MoSi_{10}$ is similar to the counterpart of the high pressure phase of a-Si (HPA-Si) although the present systems are obtained as a result of pressure relaxation. Its radial distribution function (RDF) among Si themselves is characterized by the absence of a gap between the first and second shells, which is also the case in . We further present the VDOS of a-WSi_{10}, whose curve shape is again similar to that of HPA-Si. A difference between a-$MoSi_{10}$ and a-WSi_{10} is that the W-atom displacement components extracted from the vibration eigenvectors are mainly distributed over a lower frequency range (< ~ 150 cm^{-1}) than the Mo counterpart (~ 150 cm^{-1} to ~ 300 cm^{-1}). This may be attributed to a larger atomic mass of W than Mo.

INTRODUCTION

While hydrogenated amorphous Si (a-Si:H) has been a principal material for thin-film transistors (TFTs), it suffers from degradation of its electrical properties due to the Staebler-Wronsky effect (SWE) [1]. A widely accepted remedy for this drawback is to replace a-Si:H by micro-crystalline Si (μ-Si). However, aggressive downsizing of the TFTs with μ-Si might be difficult since grains in μ-Si may induce the spatial fluctuation of the channel properties on the nano-meter scale. In order to achieve both the SWE-free nature and the atomic scale granularity, fabrication of yet another class of Si-related materials is necessary.

Recently, novel amorphous semiconductors (a-MSi_n, M=Mo or Nb, $7 \leq n \leq 16$) with relatively high carrier mobilities have been synthesized [2]. The materials were grown by the deposition of *M*-encapsulating Si clusters. The Hall measurement showed the p- and n-type semiconducting characters of a-$MoSi_n$ and a-$NbSi_n$, respectively. The Raman shift spectra of a-$MoSi_{12}$ and a-$NbSi_{13}$ suggested that a-Si-like network occurs without H. Thus the materials were assumed to be stabilized as the assembly of the deposited clusters, where the M atoms terminated the Si dangling bonds (DBs) and hence suppress the electronic disorder.

The above insight was supported by *ab initio* structure optimization of a-$MoSi_{12}$

[3]. The radial distribution function (RDF) of Si around Mo showed that each Mo atom is encapsulated with ~9 Si atoms on average ($N_c \sim 9$) and the rest of ~ 3 Si atoms per Mo are located among the clusters. The Mo d-orbitals were found to terminate the p-like DBs of the surrounding Si atoms to reduce the electronic density of states near the Fermi energy. The Si-Si RDF shape was similar to that of a-Si:H [4]. However, a clear gap was lost between the first and second peaks.

A purpose of this study is to characterize the Si network structure of a-MSi_n in more detail, base on a first-principles calculation. For that purpose, we present the vibrational density of states (VDOS) of a-$MoSi_{10}$ and a-WSi_{10}. The VDOS curve of a-$MoSi_{10}$ has a plateau around 250 cm^{-1}, where a valley occurs for a-Si:H. In the range between 400 $cm^{-1}$and 500 cm^{-1}, the a-$MoSi_{10}$ VDOS decreases monotonically while the counterpart of a-Si:H has a peak. We find that the VDOS shape of a-$MoSi_{10}$ is similar to the that of high pressure phase of a-Si (HPA-Si) found by Durandurdu and Drabold [5, 6]. In addition, the RDF among the Si atoms (Si-Si RDF) of a-$MoSi_{10}$ does not possess a clear gap between the first and second shells, which is also the case of HPA-Si. We further find a strong similarity in the VDOS and Si-Si RDF curve shapes between a-$MoSi_{10}$ and a-WSi_{10}. Judging from these findings, the structure of the Si network in a-$MoSi_{10}$ and a-WSi_{10} appears to resemble that of HPA-Si, despite that the a-$MoSi_{10}$ and a-WSi_{10} systems are obtained through relaxation of not only the atomic positions but also the pressure. A difference between a-$MoSi_{10}$ and a-WSi_{10} is that the W-atom displacement components extracted from the vibration eigenvectors are mainly distributed over a lower frequency range ($< \sim 150$ cm^{-1}) than the Mo counterpart (~ 150 cm^{-1} to ~ 300 cm^{-1}). This may be attributed to a larger atomic mass of W than Mo.

METHOD OF CALCULATION

We modeled the atomic structures of the a-$MoSi_{10}$, a-WSi_{10} and a-Si:H systems by adopting a first-principles method that calculates the total energies of the system based on the density-functional theory [7,8] for the electronic structures within the generalized gradient approximation [9]. The ultrasoft pseudopotentials [10] have been used for description of interactions between the valence electrons with the Mo, W, Si and H atomic cores. For a-$MoSi_{10}$ (a-WSi_{10}), we put twenty Mo (twenty W) and two hundred Si atoms in a supercell. For a-Si:H, 120 Si and 24 H atoms were placed in a supercell. In each case, a supercell has a rhombohedral shape subject to periodic boundary conditions in all three directions. The wavefunctions were expanded in a plane-wave basis set up to the energy cutoff of 25 Ry. The self-consistent field (SCF) iteration was performed until the difference in the total energies in the successive iterations becomes less than 1.4×10^{-7} eV per supercell. The structure was optimized until the absolute value of all components of atomic forces becomes smaller than 0.0026 eV / Å. Further, the size and shape of the supercell were optimized by calculating the stresses until the pressure on the supercell becomes smaller than 0.01 GPa. Only the Γ point was sampled in the first Brillouin zone for the structure optimization. All the calculations were done using the Quantum Espresso program suite [11].

In order to study the vibration of a-$MoSi_{10}$, a-WSi_{10} and a-Si:H within the harmonic oscillation approximation, we diagonalized the dynamical matrix that was generated from the atomic forces of the configurations, in each of which an atom is displaced in either of the six directions (x, y, z, -x, -y, or -z) from its optimized position by 0.026 Å.

RESULTS

Atomic configurations of a-$MoSi_{10}$ and a-WSi_{10}

Figure 1 shows the atomic structure of a-$MoSi_{10}$ at a local energy minimum [panel (a)] and its RDF [panel (b)], which we have obtained in our recent study [12].

Here we briefly review the procedure of structure optimization for a-$MoSi_{10}$ [12]. First, we prepared the initial configuration by putting twenty $MoSi_{10}$ clusters with random positions and orientations in a cubic supercell with the edge length of 30 Bohrs ($\simeq$ 15.88 Å). Each cluster has a cage of ten Si atoms arranged in a regular pentagonal prism (RPP) shape with a Mo atom inside of the Si cage. Random positioning of the twenty clusters was done so that they do not overlap with each other. Then we performed damped molecular dynamics followed by the BFGS optimization of the atomic positions, the size and shape of the supercell until the convergence criteria described above were all met.

As a result of optimization, the supercell shrank close to a rectangular parallelepiped with the edges being 15.42 Å, 15.49 Å and 15.77 Å long, respectively. We also summarize the structural features of a-$MoSi_{10}$ [12]. The RPP shape of Si atoms around each Mo atom was broken but the Mo atoms were kept in the Si cages. In the RDF of Si around Mo (Mo-Si RDF), a curve illustrated with a dotted line in Fig. 1(b), a clear Si shell is identified with the first peak at 2.56 Å separated from the rest part of the Mo-Si RDF for $\geq$ 3.35 Å. Integration of the Mo-Si RDF from the origin up to 3.35 Å gives rise to the average number 9.15 of the Si atoms belonging to the first shell. This establishes a picture that each Mo atom is encapsulated in a Si cage with roughly nine Si atoms with the remaining Si being located among the Si cages. As for the RDF of Si around itself (Si-Si RDF), a curve illustrated with a solid line in Fig. 1(b), no clear gap occurs between the first shell peaked at 2.41 Å and the second shell peaked at ~ 3.7 Å. The

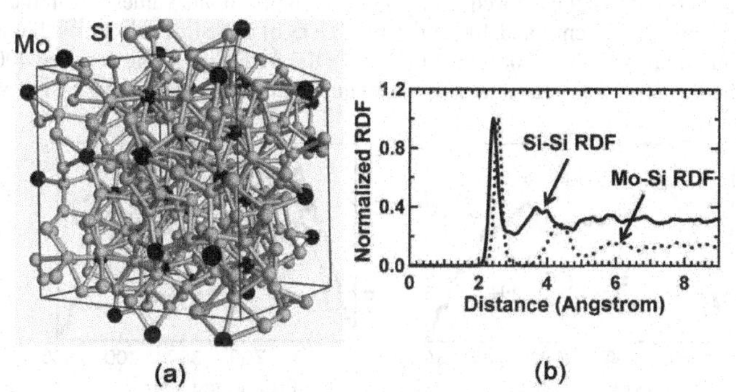

Figure 1. Atomic structure (a) and radial distribution function (RDF) (b) of a-$MoSi_{10}$ at a local energy minimum. In panel (a), black and white balls represent Mo and Si atoms, respectively. The box in panel (a) represents the supercell edges. The bars between atoms are drawn if the Si-Si (Si-Mo) distance is smaller than 2.6 (3.2) Å. In panel (b), "Mo-Si RDF" (solid curve) and "Si-Si RDF" (dotted curve) mean the RDFs of a Si atom around Mo and Si atoms as functions of the Mo-Si and Si-Si distances, respectively. The RDFs are normalized so that their maximum values are unity.

absence of the gap in Si-Si RDF suggests that, although the Si network as a whole may be regarded amorphous, the local connectivity among the Si atoms within the second nearest neighbors should be different from that for a-Si:H. The features of the Mo-Si and Si-Si RDFs are qualitatively the same as those already pointed out for a-$MoSi_{12}$ [3].

In order to investigate the effect of changing the transition metals on the Si network structure, we performed the geometry optimization for a-WSi_{10}. We used the optimized structure of a-$MoSi_{10}$ as the initial configuration by simply replacing the Mo by W atoms. The resultant structure and RDF of a-WSi_{10} is mostly unchanged from that of a-$MoSi_{10}$ mainly because the chemical environments around Mo and W are very similar due to the fact that the both elements have the same valency, six. As a matter of fact, integrating the W-Si RDF from the origin up to 3.33 Å, we obtain the average number of the Si atoms in the first shell to be 9.14. The difference in the covalent atomic radii for Mo (1.54 Å [13]) and W (1.62 Å [13]) is "absorbed" on the encapsulation in the Si cages. We are currently investigating the effect of changing the valency of the transition metal atoms on the Si network structures, which will be published elsewhere.

Vibrational properties of a-$MoSi_{10}$ and a-WSi_{10}

Figure 2 shows the VDOSs calculated for a-$MoSi_{10}$ [Fig. 2(a)] and a-WSi_{10} [Fig. 2(b)]. Our present purpose is a comparison of the VDOS *curve shape*. First of all, we notice a strong similarity in the VDOS curve shape between a-$MoSi_{10}$ and a-WSi_{10} over a whole range of frequency plotted. Combining this result with the RDF in Fig. 1, the effect of changing the transition metal element on the structure of the Si network appears to be small, if the valency of the metals is unchanged. Next, we characterize the structure of the Si network of a-MSi_{10} (M=Mo and W), by comparing their VDOSs with that of a-Si:H. There are outstanding differences in the VDOS curve shapes between a-MSi_{10} and a-Si:H depending on the frequency ranges. We note that the frequencies that bracket each frequency range are approximate values. (1) In the frequency range between 150 cm^{-1} and 400 cm^{-1}, the VDOS of a-MSi_{10} exhibits much a weaker dependence on the frequency than that for a-Si:H. (2) In the frequency range between 400 cm^{-1} and 500 cm^{-1}, the a-MSi_{10} VDOS decreases as the frequency increases while the a-Si:H VDOS

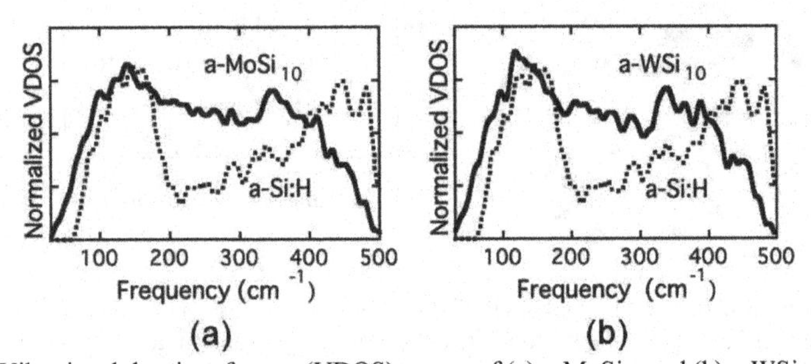

Figure 2. Vibrational density of states (VDOS) curves of (a) a-$MoSi_{10}$ and (b) a-WSi_{10} (solid curves), compared with that of a-Si:H (dotted curves). Since the number atoms in a supercell for modeling a-Si:H is different from that for a-$MoSi_{10}$ and a-WSi_{10}, the comparison is made only for the shape of the VDOS curves between a-$MoSi_{10}$ (a-WSi_{10}) and a-Si:H. For that purpose, all the VDOS curves are normalized so that they are integrated to unity. The frequency range on the abscissa is from 30 cm^{-1} to 500 cm^{-1}.

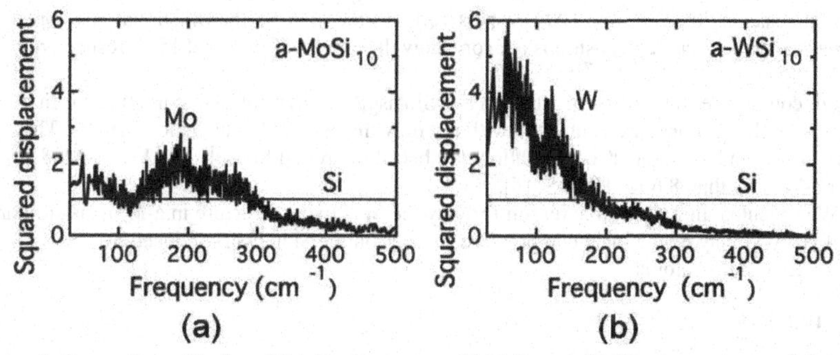

Figure 3. Squared amplitudes of the displacement of (a) Mo and (b) W atoms, extracted from the vibration eigenvectors of a-$MoSi_{10}$ and a-WSi_{10}, respectively, as functions of vibration frequency. The ordinate values are relative to the counterpart of Si. In other words, the unity in the ordinate corresponds to the squared amplitude of the Si displacement. The frequency range on the abscissa is from 30 cm^{-1} to 500 cm^{-1}.

has a peak around 450 cm^{-1}. (3) The a-MSi_{10} VDOS has a broader tail than the a-Si:H counterpart in the frequency range between 30 cm^{-1} and 150cm^{-1}.

In the present study, we do not include the eigenmodes found at frequencies lower than 30 cm^{-1} in our scope, since we could not judge whether the obtained eigenvectors are physically meaningful. It appears that the potential energy surface of the Si network in these materials might be much flatter than that of a-Si:H. Thus the numerical evaluation of the low frequency modes should require a special care, which will be an interesting future subject.

Figure 3 shows the squared amplitudes of the displacement of the M atom, extracted from the vibration eigenvectors of a-MSi_{10}, where M=Mo [Fig. 3(a)] and W [Fig. 3(b)]. For a-$MoSi_{10}$, each Mo atom moves with its surrounding Si shell for frequencies $\simeq$ 100 cm^{-1}, suggesting a "dressed Mo" picture. In the frequency range between 200 cm^{-1} and 300 cm^{-1}, the Mo atoms oscillate in their Si shells relatively freely. Since the Mo oscillation degrees of freedom is consumed up to 300 cm^{-1}, the Mo atoms cannot follow the oscillation of the Si cages for frequencies higher than 300 cm^{-1}. On the other hand, the movement of the W atom in a-WSi_{10} is localized in a narrower frequency range less than ~ 200 cm^{-1}, since W is heavier than Mo. The "dressed W" picture does not apply.

DISCUSSION

The curve shapes of the VDOS and Si-Si RDF of a-MSi_{10} (M=Mo, W) are similar to those of HPA-Si [5, 6]. Durandurdu and Drabold find an irreversible sharp transition of a-Si into the high pressure phase at 16.25 GPa on compression [5]. Their VDOS of a-Si at 16.25 GPa is mostly constant in the frequency range between 150 cm^{-1} and 300 cm^{-1}, where a valley-like sharp drop and increase occurs at 0 GPa. This feature found for HPA-Si is analogous with the VDOSs of a-MSi_{10} shown in Fig. 2. As for Si-Si RDF, the result by Durandurdu and Drabold show the absence of a gap between the first and second shells, that is the case of a-MSi_{10}, too. The bond angle distribution function (BADF) among the Si atoms of a-$MoSi_{10}$ appears to have three broad peaks at 70°, 100° and 120°, respectively. As for HPA-Si, the BADF has also three broad peaks although the peak positions are somehow different, 60°, 90° and 150°, respectively [5].

Of course, the Si network structure in a-MSi_{10} is not equivalent to that in HPA-Si. In a-MSi_{10}, most of the Si atoms are connected with not only themselves but also the M atoms. This may give rise to much a smaller coordination number of Si around Si itself, 4.33 for a-$MoSi_{10}$ (5.24 for a-WSi_{10}), than 8.6 for HPA-Si [5].

We speculate that a possible reason for why the Si network structure in a-MSi_{10} is similar to that of HPA-Si may be the local compressive stresses on the Si network induced by incorporation of the M atoms.

CONCLUSIONS

This study presents a first-principles calculation for vibrational properties of a-MSi_{10} (M=Mo, W), the assembly of the MSi_{10} clusters in amorphous phase. The calculated VDOS curves of the two systems both resemble that of the HPA-Si because they exhibit the plateau-like features over the frequency range where the valley-like shape occurs in a-Si:H. In addition, a clear gap in the Si-Si RDF curve is missing between the first and second shells for both a-MSi_{10}, which is also the feature in the HPA-Si. Judging from the results, the Si network structure in the present materials appears like that of HPA-Si despite that the structure of a-MSi_{10} is obtained by relaxation of the pressure. A possible cause of the occurrence of the HPA-Si-like network in a-MSi_{10} may be compressive stresses on the Si atoms induced by incorporation of the M atoms.

ACKNOWLEDGMENTS

This research was partially supported by the "KAKEN-HI" Grant-in-Aid for 472, 19051017, 2007. The calculations have been performed using the T2K-Tsukuba supercomputer and the AIST Supercluster. Figure 1 (a) was produced using the VESTA software[14].

REFERENCES

1. D. L. Staebler and C. R. Wronsky, *Appl. Phys. Lett.* **31**, 292 (1977).
2. N. Uchida, et al., *Appl. Phys. Express* **1**, 121502 (2008).
3. T. Miyazaki, N. Uchida, and T. Kanayama, *Phys. Stat. Solidi* C **7** 636 (2010).
4. P. Biswas, R. Atta-Fynn, and D. A. Drabold, *Phys. Rev.* B **76**, 125210 (2007).
5. M. Durandurdu and D. A. Drabold, *Phys. Rev.* B **64**, 014101 (2001).
6. M. Durandurdu and D. A. Drabold, *Phys. Rev.* B **66**, 155205 (2002).
7. P. Hohenberg and W. Kohn, *Phys. Rev.* **136**, B864 (1964).
8. W. Kohn and L. J. Sham, *Phys. Rev.* **140**, A1133 (1965).
9. J. P. Perdew et al., *Phys. Rev.* B **46**, 6671 (1992).
10. K. Laasonenn et al., *Phys. Rev.* B **47**, 10142 (1993).
11. P. Giannozzi et al., *J. Phys.: Condens. Matter* **21,** 395502 (2009). URL http://www.quantum-espresso.org
12. N. Uchida et al., submitted to *Thin Solid Films.*
13. B. Cordero et al., *Dalton Trans.*, 2832 (2008).
14. K. Momma and F. Izumi, *J. Appl. Crystallograph.* **41**, 652 (2008).

Mater. Res. Soc. Symp. Proc. Vol. 1321 © 2011 Materials Research Society
DOI: 10.1557/opl.2011.1097

Microscopic Characterizations of Nanostructured Silicon Thin Films for Solar Cells

Antonín Fejfar[1], Petr Klapetek[2], Jakub Zlámal[3], Aliaksei Vetushka[1], Martin Ledinský[1] and Jan Kočka[1]

[1]Institute of Physics of the Academy of Sciences of the Czech Republic, v.v.i, Cukrovarnická 10, 162 00 Prague 6, Czech Republic
[2]Czech Metrology Institute, Okružní 31, 638 00 Brno, Czech Republic
[2]Brno University of Technology, Technická 2, 616 69 Brno, Czech Republic

ABSTRACT

Microscopic characterization of mixed phase silicon thin films by conductive atomic force microscopy (C-AFM) was used to study the structure composed of conical microcrystalline grains dispersed in amorphous matrix. C-AFM experiments were interpreted using simulations of electric field and current distributions. Density of absorbed optical power was calculated by numerically solving the Maxwell equations. The goal of this study is to combine both models in order to simulate local photoconductivity for understanding the charge photogeneration and collection in nanostructured solar cells.

INTRODUCTION

Individual grains in silicon thin films prepared close to the border between the amorphous and microcrystalline growth have sizes from 10 to ~1000 nm or more. The grains may be aggregated in typical cones connected to each other via boundaries or surrounded by amorphous tissue. The cones end on the surface by nearly spherical caps [1,2] with a cauliflower structure, as shown in Fig. 1.

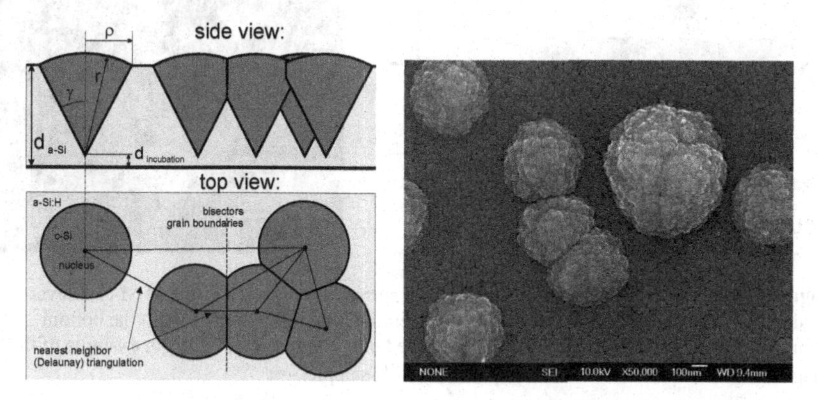

Figure 1 Left: Scheme of the typical structure of mixed phase Si layer. Right: Scanning electron microscopy image showing an example of mixed phase microcrystalline silicon layer. Spherical caps of the microcrystalline Si grains with a cauliflower structure can be seen together with surrounding columnar amorphous phase [3].

Macroscopic properties and thus also the operation of the devices based on nanostructural silicon are determined by the properties of structural components, their spatial arrangement and mutual interaction. Large differences in conductivities of the components lead to substantial redistribution of internal electrical fields [4]. Local variation of the internal fields are usually not taken into account for discussing the operation of thin film solar cells, in spite of the fact that both the thickness and spatial features are comparable to photon wavelengths, leading to pronounced near field effects [5].

Since more than 10 years we have used the tip of atomic force microscopy (AFM) cantilever in contact mode as a local contact (i.e., Conductive AFM or C-AFM [6]) to measure the structure and local electronic properties with spatial resolution down to several nanometers [7,8]. In the course of the study we discovered the differences between the conductive AFM measurements in UHV and ambient atmosphere [3] and clarified how the applied bias may change the surface properties of the samples, even leading to fundamental changes of conductivity maps [9]. This approach was later applied to other nanostructured materials, including polycrystalline silicon layers, organic bulk heterojunctions [10], carbon nanowalls and other nanostructures. The AFM measurements were complemented by maps obtained by Raman microspectroscopy. Raman spectroscopy is commonly used to obtain the crystallinity of the μc-Si:H films, which is the most important parameter for characterization of the μc-Si:H films structure [11], but it may provide much more information, including local stress in the films [12] and crystallographic orientation of individual grains [13].

In our present research we are trying to introduce additional illumination to the C-AFM in order to study local photoconductivity and related phenomena. At the same time, we develop numerical simulation of the C-AFM experiments with the goal of extracting local material parameters from the measured conductance maps.

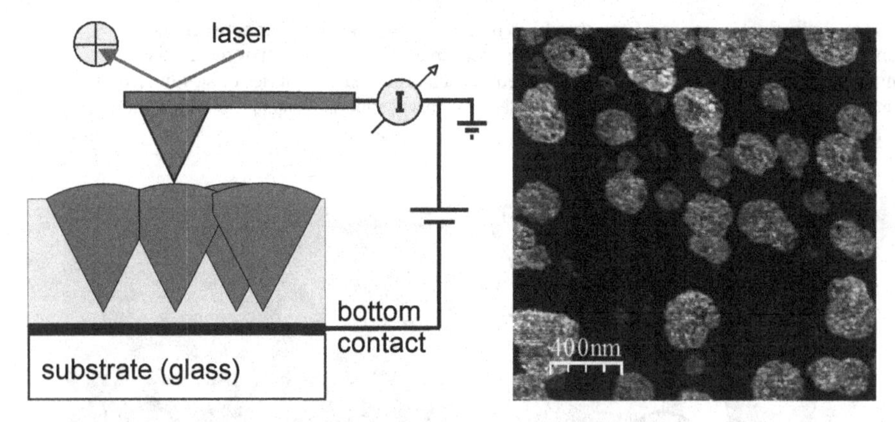

Figure 2. Left: Scheme of the C-AFM where the surface is contacted by an AFM cantilever made from doped silicon with PtIr coating. Current flowing through the film to the bottom contact is measured while the cantilever is scanned over the surface. Right: An example of the map of local conductance on a mixed phase μc-Si:H sample.

THEORY AND EXPERIMENT

An example of the local conductance map obtained by C-AFM is shown in Fig. 2. The bright circular areas in the conductance map correspond to the more conductive microcrystalline grains. It is apparent that the current on each grain is homogeneous (except for noise due to dynamic contact of the tip to the scanned surface). It can also be observed that the average value of the current increases with the diameter of the grain [14].

In order to understand the results we have used numerical simulations of the electrical current distribution in the system with a geometry corresponding to our experiment (as shown in Fig. 3). We have used the finite element method (FEM) to solve the Laplace partial differential equation for the electrostatic potential. The situation is complicated by the conical shape of the more conductive grains. At the grain tip a singularity in the electric field is expected. We have used two different approaches, either standard FEM with linear approximation (in the Comsol Multiphysics [15]) or adaptive higher-order polynomial FEM (in an open-source Agros/Hermes package [16]). In both case we simplified the solution by using the cylindrical symmetry of the system which occurs for the C-AFM tip positioned on the axis of a conical grain. The geometry of the model and the corresponding FEM meshes obtained using either linear FEM or hp-FEM are shown in the Fig. 3. The linear mesh of Comsol is colored for contrast only, but the hp-FEM mesh is color coded to show the degree of polynomial approximation within the cell. It can also be seen that the hp-FEM adaptive mesh generator creates fine elements in areas where the potential changes rapidly and leaves large elements for areas at the outside border, where the influence of the grain presence on the field distribution is small. For the hp-FEM mesh a detail is shown from the area close to the tip where rapid changes of potential are better approximated by a higher polynomial degree of the elements.

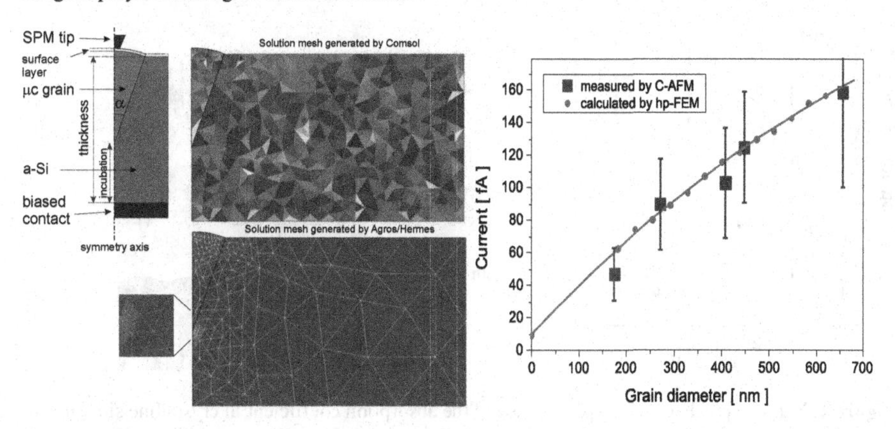

Figure 3. Left: The geometry of the modeled system in a cylindrically symmetrical case. Center: The corresponding FEM meshes (top – linear mesh in Comsol Multiphysics, bottom – adaptive mesh with higher order polynomial approximation in Agros/Hermes). Right: comparison of the average C-AFM current measured on grains (squares) with the calculated values (red points).

The graph in the right part of the image shows how the average current registered on grains depends on the grain size (represented by the grain diameter on the surface which can be easily measured by the AFM). Agreement between the experimentally observed local currents and simulations shows that the local conductivities can be understood as being the result of the grain shape and of the presence of a resistive layer on the surface [14].

The results described above were concerned with measurements of dark conductivity (or at least they were measured without intentionally introducing light). Yet the main interest lies namely in using the local contacts to study photoconductivity and related phenomena, as recently explored for organic semiconducting films [2,17].

When the FEM modeling can be used to interpret the C-AFM conductance maps without illumination, the next step would be to use it for simulation of the photoconductive AFM. Understanding the local photoconductivity requires knowledge of the optical field distribution in structures with geometry like the one shown in Fig. 1 where the dimensions are comparable to or smaller than the visible light wavelengths and the optical properties of components may be very different. This is a situation of interest not only for local photoconductivity, but also for other experiments, for example for an interpretation of the spectra of Raman scattering in mixed phase silicon samples [11,18]. The reason is illustrated in the figure 4 which shows the dependencies of absorption coefficient on the photon energy for typical crystalline, amorphous or microcrystalline silicon [19-21], together with photon energies of excitation lasers we use for measuring Raman spectra.

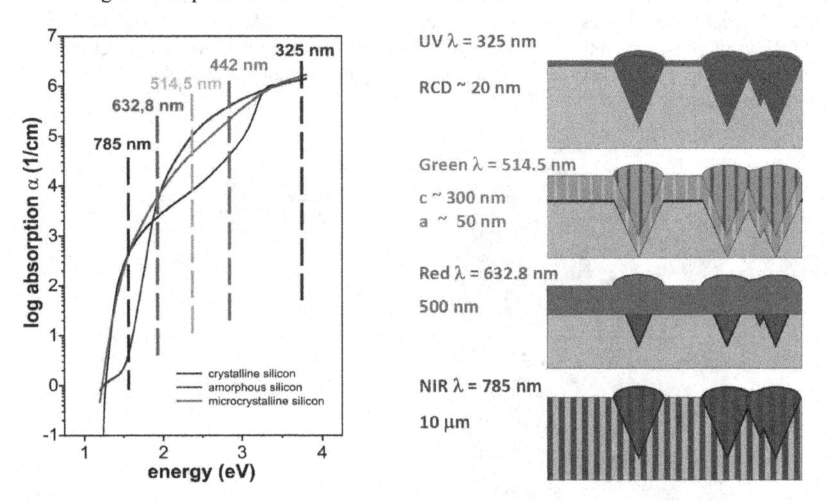

Figure 4. Left: Typical spectra dependencies of the absorption coefficient in crystalline silicon, amorphous silicon and microcrystalline silicon [21]. Energies of laser beams used by us for Raman spectra excitation are shown by vertical bars. Right: Corresponding penetration of the excitation wavelength into the films for four of the wavelengths.

The absorption depth (1/absorption coefficient) varies strongly between the materials. Raman spectra collection depth is equal to half of the absorption depth in homogeneous materials. However, for mixed phase materials where the absorption coefficients of the components may differ by more than an order of magnitude, the collection of Raman spectra becomes much more complicated. As shown in the right part of the Figure 4, the strongly absorbed UV light (top) penetrates only to the thin top layer. The green light (second row) passes through the crystalline grains easily but it is strongly absorbed in the amorphous phase. The red light (third row) is absorbed similarly in both components thanks to a coincidence of the absorption values in both amorphous and crystalline Si. Finally NIR light (bottom) is only weakly absorbed within the whole film.

In order to find the distribution of the absorbed power of the incident light, we have tried two alternative software packages for solving Maxwell equations for the geometry of the mixed phase Si thin films. The first approach used the RF module of Comsol Multiphysics 4.1 to find the field distribution for a sample illuminated by Gaussian beam. The solution is found as a combination of incident and scattered beams using appropriate boundary conditions, in this case perfectly matched layers (PML) which prevent the scattered waves reflecting back towards the sample. The solver aims at stationary solution corresponding to the time increasing towards infinity. This corresponds to a monochromatic wave, in other words, to the case when the Maxwell equations reduce to Helmholtz equation. An example of the calculated optical power loss density for weakly absorbed light ($\lambda = 785$ nm) is shown in the Figure 5. Note the concentration of the photogeneration density to a spot in the center of the grain. There is only a weak absorption in the surrounding amorphous phase and so the standing waves in the surrounding amorphous layer can be better seen in a logarithmic scale in the right.

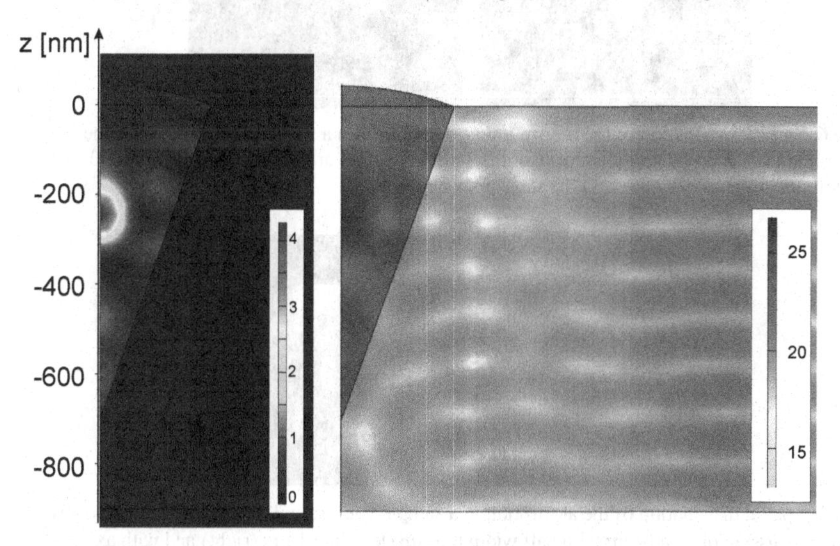

Figure 5. Electromagnetic power loss density (W/m^3) in the microcrystalline grain embedded in an amorphous layer (layer thickness ~900 nm) with incubation layer = 200 nm (distance of the grain tip from glass substrate) shown in linear color scale (left) and logarithmic scale (right).

For comparison, Finite Difference in Time Domain method (FDTD) was also used. In this case we have used a home- built custom simulator using graphics cards [22]. The simulated sample volume was divided into 500 x 500 x 180 pixels (with pixel corresponding to the smallest considered volume unit). The calculation was done either for planar wave or Gaussian beam introduced using the boundary conditions by total/scattered field (TSF) method. The calculated volume was surrounded by simple absorption boundary conditions. Density of absorption was calculated by averaging the local electric field intensity (after the calculation settled to equilibrium) and then using relationship $A = \sigma * E^2$ (using the constant volume of the "pixels") where A is the power loss density, σ is the local conductivity and E is the electric field of the optical wave.

The calculations were done for the following optical constants [21]: amorphous silicon n = 4.89, $\alpha = 256 * 10^3$ cm^{-1}, microcrystalline grain n = 4.37, $\alpha = 160 * 10^3$ cm^{-1}, which can approximate the properties for the blue light with $\lambda = 442$ nm, i.e., relatively strongly absorbed light for which we could consider the sample to be semi-infinite.

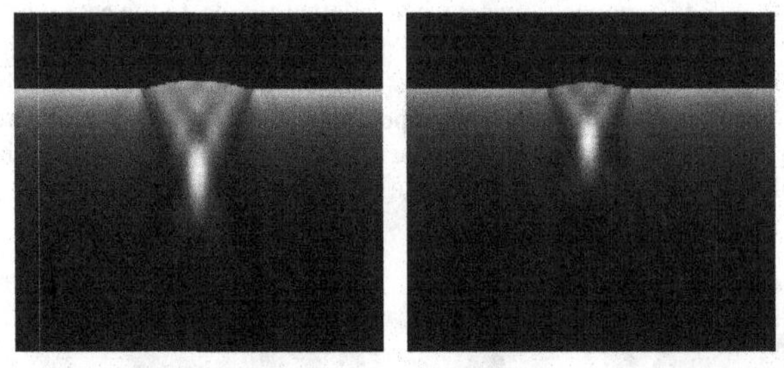

Figure 6. Spatial distribution of the absorption in a sample with a single conical grain for the case of a planar wave. Calculations for two different grain sizes are shown (grain height 800 nm in the left and 500 nm in right) with the same apex angle = 36°.

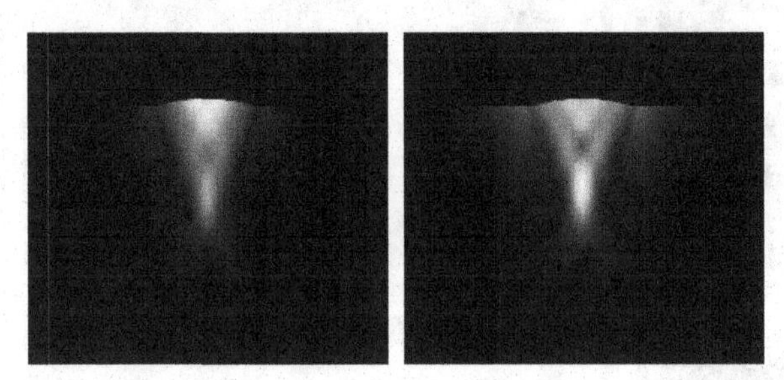

Figure 7. Spatial distribution of the absorption in a sample with a single conical grain for the two cases of a Gaussian optical beam with half width 0.5 μm (left) and 1 μm (right) and with axis coincident with the grain axis. The grain height was 800 nm and the apex angle = 36°.

While the planar wave case in Fig. 6 corresponds to the homogeneously illuminated sample, it is interesting to consider also the case when sample is illuminated locally with a focused laser beam like in the Raman microspectroscopy or in the case of scanning near-field optical microscopy (SNOM). These results are shown for two differently focused Gaussian beams on grain in Fig. 7. We could also explore the case when the illuminating Gaussian beam was moved off the grain axis, i.e. without the axial symmetry, see Fig. 8.

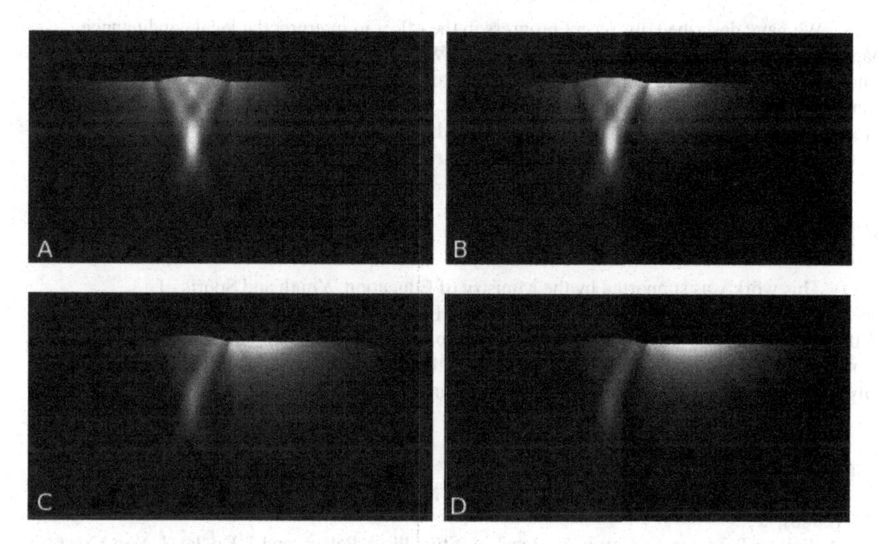

Figure 8. Spatial distribution of the absorption in a sample with a single conical grain for the case of a Gaussian optical beam with half width 1 μm for four different positions (A, B, C, D) of the Gaussian beam. Grain size was again 800 nm and the apex angle = 36°.

DISCUSSION

Large variations in both electrostatic field as well as in optical power loss density were explored by numerical simulations. These variations both have influence on the expected photogeneration and collection of charge carriers in solar cells and need to be taken into account for understanding the operation of the nanostructured solar cells based on hydrogenated silicon thin films grown at boundary between amorphous and microcrystalline structures.

The following step in our research will be the connected simulations of the local conductivity measurement using the C-AFM tip with the local co nductivity varying according to the photogeneration rate derived from the results of optical simulations. We aim at using these simulation for interpreting the photoconductive AFM, which is being recently attempted in our laboratory as well as in number of other laboratories worldwide [23,24].

However, we need to point out that this would still be a simple case of resistive calculations which do not consider non-linearities due to the barriers at interfaces, to the space charge within the films. No size effects are taken into account either. There is a number of directions into which the simulations need to expanded, but even the simpler limiting cases provide interesting insights into the operation of nanostructured solar cells.

CONCLUSIONS

We have described our recent progress in the effort to interpret the local conductance maps measured on mixed phase silicon thin films. We have presented both electrical and optical simulations of the fields in thin films with typical conical microstructure. The work on connecting both types of simulations is in progress. The simulations would then be suitable also for other solar cell designs using the nanostructured contacts for light trapping or for improving the transport of photogenerated carriers to the electrodes.

ACKNOWLEDGMENTS

This work was supported by the Ministry of Education, Youth and Sports of the Czech Republic through Institutional Research Plan AV0Z 10100521 and projects LC510, LC06040 and MEB 061012, by the Grant Agency of the Academy of Sciences of the Czech Republic through project KAN400100701 and by the European Commission through the PolySiMode project, FP7-Energy-2009-1 Grant Agreement No. 240826.

REFERENCES

1. C. W. Teplin, C.-S. Jiang, P. Stradins, and H. M. Branz, Appl. Phys. Lett. **92**, 093114-3 (2008).
2. A. Fejfar, T. Mates, O. Čertík, B. Rezek, J. Stuchlík, I. Pelant, and J. Kočka, J. Non-Cryst. Sol. **338-340**, 303-309 (2004).
3. T. Mates, P. C. P. Bronsveld, A. Fejfar, B. Rezek, J. Kočka, J. K. Rath, and R. E. I. Schropp, J. Non-Cryst. Sol. **352**, 1011-1015 (2006).
4. A. Fejfar, T. Mates, S. Honda, B. Rezek, J. Stuchlík, J. Kočka, J. Matějková, P. Hrnčiřík, and J. Valenta, in *Proceedings of the 19th Europ. Photovoltaic Solar Energy Conference, 7-11 June 2004, Paris*, edited by W. Hoffmann, J. L. Bal, H. Ossenbrink, W. Palz, and P. Helm (WIP-Munich and ETA-Florence, Munich 2004, 2004), pp. 1564 - 1566.
5. C. Rockstuhl, F. Lederer, K. Bittkau, and R. Carius, Appl. Phys. Lett. **91**, 171104 (2007).
6. C. Teichert and I. Beinik, in *Scanning Probe Microscopy in Nanoscience and Nanotechnology 2* (Springer, 2011), pp. 691-721.
7. B. Rezek, J. Stuchlík, A. Fejfar, and J. Kočka, Appl. Phys. Lett. **74**, 1475-1477 (1999).
8. B. Rezek, J. Stuchlik, A. Fejfar, and J. Kočka, J. Appl. Phys. **92**, 587 - 593 (2002).
9. A. Vetushka, A. Fejfar, M. Ledinsky, B. Rezek, J. Stuchlik, and J. Kočka, Phys. Rev. B **81**, 237301 (2010).
10. J. Čermák, B. Rezek, V. Cimrová, D. Výprachtický, M. Ledinský, T. Mates, A. Fejfar, and J. Kočka, Phys. Stat. Sol. (RRL) **1**, 193-195 (2007).
11. M. Ledinský, A. Vetushka, J. Stuchlík, T. Mates, A. Fejfar, J. Kočka, and J. Štěpánek, J. Non-Cryst. Sol. **354**, 2253–2257 (2008).

12. A. Vetushka, M. Ledinský, J. Stuchlík, T. Mates, A. Fejfar, and J. Kočka, J. Non-Cryst. Sol. **354**, 2235 - 2237 (2008).
13. M. Ledinský, A. Vetushka, J. Stuchlík, A. Fejfar, and J. Kočka, Phys. Status Solidi (c) **7**, 704-707 (2010).
14. A. Fejfar, A. Vetushka, V. Kalusová, O. Certík, M. Ledinský, B. Rezek, J. Stuchlík, and J. Kocka, Phys. Status Solidi (a) **207**, 582-586 (2010).
15. *Multiphysics Modeling and Simulation Software - COMSOL* (2005), see also http://www.comsol.com/.
16. P. Karban, Mach, F., and O. Čertík, *Agros 2D, Hpfem.org: Free Adaptive hp-FEM* (2009), see also http://hpfem.org/agros2d/.
17. B. H. Hamadani, S. Jung, P. M. Haney, L. J. Richter, and N. B. Zhitenev, Nano Letters **10**, 1611-1617 (2010).
18. M. Ledinský, L. Fekete, J. Stuchlík, T. Mates, A. Fejfar, and J. Kočka, J. Non-Cryst. Sol. **352**, 1209-1212 (2006).
19. A. V. Shah, H. Schade, M. Vanecek, J. Meier, E. Vallat-Sauvain, N. Wyrsch, U. Kroll, C. Droz, and J. Bailat, Progress in Photovoltaics: Research and Applications **12**, 113-142 (2004).
20. M. Vanecek, A. Poruba, Z. Remes, J. Rosa, S. Kamba, V. Vorllcombining acute accent.cek, J. Meier, and A. Shah, Journal of Non-Crystalline Solids **266-269**, 519-523 (2000).
21. M. Vanecek and A. Poruba, in *Properties and Applications of Amorphous Materials* (Kluwer Academic Publishers, 2001), pp. 401-433.
22. P. Klapetek, M. Valtr, A. Poruba, D. Necas, and M. Ohlídal, Applied Surface Science **256**, 5640-5643 (2010).
23. T. Gotoh, Y. Yamamoto, Z. Shen, S. Ogawa, N. Yoshida, T. Itoh, and S. Nonomura, Jpn. J. Appl. Phys. **48**, 091202 (2009).
24. M. Kawai, T. Kawakami, T. Inaba, F. Ohashi, H. Natsuhara, T. Itoh, and S. Nonomura, Current Applied Physics **10**, S392-S394 (2010).

Mater. Res. Soc. Symp. Proc. Vol. 1321 © 2011 Materials Research Society
DOI: 10.1557/opl.2011.940

Band alignment at amorphous/crystalline silicon hetero-interfaces

L. Korte, T. F. Schulze, C. Leendertz, M. Schmidt and B. Rech
Helmholtz-Zentrum Berlin für Materialien und Energie, Institut Silizium Photovoltaik,
Kekuléstr. 5, D-12489 Berlin, Germany

ABSTRACT

We present an investigation of the band offsets in amorphous/crystalline silicon heterojunctions (*a*-Si:H/*c*-Si) using low energy photoelectron spectroscopy, ellipsometry and surface photovoltage data. For a variation of deposition conditions that lead to changes in hydrogen content and the thereby the a-Si:H band gap by ~180 meV, we find that mainly the conduction band offset ΔE_V varies, while ΔE_C stays constant within experimental error. This result can be understood in the framework of charge neutrality (CNL) band lineup theory.

INTRODUCTION

Due to their high power conversion efficiency potential [1], amorphous/crystalline silicon (*a*-Si:H/*c*-Si) solar cells are currently in the focus of many research activities. The essential feature of these cells is the use of *a*-Si:H/*c*-Si heterojunctions for charge carrier extraction and wafer surface passivation. In two recent papers, we have investigated the band line-up at the *a*-Si:H/*c*-Si heterointerface. The main findings were that the valence band offset ΔE_V is independent on the doping of both *c*-Si substrate and *a*-Si:H thin film [2], and that the widening of the *a*-Si:H band gap with increasing hydrogen content in the film leads primarily to an increase in ΔE_V, while the conduction band offset ΔE_C is only slightly changed (compatible with zero change within the error margin) [3]. In the present paper, we elaborate on those results, making use of additional information on the work function, ionization energy and electron affinity in the *a*-Si:H layers. Furthermore, we use the charge neutrality level (CNL) concept [4] to discuss the physical origins of the measured band offsets. In order to obtain the CNL in *a*-Si:H, we suggest to use calculations based on the defect pool model [5].

EXPERIMENT

A set of ~10 nm thin (i)*a*-Si:H layers were deposited by RF-PECVD (f_{RF} = 13.56 MHz) on 3 Ωcm (n)*c*-Si wafers (FZ, {111}, mirror-polished). The hydrogen content in the film was varied using different deposition temperatures (T_{depo} = 130-210°C), pressures (p = 0.5, 1 and 4 mbar: "low pressure" (LP), "medium pressure" (MP) and "high pressure" (HP)) and hydrogen dilution ratios ($[H_2]/[SiH_4]$ = 0 or 10). As shown in [3], this results in a variation of the band gap E_g and structural disorder in the film. E_g was obtained from fitting spectral ellipsometry (SE) data using a Tauc-Lorentz approach and the program *rig-vm* [6]. Using surface photovoltage (SPV), we measure the band bending $e\varphi_s$ in the *c*-Si wafer and calculate the band offsets $\Delta E_V = E_{F,aSi} - (E_{F,c\text{-}Si,bulk} - e\varphi_s)$, $\Delta E_C = E_{g,aSi} - E_{g,cSi} - \Delta E_V$ (cf. Fig. 1 and [3]). Photoelectron spectroscopy in the constant final state yield (CFSYS) mode [7] was carried out in UHV at photon energies of 3 – 7 eV. The measured quantity is the internal photoelectron yield Y_{int}, i.e. the ratio between the spectrally resolved photon flux absorbed in the sample and the photoelectron flux detected in the energy analyzer. Y_{int} is related to the occupied density of states $N_{occ}(E)$ via $Y_{int}/R^2(h\nu) \propto N_{occ}(E)$ /

$h\nu R^2(h\nu)$, where $R^2 \propto h\nu^{-5}$ is the dipole matrix element for optical excitation in *a*-Si:H. We use a model-based fitting approach [8] to extract the *a*-Si:H density of states parameters from Y_{int}, cf. Fig. 2: the valence band edge E_V, valence band tail slope E_{0v} and the parameters of the dangling bond distribution: its density N_{db}, energy position E_{db} and width σ_{db}. See [8, 2] for details.

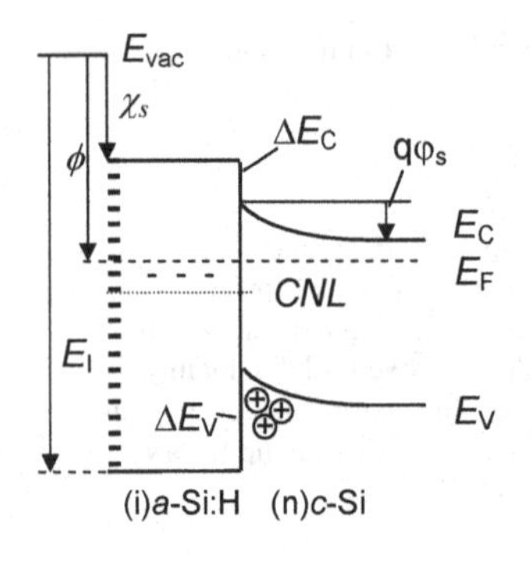

Figure 1. Electronic parameters of an *a*-Si:H/*c*-Si heterojunction: Electron affinity χ_s, ionization energy E_I, work function ϕ, band offsets ΔE_V, ΔE_C, band bending $e\varphi_s$, charge neutrality level CNL.

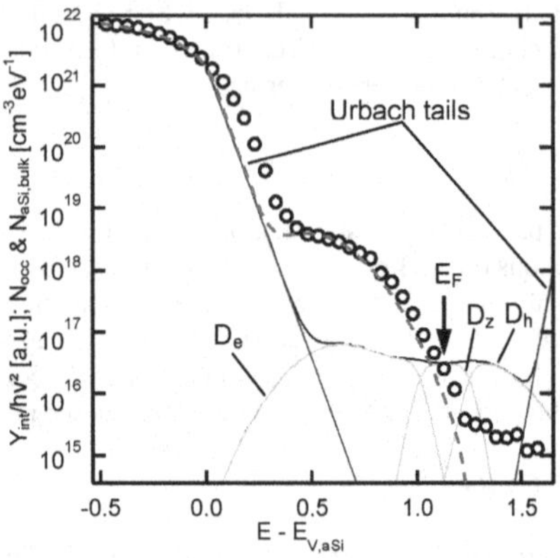

Figure 2. Measured CFSYS data Y_{int}, rescaled to an approximate occupied DOS distribution in the *a*-Si:H band gap (circles), model DOS of occupied states N_{occ} (dashed), and bulk DOS calculated from the defect pool model (full lines), notation as in [5].

RESULTS

As we have shown previously [3], for the (i)*a*-Si:H layers investigated here, the hydrogen content as measured by FTIR can serve as ordering parameter for a variety of structural and electronic parameters, among them the *a*-Si:H band gap. In Fig. 3 (left), the valence and conduction band offsets ΔE_V and ΔE_C as reported in [3] are plotted against the band gap. Using the same scaling on the ordinate, the middle panel of Fig. 3 shows the ionization energy E_I and the electron affinity χ_s of the films obtained from UPS spectra at 6.5 eV excitation energy. As depicted in Fig. 1, these quantities denote the positions of the valence and conduction band edges with respect to the vacuum level. They are a true surface property and remain unaffected by changes of the surface Fermi level brought about by charge equilibration processes between the semiconductor bulk vs. its surface [9]. Thus, a change of E_I and/or χ_s is related to either a change in the band structure of the material under investigation, or to surface charges.

Comparing the left and middle panels of Fig. 3, it is obvious that the trend that is found in the band offsets, namely an *a*-Si:H valence band edge that is receding from the *c*-Si valence band edge with increasing gap (increasing H concentration) and a nearly constant ΔE_C, is mirrored by the corresponding values of the *a*-Si:H free surface referenced to the vacuum level, i.e. E_I and χ_s. This indicates, that the main change occurring in the *a*-Si:H upon changing H content is indeed a change in the band structure. However, Fig. 3 (right) shows that there are small changes in the work function ϕ – note that the maximum difference is only ~ 100 meV – that are mainly correlated with the deposition pressure: While ϕ is almost constant for the HP and LP deposition series, being ~ 100 meV higher for LP than for HP, the MP data lie in between and additionally show a clear trend of increasing ϕ with decreasing deposition temperature (increasing H content and band gap). This indicates changes in the charge distribution of the different *a*-Si:H layers. In the following, we will analyze these trends using Powell and Deane's defect pool model [5].

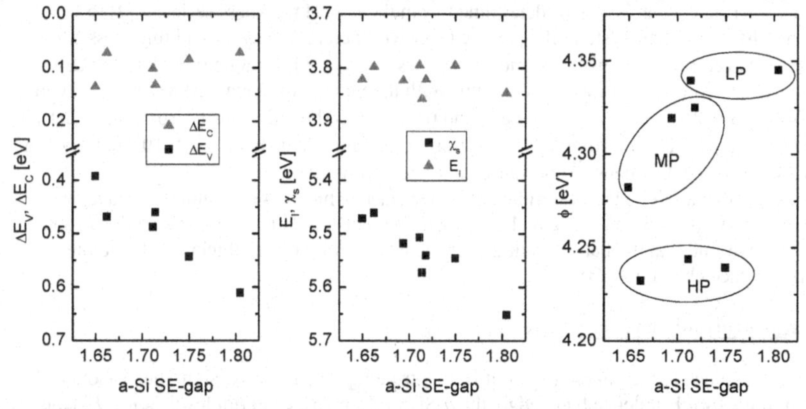

Figure 3. Electronic levels extracted from Near-UV PES on (i)*a*-Si:H/*c*-Si samples. Left: Valence- and conduction band offsets ΔE_V and ΔE_C. Middle: Electron affinity χ_s and ionization energy E_I. Right: Work function ϕ, labels: deposition pressure, see sec. "Experiment"; note the different scaling of the ordinate. Abscissae: *a*-Si:H E_g from a Tauc-Lorentz fit to SE data [3].

Briefly, the defect pool (dp) model assumes a "pool" of dangling bond defects that can potentially be realized in a given *a*-Si:H network. The energetic width of this pool – i.e. the range of energies in the band gap where it is favorable to break a Si-Si bond and create two dangling bonds (dbs) – is related to the strain in the Si-Si bonds of the disordered network. The strain is reflected in the slope of the exponential band tails, i.e. the Urbach energies E_{0V}, E_{0C}. In addition, depending on the energetic position of the amphoteric dangling bond and the background doping level, energy can be gained by occupying the db with 0, 1 or 2 electrons. Thus, the shape of the defect distribution that is finally realized in an *a*-Si:H network in thermal equilibrium depends on both the slopes E_{0v}, E_{0c} and the position of E_F.

Surface and bulk dangling bond density in the investigated (i)*a*-Si:H layers

Fig. 2 shows an the theoretical bulk defect distribution obtained with this model, using the Urbach energy E_{0v} and E_F from PES and the measured band gap E_g as input. In the same plot, also both the measured CFSYS and the model density of states fitted to this data are shown. As already pointed out by Winer *et al.* [10], the density of db states in (i)*a*-Si:H as measured with near-UV PES is typically 1-2 orders of magnitude higher than the bulk DOS: PES is a surface sensitive method, therefore the sample surface towards the vacuum gives a disproportionately high contribution to the total signal. The high density of (i)*a*-Si:H surface states has been assigned to either surface contamination – unlikely in our case since the samples were transferred in HV/UHV conditions – or a surface region of un-equilibrated *a*-Si:H [10].

One might wonder if these surface defects can influence the overall band bending in the *a*-Si:H, which would induce an error in the procedure used in [3] to calculate the band offsets. To clarify this, we have calculated the db density in the surface layer, assuming: i) that the PES signal is limited by the "exit depth" of the photoelectrons, i.e. the mean free path λ_{imfp} between inelastic scattering events. λ_{imfp} was determined from the *a*-Si:H thickness series in [2] to 1.14(6) nm. ii) a two-layer system of a surface layer with defect density N_S and thickness d_s on top of a bulk layer with db density N_B and a thickness >> λ_{imfp}. iii) a single monolayer of defect-rich material as a lower limit, i.e. $d_s = 0.23$ nm. With the maximum (apparent) defect density in Fig. 2, $N_{db,S} = 3.9\times10^{18}$ cm^{-3}eV^{-1} at ~ 0.6eV, and the expected bulk db density from the defect pool model, $N_{db,B} = 7.0\times10^{16}$ cm^{-3}eV^{-1}, this yields a maximum db density of ~2×10^{19} cm^{-3}eV^{-1} in those 0.23 nm, or a defect density concentrated in the first monolayer of ~4.2×10^{11} cm^{-2}eV^{-1}. This value is comparable to the H-terminated *c*-Si surface, which has a minimum surface db density in the low 10^{11} cm^{-2}eV^{-1}, e.g. [11]. Using numerical simulations (not shown due to space constraints), we find that the band bending over the whole *a*-Si:H layer thickness due to the additional surface charge is ≤ 50 meV.

The charge neutrality level in *a*-Si:H

Coming back to the data in Fig. 3, it is an intriguing feature that despite the pronounced recession of the *a*-Si:H valence band edge, the *a*-Si:H/*c*-Si interface Fermi level $E_{F,IF} - E_{V,cSi}$ is nearly constant (except for the two samples with the largest band gap). It appears that although the density of states distribution in the *a*-Si:H band gap changes strongly, the charge equilibration between *a*-Si:H and *c*-Si always leads to the same charge in the *a*-Si:H and thereby also to the same *c*-Si band bending, i.e. the creation of a space charge region whose charge compensates that in the *a*-Si:H. From the band bending and doping level in the *c*-Si wafer, the total charge density in the *a*-Si:H layer is easily calculated to the very low value of Q/e = $-3.5\pm1\times10^{10}$ cm^{-2} for all samples except the two with $E_g > 1.8$ eV, which have -5.4 and -6.4×10^{10} cm^{-2}; e is the (absolute) elementary charge. Assuming that the *a*-Si:H is truly intrinsic, the net charge in the film must be entirely due to charges transferred from the *c*-Si and tunnelling from *a*-Si:H gap states into *c*-Si, i.e. due to the equilibration imposed by the band offsets.

It is interesting to check the measured band offsets against established models. In [3], we have already compared them to density functional theory results by various authors. Here, we investigate whether they can also be understood in the context of the well-established charge neutrality level (CNL) theory [4, 12]. The CNL theory assumes that the band lineup in a heterojunction is determined by the continuum of interface-induced gap states. These are an intrinsic property of the materials forming the junction: As one material ends at the junction, the wave functions of its electronic states tail exponentially into the other material, just like at the semiconductor-vacuum interface. The continua of states in the band gap induced by the interface have branch points where their character changes from valence band-(donor-) to conduction band- (acceptor-)like. These branch points are thus the CNLs of the materials, and the CNL theory asserts that they should line up at the heterointerface, yielding the band offsets

$$\Delta E_V = (CNL_A - E_{V,A}) - (CNL_B - E_{V,B}) - \Delta, \quad \Delta E_C = (E_{C,A} - CNL_A) - (E_{C,B} - CNL_B) - \Delta, \quad (1)$$

where $CNL_i - E_{V[C],i}$ are the distances of the CNLs to the valence [conduction] band edges and Δ accounts for dipole effects due to an non-vanishing difference of the electronegativities of the two materials, or due to additional dipole-inducing species at the interface. These effects can be expected to be small at the *a*-Si:H/*c*-Si junction [2], as the two materials are very similar.

The CNLs of various crystalline semiconductors have been calculated using DFT, e.g. [4, 13,14]. However, to the knowledge of the authors, it has not been tried to calculate the CNL in *a*-Si:H from DFT and use this information to obtain band offsets in *a*-Si:H/*c*-Si or other *a*-Si:H-based heterojunctions. Interestingly, the amorphous nature of *a*-Si:H suggests another approach: As outlined above, the gap states at the interface between two crystalline semiconductors result from the tailing of the electronic states into the other material. They can thus be thought of as localized states at the heterointerface. In *a*-Si:H, however, localized states fill the *a*-Si:H band gap not only at the surface, but throughout the material: the tail states and dangling bonds. Just like the virtual gap states, the tail states related to the conduction band have acceptor character, those of the valence band are donors. In addition, the amphoteric dangling bonds have to be accounted for. Now, in a generic *a*-Si:H film, the free charges (electrons and holes) compensate the net charge in defects, impurities and dopant atoms. The charge balance is reflected in the Fermi level. In a (truly) intrinsic *a*-Si:H film, without any extrinsic charge carriers, E_F will move to a position where the net charge in defects – band tails and dangling bonds – is zero, i.e. the defects are neutral. Therefore, we propose that the Fermi level of such an *a*-Si:H should lie at its "intrinsic" CNL. Note, that this implies that while the CNL of a crystal is defined only at its surface, the CNL of *a*-Si:H – or other disordered semiconductors – is a bulk property.

To test our hypothesis, we would need to calculate E_F in the "truly intrinsic" *a*-Si:H corresponding to our experimentally realized films (which might have some net doping due to residual doping and impurities). To this end, we have extended our semianalytical model for recombination in *a*-Si:H/*c*-Si heterostructures [15] to include the defect pool model. Thus, we can calculate the gap state distribution in the *a*-Si:H layer; the doping of the *a*-Si:H is then adjusted to yield the measured *a*-Si:H Fermi level for the *a*-Si:H/*c*-Si system. In a next step, the equilibrium Fermi level position in the *a*-Si:H *alone*, i.e. not in contact with the *c*-Si, is calculated. Finally, the doping in the *a*-Si:H film is set to zero. Fig. 4 (left) shows the result of this procedure, triangles: the E_F-position as obtained from the PES measurements and squares: calculated E_F in the *a*-Si:H film alone, i.e. not in contact (charge balance) with the (n)*c*-Si substrate and charge set to zero. All values are referenced to the *a*-Si:H valence band edge. Note, that as we have used measured *a*-Si:H bulk properties (notably, E_{0v}, E_{0c} and E_g) for the calculations, we implicitly assume that the *a*-Si:H at the heterointerface has the same electronic properties as the bulk [16].

Now, to be able to compare this data to the CNL theory, we subtract the measured band offset ΔE_V from these values and obtain the data in the right panel of Fig. 4: If the calculated "intrinsic" E_F in the *a*-Si:H is indeed the CNL, $(E_F - E_{V,aSi}) - \Delta E_V$ should give the CNL in *c*-Si (Eq. (1); Fig. 1). It is encouraging that for all films except one, the calculated values lie within a narrow band, with a mean value of 0.63(2) eV (calculated over all values except the one at 0.52 eV): The *c*-Si CNL should be a fixed value. For CNL_{cSi}, different values have been published, e.g. [12]. Tersoff's data based on calculations with the augmented plane-wave method is generally considered very reliable [4], he finds $(CNL-E_V)_{cSi} = 0.36$ eV. However, a considerable spread in the published values is found, e.g. Cardona *et al.* report 0.23 eV [13]. Interestingly, Flores *et al.* conclude from the calculation of intrinsic CNLs and the comparison to Schottky barriers that $(CNL-E_V)_{cSi} = 0.63$ eV [14]. This is identical to our averaged value.

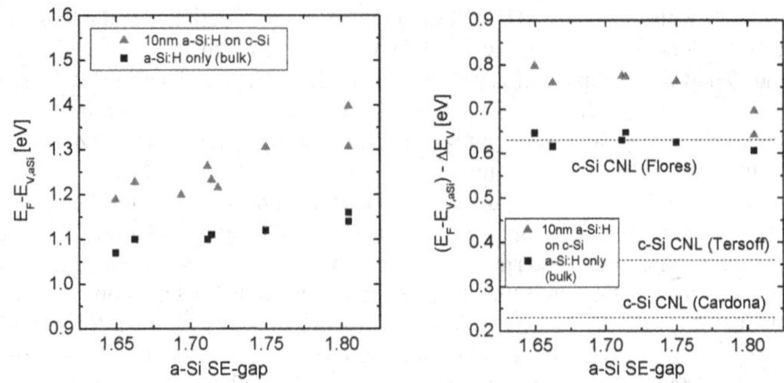

Figure 4. Left: Position of the valence band edge relative to E_F in *a*-Si:H from PES (triangles) and calculated E_F in the corresponding fully intrinsic *a*-Si:H layer *not* in contact with the *c*-Si (squares). Right: Subtracting ΔE_V (Fig. 3, left), an energy level is calculated that we propose to identify with the *a*-Si:H CNL (see text). Dashed lines: Calculated CNLs in *c*-Si [4,13,14].

CONCLUSIONS

Near-UV photoelectron spectroscopy and surface photovoltage were combined to extract parameters of the *a*-Si:H density of states, evaluate the impact of *a*-Si:H surface states and gain insight into the mechanisms that govern band lineup at the *a*-Si:H/*c*-Si heterointerface. A new approach to understand the *a*-Si:H/*c*-Si band offsets in the context of charge neutrality level considerations combined with the *a*-Si:H defect pool model has been proposed. While a more detailed discussion is beyond the scope of this paper, we find encouraging first results: the correct trends of ΔE_V, ΔE_C vs. *a*-Si:H band gap are reproduced, and the agreement to theoretical considerations (that show a large scatter) is satisfying.

REFERENCES

1. H. Sakata *et al.*, *Proc. 25th Eur. Photovolt. Sol. Ener. Conf.* (WIP, 2010) pp. 1102-5.
2. L. Korte and M. Schmidt. *J. Appl. Phys.* **109**, 063714(2011).
3. T. F. Schulze, F. Ruske, B. Rech, and L. Korte, *Phys. Rev. B*, in press (2011).
4. J. Tersoff, *Phys. Rev. B* **30**, 4874 (1984).
5. M. J. Powell and S. C. Deane, *Phys. Rev. B* **48**, 10815 (1993).
6. A. Pflug, RIG-VM simulation system, Fraunhofer IST, http://www.simkopp.de/rvm/.
7. M. Sebastiani et al. *Phys. Rev. Lett.* **75**, 3352 (1995).
8. L. Korte and M. Schmidt, *J. Non-Cryst. Sol.* **354**, 2138 (2008).
9. D. Cahen and A. Kahn, *Adv. Mat.* **15** 271 (2003).
10. K. Winer and L. Ley. *Phys. Rev. B* **36**, 6072 (1987).
11. H. Angermann *et al.*, *phys. stat. sol. (c)* **8**, 879 (2011).
12. W. Mönch, *Electronic Properties of Semiconductor Interfaces*. (Springer, Berlin, 2004).
13. M. Cardona and N.E. Christensen, *Phys. Rev. B* **35**, 6182 (1987).
14. F. Flores, A. Munoz, and J.C. Duran, *Appl. Surf. Sci.* **41-42**, 144 (1989).
15. C. Leendertz *et al.*, *Proc. 25th Eur. Photovolt. Sol. Ener. Conf.* (WIP, 2010) pp.1377-81.
16. T. F. Schulze *et al.*, *Appl. Phys. Lett.* **96**, 252102 (2010).

Mater. Res. Soc. Symp. Proc. Vol. 1321 © 2011 Materials Research Society
DOI: 10.1557/opl.2011.1229

Electron emission from deep traps in hydrogenated amorphous silicon and silicon-germanium: Meyer-Neldel behavior and ionization entropy

Qi Long[1], Steluta Dinca[1], Eric A. Schiff[1], Baojie Yan[2], Jeff Yang,[2] and Subhendu Guha[2]
[1]Department of Physics, Syracuse University, Syracuse, New York 13244-1130, U.S.A.
[2]United Solar Ovonic LLC, Troy, Michigan 48084, U.S.A.

ABSTRACT

We have measured electron drift in amorphous silicon-germanium *nip* photodiodes using the photocarrier time-of-flight technique. The samples show electron deep-trapping shortly after photogeneration, which is generally attributed to capture by a neutral dangling bond (D^0) to form a negatively charged center (D^-). An unusual feature is that electron re-emission from the trap is also clearly seen in the transients. Temperature-dependent measurements on the emission yield an activation energy of about 0.8 eV and the remarkably large value of 10^{15} Hz for the emission prefactor frequency. We also compiled results on electron emission from deep traps in a-Si:H, a-SiGe:H, and a-SiC:H from six previous publications. Collectively, these measurements exhibit "Meyer Neldel" behavior for electron emission over a range of activation energies from 0.2-0.8 eV and a prefactor range extending over nine decades, from 10^6 to 10^{15} Hz. The Meyer-Neldel behavior is consistent with the predictions of the multi-excitation entropy model. We extract a ionization entropy of $20k_B$ from the measurements, which is very large compared to crystal silicon. We discuss this result in terms of a bond charge model.

INTRODUCTION

The capture and emission of photogenerated electrons has been studied in amorphous silicon for over thirty years, typically with transient photocurrent techniques such as time-of-flight [1]. In this paper we focus on measurements of the emission time t_E of electrons from deep levels in a-SiGe:H (i.e. the (0/-) transition of a dangling bond). These obey the temperature dependence of the general Arrhenius equation:

$$1/t_E = \nu_0 \exp(-E_a/k_B T)\,. \tag{1}$$

Several aspects of these measurements were unusual. Deep level emission is often difficult to observe in time-of-flight measurements. We see it readily because in our samples the activation energies are unusually large (about 0.8 eV), as are the emission prefactors ν_0 (10^{15} s^{-1}).

When we place these Arrhenius parameters on a scatter plot with all other measurements for amorphous silicon based materials, we find a single "Meyer-Neldel" line extending from 0.2-0.8 eV and from 10^6 and 10^{15} s^{-1}. The slope is consistent with the multiple-excitation entropy model proposed by Yelon and Movaghar [2,3], which has been applied with some success to a remarkably wide range of processes and materials [4]. We speculate that our measurements also address the huge range of defect energies and emission rates that are simply assigned to a single defect, the dangling bond. Such variability may be unique to amorphous silicon and related materials; it is certainly unexpected from the experience of defects in single-crystal materials. The emission measurements also indicate that the ionization entropy of this defect level is about five times larger than it is for defects in crystalline silicon and other crystals, which suggests that charge state transitions affect nearby chemical bonds much more strongly than in crystals. We suggest that this larger interaction volume is essential to the wide variation in level positions of the dangling bond.

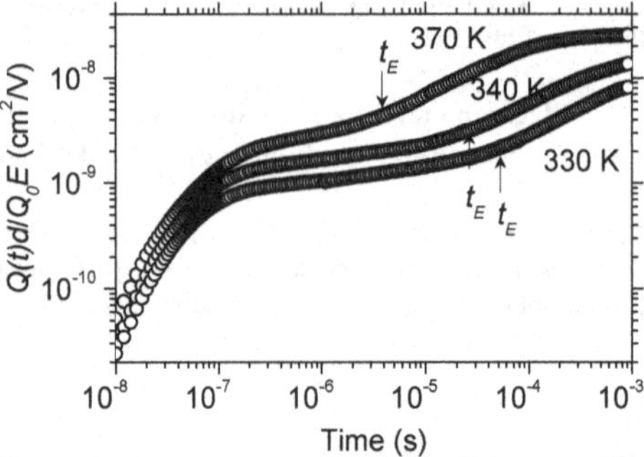

Figure 1. Temperature dependence of the normalized transient photocharge in an a-SiGe solar cell. Q_0 is total photocharge generated by a laser pulse, E is the electric field across the sample, and d is the thickness. We interpret the times t_E as the emission times for an electron from a deep-trap.

EXPERIMENTAL RESULTS

Hydrogenated amorphous silicon germanium a-SiGe:H *nip* structures were made with radiofrequency (RF) and very high frequency (VHF) plasma deposition [5]. Some of the specimens' properties are shown in Table 1. The open-circuit voltage was measured under a solar simulator. The *i*-layer thicknesses were inferred from dark capacitance measurements using the infrared dielectric constants of similar a-SiGe:H samples; the latter were obtained from interference fringe measurements, which yielded 1.21×10^{-12} F/cm.

For the experiment, the sample was illuminated through the *p*-layer by a 3 ns pulsed nitrogen-laser pumped dye laser. The wavelength we used in the experiment was 640 nm; for this wavelength, the absorption depth is 0.12 μm [6], which is much less than the thickness of the *i*-layer. The samples were either under short-circuit or pulsed reverse bias of magnitude V.

We present normalized short-circuit photocharge transients $Q(t)$ for different temperatures in Fig.1. Q_0 is the total photocharge; Q_0, the built-in potential V_{BI}, and the "electron deep-trapping mobility-lifetime product" $\mu\tau_{e,t}$ were determined by a "Hecht analysis" of the photocharge measured at 10^{-5} s as a function of the reverse bias voltage [7,8].

The electron deep-trapping mobility-lifetime product $\mu\tau_{e,t}$ is indicative of dangling bond densities. For the a-Si:H, the correlation of $\mu\tau_{e,t}$ with spin density measurements is consistent with the identification of deep trapping with $e^- + D^0 \rightarrow D^-$ [9,10].

For a-SiGe:H sample, we don't have strong evidence to show the defects are mostly neutral dangling bonds, we do know that the defects in a-SiGe:H are mostly Ge-dangling bonds, even though more than half of the atoms are still Si [11]. The similar values for $\mu\tau_{e,t}$ in Table 1

Table 1. Specifications of the two samples at room temperature.

Sample Code	Deposition frequency	Deposition Rate(Å/s)	V_{oc} (V)	d (μm)	$\mu\tau_{e,t}$ (cm^2/V)	ν_0 (s^{-1})	E_a(eV)
a	RF	4.4	0.64	1.4	5×10^{-10}	2.2×10^{15}	0.71
b	VHF	5.5	0.62	1.3	1×10^{-9}	1.3×10^{15}	0.74

for the VHF sample (at high deposition rate) and the RF sample (lower rate) suggests that the defect densities in the two samples were similar.

We now return to discussing the form of the photocharge transients in Fig. 1. These essentially record the displacement of the electron photocarrier generated at t=0. As can be seen, they consist of several segments. The early time photocharge (before 10^{-7} s) is the motion of the electron in the conduction band. At about 10^{-7} s the electron's motion is arrested by its capture by a "deep trap", which is generally understood to be a neutral dangling bond [9,10]. Note that, by definition, the value of the normalized photocharge at 10^{-5} s is $\tau_{e,t}$. The upward bending of the transient marked "t_E" indicates re-emission of the electron. For the highest temperature transient, the electron is finally collected at the bottom electrode at about 10^{-4} s.

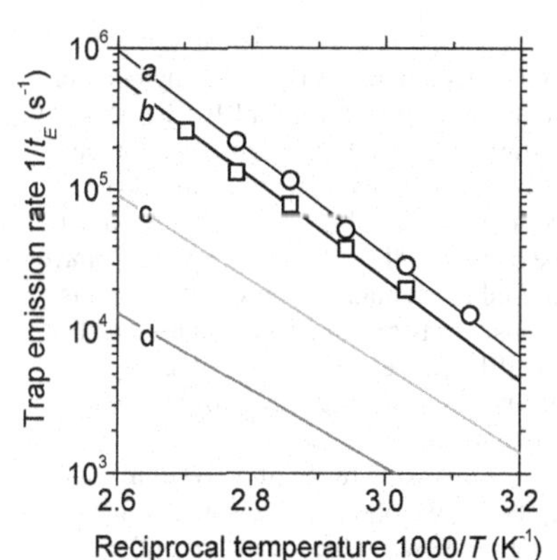

Figure 2. Temperature-dependence of electron emission from a trap for two a-SiGe:H samples. The straight lines shown are standard Arrhenius fits. Lines c and d are results for two samples of a-Si:H

As can be seen in Fig. 1, the apparent emission time shortens as the temperature increased. In Fig. 2 we graph the inverse of the emission time $1/t_E$ versus the reciprocal temperature $1000/T$. The straight lines shown on the graph are based on the standard Arrhenius model for emission from a trap:

$$\frac{1}{t_E} = \nu_0 \exp[-E_a/k_B T], \quad (2)$$

where ν_0 is the thermal emission prefactor ("attempt-to-escape" frequency) for the trap and E_a is the activation energy.

DISCUSSION

In Fig. 3 we present a correlation plot for measurements of ν_0 and E_a for electron traps in a-Si:H, and a-SiGe:H, and a-SiC:H. Table 2 shows that the original measurements involve several different transient and modulated photocurrent techniques, but we don't think this is important. Most of the points represent a-Si:H. Overall, a-SiGe:H and a-SiC:H samples occurred at the higher activation energies, but otherwise there is no sharp distinction between a-Si:H and

Table 2. List of D^- emission measurements

Publication	a	b[10]	c[12]	d[13]	e[14]	f[14]	g[15]	h[16]
Material	aSiGe:H	aSi:H	aSiGe:H	aSiGe:H	aSiGe:H	aSi:H	aSiC:H	aSi:H
Method	TOF*	TOF	DLCP	DLCP	MPCS	MPCS	TOF	TOF

*TOF: time of flight. DLCP: drive-level capacitance profiling [17]
MPCS: modulated photocurrent spectroscopy [14].

the alloy samples. Most evidence indicates that the defect involved is a dangling bond on a silicon or germanium atom [9]. The range of activation energies in this figure is remarkable, and suggests that level positions for the (0/-) transition of a dangling bond vary over most of the bandgap. "Defect pool" arguments have been used to account for the variation in this level position between doped and undoped a-Si:H [9], but we are unaware of an explanation for the wide range of Fig. 3 in undoped materials.

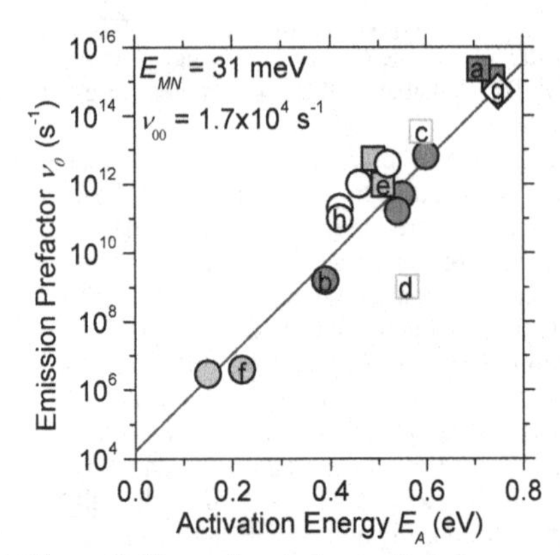

Figure 3. Thermal emission prefactors ν_0 and activation energies E_a for electron emission from the D^- center in a-Si:H, a-SiGe:H, and a-SiC:H samples. The letter codes for each symbol are defined in Table 2; the symbol shape indicates the material (circle: a-Si:H, square: a-SiGe:H, and

Over this wide range of activation energies E_a, we observe a reasonably linear correlation between the logarithm of the emission prefactor and E_a. The correlation is an example of a "Meyer-Neldel" relation, which are ubiquitous (see [4] for a recent review). We write:

$$\nu_0 = \nu_{00} exp\left(\frac{E_a}{E_{MN}}\right) \quad (3)$$

where E_{MN} is the Meyer-Neldel energy and ν_{00} is MNR prefactor which we will discuss later in the paper. In Fig. 3 we have shown a best fit yielding E_{MN} = 31 meV and $\nu_{00} = 2 \times 10^4$ s^{-1}. Similar parameters were previously reported based on only 4 of the specimens in Fig. 3 [10].

There have also been estimates for Meyer-Neldel parameters from electron and hole drift mobility measurements. These involve photocarrier emission from bandtail traps, as opposed to dangling bonds. These give quite different values for the Meyer-Neldel parameters: E_{MN} = 40 meV and $\nu_{00} = 5 \times 10^9$ s^{-1} for electrons in the conduction bandtail, and E_{MN} = 70 meV and $\nu_{00} = 7 \times 10^{10}$ s^{-1} for holes in the valence bandtail [18,19].

One mechanism that accounts for a Meyer-Neldel relation is the multiple excitation entropy (MEE) model [2,3]. This model presumes that the activation energy is larger than a typical thermal excitation of the system such as an optical phonon. Several excitations must merge to permit a carrier to be emitted from the trap. In the simplest case, consideration of the entropy of these multiple excitations leads to Meyer-Neldel behavior with $E_{MN} \approx E_{ex}$, where E_{ex} is the typical value for the excitation. This appears to apply well to hole emission from bandtail traps, for which E_{MN} is close to the energy of an optical phonon in silicon at 63 meV.

Interestingly, the smaller value for E_{MN} that we measured for electrons in dangling bonds is more typical of a-Si:H than is the value for hole emission from the bandtail. As summarized in a recent review [4], defect annealing processes and dark conductivity typically have a value $E_{MN} \approx$ 40 meV. Yelon, Movaghar, and Branz proposed that polaron effects, which are an interaction term between the local electrons with optical phonons, can lower E_{MN} below 60 meV even when optical phonons are the key excitation; they also conclude that acoustic phonons yield values for E_{MN} that are much too small to be consistent with experimental results [3]. Presuming that the polaron mechanism is involved, we would speculate that the proximity of the dangling bond is modifying the nearby conduction band states and the corresponding polaron correction.

The other fitting parameter from Fig. 3 is ν_{00}. There are several perspectives on this parameter; we shall use one based on "ionization entropy" concepts. In this perspective, $\nu_{00}$is thought to be related to a fundamental Eyring frequency $\nu_{Eyring} = k_B T/h$ (5.99×10^{12} s^{-1} near room-temperature) [20]:

$$\nu_{00} = \frac{k_B T}{\hbar} \exp(-\Delta S_{ion}/k_B) \tag{4}$$

where ΔS_{ion} is the entropy change due to the ionization process. In this case the change involves the entropy of the defect in the 2 charge states, as well as the entropy associated with a free carrier in the conduction band.

Ionization entropies are a source of the temperature-dependence of level positions through $\Delta S_{ion} = -\frac{\partial \Delta E}{\partial T}$ [21], where ΔE is the energy required to detach an electron from a D^-, creating a D^0 and an electron at the bandedge. For $\nu_{00} = 1.7 \times 10^4$ s^{-1} we obtain $\Delta S_{ion} \cong 20k_B$ (-1.7 meV/K). This can be compared with the "ionization entropy" associated with an electron-hole pair. For $\frac{\partial E_G}{\partial T} = -0.47$ meV/K, we obtain $\Delta S_{ion} = 5.5\ k_B$ [22]. For reference, the ionization entropy for defect annealing in a-Si:H reported long ago by Crandall was $37k_B$ [23]. We've listed the ionization entropies for several types of silicon materials in Table 3.

These entropy changes are several times larger than expected from the simplest model, which is that removal or addition of a charge to a bond is roughly equivalent to removing the bond itself. The latter changes the number of degrees of freedom in the solid, slightly changes the entire vibrational spectrum, and thereby modifies the entropy (by roughly $0.75k_B$) [24]. As Heine and van Vechten emphasized for crystalline silicon, the measured temperature-dependences are several times larger than expected for breaking of a bond; they comment that "one hole in the most bonding states at the top of the valence band could destroy the bond charge Z_B in six bonds, which is our qualitative explanation for the large entropy change" [24]. For c-Si, the ionization entropy change of several deep level defects is approximately the same as for electron-hole pair creation [25], and a similar discussion applies.

It is intriguing that, while the ionization entropy associated with the temperature-dependence of the bandgaps is about the same for c-Si and for a-Si:H, ΔS_{ion} for the (0/-) transition of the D-center is several times larger ($20k_B$) than for several deep defects in c-Si. In Heine and van Vechten's language, ionization of the defect is "destroying the bond charge" in more than two dozen nearby bonds. The large magnitude of the ionization entropy is perhaps the single most dramatic distinction that has been drawn between defects in crystals and in non-crystalline solids. Of course, the extraordinary range of binding energies for the *D*-center in Fig. 3 is also unexpected in the context of defects in crystals.

Table 3. Ionization entropy for different types of silicon. C.B is for the conduction band; V.B is for the valence band, and E_g is measurement for the band gap.

Silicon	a-Si	a-Si	a-Si	a-Si	c-Si	c-Si	Bond
Defect	$D^{0/-}$	C. B.	V. B.	E_g	E_g	Au^-	n-p
ΔS_{ion}	$20k_B$	$7.1k_B$[18]	$4.5k_B$[18]	$5.5k_B$[22]	$2.9k_B$[25]	$2.9k_B$[25]	$3/4k_B$[24]
Methods	MNR	MNR	MNR	$-\partial E/\partial T$	$-\partial E/\partial T$	$-\partial E/\partial T$	$-\partial E/\partial T$

CONCLUSION

Trap emission prefactor frequencies for many amorphous silicon type materials are often in the range 10^{11}-10^{13} s^{-1}, which is similar to the fundamental Eyring frequency. In this paper we've shown that a much larger range of prefactors (10^9) has been measured, including some very large values in device-grade materials. There is a fairly good "Meyer-Neldel" correlation with the emission activation energy, which likely reflects "multi excitation entropy" of phonons. A less familiar aspect is the enormous ionization entropy implied by the Meyer-Neldel graph for dangling bond emission when compared with related processes such as bandtail trap emission. We are not aware of any satisfactory explanation for this aspect of the measurements.

This research has been supported by the U. S. Department of Energy through the Solar America Initiative (DE-FC36-07 GO 17053). Additional support was received from the Empire State Development Corporation through the Syracuse Center of Excellence in Environmental and Energy Systems.

REFERENCES

1. E. A. Schiff, *J. Phys.: Condens. Matter* **16**, S5265-5275 (2004).
2. A. Yelon, B. Movaghar, *Phys. Rev. Lett* **65**, 618 (1990).
3. A. Yelon, B. Movaghar, and H. M. Branz, *Phys. Rev. B* **46**, 12244 (1992).
4. A. Yelon, B. Movaghar, and R. S. Crandall, *Rep. Prog. Phys.* **69**, 1145 (2006).
5. J. Yang, B. Yan, J. Smeets, and S. Guha, *Mater. Res. Soc. Symp. Proc.* **664**, A11.3 (2001).
6. Y. Xu, A. H. Mahan, L. M. Gedvilas, R. C. Reedy, and H. M. Branz, *Thin Solid Films* **501,** 198 (2006).
7. Q. Wang, H. Antoniadis, E. A. Schiff, and S. Guha, *Phys. Rev. B* **47**, 9435 (1993).
8. S. A. Dinca, E. A. Schiff, B. Egaas, R. Noufi, D. L. Young, and W. N. Shafarman, *Phys. Rev. B* **80**, 235201 (2009).
9. R. A. Street, *Hydrogenated Amorphous Silicon* (Cambridge University Press, 1991).
10. H. Antoniadis, E. A. Schiff, *Phys. Rev. B.* **46**, 9482dd (1992).
11. J.-K. Lee and E. A. Schiff, *J. Non-Cryst. Solids* **114**, 423-425 (1989).
12. S. Datta, J. D. Cohen, *J. Non-Cryst. Solids* **354**, 2126. (2008).
13. F. Zhong, C. -C. Chen, J. D. Cohen, *J. Non-Cryst. Solids* **198-200**, 572 (1996).
14. Y. Tsutsumi, *Phil. Mag. B* **60**, 695 (1988).
15. M. Brinza, G. J. Adriaenssens, *J. Mat. Sci.* **14**, 749 (2003).
16. B. Yan and G. J. Adriaenssens, *J. Appl. Phys.* **77**, 5661 (1995).
17. C. E. Michelson, A. V. Gelatos and J. D. Cohen, *Appl. Phys. Lett* **47**, 412 (1985).
18. W.-C. Chen, L. -A. Hamel, and A. Yelon, *J. Non-Cryst. Solids.* **200**, 254 (1997).
19. E. A. Schiff, *Phil. Mag. B* **89** 2505 (2009).
20. R. S. Crandall, *J. Appl. Phys.* **108**, 103713-1 (2010).
21. J. A. VanVechten, in *Handbook of Semiconductors* (edited by S. P. Keller), **Vol. 3**, Chap. 1, North Holland, Amsterdam (1980).
22. G. D. Cody, T. Tiedje, B. Abeles, B. Brooks, and Y. Goldstein, *Phys. Rev. Lett.* **47**, 1480 (1982).
23. R. S. Crandall, *Phys. Rev. B* **43**, 4057 (1991).
24. V. Heine and J. A. VanVechten, *Phys. Rev. B* **13**, 1622 (1976).
25. J. A. Van Vechten and C. D. Thurmond, *Phys. Rev. B* **14**, 3539 (1976).

Nanostructures

Mater. Res. Soc. Symp. Proc. Vol. 1321 © 2011 Materials Research Society
DOI: 10.1557/opl.2011.936

Opto-electronic properties of co-deposited mixed-phase hydrogenated amorphous/nanocrystalline silicon thin films

James Kakalios,[1] U. Kortshagen,[2] C. Blackwell,[1] C. Anderson,[2] Y. Adjallah,[1] L. R. Wienkes,[1] K. Bodurtha,[1] and J. Trask[2]

[1]School of Physics and Astronomy,
[2]Department of Mechanical Engineering
University of Minnesota, Minneapolis, MN 55455

ABSTRACT

Mixed-phase thin film materials, consisting of nanocrystalline semiconductors embedded within a bulk semiconductor or insulator, have been synthesized in a dual-chamber co-deposition system. A flow-through plasma reactor is employed to generate nanocrystalline particles, that are then injected into a second, capacitively-coupled plasma deposition system in which the surrounding semiconductor or insulating material is deposited. Raman spectroscopy, X-ray diffraction and high resolution TEM confirm the presence of nanocrystals homogenously embedded throughout the a-Si:H matrix. In undoped nc-Si within a-Si:H (a/nc-Si:H), the dark conductivity increases with crystal fraction, with the largest enhancement of several orders of magnitude observed when the nanocrystalline density corresponds to a crystalline fraction of 2 – 4%. These results are consistent with the nc donating electrons to the surrounding a-Si:H matrix without a corresponding increase in dangling bond density for these films. In contrast, charge transport in n-type doped a/nc-Si:H films is consistent with multi-phonon hopping, possibly through extended nanocrystallite clusters with weak electron-phonon coupling. The flexibility of the dual-chamber co-deposition process is demonstrated by the synthesis of mixed-phase thin films comprised of two distinct chemical species, such as germanium nanocrystallites embedded in a-Si:H and Si nanocrystallites embedded within an insulating a-SiN_x:H film.

INTRODUCTION

The opto-electronic properties of semiconducting materials are primarily determined by their chemical composition and atomic bonding arrangements. In contrast, in nanostructured systems the low-dimensionality and short length scales can have a significant influence on the material's characteristics, such as the energy gap [1]. The ability to synthesize mixed-phase materials, consisting of nano-scale materials embedded within a bulk semiconductor or insulator has enabled the development of novel materials with properties not easily realized in homogeneous thin films. Solar cells utilizing silicon nanocrystallites embedded within an hydrogenated amorphous silicon matrix (a/nc-Si:H) as the photovoltaic material have recently attracted considerable attention due to their high solar conversion efficiencies, high deposition rates and improved resistance to light-induced defect creation [2-4].

These mixed phase films are typically synthesized in a capacitively-coupled single-chamber plasma system, using high gas pressures and a heavily hydrogen-diluted silane precursor [5-9]. This single-chamber process poses limitations for the production of a/nc-Si:H, since (1) the properties of the crystalline particles produced in the gas phase are not easily controlled and (2) gas-chamber conditions are not conducive to growing high quality amorphous material. To address these issues, we have recently developed a dual-chamber co-deposition plasma enhanced chemical vapor deposition (PECVD) system, shown in Figure 1, that enables the growth of a wide variety of mixed-phase thin film materials. Nanocrystalline particles are synthesized in an upstream flow-through tube plasma reactor (Fig. 1a) [10], where the size of the nanoparticles is determined by the plasma power and gas residence time in the reactor. These nanoparticles are then injected into a second capacitively-coupled plasma deposition system (Fig. 1b) in which the surrounding matrix material is grown. The synthesis of the nanocrystals is decoupled from the deposition of the host semiconductor matrix, so that the growth conditions can be independently optimized for each component of the mixed-phase material. This paper will review some of the novel optical and electronic properties observed in mixed-phase films synthesized in this dual-chamber co-deposition system.

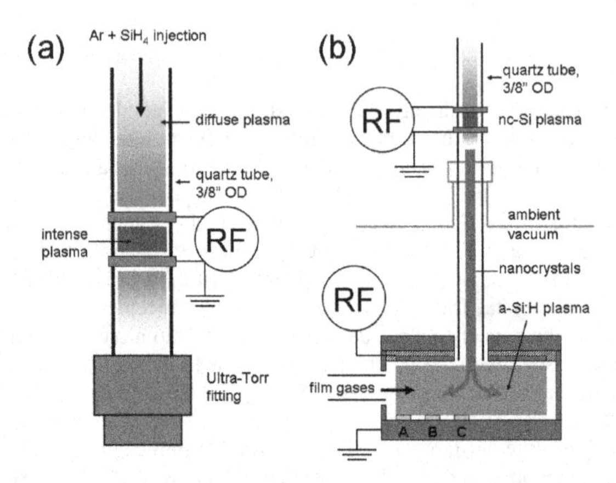

Figure 1: (a) Schematic of particle synthesis chamber, and (b) the dual plasma co-deposition system.

SAMPLE PREPARATION

The particle synthesis reactor, shown in Figure 1a, consists of a 3/8-inch diameter quartz tube fitted with two ring electrodes [10]. Using an Advanced Energy RFX-600 signal generator/amplifier and a p-type matching network, the excitation voltage at 13.56 MHz was applied to the upper ring electrode in Figure 1a, while the lower ring electrode is grounded. By varying the deposition conditions, in particular the gas chamber pressure and RF power, the diameter of the silicon nanocrystals can be controlled. The gas flow of silane is set at 10 sccm of 5% silane pre-diluted in 95% helium. As a carrier gas to entrain the nanoparticles and deliver them into the second plasma deposition chamber, 50 sccm of argon was used, for a total pressure of 1.5 Torr. The argon promotes silicon nanocrystal formation, and inhibits diffusion losses in the narrow particle synthesis tube, compared to when helium is employed as a carrier gas, enabling the formation of nanocrystallites at lower pressures. Using ~50 W of RF power, a discharge is produced in the tube that extends for a length of around 5 cm, and is extinguished as the tube feeds into the grounded, second deposition chamber (Figure 1b). The size of the particles is directly related to the gas residence time in the plasma [10]. A typical residence time of 3 ms results in nanocrystallite formation with an average diameter of between 5-6 nm. Details of the nanocrystal diameter dependence on chamber gas pressure and plasma residence time have been published previously [11,12].

Concurrent with the continuous injection of silicon nanocrystals, a-Si:H is deposited in the lower, capacitively-coupled plasma (CCP) chamber, shown in Figure 1b. The flow of silane in the second chamber is 20 sccm of 5%-silane in helium, introduced through a separate gas feedthrough from the side of the chamber (not shown in fig.1b). Helium is employed for the second deposition chamber, as the lower mass noble gas ion induces less surface damage of the growing thin film semiconductor. The total pressure in the second chamber is 600 mTorr, and the RF (13.56 MHz) power applied across the two electrodes (the R.F. powered electrode has an area of 315 cm^2 while the grounded electrode (on which the deposition substrates reside) has an area of 1338 cm^2, with a fixed separation of 5 cm) is 5 W. For the films described here the grounded electrode is kept at 250°C while the top RF electrode is unheated.

Doped mixed-phase films have also been deposited in the dual-chamber system, similar in design but slightly different in construction to the chamber used for the growth of the undoped a/nc-Si:H films, when silane (SiH_4), argon and phosphine (PH_3) are used in the particle synthesis chamber [13]. The doping level was set at $[PH_3]/[SiH_4] = 6 \times 10^{-4}$ for all of the films studied here. With the pressure at 1.7 Torr and RF power of 70W for the particle reactor, the resulting crystalline nanoparticles are then entrained by the argon and injected into the second PECVD chamber where a-Si:H is deposited using the residual silane/dopant gas mixture from the first chamber. The a-Si:H matrix is deposited in the second chamber at a total pressure of 700 mTorr and an RF power of 4W (electrode area of 53.5 cm^2). The total gas flow was held at 90 sccm for all films described here, but the ratio of silane/dopant mixture to argon was adjusted in order to vary the crystalline content in the mixed-phase films. The substrate electrode is heated to 250°C while the RF electrode is unheated.

The concentration of silicon nanocrystals incorporated into an amorphous silicon matrix can be controlled by either a thermophoresis force induced by a thermal gradient applied across the silane plasma [14] or via the convection of the carrier gas as the nanoparticles are injected into the capacitively-coupled chamber from the particle synthesis reactor [11,12,15]. In the latter configuration, substrates are placed radially outwards from the center position of the tube exit (located 2.5 cm above the lower electrode), labeled alphabetically from the electrode periphery inwards, as shown in Figure 1. This reactor geometry takes advantage of the fact that the deposition rate of particles decreases radially outwards from the center position of the electrode. We thus have a unique way of studying the effect of crystalline fraction on the film properties.

STRUCTURAL CHARACTERIZATION

High-resolution transmission electron microscopy (HRTEM) studies of cleaved films verify that silicon nanocrystals are indeed introduced into the amorphous silicon matrix, and reside homogenously throughout the film's thickness [11]. Comparison of films deposited with the particle synthesis chamber turned on (leading to a/nc-Si:H films) and turned off (in which a-Si:H is deposited) confirms that the crystallites are synthesized in the particle synthesis reactor, rather than forming within the silane plasma in the capacitively-coupled plasma chamber. Focal series imaging confirms that the lattice fringes in Figure 2b are not microscope artifacts [16].

Additional evidence that the nanoparticles synthesized in the first plasma chamber are crystalline is provided by X-ray diffraction (XRD) studies. Fig. 2 shows an XRD spectra taken on a Bruker-AXS Microdiffractometer at room temperature. Silicon nanocrystals were collected on a stainless steel mesh and then transferred to a glass slide. The background diffraction signal from the glass substrate has been subtracted from the data. The average size of the crystallites was calculated from the observed peaks to be 3.4 nm, correcting for instrument broadening.

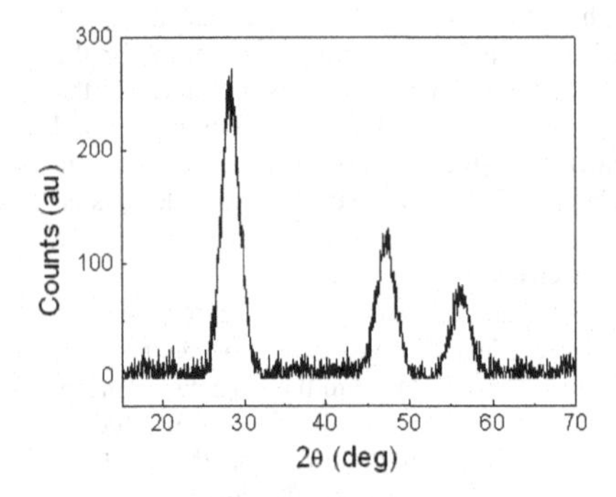

Figure 2: X-Ray diffraction spectrum for silicon nanocrystallites synthesized in the plasma reactor shown in Fig. 1a. The free-standing nanoparticles are collected on a wire mesh and transferred to a glass substrate for diffraction studies (background diffraction from glass slide subtracted). The average crystallite size for the particles in this figure is 3.4 nm.

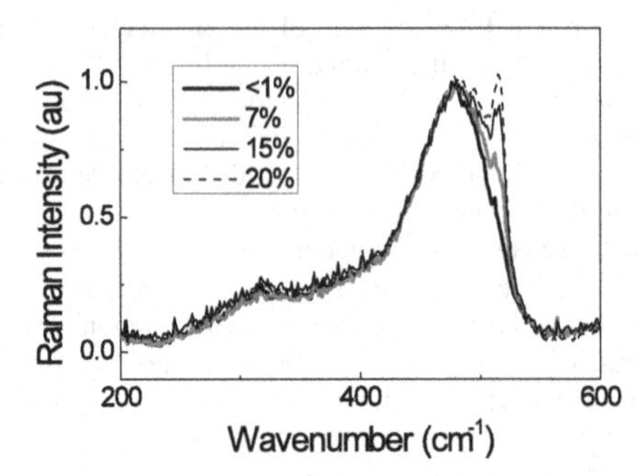

Figure 3: Raman spectra for a series of n-type doped a/nc-Si:H films deposited in the dual chamber co-deposition system. The growth conditions were varied so that the concentration of Si nc in the a-Si:H matrix is increased from < 1% to 20%. Comparison of the relative intensity of the c-Si and a-Si:H TO modes at 518 cm^{-1} and 480 cm^{-1} respectively, provides an estimate of the crystal fraction X_C.

The concentration of nanocrystals is ascertained from Raman spectroscopy, using a Witec Alpha 300R confocal Raman microscope, equipped with an UHTS 200 spectrometer, using an argon ion excitation laser of wavelength 514.5 nm at a power of 5 mW focused to an area of ~ 2-3 μm diameter. By comparing the sharp TO mode for crystalline silicon at 518 cm^{-1} (this peak may be shifted to lower wavenumbers due to quantum confinement in the nanoparticles) to the broader TO signal centered at 480 cm^{-1} for the amorphous silicon matrix, a quantitative estimate of the crystal fraction X_C can be obtained [12,17]. Figure 3 shows the Raman spectra for a series of n-type doped a/nc-Si:H films, where the nanoparticle synthesis chamber is side injected into the capacitively-coupled amorphous silicon deposition system.

We next describe measurements of the electronic and optical properties of undoped and n-type doped mixed-phase thin films consisting of silicon nanocrystallites embedded within an a-Si:H matrix, and some recent results on novel mixed phase films such as germanium nanocrystals in an a-Si:H film and silicon nanocrystallites in an insulating silicon nitride layer that are synthesized in the dual chamber co-deposition system.

UNDOPED a/nc-Si:H FILMS

One of the initial motivations for studying the properties of a/nc-Si:H mixed-phase thin films was the report that these materials exhibited enhanced resistance to light-induced defect creation (the Staebler-Wronski effect) without a concomitant degradation in the opto-electronic properties. The undoped mixed-phase a/nc-Si:H films synthesized in the dual chamber co-deposition system exhibit an enhancement in the dark conductivity without a significant increase in mid-gap optical absorption coefficient, a measure of the defect density [12,18].

When the nanoparticles grown in the first plasma reactor are injected into the second capacitively-coupled plasma (CCP) chamber from the top through a showerhead RF electrode, as illustrated in Fig. 1, they are subject to convection of the inert carrier gas (typically argon) in the CCP chamber. The deposition rate of particles decreases radially from the center position of the electrode, where the outlet of the particle injection tube is located, as confirmed by tapping-mode Atomic Force Microscopy measurements [12]. By placing substrates radially outwards from the center position of the tube exit, we have a unique way of studying the influence of crystalline fraction on the properties of a series of a/nc-Si:H deposited simultaneously [11,12]. Raman studies for films synthesized in a single deposition run, at substrate locations marked A, B and C in Fig. 1b, find that the crystalline fraction is less than 1% for the substrate the greatest distance from the particle injection tube (position A), while the film deposited directly underneath the outlet of the particle synthesis chamber outlet (position C) has a crystal fraction of 10 - 20%. Films deposited at the position labeled B in Fig. 1b typically have crystal fractions of 2 – 4%, as measured by Raman spectroscopy. As these films are all deposited at the same time in a single run, the quality of the a-Si:H matrix should be identical for these three films, and any changes in the films' properties should be primarily due to the varying concentration of nanocrystalline inclusions.

Figure 4 shows an Arrhenius plot of the dark conductivity for three films grown in such a manner in a single deposition run. There is a striking non-montonic dependence of the dark conductivity with nanocrystal content. As expected, the film grown the furthest from the particle injection tube outlet with the lowest nanocrystal concentration exhibits a conductivity not unlike that of nominally homogeneous a-Si:H. The addition of nanocrystallites corresponding to a crystal fraction of ~ 2% leads to a several order of magnitude enhancement of the dark conductivity. However, the conductivity for $X_C = 18\%$ is intermediate to samples between 0 and 2% over the range of temperatures investigated. In order to confirm that the dark conductivity is indeed non-monotonic with nanocrystalline concentration, this deposition series was repeated five times. While there is a run to run variation in the exact concentration of silicon nanocrystals embedded within the a-Si:H matrix, in all trials the general behavior of an enhancement of the dark conductivity, with a decrease in the measured activation energy for the a/nc-Si:H film containing a crystalline fraction of $X_C \sim 2 – 4\%$, was observed [12].

These results have been attributed to charge donation from the silicon nanocrystals into the surrounding a-Si:H matrix. Studies [19] indicate that for Si nanocrystals in a-Si:H, the conduction band offset is approximately 0.1 eV, while the valence band offset is closer to 0.3 eV (assuming a bandgap of 1.4 eV for the nanocrystal due to confinement effects, and a bandgap of 1.8 eV for a-Si:H). In this case a thermally excited electron in the nanocrystal can easily escape to the surrounding a-Si:H, while the hole is less likely to leave the nanocrystallite. The excess electrons donated by the nanocrystals to the a-Si:H would fall into dangling bond defects in the

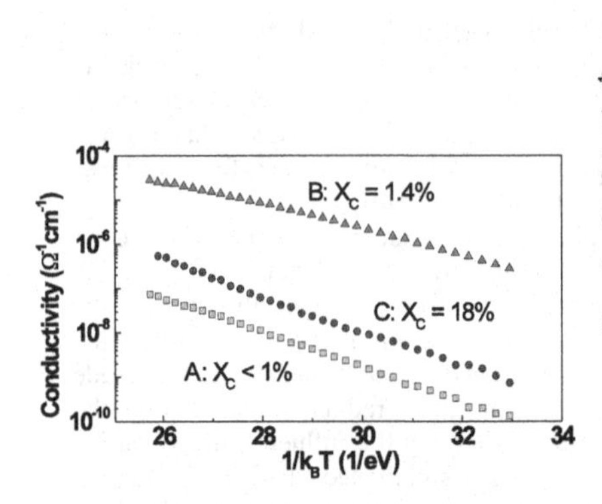

Figure 4: Arrhenius plot of the conductivity of three a/nc-Si:H films grown in a single run. The dark conductivity exhibits a non-monotonic dependence on crystal fraction. The activation energies were 0.90, 0.64, and 0.91 eV for films A, B, and C respectively.

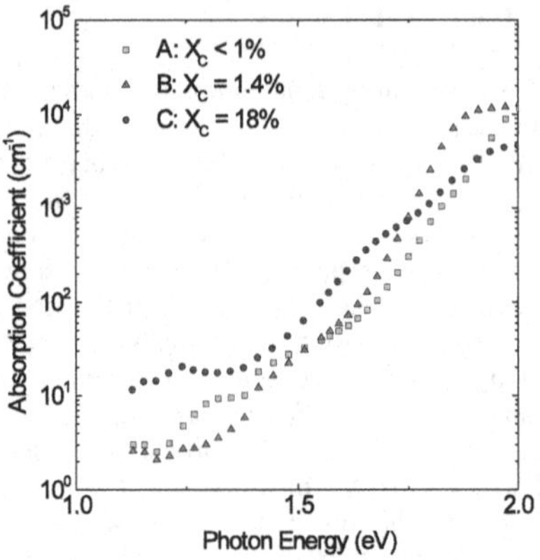

Figure 5: Plot of the spectral dependence of the optical absorption coefficient as determined by CPM for the three mixed-phase films in Fig. 4.

middle of the mobility gap, shifting the Fermi energy and raising the conductivity of the a-Si:H. Constant Photocurrent Measurements of the optical absorption spectra of these films, shown in Figure 5, find that the midgap defect density is unchanged for the films with $X_C \sim 2$ -4 % and those without nanocrystals.

However, the film deposited directly underneath the particle injection tube ($X_C \sim 18\%$) displays an order of magnitude higher dangling bond concentration than the substrates displaced radially from the outlet of the injection tube. Tapping mode AFM measurements confirm that the films deposited directly underneath the particle injection tube have a higher concentration of nanoparticle inclusions than would be expected by a simple extrapolation from the densities found in the substrates located at greater separations. Moreover, it is likely that films with high nanocrystal concentration contain more small nanocrystal agglomerates, which may form either in the gas phase or by crystals depositing onto or next to other crystals that are already residing on the substrate. These agglomerates are more likely to lead to small voids due to shading effects during the amorphous film deposition. It is thus probable that a larger density of dangling bonds is associated with the higher nanoparticles densities in these films, consistent with the observed lower photosensitivity in these films. The excess dangling bond density is higher than the density of excess charge donated by the nanocrystals in these films, and the conductivity activation energy returns to a larger value, decreasing the conductivity [12]. These results suggest that the optimal dangling bond density, photosensitivity and dark conductivity result from a/nc-Si:H films with Xc ~ 2 – 4%.

N-TYPE DOPED a/nc-Si:H FILMS

Nearly all applications of semiconductors derive from the fact that these materials can be readily doped through the introduction of chemical impurities. While the doping mechanism is well understood for both single crystalline and amorphous semiconductors [20, 21], the effect of dopants on the electronic properties of semiconductor nanocrystals, and on the surrounding amorphous semiconductor in mixed-phase a/nc-Si:H films, has not been thoroughly investigated.

By dynamically mixing phosphine with silane in the particle synthesis chamber of Fig. 1, we have recently synthesized n-type doped mixed-phase a/nc-Si:H films where the crystalline fraction, as determined by Raman spectroscopy, varies from < 1% up to 60%. Fig. 3 shows the Raman spectra for this series of films, up to X_C = 20% for clarity. The remaining Raman curves are shown in [13]. The doping level was moderate ($[PH_3]/[SiH_4] = 6 \times 10^{-4}$), and any unreacted phosphine is available for incorporation into the surrounding a-Si:H matrix in the CCP chamber. An Arrhenius plot of the temperature dependence of the dark conductivity (Fig. 6), indicates that transport in these films is sensitive to the Si nanocrsytalline concentration. Doped films with X_C = 0 (that is, a film deposited with the particle synthesis reactor turned off and no nanocrystallites incoproated into the a-Si:H film) exhibit thermally activated behavior, characterized by a single activation energy over the entire temperature range investigated (450 to 100 K). In fact these films exhibit a high temperature kink near 400 K, with the conductivity activation energy increasing at higher temperatures. This is the well known "thermal equilibration effect" observed in doped a-Si:H [22,23]. The presence of the nanocrystallites has been found to lead to slower stretched exponential relaxation and a reduction in the difference in the conductivity between the slow cool and rapid quench states, consistent with slower hydrogen diffusion in the a/nc-Si:H films [24].

The dark conductivity of n-type doped a/nc-Si:H films increases monotonically with crystal fraction, unlike for undoped a/nc-Si:H (Fig. 4), and displays marked deviations from simple thermally activated temperature dependence [13]. In order to ascertain the correct functional form of the conductivity's temperature dependence for these films, an analysis technique described by Zabrodskii [25] and Hill [26], involving the temperature dependence of the conductivity activation energy E_a was employed. Given that the conductivity σ can be written in the form $\sigma = \sigma_o \exp[-\beta E_a]$ where σ_o is the conductivity pre-exponential factor, $\beta = (k_B T)^{-1}$, and k_B is Boltzmann's constant, then from the measured dark conductivity as a function of temperature, we calculate $- d\ln(\sigma)/d\beta$ and plot this quantity against temperature, on a log-log plot. For simple thermally activated behavior such a plot yields a horizontal line reflecting the value of the single activation energy, while for Mott variable range hopping [27] this analysis displays a power-law temperature dependence, that is, $E_a \sim T^m$ where m = 0.75 [13].

The analysis of the temperature dependence of the activation energy for n-type doped a/nc-Si:H containing X_C = 15% or higher nanocrystalline content indicates that the dark conductivity exhibits a power-law temperature dependence, $\sigma = \sigma_o (T/T_o)^n$ where n = 4 – 2 outside of any experimental uncertainty [13]. Fig. 7 shows a log-log plot of σ against T for the n-type doped a/nc-Si:H films, and indeed a power-law temperature dependence does provide a superior fit to the data over most of the temperature range studied. Such a temperature dependence has been observed in unhydrogenated evaporated a-Ge [28], sputtered amorphous

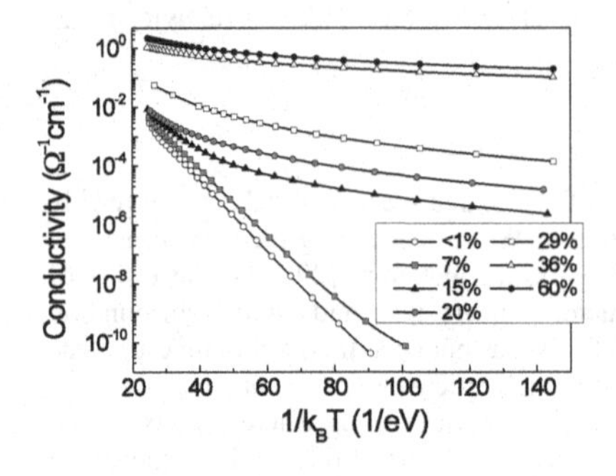

Figure 6: Arrhenius plot of the dark conductivity of n-type doped a/nc-Si:H with increasing nanocrystalline concentration.

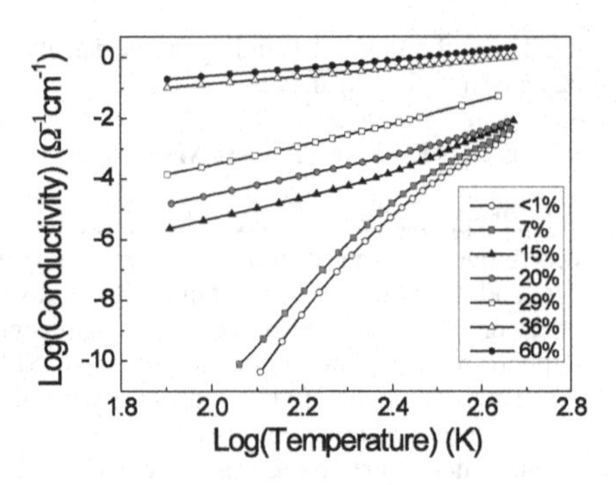

Figure 7: log-log plot of the dark conductivity against temperature of n-type doped a/nc-Si:H with increasing nanocrystalline concentration.

carbon [29], lightly doped $(CH)_x$ [30], semiconducting strontium vanadate glasses [31], Cr containing multicomponent oxide glasses with nanocrystalline particles and clusters [32], and vanadium oxide glasses [33]. In the polyacetylene films hopping is believed to occur through a solition band [30], while in the transition metal oxide glass systems it is proposed that above a given concentration, the wavefunctions for the metallic or nanocrystalline clusters overlap sufficiently to produce percolating microclusters with weak electron-phonon coupling between the clusters [34-36].

Figure 7 indicates that for doped a/nc-Si:H films with $X_C \geq 15\%$, the low temperature electronic transport is dominated by multiphonon hopping between connected clusters of nanocrystals. In order to hop from one extended cluster to another, the electron should couple with a phonon whose wavelength is comparable to the radius of the electron's wavefunction [37]. If the electron is delocalized over the long length scale comprising several connected nanocrsytallites, then this would suggest that hopping requires many such long-wavelength, low-energy phonons. The power-law temperature dependence of the dark conductivity arises from expanding the product of n Bose-Einstein occupation functions, which describe the likelihood of n such low energy phonons being available for the hopping event. The distribution of hop energies would then account for the non-integer power-law exponent. Experiments are underway to ascertain whether this power-law temperature dependence persists to lower temperatures (< 10K), or whether a transition to Efros-Shklovskii Variable Range Hopping [38] occurs.

NOVEL MIXED PHASE FILMS

The flexibility of the dual-chamber co-deposition process is demonstrated by the ability to synthesize mixed-phase thin films comprised of two distinct chemical species. As a proof of concept, we have fabricated 5 nm diameter germanium nanocrystallites in the particle synthesis chamber in Fig. 1, that are then injected into the second CCP chamber in which a surrounding a-Si:H film is deposited. Raman spectroscopy, shown in Fig. 8, confirms the presence of Ge nanocrystallites, which are 5 nm in diameter based upon XRD data. The sharp peak near 300 cm^{-1} is due to the crystalline Ge TO mode, while the signal at 400 cm^{-1} indicates the presence of Si-Ge bonding [39, 40]. There is no indication of a reduction of the optical absorption gap [41], as suggested by the optical transmission measurements, in these nc-Ge /a-Si:H films, suggesting no significant alloying with germanium in the amorphous silicon matrix [42]. The Raman mode at 380 cm^{-1} may therefore reflect bonds at the interface of the Ge nanocrystal in the a-Si:H matrix. It is difficult to imagine how such a mixed-phase film could be synthesized in a single chamber plasma reactor.

As described above, when silicon nanocrystallites are embedded within an undoped a-Si:H matrix, the nanoparticles appear to donate electrons to the surrounding amorphous film. The excess charge carriers are believed responsible for the enhancement of the conductivity, and decrease in the activation energy, shown in Fig. 4. The fact that electrons are donated to the a-Si:H matrix is confirmed by thermopower measurements, which indicate that the majority carriers are electrons. Fig. 9 shows an Arrhenius plot of the conductivity of the nc-Ge/a-Si:H film from Fig. 8, along with that for pure a-Si:H (grown with the particle synthesis chamber off)

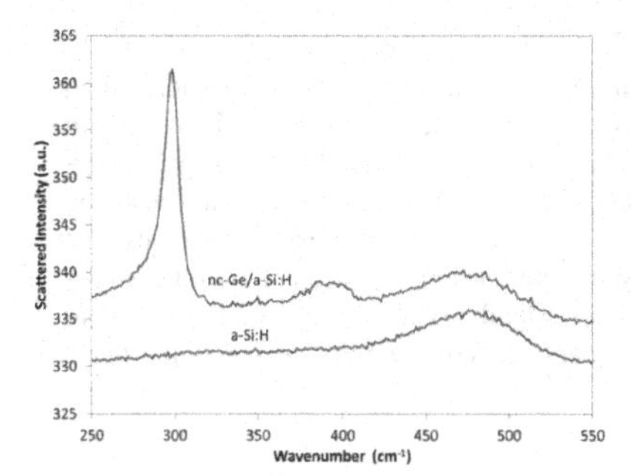

Figure 8: Raman spectra of pure a-Si:H and a mixed phase a-Si:H film containing nc-Ge.

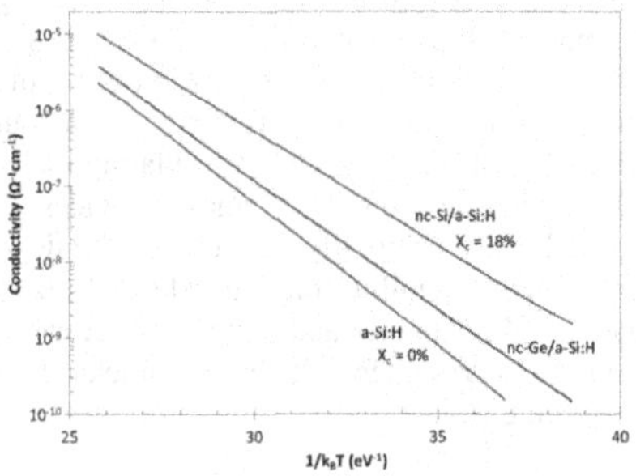

Figure 9: Arrhenius plot of the dark conductivity of a/nc-Si:H, a-Si:H and a mixed phase a-Si:H film containing nc-Ge.

and an a/nc-Si:H (that is, a-Si:H with Si nc inclusions) for comparison. The nc-Ge/a-Si:H also exhibits an enhancement of the dark conductivity, but not as large as for the a/nc-Si:H film. However, thermopower measurements of this nc-Ge/a-Si:H film find a positive Seebeck coefficient, with a temperature dependence similar to that observed in p-type a-Si:H, while other films with a smaller Ge crystalline fraction detectable by Raman spectroscopy have exhibited an n-type thermopower. P-type conduction has been observed in single-crystal germanium nanowires [43], Ge/Si core-shell nanowires [44] and Ge nanocrystals embedded within a SiO_2 matrix, prepared by co-sputtering [45]. The hole conduction in these nanoscale systems is attributed to the formation of a hole accumulation layer in the germanium nanocrystallites due to acceptor-like surface states [43-45]. By growing a series of nc-Ge/a-Si:H films for which the germanium crystal fraction is held fixed, but the average diameter and density of the nanoparticles are systematically varied, the relative contribution of surface and bulk defect states can be examined.

The addition of an NH_3 gas line to the CCP chamber in Fig. 1 enables the synthesis of a-SiN_x:H in the dual chamber co-deposition system. We have begun fabricating thin film transistor structures using the nitride as the insulator and a/nc-Si:H as the semiconductor material. Studies of the electron mobility in these mixed-phase TFT devices should provide important information on the nature of charge transport in these materials. Ellipsometry and FTIR measurements verifies the deposition of thin insulating a-SiN_x:H layers, and Raman spectroscopy (Figure 10) confirms that we can also inject Si nanocrystals into the nitride from the particle synthesis reactor. This Raman spectrum is consistent with measurements of free-standing silicon nanocrystals synthesized in the plasma reactor of Fig. 1a [46]. If the concentration of Si nc exceeds the percolation threshold, then transport through the nanoparticles can be studied. Three-dimensionally confined silicon nanocrystals embedded within an insulating dielectric have been investigated for their promising application in non-volatile memory devices [47,48]. Layers of nanocrystals can also be synthesized by pulsing the particle synthesis reactor. With our dual-plasma co-deposition system, Si nanocrystallite floating-gate TFTs can be fabricated at lower processing temperatures and the particle diameter and concentration can be independently controlled, providing an ideal platform to investigate Coulomb blockade and quantum confinement effects in these nanoelectronic devices.

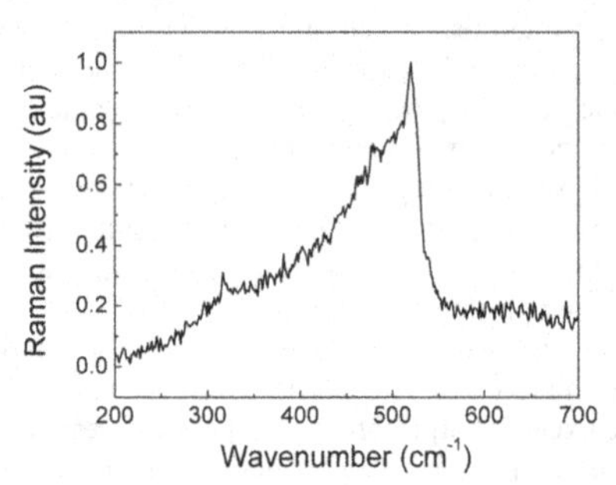

Figure 10: Raman spectrum of silicon nanocrystals embedded within a-SiN_x:H insulating film.

CONCLUSIONS

A dual chamber co-deposition system has been developed in order to synthesize mixed phase amorphous semiconductor thin films containing nanocrystalline inclusions, enabling the independent optimization of the growth conditions for each phase. Studies of undoped a/nc-Si:H films find that the dark conductivity of films with a crystal fraction is enhanced by several orders of magnitude compared the films with lower or higher crystal fractions, which has been ascribed to charge donation from the nanocrystalline inclusions to the surrounding a-Si:H matrix. In contrast, the conductivity of n-type doped a/nc-Si:H films increases monotonically with crystal fraction, and in fact there is a transition for Xc ~ 15% from Arrhenius behavior to a power-law dependence of the conductivity on temperature, consistent with charge transport via multi-phonon hopping. The dual chamber system's flexibility is illustrated by the successful fabrication of a-Si:H films containing germanium nanocrystalline inclusions, and nc-Si embedded within an insulating silicon nitride film. This system opens up the possibility of study of a wide range of mixed-phase materials that can not be readily deposited in a single PECVD chamber system.

ACKNOWLEDGEMENTS

The assistance of C. R. Perrey, J. Deneen and C. Barry Carter with high-resolution TEM imaging is gratefully acknowledged, as are insightful conversations with B. I. Shklovskii and P. Stradins. This work was partially supported by NSF grants NER-DMI-0403887, DMR-0705675, IGERT grant DGE-0114372, in part by the MRSEC Program of the NSF-DMR-0212302, NSF grant DMR-0705675, the NINN Characterization Facility, the Xcel Energy grant under RDF contract #RD3-25, NREL Sub-Contract XEA-9-99012-01 and the University of Minnesota.

REFERENCES

1. A. I. Gusev and A. A. Rempe, *Nanocrystalline Materials* (Cambridge International Scence Publishing, 2004).
2. R. Butte, R. Meaudre, M. Meaudre, S. Vignoli, C. Longeaud, J. P. Kleider, and P. R. Cabarrocas, *Phil. Mag. B* **79,** 1079 (1999).
3. Y. Lubianiker, J. D. Cohen, H. C. Jin, and J. R. Abelson, *Phys. Rev. B* **60,** 4434 (1999).
4. T. Kamei, P. Stradins, A. Matsuda, *Appl. Phys. Lett.*, **74**, 1707, (1999).
5. A. Fontcuberta i Morral, H. Hofmeister, and P. Roca i Cabarrocas, *J. Non-Crystal. Solids*, **299-302**, 284 (2002).
6. J. Yang, Lord, K., Guha, S., Ovshinski, S.R., *Materials Research Society Symposium - Proceedings*, **609**, pp. A15.4.1, (2000).
7. R. W. Collins, Ferlauto, A.S., Ferreira, G.M., Joohyun, K., et al., *Materials Research Society Symposium - Proceedings*, **762**, pp. A.10.1, (2003).
8. A. S. Ferlauto, Ferreura, G.M., Koveal, R.J., Pearce, J.M., Wronski, C.R., Collins, R.W., *et al.*, *Materials Research Society Symposium - Proceedings*, **762**, pp. A5.10, (2003).
9. C. R. Wronski, J. M. Pearce, R. J. Koval, X. Niu, A. S. Ferlauto, J. Koh, and R. W. Collins, *Mater. Res. Soc. Symp. Proc.* **715**, 459 (2002).
10. L. Mangolini, E. Thimsen, U. Kortshagen, *Nano Lett.* **5**, 655 (2005).
11. C. Anderson, C. Blackwell, J. Deneen, C. B. Carter, U. Kortshagen, and J. Kakalios, *Materials Research Society Symposium - Proceedings*, **910**, 79 (2006).
12. Y. Adjallah, C. Anderson, U. Kortshagen, and J. Kakalios, *J. Appl. Phys.* **107**, 43704 (2010).
13. L. R. Wienkes, C. Blackwell, and J. Kakalios, this proceedings.
14. C. Blackwell, C. Anderson, J. Deneen, C. B. Carter, U. Kortshagen, and J. Kakalios, *Materials Research Society Symposium - Proceedings*, **910,** 181 (2006).
15. N. A. Fuchs, *The Mechanics of Aerosols* (Dover, New York, 1964).
16. D. B. Williams and C. B. Carter, *Transmission Electron Microscopy* (Plenum, New York, 1996).
17. H. Richter, Z. P. Wang and L. Ley, *Solid State Commun.* **39**, 625 (1981); V. Paillard, P. Puech, M. A. Laguna, R. Carles, B. Kohn and F. Huisken, *J. Appl. Phys.* **86**, 1921 (1999); G. Viera, S. Huet and L. Boufendi, *J. Appl. Phys.* **90**, 4175 (2001); R. Meyer and D. Comtesse, *Phys. Rev. B* **83**, 014301 (2011).
18. L. R. Wienkes, A. Besaws, C. Anderson, D. Bobela, P. Stradins, U. Kortshagen, and J. Kakalios *Materials Research Society Symposium - Proceedings*, **1245**, 201 (2010).
19. D. C. Marra, E. A. Edelberg, R. L. Naone, and E. S. Aydil, *J. Vac. Sci. Technol.* A 16, 3199 (1998).

20. M. Stutzmann and R. A. Street, *Phys. Rev. Lett.* **54**, 1836 (1985).
21. M. Stutzmann, D. K. Biegelsen and R. A. Street, *Phys. Rev. B* **35**, 5666 (1987).
22. J. Kakalios and R. A. Street, *Phys. Rev. B* **34**, 6014 (1986).
23. J. Kakalios, R. A. Street and W. B. Jackson, *Phys. Rev. Lett.* **59**, 1037 (1987).
24. L. R. Wienkes, T. Hutchinson, C. Blackwell, and J. Kakalios, *J. Appl. Phys.* (submitted).
25. A. G. Zabrodskii and I. S. Shlimak, *Sov. Phys. Semicond.* **9**, 391 (1975); A. G. Zabrodskii, *Sov. Phys. Semicond.* **11**, 345 (1977).
26. R. M. Hill, *Phys. Stat. Sol. (a)* **35**, K29 (1976).
27. N. F. Mott and E. A. Davis, *Electronic Processes in Non-Crystalline Materials*, 2nd ed. (Oxford University Press, 1979); N. F. Mott, *Philos. Mag.* **19**, 835 (1969).
28. K. Shimakawa, *Phys. Rev. B*, **39**, 12933, (1989).
29. K. Shimakawa, K. Miyake, *Phys. Rev. Lett.* **61**, 994, (1988); K. Shimakawa, K. Miyake, *Phys. Rev. B*, **39**, 7578, (1989).
30. S. Kivelson, *Phys. Rev. Lett.* **46**,1344 (1981); S. Kivelson, *Phys. Rev. B*, **25**, 3798, (1982).
31. S. Sen and A. Ghosh, *J. Phys.:* Condens. Matter **11**, 1529, (1999).
32. S. Bhattacharya, B. K. Chaudhuri and H. Sakata, *J. Appl. Phys.* **88**, 5033 (2000).
33. K. Shimakawa, *Philos. Mag. B* **60**, 377 (1989).
34. G P Triberis, L R Friedman, *J. Phys. C: Solid State Phys.* **14**, 4631, (1981).
35. Mahi R. Singh, Graeme Bart, Martin Zinke-Allmang, *Nanoscale Res. Lett.* **5**, 501, (2010).
36. E. Mansour, K. El-Egili, G. El-Damrawi, *Physica B* **389**, 355, (2007).
37. D. Emin, *Phys. Rev. Lett.* **32**, 303 (1974).
38. A. L. Efros and B. I. Shklovskii, *J. Phys. C* **8**, L49 (1975).
39. S. Hayashi, M. Ito and H. Kanamori, *Solid State Comm.* **44**, 75 (1982).
40. A. Picco, E. Bonera, E. Grilli, M. Guzzi, M. Giarola, G. Mariotto, D. Chrastina and G. Isella, *Phys. Rev. B* **82**, 115317 (2010).
41. K. D. Mackenzie, J. H. Burnett, J. R. Eggert, Y. M. Li and W. Paul, *Phys. Rev. B* **38**, 6120 (1988).
42. R. A. Street, C. C. Tsai, M. Stutzmann and J. Kakalios, *Philos. Mag. B* **56**, 289 (1987).
43. T. Hanrath and B. A. Korgel, *J. Phys. Chem.* B **109**, 5518 (2005); S. Zhang, E. R. Hamesath, D. E. Perea E. Wijaya, J. L. Lensch-Falk and L. J. Lauhon, *Nano Lett.* **9**, 3268 (2009).
44. J.-S. Park, B. Ryu, C.-H. Moon and K. J. Chang, *Nano Lett.*, **10**, 116 (2010).
45. B. Zhang, S. Shrestha, M. A. Green and G. Conibeer, *Appl. Phys. Lett.* **97**, 132109 (2010).
46. R. Anthony and U. Kortshagen, *Phys. Rev. B* **80**, 115407 (2009).
47. S. Tiwari, F. Rana, H. Hanafi, A. Hartstein, E. F. Crabbé, and K. Chan, *Appl. Phys. Lett.*, **68**, 1377 (1996).
48. T. Z. Lu, M. Alexe, R. Scholz, V. Talelaev and M. Zacharias, *Appl. Phys. Lett.*, **87**, 202110 (2005).

Mater. Res. Soc. Symp. Proc. Vol. 1321 © 2011 Materials Research Society
DOI: 10.1557/opl.2011.813

Mixed phase silicon oxide layers for thin-film silicon solar cells

Peter Cuony[1], Duncan T.L. Alexander[2], Linus Löfgren[1], Michael Krumrey[3], Michael Marending[1], Mathieu Despeisse[1], Christophe Ballif[1]

[1] Ecole Polytechnique Fédérale de Lausanne (EPFL), Institute of Microengineering (IMT), Photovoltaics and Thin Film Electronics Laboratory (PV-Lab), Rue Breguet 2, 2000 Neuchâtel, Switzerland

[2] Ecole Polytechnique Fédérale de Lausanne (EPFL), Interdisciplinary Centre for Electron Microscopy (CIME), 1015 Lausanne, Switzerland

[3] Physikalisch-Technische Bundesanstalt (PTB), Abbestr. 2-12, 10587 Berlin, Germany

ABSTRACT

Lower absorption, lower refractive index and tunable resistance are three advantages of doped silicon oxide containing nanocrystalline silicon grains (nc-SiO_x) compared to doped microcrystalline silicon, for the use as p- and n-type layers in thin-film silicon solar cells. In this study we show how optical, electrical and microstructural properties of nc-SiO_x layers depend on precursor gas ratios and we propose a growth model to explain the phase separation in such films into Si-rich and O-rich regions as visualized by energy-filtered transmission electron microscopy.

INTRODUCTION

The tandem configuration with a hydrogenated amorphous silicon (a-Si) top cell and a hydrogenated microcrystalline silicon (μc-Si) bottom cell, also called the Micromorph configuration is a promising candidate for future large scale deployment of photovoltaics for electricity generation due to abundant source materials and scalable and low-cost deposition processes. Hydrogenated silicon oxide containing nanocrystalline silicon grains (nc-SiO_x) has attracted much interest in the last years because of different applications in thin-film silicon solar cells: First, p-type nc-SiO_x is an excellent window and anti-reflection layer, due to lower absorption coefficient and lower refractive index when compared to p-type μc-Si layers [1]. Second, n-type nc-SiO_x can be used as intermediate reflecting layer, when inserted between two sub-cells of a tandem configuration, allowing for advanced light-trapping schemes [2-5]. Third, tunable resistance of p- and n-type nc-SiO_x can help to reduce the impacts of shunts on the electrical cell parameters [1, 6].

EXPERIMENTAL DETAILS

The nc-SiO_x layers are deposited at 200 °C from a gas mixture of SiH_4, H_2, and CO_2 by plasma enhanced chemical vapor deposition, and p- and n-type doping is achieved by adding $B(CH_3)_3$ and PH_3, respectively. Fourier Transform Infrared (FTIR) absorption measurements are performed with a Nicolet 8700 system from Thermo on samples deposited on silicon wafers and absorption spectra are normalized with the layer thickness. Rutherford backscattering (RBS) and hydrogen forward scattering (HFS) measurements to determine Si, O, and H contents are carried out by the Evans Analytical Group. The density is deduced from X-ray reflectometry performed with synchrotron radiation at the PTB four-crystal monochromator beamline at BESSY II [7]. The absorption coefficient (α) and refractive index (n) are determined from fitting spectroscopic ellipsometry measurements to a Tauc-Lorentz dispersion model including a surface roughness layer. Two aluminum contacts are evaporated onto the samples, in order to measure the electrical in-plane dark conductivity (σ) after 90 minutes annealing at 180 °C in vacuum. The crystalline fraction of the Si-rich phase (hence not taking into account the O-rich phase) is evaluated by Raman spectroscopy [8].

RESULTS AND DISCUSSION

Figure 1a shows infrared absorption peak of Si-O-Si stretching mode at 1050 cm^{-1} [9, 10], which is increasing with increasing CO_2/SiH_4 precursor gas ratio due to increased oxygen content in the nc-SiO_x films. From this absorption peak one can estimate the oxygen content and we found the best correlation to RBS measurements when using the proportionality constant $A_{SiO} = 1.48 \cdot 10^{19}$ cm^{-2} proposed by He et al. [11]. Figure 1b demonstrates that oxygen incorporation is also enhanced by increasing H_2 source gas dilution. To explain this effect, Iftiquar et al. [10] argue that in the plasma, atomic hydrogen inhibits the backward reaction of $CO + O \rightarrow CO_2$ by forming OH. Because highly reactive and electronegative O hardly leaves the plasma it is speculated that oxygen mainly reaches the growing nc-SiO_x film in the form of OH. In this context it is interesting to note that the hydrogen content does not correlate with H_2 dilution in the plasma but shows a correlation to silicon content of the nc-SiO_x films which can be explained by free energy models predicting that H in SiO_x films will be only bound to Si with an absence of O-H bounds [12].

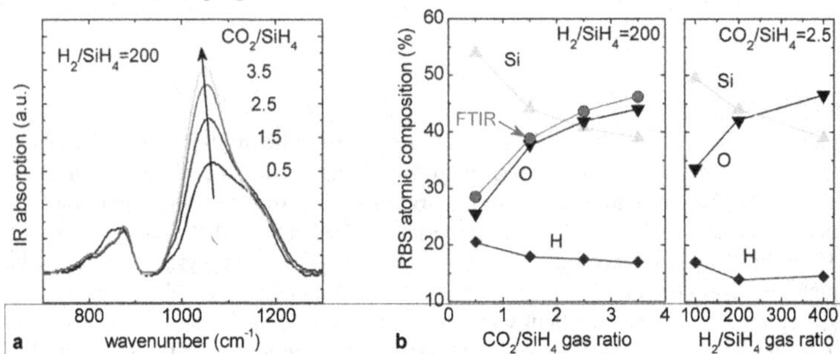

Figure 1. a) Infrared absorption peak of Si-O-Si stretching mode at 1050 cm^{-1} of 100 nm thick nc-SiO_x layers. **b)** Silicon, oxygen and hydrogen content measured by RBS and HFS for different source gas ratios and oxygen content calculated from FTIR measurements for comparison.

Figure 2a shows dark conductivity (σ), silicon crystalline fraction ($Si\ Rc$) and absorption coefficient (α) of p- and n-type nc-SiO_x layers as a function of CO_2/SiH_4 gas flow ratio. In this experiment, the H_2-dilution is set to 200 for p-type layers and 100 for the n-type layers, because in the p-i-n configuration of μc-Si cells it is common to use higher H_2 dilution for the p-layer because it grows on a ZnO substrate, with longer incubation phase compared to n-layer which grows directly on a μc-Si layer, favoring the growth of microcrystalline material. For both p- and n-type layers, the dark in-plane conductivity is around 10 $S \cdot cm^{-1}$ for an oxygen free sample, and decreases rapidly with higher CO_2/SiH_4 gas ratio, due to an increasing amount of oxygen incorporated in the nc-SiO_x film. Furthermore, increasing CO_2/SiH_4 gas ratio reduces the Raman crystalline fraction of the Si-rich phase in the nc-SiO_x layers, and decreases the optical absorption coefficient α. Figure 2b shows the results obtained for different doping concentrations, ranging from 0 ‰ to 50 ‰, with a constant CO_2/SiH_4 gas ratio of 0.5. p-type doping with trimethylboron (TMB) rapidly reduces the crystalline fraction of the silicon phase, and a maximum dark conductivity is obtained at TMB/SiH_4 = 8 ‰. This is different for n-type doping with PH_3, where dark conductivity is continuously increasing with increasing doping gas concentration as phosporous does not reduce the crystalline fraction of the silicon phase. This amorphizing effect of TMB limits the oxygen content in nc-SiO_x films to x~0.5 before the films become too resistive to be used in thin-film silicon solar cells. For n-type nc-SiO_x layers, where high phosphorous doping concentrations can partially make up for lower conductivity due to oxygen incorporation, we achieved nc-SiO_x films with x~1 which have still a transverse conductivity high enough for the use in thin-film silicon solar cells.

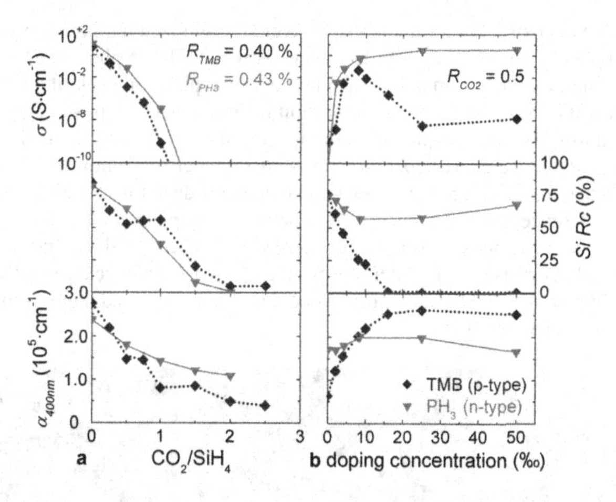

Figure 2. Evolution of dark conductivity (σ), silicon Raman crystalline fraction (*Si Rc*) and absorption coefficient (α) of p- and n-type nc-SiO_x layers deposited with different CO_2/SiH_4 gas ratios (**a**) and doping concentrations (**b**).

Figure 3a shows the decrease in dark conductivity due to air exposure of two nc-SiO_x layers with different oxygen contents. The initial conductivity can be recovered by annealing the samples during 90 minutes at 180° in vacuum. Water vapor penetrating into nano-pores of nc-SiO_x films is believed to be the major reason for this reversible in-plane conductivity degradation. Figure 3b shows X-ray reflectometry measurements revealing a lower critical angle for nc-SiO_x than for a-Si suggesting a lower density [13] and strengthening the hypothesis of nano-scale porosity in nc-SiO_x layers.

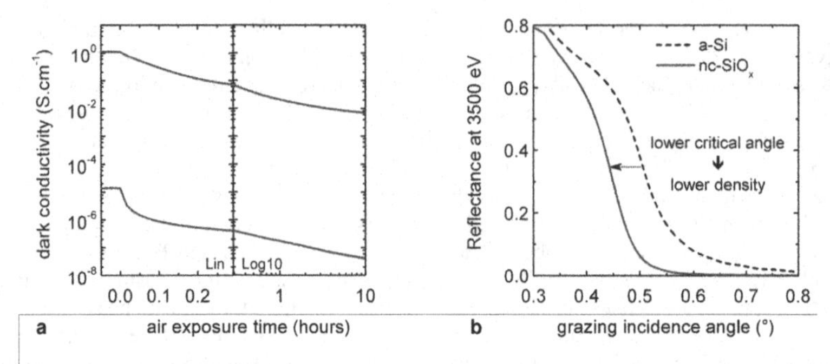

Figure 3. a) Decreasing dark conductivity due to air exposure of two different 100 nm thick nc-SiO_x layers. **b)** X-ray reflectometry measurements at 3500 eV (far away from the Si absorption edge) indicating significantly less dense material for nc-SiO_x when compared to a-Si material.

Figure 4 shows the nanostructure of the nc-SiO_x layer as visualized in plan-view by energy-filtered transmission electron microscopy (Si-rich phase = white; O-rich phase = dark). The image is obtained using electrons from a 4 eV window over the plasmon peak of silicon at ~17.5 eV, after removal of the background contribution from the plasmon peak of SiO_2[14]. While layers produced with low H_2 dilution appear as homogenous SiO_x mixture (Figure 4a), this technique reveals a pronounced phase separation on the nanometer scale into Si-rich regions surrounded by an O-rich material for samples that have been produced with a high H_2 dilution (Figure 4b-c). The difference in silicon particle sizes between sample b and c is due to the different silicon content of the nc-SiO_x films. It is interesting to note that the homogenous $SiO_{x\sim1.4}$ (sample a) and the phase-separated nc-$SiO_{x\sim1}$ (sample b), both have a refractive index of ~ 1.8 but different SiO_x stoichiometry, which could be due to the increased nano-porosity in nc-SiO_x samples with phase separation.

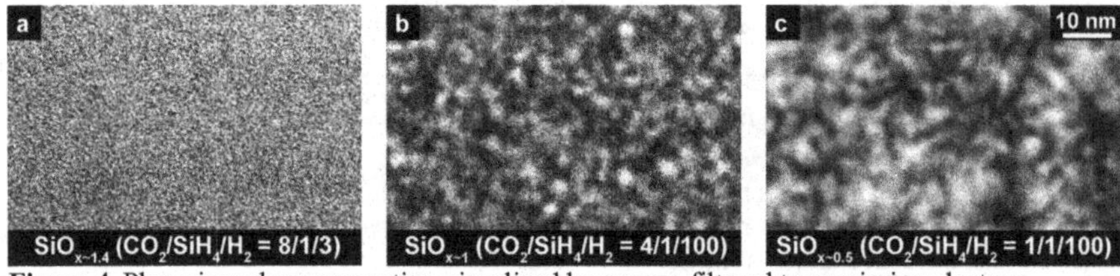

Figure 4. Plan-view phase separation visualized by energy-filtered transmission electron microscopy (Si phase = white) in 50-70 nm thick nc-SiO_x layers produced with different $CO_2/SiH_4/H_2$ gas ratios.

Mixed-phase nc-SiO_x films have been studied previously, but the energetically favorable phase separation is usually obtained after high-temperature annealing of initially homogenous SiO_x films [14-16]. In our case, we show that the phase separation is also possible at low substrate temperatures of 200 °C, if produced with highly H_2 diluted plasmas. This growth mechanism can be explained with the surface diffusion model [17], as developed for mixed-phase a-Si/μc-Si films. For low H_2 dilution all the Si and O atoms arriving from the plasma stick to the surface leading to a homogenous SiO_x film (Figure 5a), whereas high H_2 dilution enables phase-separation by increasing adatom mobility via surface heating and passivation of surface dangling bonds (Figure 5b).

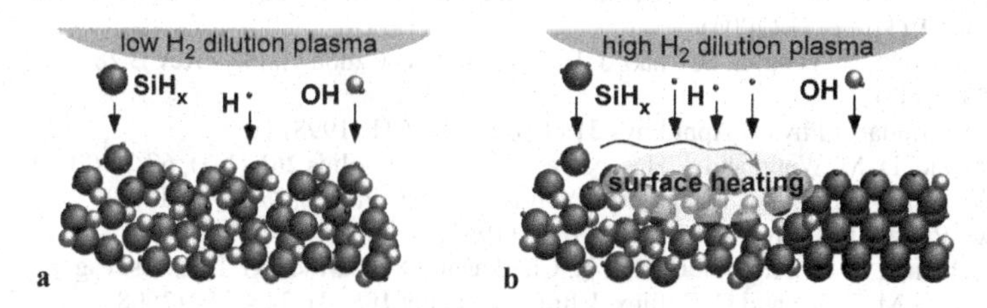

Figure 5. Surface diffusion model [17] with low adatom mobility in the case of low H_2 dilution leading to homogenous SiO_x material (**a**) and high adatom mobility due to surface heating and H-bonding of surface dangling bonds in the case of high H_2 dilution leading to Si/SiO_2 mixed-phase structure (**b**).

CONCLUSIONS

In this contribution we have related precursor gas ratios to optical, electrical and microstructural characteristics of nc-SiO_x films. We evidenced a possible nano-porosity in these films which makes them sensitive to air exposure. With energy-filtered transmission electron microscopy we demonstrated a phase separation within nc-SiO_x films into Si-rich and O-rich regions. H_2 source gas dilution has been identified as a key parameter for the phase separation, whereas the size of the Si-rich regions depends mostly on the silicon content in nc-SiO_x films. To explain the phase-separation we used a modified surface diffusion model, originally developed for mixed-phase a-Si/μc-Si materials.

ACKNOWLEDGMENTS

The authors acknowledge support by the Swiss Federal Energy Office (Grant No. 101191).

REFERENCES

1. P. Cuony, M. Marending, D. T. L. Alexander, M. Boccard, G. Bugnon, M. Despeisse and C. Ballif, Appl Phys Lett **97** (21), 213502 (2010).
2. P. Buehlmann, J. Bailat, D. Domine, A. Billet, F. Meillaud, A. Feltrin and C. Ballif, Appl Phys Lett **91**, 143505 (2007).
3. T. Grundler, A. Lambertz and F. Finger, physica status solidi (c) **7** (3-4), 1085–1088 (2010).
4. D. Dominé, P. Buehlmann, J. Bailat, A. Billet, A. Feltrin and C. Ballif, Phys Status Solidi-R **2** (4), 163–165 (2008).
5. P. Buehlmann, J. Bailat, A. Feltrin and C. Ballif, Photovoltaic Materials and Manufacturing Issues **1123** (2009).

6. M. Despeisse, G. Bugnon, A. Feltrin, M. Stueckelberger, P. Cuony, F. Meillaud, A. Billet and C. Ballif, Appl Phys Lett **96**, 073507 (2010).
7. M. Krumrey, M. Hoffmann, G. Ulm, K. Hasche and P. Thomsen-Schmidt, Thin Solid Films **459** (1-2), 241-244 (2004).
8. C. Droz, E. Vallat-Sauvain, J. Bailat, L. Feitknecht, J. Meier and A. Shah, Sol Energ Mat Sol C **81** (1), 61–71 (2004).
9. G. Lucovsky, J. Yang, S. S. Chao, J. E. Tyler and W. Czubatyj, Phys Rev B **28** (6), 3225-3233 (1983).
10. S. M. Iftiquar, J Phys D Appl Phys **31** (14), 1630–1641 (1998).
11. L. N. He, D. M. Wang and S. Hasegawa, J Non-Cryst Solids **261** (1-3), 67-71 (2000).
12. F. W. Smith and Z. Yin, J Non-Cryst Solids **137-138** (Part 2), 871-874 (1991).
13. L. G. Parratt, Phys Rev **99** (5), 1635-1635 (1955).
14. S. Schamm, C. Bonafos, H. Coffin, N. Cherkashin, M. Carrada, G. Ben Assayag, A. Claverie, M. Tence and C. Colliex, Ultramicroscopy **108** (4), 346–357 (2008).
15. G. Nicotra, S. Lombardo, C. Spinella, G. Ammendola, C. Gerardi and C. Demuro, Appl Surf Sci **205** (1–4), 304–308 (2003).
16. F. Iacona, C. Bongiorno, C. Spinella, S. Boninelli and F. Priolo, J Appl Phys **95** (7), 3723–3732 (2004).
17. A. Matsuda, Thin Solid Films **337** (1–2), 1–6 (1999).

Mater. Res. Soc. Symp. Proc. Vol. 1321 © 2011 Materials Research Society
DOI: 10.1557/opl.2011.1096

Silicon Thin-Films from Nanoparticle dispersion: Tailoring Morphological, Electrical and Optical Characteristics.

Etienne Drahi, Sylvain Blayac and Patrick Benaben
Centre Microélectronique de Provence – Georges Charpak, Ecole Nationale Supérieure des Mines de Saint Etienne, 13541 Gardanne, France

ABSTRACT

Amorphous and microcrystalline silicon are currently used for electronic devices such as solar cells and thin-film transistors. This paper shows that silicon nanoparticle dispersion has the potential to be used as source material for polycrystalline silicon thin-film thus opening a route to solution processed silicon devices. After deposition, a classical thermal or microwave annealing step is used to induce the coalescence of the silicon nanoparticles. Both sintering techniques are studied in terms of morphology, electrical and optical properties.

INTRODUCTION

Oil exhaustion as well as growing needs in energy guarantee a bright future for renewable energies. Solar energy is very promising for heat and electricity generation. Nevertheless technological efficiency improvements are necessary in order to make solar energy competitive. This motivated the development of new generations of solar cells with the main objective to lower the cost of photovoltaic energy.

Amorphous (a-Si) and microcrystalline (μc-Si) silicon allow much thinner and cheaper devices than crystalline silicon (c-Si) solar cells technology. Nevertheless, the use of vacuum processes raises the cost of large area devices.

Solution processed technologies are already used to lower the cost of the processes for organic or chalcogenide solar cells. Recently, silicon nanoparticle dispersion has been used to lower the contact resistance between the contact layer and the metallic fingers [1] or to enhance the UV absorption [2] of a solar cell. Thus, the idea of an entire printed silicon solar cell is emerging. Annealing is mandatory for post deposition film properties as a morphological and functional properties restoring step. In an industrial point of view both thermal budget and processing time are key evaluation parameters. Laser annealing [3,4] or ALuminum Induced Layer Exchange (ALILE) [4] are two promising candidates. Nevertheless, laser annealing is not compatible with very large area processes, while ALILE process applied to nanoparticles lowers the process speed by two orders of magnitude. Microwave annealing shows a good potential for both time and temperature reduction [5]. This paper reports a comparison between thermal versus microwave annealing.

EXPERIMENT

This part explains the experimental settings from suspension fabrication from dry silicon nanoparticles, deposition of this suspension and annealing processes and the characterizations made on the samples.

Spherical undoped silicon nanoparticles (99% metals basis) with BET (Brunauer-Emmett-Teller) diameter of 32.19 nm and size dispersion between 20 and 200 nm were used.

Native oxide shell thickness was estimated to about 0.3-0.5 nm by thermogravimetric analysis (TGA) coupled with differential thermal analysis (DTA) experiment. Nanoparticles were dispersed in ethanol with a concentration of 10 % weight. They were then sonicated with an ultrasonic probe (VCX500) during 15 min (1 s pulse on – 1 s pulse off) in a refrigerated bath.

Two types of substrates have been used: thermally oxidized silicon wafer (for high temperature processes > 1170 °C) and quartz (for samples annealed below 1200 °C). The substrate was cleaned with ethanol and dried under N_2 flux. The dispersion was then spin coated at 2000 rpm during 30 s with no specific acceleration ramp. After deposition, the sample was dried 15 min in an atmospheric oven at 90 °C.

Two types of annealing techniques have been studied: thermal (SETARAM 1612 dilatometer) in argon and vacuum and microwave (multimode 2.45 GHz) annealing in nitrogen (argon is prohibited due to possible creation of plasma when exposed to microwaves).

Microwave annealing system and temperature measurement procedure of the silicon nanoparticle layer by pyrometry have been described extensively in [6]. For microwave, target temperatures of: 800 (M800), 900 (M900), 1000 (M1000), 1100 (M1100) and 1170 °C (M1170) in N_2 atmosphere were used with an average heating ramp around 20 °C/min. As a first approach, no constant temperature step was applied given the complexity of the temperature regulation with this kind of material.

For thermal annealing, ramps of 50 °C/min have been applied to reach temperatures of 1000 (T1000), 1100 (T1100), 1150 (T1150) and 1200 °C (T1200) followed by a 1 h constant temperature step. In a second process after reaching at 1200 °C for 5 min, the samples were maintained at 1100 °C for 1 h (T1200-1100). Temperature of the ambient atmosphere was measured by a thermocouple placed in the oven cavity.

The morphology of the layers was characterized with a Carl Zeiss ultra 55 Scanning Electron Microscope (SEM) and a Jobin-Yvon LabRam HR800 Raman microscope (around 2 µm diameter spot). For Raman microscopy, a 488 nm argon laser was used for its low penetration depth allowing a probing of the sole coalesced nanoparticles thin-film. A particular attention has been brought to laser power in order to prevent laser-induced recrystallization of the layer during the measurement.

Electrical measurements were made with a Keithley 4200 semiconductor parameter analyzer using large spring probes to guarantee a large contact surface and less damage of the nanoparticles layer.

Optical measurements in the range of 300 nm to 2000 nm were made using Perkin Elmer Lambda 950 UV/VIS optical spectrometer. A quartz substrate without nanoparticles layer was used as reference.

DISCUSSION

Layers morphology

a- Impact of native oxide on coalescence of silicon nanoparticles

DTA/TGA experiment shows an important mass loss from 1200 °C conjugated with an important variation of heat flow showing the beginning of melting of the smaller nanoparticles. Nevertheless, melting of silicon nanoparticles happens at higher temperatures than expected [7,8]. This effect is attributed to the native oxide around the nanoparticles which inhibits the sintering under 1100 °C, where the reduction of silica by silicon is allowed [9,10] given the following reaction:

$$Si_{(Solid)} + SiO_{2(Solid)} \leftrightarrow 2SiO_{(Gas)}$$

Coblenz showed that the presence of an oxide around the silicon nanoparticles inhibits surface diffusion but also allows the densification of the layer [11].

b- Thermal annealing vs. microwave annealing: morphology comparison

Qualitative SEM observations show very high roughness of the layers even for high annealing temperatures (Figure 1). Figure 1a shows morphology of dried layer. The average thickness of the layer is around 10 µm with a very high porosity due to agglomerates of nanoparticles in the suspension and after spin-coating process.

For classical thermal annealing the first sintering stage, neck formation [12], is observed for T1000. For T1100 under vacuum, sintering is at an advanced stage with a well grown neck while T1100 in argon shows no clear evolution. T1200 process in vacuum (Figure 1b) exhibits faceted silicon nanocrystals on a rough silicon layer.

Microwave sintering does not follow the same evolution than for thermal annealing. Neck formation is observed at temperature as low as 800 °C. SEM observations show no significant evolution of the nanostructure of the layer between 800 °C and 1100 °C. Nevertheless, high stage of sintering is reached for M1170 (Figure 1c): a continuous layer with low open porosity is obtained.

Figure 1. Secondary electrons SEM pictures of: a) as deposited silicon nanoparticles layer and sintered silicon nanoparticles layers: b) T1200 process under vacuum and c) M1170 process.

Raman studies were carried out in order to quantify crystallinity of the layers. Stoke peak positions were extracted from the spectra. Lorentzian fit was used to extract the full width at half maximum (FWHM). These two quantities as a function of annealing temperature are reported on the Figure 2.

For thermal annealing (Figure 2a), the experiment confirms that under 1100 °C there is no coalescence or grain growth. For T1100 process in argon no evolution is seen from the as dried layer. T1100 in vacuum exhibits an important shift of the peak position and a narrowing of the peak showing that thermal annealing under vacuum allows coalescence at lower temperatures than in argon atmosphere. At 1200 °C, the peak position and FWHM are c-Si-like (dash and dot lines representing respectively the position of the peak and FWHM).

In Figure 2b microwave and thermal annealing under vacuum are compared. Thermal annealing under vacuum is chosen for comparison because of its better crystallinity at lower temperature than thermal annealing in argon. The experiment shows that for microwave annealing, a change of crystallinity appears at temperatures as low as 800 °C and crystallinity improves almost linearly. This lowering of sintering temperature is referred in literature as "microwave effect". Nevertheless, even at high temperature FWHM is still broad compared with thermal annealing, which means there is a larger distribution of crystalline sizes in the layer.

This difference in crystallinity is consistent with SEM observations shown in Figure 1. T1200 process under vacuum shows faceted crystals while M1170 process shows a continuous layer with no particular crystalline structures.

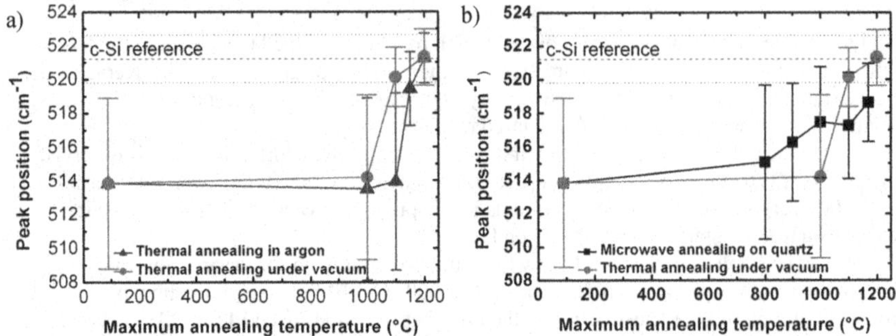

Figure 2. Maximum peak position and FWHM (vertical bars) of the Raman spectra as a function of maximum annealing temperature: a) comparison of thermal annealing under vacuum and in argon atmosphere; b) comparison of microwave (on quartz) and vacuum thermal annealing.

In this paper, the nanoparticle layers are very thin in comparison with the thickness of the substrate. Depending of the substrate, it could absorb, reemit or be transparent to microwaves. Thus, its coupling effect on sintering processes has to be taken into account. Its effect will be further studied in the future.

Layers electrical properties

I(V) measurements under dark made with 2 mm contacts separation are shown in Figure 3. For measurements made on oxidized silicon substrates (T1200-1100 and T1200), the leakage current through the oxide and the wafer was also measured.

As-deposited nanoparticles and annealed layers until 1100 °C show purely dielectric behavior with only leakage current measured around 0.1 pA. This insulating behavior can be attributed to a low coalescence stage limiting charge transfer from a nanoparticle to another.

M1170 and T1200-1100 under vacuum processes exhibit a very high resistivity (Figure 3a and 3b). Nevertheless, for T1200-1100 under vacuum, leakage current through the substrate (oxidized silicon wafer) is higher than the current travelling through the nanoparticles layer.

In agreement with the microstructure observed with SEM (Figure 1b), T1200 process under vacuum shows a low resistivity with a sheet resistance of 70 ohm. For this sample, substrate leakage current through the substrate (oxidized silicon wafer) much smaller than the current travelling through the layer has been measured.

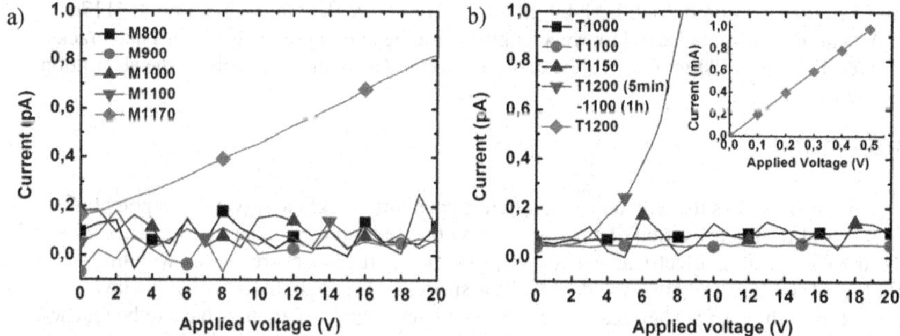

Figure 3. I(V) curves of: a) microwave annealed layers on quartz (contacts separation: 2 mm) under dark, b) thermal annealed layers under dark (contacts separation: 2 mm)

These results show that recovering of electrical properties is possible for solution processed silicon nanoparticles layers. Thermal annealing under vacuum allowed the formation of a low resistivity thin-film. The measured current of microwave annealed layers is very small in spite of better crystallinity at lower annealing temperatures than thermal annealing. Up to date, because of technical limitations, a constant temperature step could not be applied and is still under development. This stabilization step should help to improve both crystallinity and resistivity.

Layers optical properties

Transmission intensity was measured on Si/quartz samples versus bare quartz reference (Figure 4). No integrating sphere was used, thus the reflective intensity at the air/semiconductor interface was not measured. This might explain low transmittance at high wavelengths.

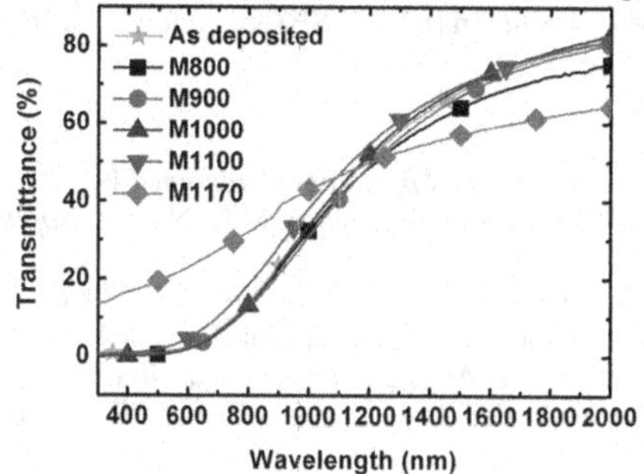

Figure 4. Transmittance spectra of the microwave annealed samples.

For annealing temperatures lower than 1100°C, no fundamental change in the transmission spectrum can be seen. This behavior is consistent with SEM observations. For M1100 and M1170 processes, a change of visual opacity of the layer is seen which is coherent with the transmittance measurements, microstructure changes and thickness variation from around 10 µm (as deposited, M800, M900, M1000, M1100 samples) to around 2 µm (M1170 sample). Gain in transmittance at low wavelengths could be due to presence of holes and cracks in the layer. Further investigations will be carried out in order to qualify precisely the absorption coefficient.

CONCLUSIONS

This study shows that silicon nanoparticle dispersion in organic solvents is a potential source material for the formation of silicon thin-film for electronic applications and especially for photovoltaic devices. Electrical and optical properties of the layers are related with the morphology. First stages of sintering do not allow sufficient charge transfer within the layer. Because of the native oxide shell around the nanoparticles, high temperatures have to be reached in order to activate high coalescence stage such as densification of the layer. At the present time even though electrical measurements do not show an advantage of using microwave annealing, Raman study clearly shows that nanoparticles coalescence takes place at lower temperature. Microwave annealing is thus a promising candidate to lower the annealing temperatures and processing time even with the presence of native oxide. In order to assess these observations, further investigations will be made: the maximum temperature will be maintained and different heating ramps applied in order to evaluate the coalescence dynamics. Sintering of oxide free nanoparticles will take place at lower temperature and coupling effect of the substrate will also be further investigated.

ACKNOWLEDGMENTS

This work is financially supported by the French National Research Agency (ANR) through the INXILICIUM project. Authors are particularly thankful to S. Saunier, F. Valdivieso and D. Zymelka (ENSM-SE) for their help with the microwave oven, the dilatometer and their background of the sintering processes. Helpful discussions with P. Grosseau (ENSM-SE) and the experimental assistance of M-C. Bartholin (ENSM-SE) are gratefully acknowledged.

REFERENCES

1. H. Antoniadis, *34th IEEE Photovoltaic Specialists Conference* **1-3**, 2091-2095 (2009).
2. M. Stupca, M. Alsalhi, T. Al Saud, A. Almuhanna, M.H. Nayfeh, *Applied Physics Letters* **91**, (2007).
3. Bet, Kar, *Materials Science and Engineering B* **130**, 228–236 (2006).
4. R. Lechner, PhD. Thesis, Technische Universität München 2009
5. J.H. Ahn, J.N. Lee, Y.C. Kim, B.T. Ahn, *Current Applied Physics* **2**, 135-139 (2002).
6. D. Zymelka, S. Saunier, J. Molimard, D. Goeuriot, *Advanced Engineering Materials* **13** (2011).
7. M. Wautelet, *Solid State Communications* **74**, 1237-1239 (1990).
8. A.N. Goldstein, *Applied Physics A: Materials Science & Processing* **62**, 33-37 (1996).
9. C. Greskovich, *Journal of Materials Science* **16**, 613-619 (1981).
10. N.J. Shaw & A.H. Heuer, *Acta Metallurgica* **31**, 55-59 (1983).
11. W.S. Coblenz, *Journal of Materials Science* **25**, 2754-2764 (1990).
12. J. Frenkel, *Journal of Physics* **9**, 385-391 (1945).

Mater. Res. Soc. Symp. Proc. Vol. 1321 © 2011 Materials Research Society
DOI: 10.1557/opl.2011.1250

Electric field effect in amorphous semiconductor films assembled from transition-metal-encapsulating Si clusters

N. Uchida [1], T. Miyazaki[2], Y. Matsushita[1, 3], K. Sameshima[1, 3], and T. Kanayama [3, 4]

[1] Nanodevice Innovation Research Center, National Institute of Advanced Industrial Science and Technology, 1-1-1 Higashi, Tsukuba, Ibaraki 305-8562, Japan
[2] Nanosystem Research Institute, National Institute of Advanced Industrial Science and Technology,, 1-1-1 Umezono, Tsukuba, Ibaraki 305-8568, Japan
[3] Institute of Applied Physics and Doctoral Program in Applied Physics, University of Tsukuba, 1-1-1 Tenoudai, Tsukuba, Ibaraki 305-8573, Japan
[4] National Institute of Advanced Industrial Science and Technology,, 1-1-1 Umezono, Tsukuba, Ibaraki 305-8568, Japan

ABSTRACT

We synthesized amorphous semiconductor films composed of Mo-encapsulating Si clusters ($MoSi_n$: n~10) on solid substrates. The $MoSi_{10}$ films had Si networks similar to hydrogenated amorphous Si and an optical gap of 1.5 eV. Electron spin resonance signals were not observed in the films indicating that dangling bonds of Si were terminated by Mo atoms. We fabricated thin-film-transistors using the $MoSi_{10}$ film as a channel material. The electric field effect of the film was clearly observed. This suggests that the density of mid-gap states in the film is low enough for the field effect to occur.

INTRODUCTION

Carrier transport properties of amorphous semiconductors are sensitive to structural disorders. Hydrogenated amorphous Si (a-Si:H) is useful for a wide range of applications, e.g., thin film transistors (TFT), sensors, and solar cells, however structural disorders are present in the Si sp^3 bonding network in a-Si:H [1]. As an alternative way to compose Si-based amorphous semiconductor films with reduced disorders, we have demonstrated the synthesis of amorphous films by using deposition of transition metal (M= Mo, Nb and W) encapsulating Si clusters, MSi_n, where n=7–16 on solid substrates [2, 3]. The MSi_n and the hydrogenated clusters (MSi_nH_x) were actually synthesized through the reaction between the M vapor and the silane (SiH_4) gas [4, -7]. The Si cage structures were observed for the MSi_n ($n \geq 10$) and stabilized by covalent bonding between the M atoms and the Si cages. We confirmed for M=W by X-ray absorption spectroscopy that the MSi_n films were actually composed of random arrangement of unit MSi_n clusters with n= 8 – 10 [8]. The MSi_n films have tunable optical gap (E_{og}) of 0.4 -1.8 eV by changing unit clusters. Raman spectra show that MSi_n films have amorphous Si (a-Si) networks, but do not contain hydrogen. The M atoms stabilize a-Si networks and terminate the Si dangling bonds due to formation of MSi_n clusters as unit structures. The aspect of the cluster-assembled material was supported by *ab-initio* structural modeling of the MSi_n film [9].While the MSi_n film has amorphous Si networks, we observed superior properties owing to a reduction in electronic disorder in Si networks as a result of the MSi_n clusters being used as the buildig blocks. For example, The MSi_n film has higher carrier mobility, which is 1.6－32 cm^2/Vs for hole and 0.69

$-18\ cm^2/Vs$ for electron, than a-Si:H films (hole:<~0.1, electron:<~1cm^2/Vs) [2, 3]. Therefore, the MSi$_n$ cluster film is an attractive material for TFT applications.

In this paper we demonstrate the possibility of MSi$_n$ films as a channel material of TFT. The MoSi$_n$ film (n~10) is selected as a first trial of TFT fabrication, because it is expected that the film has relatively larger energy gap reflecting the highest occupied and the lowest occupied orbital gap of free $MoSi_{10}$ clusters, which is 1.6 eV by *ab initio* calculation.

EXPERIMENT

The MoSi$_n$ cluster deposition system used in this work was reported previously [2]. A key issue for obtaining the MoSi$_n$ films is to prevent the transition metals from coalescing with each other but to synthesize the materials composed of MoSi$_n$ clusters. For this purpose, we deposited $MoSi_nH_x$ clusters onto solid substrate followed by annealing at 500°C of 10 minutes for dehydrogenation. The $MoSi_nH_x$ clusters were formed by laser ablation of Mo atoms from Mo disilicide targets in the presence of silane SiH_4 gas of 50 Pa. Composition of the MoSi$_n$ film was estimated to be n~10 by using Rutherford back scattering and X-ray photoelectron spectroscopy. The XPS spectra were obtained with a system using MgKα x-ray source (1253.6 eV).We also used the spectra to investigate bonding states of Si and Mo atoms. To observe the structural alteration of the $MoSi_{10}$ film due to the thermal annealing, we measured Raman spectra of the film using 514 nm Ar CW-laser with a power of 0.14 mW focused at ~2 μm in diameter. To observe electronic structures, we measured optical absorption spectra of $MoSi_{10}$ films in the range from ultraviolet (240 nm) to infrared (2500 nm). In order to measure the field-effect characteristic of $MoSi_{10}$-film TFTs, the $MoSi_{10}$ films were deposited on thermally oxidized Si substrates and contacted with Al electrodes.

RESULTS & DISCUSSION

Figures 1 shows XPS spectra of Si 2p and Mo 3d from a $MoSi_{10}$ film before and after annealing at 500℃. The Si 2p signals were observed at 101.7 eV for the as-deposited sample and 100.6 eV for the annealed sample. These peaks show higher binding energy than those of the bulk Si (99.8 eV) and $MoSi_2$ (99.36 eV) [10].The Mo $3d_{5/2}$ peaks, which were observed at 229.4 eV for the as-deposited sample and 228.6 eV for the annealed sample, also show higher binding energy shift than those of Mo (227.9 eV) and $MoSi_2$ (227.68 eV) [10]. This indicates that Si and Mo atoms form particular Mo-Si bonds in the $MoSi_{10}$ film. This view is supported by *ab initio* calculation [9]. We found that the Si 2p and Mo 3d peaks show lower shift of -1.1 and -0.8 eV by the annealing of 500℃. This suggests that the structural alteration of the $MoSi_{10}$ film is induced by the annealing.

To investigate the structural alteration of the $MoSi_{10}$ film by the annealing, we measured Raman spectra of the films, as shown in Fig. 2. The transverse optical (TO)-phonon peak position is at 463 and 473 cm^{-1} in the $MoSi_{10}$ films, lower than that in hydrogenated a-Si (480 cm^{-1}). This lower shift is commonly observed in Raman spectra of other MSi$_n$ films [2, 3] due to containing a heavy M atom in the unit cluster. After annealing of the $MoSi_{10}$ film, we found narrowing and higher frequency shift of the TO phonon peak. This narrowing, which is estimated at 6 cm^{-1} from difference of the FWHM value, suggests that the a-Si networks obtained an ordering of Si bond angles by the annealing from analogy to the hydrogenated a-Si [11]. The higher frequency shift of 10 cm^{-1} suggests the formation of more tight Si-Si bonds.

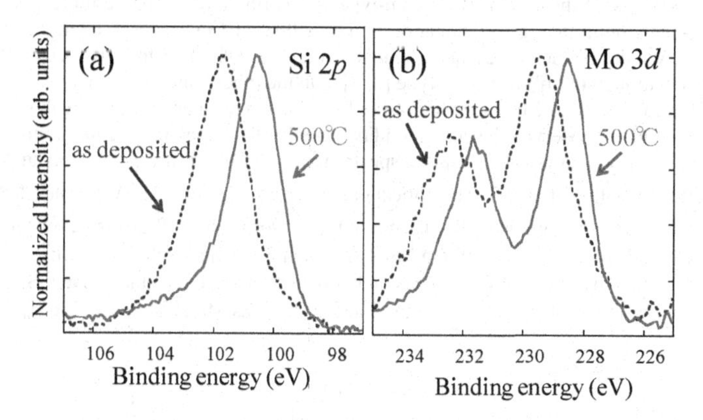

Figures 1. XPS spectra of (a) Si 2*p* and (b) Mo 3*d* from the $MoSi_{10}$ film before and after annealing at 500℃.

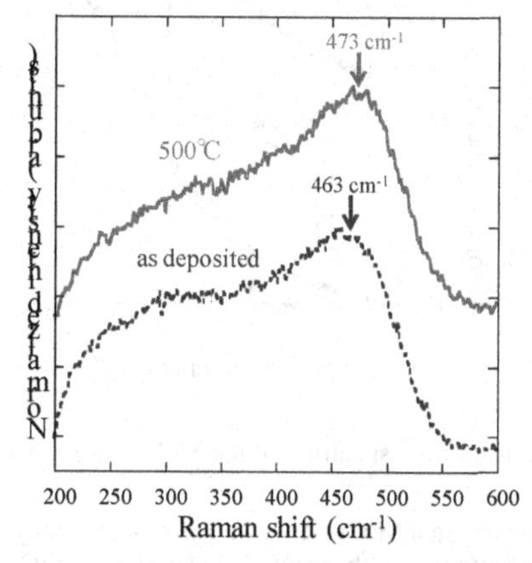

Figure 2. Raman spectra of the film before and after annealing of 500℃.

To use the $MoSi_{10}$ film as a TFT channel, density of mid-gap states, such as Si dangling bonds, must be reduced, because the carrier transport properties in a-Si networks are constrained by the trapping of carriers in mid-gap states [12]. Dangling bonds in usual a-Si are terminated by hydrogen atoms. The $MoSi_{10}$ films did not contain Si-H bonds because Si-H vibrations, typically observed at 2000–2200 cm^{-1} in a-Si:H films by the Raman and also Fourier transform infrared spectroscopy [12], were not detected in the spectra of $MoSi_n$ films [2]. However, neutral dangling bond defects (T_3^0) were not observed in X-band electron spin resonance spectra (ESR) of $MoSi_n$ (n= 7-16) films at room temperature (RT), while the T_3^0 center was observed at g= 2.0055 in a-Si films. In the case of B doped p-type a-Si:H films, when the Fermi level is close to the valence-band mobility edge, a resonance with g =2.008-2.013 was observed in the ESR spectra [13] . We did not detect resonance signals with g= 2.008-2.013 in ESR spectra of the $MoSi_n$ films, although the films were p-type [2, 3]. On the other hand, Stuzmann et al. reported that Mo impurities, such as Mo^{5+} ions, in a-Si:H films have a resonance signals with g= 1.92-1.93 [15]. We did not observed the corresponding signals of Mo ions in the $MoSi_n$ films.

Figure 3 shows an optical absorption spectrum of a $MoSi_{10}$ film deposited on an Al_2O_3 substrate after annealing at 500°C. The optical gap was estimated to 1.51 eV for the $MoSi_{10}$ film from the absorption spectrum using the Tauc plot [2]. Absorption coefficients below the optical gap of 1.5 eV is $3.0\times10^3 - 2.0\times10^4$ cm^{-1} is higher than the typical values of a-Si:H ($<10^3 cm^{-1}$) [16]. This indicates that the $MoSi_{10}$ film has higher density of mid-gap states than a-Si:H. The source for the mid-gap states may be structural defects such as Mo clusters and $MoSi_n$ ($n\neq10$) clusters.

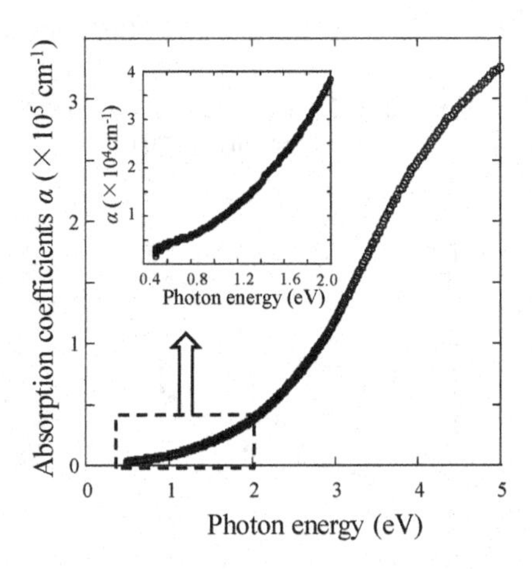

Figure 3. Optical absorption spectrum of the $MoSi_{10}$ film after annealing at 500℃.

The structure of the fabricated $MoSi_{10}$-film TFT is schematically drawn in Fig. 4(a). We deposited $MoSi_nH_x$ clusters onto thermally oxidized Si substrates (SiO_2= 200 nm thick, N-type, 0.01 Ωcm, cleaned by SPM and HPM treatment) followed by annealing at ~500°C of 10 minutes in an ultra high vacuum for dehydrogenation. The thickness of the $MoSi_n$ film was ~15 nm by deposition of 3 hours. Aluminum thin films were deposited on the $MoSi_{10}$ film to form Source/Drain electrodes, leaving the channel width W= 300 um and the channel length L= 100 um. The Si substrate was used as the gate electrode. Figure 4(b) shows drain current (I_d)-drain voltage (V_d) characteristics of the $MoSi_{10}$-film TFT at RT. We observed the field-effect characteristic of p-channel enhancement mode. This suggests that the density of mid-gap states in the $MoSi_{10}$ film is low enough for the field effect to occur. The threshold voltage (V_{th}) was estimated to be -3.0 V from I_d-V_g characteristics at V_d= 0.2 V, and the field-effect mobility was calculated from the I_d value of linear region using the equation $I_d = WC_i/L \times \mu (V_g - V_{th}) V_d$, where μ is the field-effect mobility, C_i is the capacitance of the gate SiO_2. We obtained the μ value of 3x10^{-3} cm^2/Vs at V_g = -5 V and V_d= -1.0 V.

We measured the carrier mobility to check the quality of the $MoSi_{10}$ film by Hall effect measurements at RT. Electrodes for the 4-probe method were formed beside the TFT structures simultaneously with the deposition of Al films for source/drain electrodes. The $MoSi_{10}$ film was p-type with a mobility of 1.3 cm^2/Vs and carrier density of 4.2×10^{16} cm^{-3}. The observed mobility was higher than that of hydrogenated a-Si (< 0.1 cm^2/Vs), but slightly smaller than the reported value [2, 3]. The field-effect mobility was lower than that of Hall effect measurements [2, 3]. This may be caused by carrier trapping at interface states between SiO_2 and the $MoSi_{10}$ film and at residual mid-gap states in the film. Actually, the absorption spectrum of the $MoSi_{10}$ film suggests that the film has higher density of mid-gap states than a-Si:H, as shown in Fig. 3.

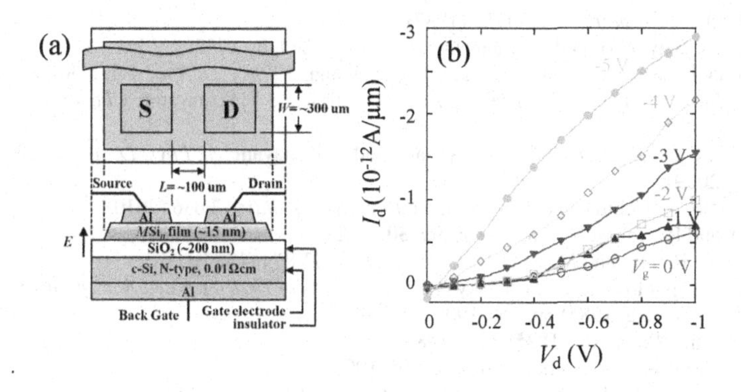

Figure 4. (a) The structure of the MSi_n TFT. (b) I_d-V_d characteristics of a $MoSi_{10}$ TFT.

CONCLUSIONS

The $MoSi_n$ films were synthesized by deposition of $MoSi_nH_x$ clusters onto solid substrates followed by annealing at 500°C for dehydrogenation. The XPS and Raman spectra suggested that the films were composed of $MoSi_{10}$ clusters. The $MoSi_{10}$ films had a-Si networks and an optical gap of 1.5 eV. We demonstrated that the $MoSi_{10}$- channel TFT operated with the p-channel enhancement mode. The hole mobility was estimated to ~3x10^{-3} cm^2/Vs. ESR signals due to Si dangling bonds were not observed. These facts show that the concept of Si dangling bonds termination by Mo atoms works well to reduce the density of dangling bond defects in the $MoSi_{10}$ film.

ACKNOWLEDGMENTS

This research was supported by an international cooperative program of Japan Science and Technology Agency (JST) and KAKENHI (Grant-in-Aid for Scientific Research for 472, 19051017, 2007.) on Priority Areas "New Materials Science Using Regulated Nano Spaces-Strategy in Ubiquitous Elements" from the Ministry of Education, Culture, Sports, Science and Technology of Japan.

REFERENCES

1. For example, K. Morigaki: *Physics of Amorphous Semiconductors* (Imperial College Press and World Scientific Publishing, London, 1999) pp. 99–136, 317– 366.
2. N. Uchida, H. Kintou, Y. Matsushita, T. Tada and T. Kanayama, *Appl. Phys. Express* **1** 121502 (2008).
3. N. Uchida, T. Miyazaki, Y. Matsushita, K, Sameshima and T. Kanayama, *Thin Solid Films, in press.*
4. S. M. Beck, *J. Chem. Phys.* **87** 4233 (1987).
5. H. Hiura, T. Miyazaki, and T. Kanayama, *Phys. Rev. Lett.* **86** 1733 (2001).
6. N. Uchida, L. Bolotov, T. Miyazaki and T. Kanayama, *J. Phys. D: Appl. Phys.* **36** L43 2003.
7. A. Negishi, N. Kariya, K. Sugawara, I. Arai, H. Hiura, and T. Kanayama: *Chem. Phys. Lett.* **388** 463 (2004).
8. Z. Sun, H. Oyanagi, N. Uchida, T. Miyazaki, and T. Kanayama: *J. Phys. D: Appl. Phys.* **42** 015412 (2009).
9. T. Miyazaki, N. Uchida, and T. Kanayama, *Phys. Stat. solidi C* **7** 636 (2010).
10. L. Shaw and R. Abbaschian, *J. Mater. Sci.* **30** 5272 (1995).
11. D. Beeman, R. Tsu, and M. F. Thorpe, *Phys. Rev. B* **32** 874 (1985).
12. J. D. Joannopoulos and G. Lucovsky ed.: *The Physics of Hydrogenated Amorphous Silicon II* (Springer-Verlag, 1984) pp. 169-190, 261-297.
13. M. Stutzmann, Phys. Rev. B **35** 5666 (1987).
14. L. Skuja, J. of Non-Crystalline Solids **239** 16 (1998).
15. M. Stutzmann and J. Stuke, Philos. Mag. B **63** 151 (1991).
16. H. Okamoto and Y. Hamakawa, J. Non-Cryst. Solids **77/78** 1441 (1985).

Mater. Res. Soc. Symp. Proc. Vol. 1321 © 2011 Materials Research Society
DOI: 10.1557/opl.2011.949

Optical characterization using ellipsometry of Si nanocrystal thin layers embedded in silicon oxide

E. Agocs[1,2], P. Petrik[1], M. Fried[1], A. G. Nassiopoulou[3]
[1]Research Institute for Technical Physics and Material Science, 1121 Budapest, Konkoly Thege u. 29-33, Hungary.
[2] University of Pannonia, 8200 Veszprem, Egyetem u. 10, Hungary
[3] Institute of Microelectronics (IMEL), NCSR Demokritos, Aghia Paraskevi, 153 10 Athens, Greece.

ABSTRACT

We have developed optical models for the characterization of grain size in nanocrystal thin films embedded in SiO_2 and fabricated using low pressure chemical vapor deposition of Si from silane on a quartz substrate, followed by thermal oxidation. The as-grown nanocrystals thin film on quartz was composed of a two-dimensional array of Si nanocrystals (Si-NC) showing columnar structure in the z-direction and touching each other in the x-y plane. The nanocrystal size in the z-direction was equal to the Si nanocrystal film thickness, changing by the deposition time, while their x-y size was almost equal in all the samples, with small size dispersion. After high temperature thermal oxidation, a thin silicon oxide film was formed on top of the nanocrystals layer. The aim of this work was to measure the grain size and the nanocrystallinity of the Si nanocrystal thin films, a quantity related to the change of the dielectric function. We used a definition for the nanorcystallinity that is related to the effective medium analysis (EMA) of the material. The optical technique used for the investigations was spectroscopic ellipsometry. To measure the above sample properties the thickness and composition of several layers on a quartz substrate had to be determined by proper modeling of this complex system. We found that the nanocrystallinity (defined as the ratio of nc-Si/(c-Si+nc-Si) decreases systematically with increasing the Si-NC layer thickness. Using this approach we are sensitive to the lifetime broadening of electrons caused by the scattering on the grain boundaries, and not to the shift of the direct interband transition energies due to quantum confinement.

INTRODUCTION

Si nanocrystals (Si-NCs) embedded in dielectrics have been widely studied recently for use in non-volatile memory devices as charge storage medium, in Si-based light emitting diodes, and sensors [1-4]. Due to the quantum-confinement effect, the band structure of Si-NCs is different from that of bulk silicon and shows discrete energy levels in the conduction and valence bands that are strongly dependent on the Si-NC size. In a system containing a large number of Si-NCs, those with sizes larger than a few nanometers show a bandgap similar to that of bulk Si, while those with sizes below ~3-5 nm show a size dependent bandgap. Therefore the electric, the transport, and the charging properties change, and they are dependent hard on grain sizes in this range [5,6].

In this study we have investigated Si nanocrystal (Si-NC) thin layers in 5 different grain sizes embedded in silicon oxide with spectroscopic ellipsometry (SE). The preparation method provides uniform properties with well-defined grain sizes controlled by the thickness of the layers that makes these samples suitable for SE model development. We show that effective medium theory can be used to determine a range of sample properties [7], however the method has limitations. Besides the layer thicknesses and composition we can also sensitively measure the nanocrystallinity (defined based on the effective medium models as described below) depending on the grain size.

EXPERIMENT

The samples were composed of two-dimensional arrays of Si-NCs with columnar structures in the z-direction that touch each other in the x-y plane. Their size in the z-direction was equal to the Si nanocrystal film thickness, changing by the deposition time, while their x-y size was almost equal in all the samples, with small size dispersion. By high temperature thermal oxidation (900°C) the size of the nanocrystals in the z-direction was reduced and they were embedded in SiO_2 (thermal oxide on top, quartz substrate at the bottom). The structures of samples have been investigated by TEM [8]. We used samples with Si-NC thin films of different thicknesses ranging from 5 to 30 nm that were fabricated by low pressure chemical vapor deposition (LPCVD) from silane at 610°C and a pressure of 300 mTorr on quartz substrates. We had a clean quartz sample too, to measure and calculate the refractive index of that easily.

Table 1. Name of samples and their nominal Si nanocrystal film thickness

Name of sample	Nominal Si nanocrystal film thickness [nm]
Sample_2	2
Sample_3	7
Sample_4	12
Sample_5	17
Sample_6	22

The ellipsometric measurements were performed by a Woollam M-2000DI rotating compensator ellipsometer in the wavelength range of 190-1690 nm in 706 points at angle of incidence ranging from 50° to 60° by 2.5°. Evaluation of the SE data was carried out using an appropriate optical model. Calculated spectra are fitted to the measured ones by varying the wavelength-independent model parameters using linear regression analysis. The best fit model parameters are obtained in terms of their 95% confidence limits by minimizing the following unbiased estimator σ of the mean square deviation (MSE):

$$\sigma = \sqrt{\frac{1}{N-P-1}\sum_{i=1}^{N}\left(\left(\cos\Delta_i^{meas} - \cos\Delta_i^{calc}\right)^2 + \left(\tan\Psi_i^{meas} - \tan\Psi_i^{calc}\right)^2\right)}$$, where Ψ and Δ are

the ellipsometric angles, N is the number of independently measured values corresponding to different wavelengths and P is the number of unknown model parameters, *meas* and *calc* refer to measured and calculated values, respectively.

The optical model consists of the substrate and two layers (Fig. 1). The refractive index of the substrate is calculated using the Cauchy parameterization of $n = A + B/\lambda^2 + C/\lambda^4$ and $k = De^{-E\lambda/\lambda 0}$, where n and k are the refractive index and the extinction coefficient, respectively. λ denotes the wavelength, whereas A, B, C, D, and E are the Cauchy parameters. Layer 1 represents the nanocrystalline silicon film. It has been described with the effective medium approximation (EMA) using 3 components: a nanocrystal silicon reference (nc-Si [9]), a single crystalline silicon reference (c-Si) and a silicon dioxide reference (SiO_2). Layer 2 represents the silicon oxide film. It is another EMA layer, which consist of SiO_2, the amorphous silicon reference (a-Si) and nc-Si.

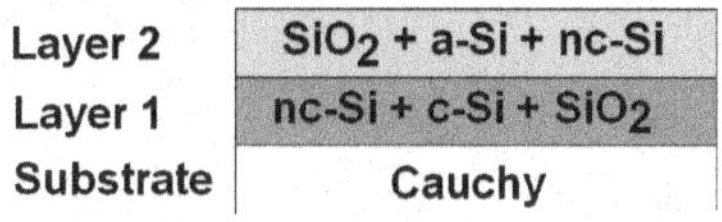

Layer 2	SiO_2 + a-Si + nc-Si
Layer 1	nc-Si + c-Si + SiO_2
Substrate	Cauchy

Figure 1 The model of Si nanocrystal thin layers embedded in silicon oxide

The model consists of 6 free parameters: four volume fraction values and two thickness values. The Cauchy parameters of substrate were calculated from the independent measurement on the clean quartz sample, and they were fixed in the Si-NC evaluations.

The silicon-based compositions are optically a kind of transition between a-Si and c-Si. It is possible to represent the Si based-layers with the mixture of c-Si and a-S. In our case these layers are poly(nano)crystalline, which means that they consist of single crystalline silicon grains in size of nanoscale. Because of the nanosize, the electronic band structure of single crystalline silicon is changing and the influence increases with increasing amount of grain boundaries. These effects are better described when using a nanocrystalline reference material (nc-Si) instead of (or in addition to) the amorphous component, where nc-Si is the measured dielectric function reference on very fine-grained polycrystalline silicon [9]. This reference includes the most characteristic behavior of the dielectric function as a result of decreasing grain boundaries. Therefore, the fit quality can be improved significantly. The ratio of nc-Si and c-Si informs us about the grain size. The nc-Si / (c-Si + nc-Si) function was used to represent the nanocrystal silicon samples with numeric values.

DISCUSSION

The fitted parameters of the optical model described in the Experimental section (Fig. 1) are compiled in Table 2. The five measured ellipsometric angles (dot), and the fitted Ψ and Δ curves (lines) are plotted in Fig. 2 for Sample 3. The Cauchy parameters of the substrate were calculated from the ellipsometric spectrum of a clean quartz sample. The fitted values are $A = 1.45 \pm 0.001$ $B = 0.0036 \pm 0.0003$ and $C = 0.000038 \pm 0.000015$.

With increasing the deposition time the thicknesses of Layer 1 are increasing. In Layer 1 the volume fraction of SiO_2 decreases with increasing thickness, in agreement with our expectation. The time of oxidation process was similar in all samples, revealed by the similar thicknesses of Layer 2 ($\approx$25-26 nm).

In case of sample 2, the total thickness is about 16 nm, which is lower than the thickness of Layer 2 of other samples. It means that during the oxidation process the Si-NC layer was almost oxidized (the remaining Si-NC size was below ~1nm). In this case Layer 1 can only be considered as an interface layer.

The volume fraction of c-Si increases, whereas that of nc-Si is more or less constant over Samples 2 to 6. In other words, the relative volume fraction of nc-Si increases with decreasing layer thickness (decreasing grain size), as shown in Fig. 3. A good fit can also be obtained by describing Layer 2 with only a SiO_2 component. The 3-component EMA, as shown in Table I, increases the fit quality only to a small extent.

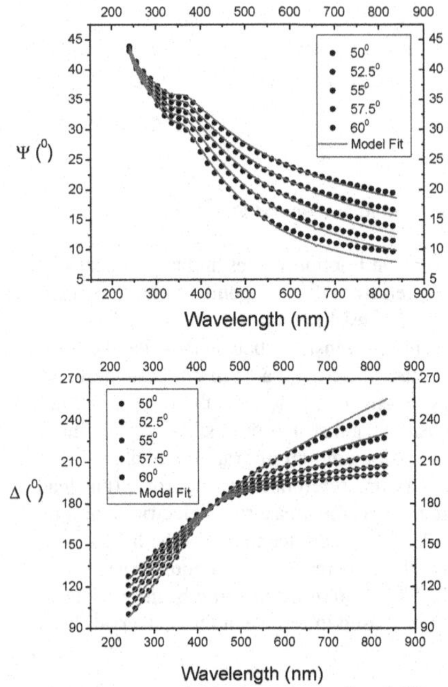

Figure 2. Measured (dots) and calculated (lines) Ψ (top) and Δ (bottom) spectra of Sample 3 in range from 240 nm to 840 nm in five different incident angles. The MSE of the fit is 32.

In some cases the volume fraction of certain components is below zero. It means, that the films are not described perfectly with the given two or three components. The negative volume fraction is usually an "overcompensation" of the effect of the particular component, but to the other direction, like negative void fraction usually means a higher density than the reference. Another example is the negative volume fraction of nc-Si in Layer 1 of Sample 2 (see Table I). In this case the negative volume fraction of nc-Si shows a grain size that is smaller than that of the nc-Si reference. In this model most of the negative values are just below zero, so fixing the values at zero doesn't affect much the dielectric function of the layer (and the fit quality). However, these discrepancies point out the necessity of using more complex parameterization, like that of Adachi [9-12].

Table 2. Parameters of optical model fitted onto the samples with Si-NC thin films of different thicknesses. MSE means mean square error; Layer 2 represents the silicon oxide layer, Layer 1 represents the Si-NC layer, and the substrate represents the quartz substrate, which is described by Cauchy formula. "T" denotes the thickness in nanometers, "SiO_2", "a-Si", "nc-Si" and "c-Si" are given in volume fractions. The errors of thicknesses and volume fractions are under 1 nm and 1%, respectively. The evaluation was performed in the wavelength range of 240-840 nm.

Name of Sample		2	3	4	5	6
MSE		17	32	23	17	17
Layer 2	T [nm]	**16.2**	**26.3**	**25.7**	**25.8**	**25.2**
	SiO2	100.3	102.5	100.6	98.9	97.3
	a-Si	15.2	-1.2	-1.6	4.3	3.6
	nc-Si	-15.5	-1.3	1.0	-3.2	-1.0
Layer 1	T [nm]	**0.2**	**8.3**	**13.2**	**16.5**	**20.8**
	nc-Si	130.7	69.7	76.3	79.0	77.6
	c-Si	-35.5	3.4	9.8	19.4	29.6
	SiO2	4.8	26.9	13.9	1.6	-7.2
Substrate Cauchy	A	1.45				
	B	0.0036				
	C	0.000038				

We can define a characteristic quantity called 'nanocrystallinity' described by nc-Si/(c-Si+nc-Si) (see Fig. 3). We obtained a systematically increasing value with decreasing grain size (thickness). In case of Sample 2, this value is higher than 1, because the grain size of nanocrystalline silicon is smaller than that of the nc-Si reference.

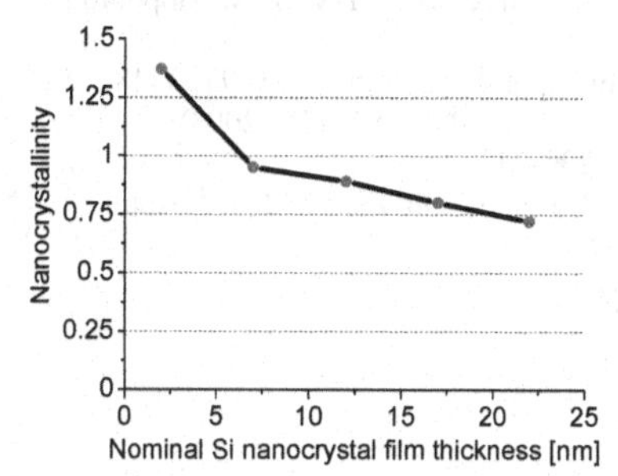

Figure 3. Nanocrystallinity of the samples (defined as the ratio of nc-Si/(c-Si+nc-Si)).

CONCLUSIONS

We developed EMA models for the ellipsometric measurement of LPCVD-deposited and oxidized thin Si-NC films on quartz. We found that the models describe the layer structure fairly well, and the nanocrystallinity (defined as the ratio of nc-Si/(c-Si+nc-Si) decreases systematically with increasing the Si-NC layer thickness. This is attributed to the decreasing percentage of grain boundaries per unit volume (increasing grain size) with increasing layer thickness. Using this approach we are sensitive to the lifetime broadening of electrons caused by the scattering on the grain boundaries, and not to the shift of the direct interband transition energies due to quantum confinement.

ACKNOWLEDGMENTS

This work has been supported by the European Commission Research Infrastructure Action under the FP6-Program contract no.026134 (RII3) ANNA, the Hungarian Research Fund (OTKA K81842), and the Hungarian NKTH ICMET07 project.

REFERENCES

1. J. U. Schmidt, and B. Schmidt, Mater. Sci. Eng. B **101** (2003) 28.
2. D. Pacifici, A Irrera, G. Franzó, M. Miritello, F. Iacona, and F. Priolo, Physica E **16** (2003) 331-340.
3. A. Irrera, M. Miritello, D. Pacifici, G. Franzó, F. Priolo, F. Iacona, D. Sanfilippo, G. Di Stefano, and P. G. Fallica, Nucl. Instr. And Meth. B **216** (2004) 222-227.
4. D. N. Pagonis, A. G. Nassiopoulou, and G. Kaltsas, J. Electrochem. Soc. **151**(8), H 174-H179 (2004).
5. E. Lioudakis, A. Othonos, A. G. Nassiopoulou, Ch. B. Lioutas, and N. Frangis, Appl. Phys. Lett. **90**, 191114 (2007).
6. E. Lioudakis, A. Othonos, and A. G. Nassiopoulou, Appl. Phys. Lett. **90**, 171103 (2007).
7. D. E. Aspnes, Thin Solid Films, **89** (1982) 249-262.
8. Ch. B. Lioutas, N. Vouroutzis, I. Tsiaoussis, N. Frangis, S. Gardelis, and A. G. Nassiopoulou, Phys. Stat. Sol. (A) **205**, No.11, 2615-2620 (2008).
9. G. E. Jellison Jr., M. F. Chisholm, and S. M. Gorgatkin, Appl. Phys. Lett. **62** (1993) 3348.
10. S. Adachi, Hirofumi Mori, and Mitsutoshi Takahashi, J. Appl. Phys. **93**, 115 (2003).
11. A. B. Djurisic, E. H. Li, Thin Solid Films **364** (2000) 239-243.
12. P. Petrik, M. Fried, E. Vazsonyi, P. Basa, T. Lohner, P. Kozma, and Zs. Makkai, J. Appl. Phys. **105**, 024908 (2009).

Growth Mechanism

Mater. Res. Soc. Symp. Proc. Vol. 1321 © 2011 Materials Research Society
DOI: 10.1557/opl.2011.1247

Control of Materials and Interfaces in μc-Si:H-based Solar Cells Grown at High Rate

Yasushi Sobajima[1,2], Chitose Sada[1,2], Akihisa Matsuda[1,2], Hiroaki Okamoto[1,2]
[1]Department of Systems Innovations, Graduate School of Engineering Science Osaka University, Toyonaka, Osaka, 560-8531, Japan
[2]The Japan Science and Technology Agency (JST) - Core Research for Evolutional Science and Technology (CREST)

ABSTRACT

Growth process of microcrystalline silicon (μc-Si:H) using plasma-enhanced chemical-vapor-deposition method under high-rate-growth condition has been studied for the control of optoelectronic properties in the resulting materials. We have found two important things for the spatial-defect distribution in the resulting μc-Si:H through a precise dangling-bond-density measurement, e. g., (1) dangling-bond defects are uniformly distributed in the bulk region of μc-Si:H films independent of their crystallite size and (2) large number of dangling bonds are located at the surface of μc-Si:H especially when the film is deposited at high growth rate. Starting procedure of film growth has been investigated as an important process to control the dangling-bond-defect density in the bulk region of resulting μc-Si:H through the change in the electron temperature by the presence of particulates produced at the starting period of the plasma. Deposition of Si-compress thin layer on μc-Si:H grown at high rate followed by thermal annealing has been proposed as an effective method to reduce the defect density at the surface of resulting μc-Si:H. Utilizing the starting-procedure-controlling method and the compress-layer-deposition method together with several interface-controlling methods, we have demonstrated the fabrication of high conversion-efficiency (9.27%) substrate-type (n-i-p) μc-Si:H solar cells whose intrinsic μc-Si:H layer is deposited at high growth rate of 2.3 nm/sec.

INTRODUCTION

Microcrystalline silicon (μc-Si:H) films prepared by plasma-enhanced chemical-vapor-deposition (PECVD) method are recognized as promising materials for their applications to optoelectronic devices [1,2]. Thinking of the application of μc-Si:H to thin film solar cells, thickness of intrinsic μc-Si:H active layer is required to be more than 2 microns to absorb sufficient visible sun light [1]. Therefore, to reduce the marketing cost, high-rate-growth process of high quality μc-Si:H by PECVD method is an essential issue to be realized. In previous studies, growth rate of ~2 nm/sec has been obtained by very high frequency (VHF as power-source frequency)-PECVD method [3,4]. High VHF-power density and high working pressure conditions are effective to obtain high electron-density and low electron-temperature in the plasma, being useful to achieve high-SiH_4-depletion (HSD) conditions [2,3,5].

However, when increasing the film-growth rate, the optoelectronic properties of resulting μc-Si:H films are severely deteriorated owing to an increase of dangling-bond-defect density [5-7]. Therefore, microscopic understanding of film-growth process and dangling-bond defect-formation process during μc-Si:H-film growth under high-rate-growth condition are quite important to improve solar-cell-device performance through the reduction of defect density in the bulk region of μc-Si:H as well as at n/i and p/i interfaces. Moreover, we have few knowledge about the spatial distribution and location of dangling-bond defects in μc-Si:H.

In this paper, we describe our model for the growth of μc-Si:H films prepared from SiH_4/H_2 plasma and propose how to control the optoelectronic properties of resulting μc-Si:H under high-rate-growth conditions. We also show how to obtain the information about the spatial distribution of dangling-bond defects in the bulk region and at the surface region in μc-Si:H by means of precise electron-spin-resonance (ESR) measurements.

We propose an approach to control the defect density in the bulk region of μc-Si:H by adjusting the starting procedure of film growth. We also propose a novel technique to reduce the surface-defect density by additional silicon-thin-layer deposition on μc-Si:H with post-deposition-thermal annealing process. Utilizing these two defect-density-controlling techniques in the bulk and at the surface of resulting μc-Si:H films together with several interface-(i/p and p/TCO)-properties-controlling methods, we demonstrate the fabrication of high efficiency (~9.3%) n-i-p-type single-junction μc-Si:H solar cells whose intrinsic layer is deposited at high rate of 2.3 nm/sec.

GROWTH PROCESS OF μc-Si:H UNDER HIGH-RATE-GROWTH CONDITION

Initial event in the growth process of μc-Si:H is electron-impact-excitation dissociation of source gas materials in monosilane (SiH_4) / hydrogen (H_2) glow-discharge plasma [2]. A variety of radicals, reactive species, emissive species, and ions (positive and negative) are produced almost simultaneously in the plasma [2]. These species collide and react mostly with parent molecules, reaching steady state. Steady-state densities of reactive species, being determined using a variety of diagnostic techniques in realistic plasmas used for μc-Si:H growth, are summarized in Fig. 1.

In the film growth process of μc-Si:H, silyl radical (SiH_3), showing no reactivity with SiH_4 and H_2, exhibits the highest steady-state density among a variety of reactive species produced in the plasma including short-lifetime reactive species (SLS) such as silylene (SiH_2), silylidine (SiH), and silicon (Si). In the deposition process of μc-Si:H, low partial-pressure condition of SiH_4 is popularly used for the survival of atomic hydrogen (H) due to its important reaction for the formation process of μc-Si:H [2] on the film-growing surface, because atomic H shows high reactivity with SiH_4 as

$$H + SiH_4 \rightarrow H_2 + SiH_3. \quad (1)$$

Atomic H is produced mainly by electron-impact dissociation of H_2 in the plasma. Therefore, steady-state density of atomic H in the plasma is easily adjusted by changing the hydrogen-

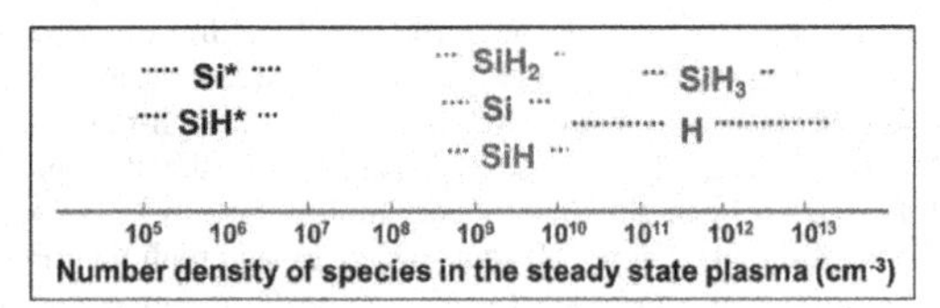

Figure 1. Number density of neutral reactive species and emissive species in the steady-state plasma for μc-Si:H growth.

dilution ratio R (H_2/SiH_4) in the starting-source-gas materials. In general, crystalline-volume fraction (X_C) in the resulting μc-Si:H is increased with increasing R [2]. In other words, atomic H is survived under low SiH_4-partial-pressure condition.

On the other hand, SLS also react well with parent SiH_4 molecule as follows [2,8];

$SiH_x + SiH_4 \rightarrow Si_2H_{4+x}$ or $2SiH_3$. (2)

Therefore, SLS tend to survive and their contribution ratio to film growth (ϕ_{SLS} / ϕ_{SiH3}) is increased under low SiH_4- partial-pressure condition, e. g., μc-Si:H formation condition. In order to increase the growth rate of μc-Si:H, generation rate of SiH_3-film precursor should be increased, being usually achieved by elevating the VHF-power density applied to the plasma (high electron density) under conventional low pressure condition. When high power density is applied to the plasma under limited SiH_4-flow-rate condition for the high rate growth of μc-Si:H, SiH_4 molecule is highly depleted, giving rise to the reduction of gas-phase-reaction probabilities of SLS with SiH_4 molecules, leading to an increase in the contribution ratio of SLS to film growth (ϕ_{SLS} / ϕ_{SiH3}), which causes high production rate of dangling bonds on the film-growing surface, being main cause of high-dangling-bond density in the bulk region of resulting μc-Si:H under high-growth-rate condition.

For the growth of high-quality μc-Si:H films, contribution ratio of SLS to SiH_3 (ϕ_{SLS} / ϕ_{SiH3}) should be minimized even under high-growth-rate condition. However, under high-generation-rate condition of SiH_3 (high-growth-rate condition), SLS-generation rate is also high in the plasma. To reduce the steady-state density of SLS [SiH_x] in the plasma, even under SiH_4-depletion condition, following reaction of SLS with H_2 molecules could be helpful and utilized.

$SiH_x + H_2 \rightarrow SiH_{2+x}$. (3)

However, the reaction rate constant for the reaction (3) is almost two orders of magnitude smaller than that for the reaction (2). To overcome this problem, the density of H_2 [H_2] in the plasma is set at high to scavenge SLS with H_2 sufficiently even under SiH_4-depletion condition, i. e., high-total pressure and SiH_4-depletion condition is popularly used recently [2,3,5]. It should be noticed that total pressure used recently for the growth of high quality μc-Si:H at high rate is in the range between 7 and 10 Torr [3-5,9], which is almost two orders of magnitude higher than that used before [2], being corresponding to the difference of reaction-rate constants between reactions (2) and (3).

Further possibility to reduce the contribution ratio of SLS to film growth is to decrease the generation rate of SLS in the plasma with respect to that of SiH_3. For this purpose control of electron temperature (T_e) in the plasma is one of the most important key issues for the growth of high quality μc-Si:H, since the threshold electron energy for the generation of SiH_3 from SiH_4 is 8.75 eV, while that of SLS from SiH_4 is higher than 9.47 eV [10] depending on the sort of SLS. Therefore, the generation-rate ratio of SLS to SiH_3 could be reduced when decreasing the electron temperature (T_e) in the plasma. However, it is well known phenomenon that T_e is increased very much by the presence of particulates produced in the plasma through an electron-attachment event. Particulates and/or higher-order-silane molecules are known to be initiated by the reaction (2) followed by successive reactions with SLS, therefore, high electron-density condition with high SiH_4 density, being used to see at the initial stage of film growth before the SiH_4-depletion circumstance during steady-state-film growth, should be avoided to realize low

T_e throughout film growth. Moreover, a part of particulates produced at the initial stage of film growth are incorporated, forming partially photo-insensitive layer in the resulting film.

EXPERIMENT

Intrinsic μc-Si:H layer was prepared by capacitively-coupled VHF-(100 MHz)-PECVD system with conventional shower-head cathode at the substrate temperature of 180 °C. Gas purifiers were set just before each (H_2 and SiH_4) mass-flow controller. Deposition rate of intrinsic μc-Si:H layer was adjusted to be 0.26 to 6.3 nm/sec by the control of growth parameters [6].

ESR measurement (JEOL JES-TM110) was performed with specially designed TM-mode cavity where planar sample (19.6 x19.6 mm^2) can be measured. This ESR-measurement system can evaluate the dangling-bond-defect density without peeling the sample from its substrate. All the μc-Si:H samples for ESR measurements were prepared on quartz-glass substrate. ESR measurements were performed in X-band (9.4 GHz).

Raman system with 514.5 nm line of Ar^+-ion laser was used (Renishaw System 1000) to estimate the crystalline-volume fraction, X_C, in μc-Si:H with Raman-scattering-intensity ratio of $I_c/(I_c + I_a)$ where I_c and I_a are the integrated intensities of spectral components at around 520 and 480 cm^{-1}, respectively [11].

Optical-emission spectroscopy (OES) was used as the monitoring method of plasma parameters [12,13]. The optical-emission-intensity ratio of I_{Si^*}/I_{SiH^*} was used as the measure of electron temperature (T_e) in the H_2 and SiH_4-gas-mixture plasma where I_{Si^*} and I_{SiH^*} are optical-emission intensities from Si^* ($\lambda = 288$ nm) and SiH^* ($\lambda = 414$ nm), respectively [14,15]. The excitation energy for the formation of Si^* (10.53-11.5 eV) is higher than that of SiH^* (10.33 – 10.55 eV) [10], both of which are formed by one-electron-impact process directly from SiH_4 molecules in the plasma and the electron-energy dependence of Si^*- and SiH^*- generation cross sections exhibits a similar trend [10,16-18]. Assuming that the electron-energy-distribution function (EEDF) obeys Maxwell-Boltzmann distribution, I_{Si^*}/I_{SiH^*} provides the information of the slope in the high-energy-tail region of the EEDF [18], being a proportional function of electron temperature. Namely, when I_{Si^*}/I_{SiH^*} takes a smaller value, the slope of the high-energy tail in the EEDF is steeper, meaning a low electron temperature

Surface morphology of μc-Si:H films was observed by topographic images taken by atomic force microscopy (AFM; SII SPA400/SPI3800N) in tapping-contact mode whose scanning range was 5 μm.

Structure of μc-Si:H solar cells fabricated in this study was; textured substrate (glass + textured Al-doped ZnO (AZO), or Asahi-type U) / Ag / transparent-conducting oxide (TCO) / n-i-p μc-Si:H / TCO/ Ag grid. Textured AZO for light trapping was prepared by radio-frequency-(RF: 13.56 MHz)-magnetron sputtering method followed by chemical etching in hydrochloric acid (HCl) aqueous solution [4]. Ag layer and TCO films (AZO, ITO and InTiO) were also prepared by RF-magnetron sputtering method. After the formation of top silver grid, solar cells were thermally annealed in N_2 atmosphere at 200 °C for 1 hour after the cell isolation by reactive-ion-etching (RIE) process. Active area of the solar cell was 0.23 cm^2.

Solar-cell parameters, short-circuit current (J_{SC}), open-circuit voltage (V_{OC}) and fill factor (FF), were evaluated from the current density-voltage characteristics under Air Mass 1.5

illumination (100 mW/cm^2). Spectral responses (external quantum efficiencies; QE) were measured with the conventional lock-in-detection system.

RESULTS AND DISCUSSION

Defect-density distribution and location in μc-Si:H prepared at high rate

Figure 2 shows areal defect density measured by precise ESR measurement mentioned above for μc-Si:H prepared under different growth-rate conditions as a function of film thickness [19]. As is clearly seen in the figure, areal defect density increases linearly with film thickness in all the samples. This result indicates that dangling-bond defects in the bulk region of μc-Si:H are uniformly distributed over the film-growth direction. This result is not easily explainable fact when thinking that dangling bonds are preferentially located at the grain boundaries of crystallites as has been believed, because the total surface area of crystallites is decreased due to an increase of crystallite size with film thickness. Therefore, it may be natural to think that the dangling-bond defect is uniformly distributed not only toward the film-growth direction but also toward lateral direction, i.e., dangling-bond defects are located even in the crystallites, at least in macroscopically large crystallites in which small fractal-like structure is observed under AFM images on the surface of μc-Si:H films [20].

Defect density at the interface region estimated from y-intercept in Fig. 2 is very high as high as middle of 10^{12} cm^{-2} which corresponds to > 10^{18} cm^{-3} in terms of volume-defect density, while volume-defect density in the bulk region of film (slope of the fitting line in Fig. 2) is as low as 10^{15} – 10^{16} cm^{-3}, being controllable by adjusting the film-growth conditions as discussed above. High surface-defect density in μc-Si:H grown at high rate gives a crucial issue to be solved for a drastic improvement of n-i-p solar cell performance through the control of defect density at the interface between i layer and p layer (between n layer and i layer for p-i-n solar cells). The presence of high surface-defect density is tentatively explained as follows; film-growing surface is heated not only by the heat radiation from high density plasma but also by the presence of many exothermic chemical reactions due to hydrogen-exchanging reaction near the film-growing surface under μc-Si:H growth conditions at high rate together with H-abstraction reaction with SiH_3, being the causes of removal process of surface-covering-bonded hydrogen leaving many dangling bonds on the film-growing surface. Many dangling bonds produced on

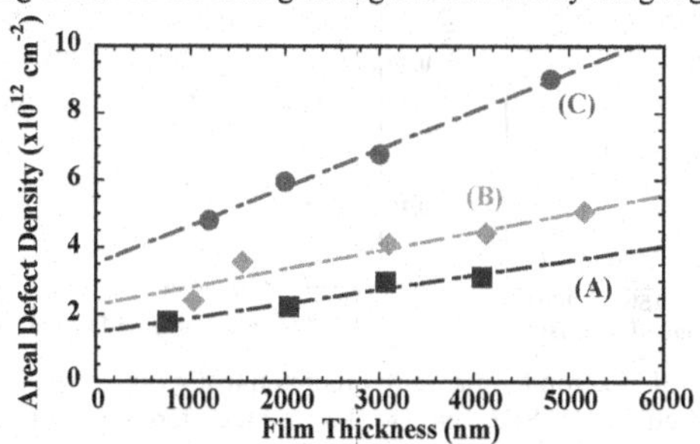

Figure 2. Areal-defect density plotted against film thickness of μc-Si:H prepared at different growth rate (A: 0.28, B: 2.15, C: 6.73 nm/sec).

the film-growing surface are usually saturated with many surface-diffusing SiH_3 during steady-state-film growth, but no coverage by SiH_3 is available anymore at the end of film growth.

Control of electron temperature during film growth by changing the starting procedure

As discussed above, to obtain high quality μc-Si:H with low dangling-bond-defect density at high growth rate, the contribution ratio of SLS to film growth should be minimized [21, 22]. For this purpose, T_e in the plasma should be reduced through the control of starting procedure of film growth. At the start of H_2/SiH_4 plasma, since the partial pressure of SiH_4 is still high as compared to that of steady-state SiH_4-depletion circumstance, particulate formation occurs easily so that T_e is affected sensitively by the starting procedure of film growth. In this section, two types of starting procedure for μc-Si:H growth were tested. First trial is SiH_4-flow-rate-control procedure at the initial stage of film growth, where the SiH_4-introduction time (t_1), the time to reach SiH_4-flow rate of 3 sccm, was changed from 15 second (fast, case A), to 50 second (medium, B) and 120 second (slow, C) under fixed power-density (0.33 W/cm^2) and H_2-flow-rate (168 sccm) condition. Second trial is power-density-control procedure at the start of film growth. In this method, SiH_4-flow rate and H_2-flow rate were fixed at 3 and 168 sccm, respectively, and the initial power density was set at 0.15 W/cm^2 for the ignition of plasma. The increasing time (t_2) of power density from 0.15 W/cm^2 to 0.33 W/cm^2 was varied from 0 second (fast, case D), to 10 second (medium, E) and 20 second (slow, F). The growth rate, being proportional to the steady-state density of main film precursor [SiH_3], is almost constant at 2.4 nm/sec, since [SiH_3] under HSD condition is limited by the SiH_4-flow rate fixed in this study.

Figure 3(a) illustrates the time evolution of T_e monitored by optical-emission-intensity ratio of I_{Si*}/I_{SiH*} after turning on the H_2-based plasma for different SiH_4-introduction times (t_1). As shown in Fig. 3(a), T_e in the plasma gradually increases after turning on the plasma, reaching saturation, and it remains the same until the end of film growth. The transition behavior of T_e is in good agreement with the SiH_4-introduction scheme, i.e., the slope of T_e becomes less steeper with elongating SiH_4-introduction time (t_1). In addition, the saturation value of T_e, is decreased with increasing SiH_4-introduction time (t_1). The electron temperature is considered to be affected by electron-attachment events to particulates formed in the plasma. The formation rate of

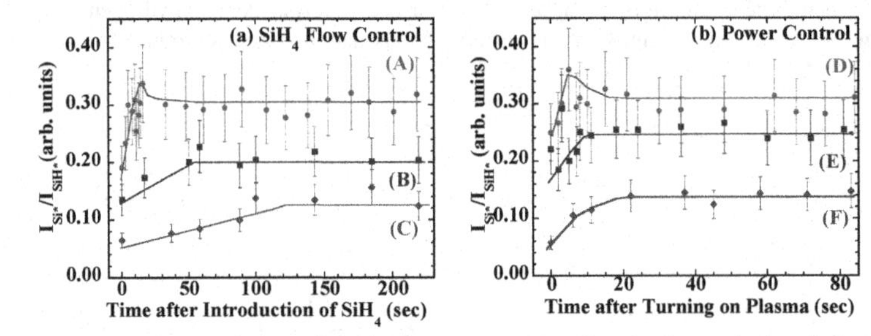

Figure 3. (a): Time evolution of electron temperature predicted using optical-emission-intensity ratio of I_{Si*}/I_{SiH*} after the introduction of SiH_4 into H_2 plasma for different introduction times of t_1 [t_1 = 15 sec (A), 50 sec (B), 120 sec (C)]. (b) Time evolution of electron temperature after application of VHF-power to H_2/SiH_4 plasma for different power- increasing times of t_2 [t_2 = 0 sec (D), 10 sec (E), and 20 sec (F)].

particulates is roughly given by the product of SiH_4-molecule density and electron density in the plasma [14,15,21]. Therefore, T_e is decreased, as demonstrated in Fig. 3 (a), when the SiH_4-introduction time (t_1) is elongated through a successful depletion of SiH_4 molecules from the start.

Figure 3 (b) shows the time evolution of T_e monitored by I_{Si*}/I_{SiH*} for different power-density-application times (t_2). The slope of T_e also becomes less steeper and the saturation value of T_e takes a lower value with increasing power-density-application time (t_2). The reason for the change in T_e is considered to be the same as that mentioned in the case of SiH_4-introduction-time series. Namely, the particulate-formation situation, being determined by the coexistence of high SiH_4 density and high electron density, is also avoided by the slow application of power density.

These results indicate that T_e during film growth is successfully controlled by the gradual introduction of SiH_4 and by the slow application of power density. A decrease in the initial slope and reduction of the saturation value of T_e are good trend for obtaining a high performance in μc-Si:H n-i-p solar cells through the improvement of optoelectronic properties at the n/i interface and the reduction of the bulk-defect density in the intrinsic layer. It is noted here that X_C in the resulting μc-Si:H is almost constant in the range of 58 to 65%, although X_C is affected by T_e. In addition, the film-growth rate is unchanged by T_e under the SiH_4-depletion conditions.

Figure 4 shows photo-J-V characteristics of μc-Si:H single-junction solar cells fabricated on Asahi type-U glass substrate for three solar cells whose starting film-growth procedure of i-layer is different. Procedures (C), (D), and (F) in the figure denote the gradual SiH_4-introduction method, the fast, and the slow power-density-application methods, respectively. As seen in the figure, the photovoltaic performance is strongly affected by the starting procedure of intrinsic μc-Si:H film growth in n-i-p solar cells. The worst photovoltaic performance (J_{SC} = 20.4 mA/cm^2, V_{OC} = 0.47 V, and FF = 59.3%) among solar cells fabricated here is obtained in the case of starting procedure (D) where severe particulate formation is seen in the starting stage of film growth and the highest T_e is observed in the steady-state plasma during film growth. Better photovoltaic performance (J_{SC} = 22.8 mA/cm^2, V_{OC} = 0.49 V, and FF = 65.6%) is seen in the case of starting procedure (F) where the formation of particulates is reasonably suppressed and a low T_e is realized. The best conversion efficiency of 8.97% (J_{SC} = 23.6 mA/cm^2, V_{OC} = 0.54 V, and FF = 70.4%) in this study is obtained in the case of starting procedure (C) where the particulate formation is minimized and the lowest T_e is achieved. Figure 5 shows the cross-

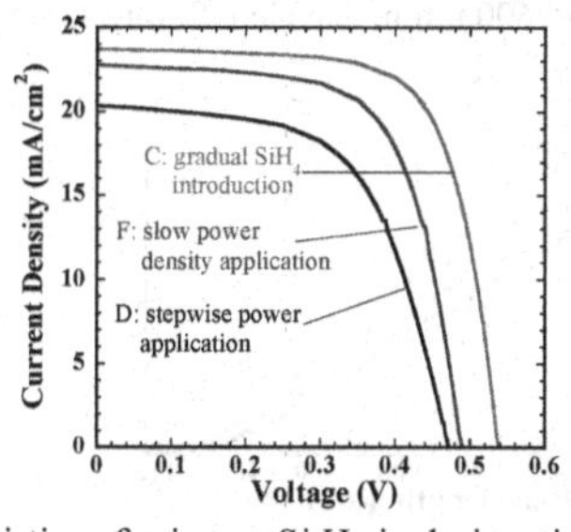

Figure 4. Photo-J-V characteristics of n-i-p μc-Si:H single-junction solar cells having 2.0-μm thick i layers prepared with different starting procedures (C), (D), and (F), measured under AM 1.5, 100 mW/cm^2 condition.

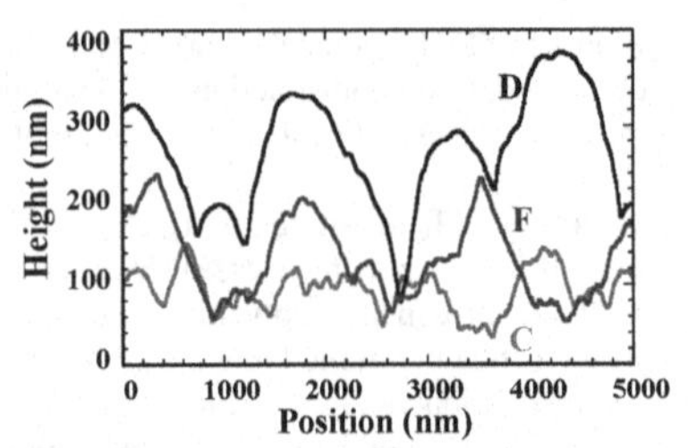

Figure 5. Cross-sectional line shape of surface roughness of n-i-p μc-Si:H solar cells fabricated with different starting procedures for i-layer growth. RMS-surface roughness for procedure (C), (F), and (D) is 35.32 nm, 52.3 nm, and 64.76 nm, respectively.

sectional line shape of the surface roughness of μc-Si:H solar cells. The highest RMS-surface roughness (64.76 nm) is seen in the case of (D) due to the incorporation of large amount of particulates at the starting period of film growth. The RMS-surface roughness (52.3 nm) in the case of (F) is higher than that (35.32 nm) in the case of (C), although T_e in the steady state shows almost the same value in these two cases. This could be explained as follows: in the case of (F), particulates are formed and negatively charged, however, they can escape from the plasma owing to rather low plasma potential at the starting period, resulting in the enhancement of surface roughness but low T_e in the steady state. On the other hand, in the case of (C), particulate formation is suppressed due to the successful depletion of SiH_4 molecules from the start and the T_e shows the lowest value. Drop in V_{OC} in the case of (F) and further drop in V_{OC} in the case of (D) in Fig. 4 are attributed to the enhancement of surface roughness at the i/p interface due to the incorporation of large size particulates at the initial stage of film growth, being confirmed by AFM observations, as shown in Fig. 5. The large surface roughness at the i/p interface becomes an origin of the TCO/intrinsic μc-Si:H direct-contact pathway owing to the poor coverage of the thin p-type μc-Si:H layer.Large amount of reduction in J_{SC} in the case of (F) and (D) has been analyzed by quantum efficiency (QE) measurements and their reverse-bias-voltage recoveries as demonstrated in Fig. 6. When comparing QE data and their reverse-bias-recovery behaviors for solar cells fabricated by staring procedure (C) and (D), the cause of J_{SC} reduction is considered as follows; the large drop in QE at around 600 nm in wavelength for the solar cell fabricated by (D)

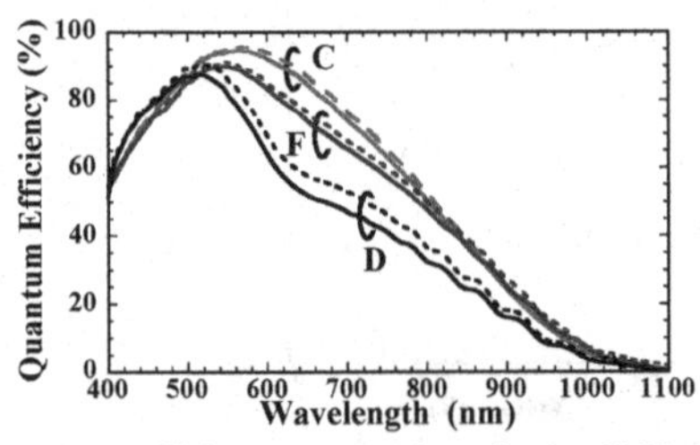

Figure 6. External quantum efficiency spectra for n-i-p μc-Si:H solar cells fabricated with different starting procedures for i-layer growth. Solid lines and dashed lines represent spectra measured under short circuit condition and reverse bias-voltage-application condition of 1V.

is due to the presence of partially photo-insensitive particulates near the n/i interface deeper part from the top surface of the solar cell. Rather large recovery for the QE spectrum in the longer wavelength side by reverse-bias-voltage application for the solar cell fabricated by (D) is due to the presence of flat band in the solar-cell-band diagram caused by higher defect density in the whole area of i layer, which is produced by higher T_e during film growth.

Improvement of photovoltaic performance in μc-Si:H solar cells by means of compress layer

As mentioned above, areal defect density on the surface (including interface) of μc-Si:H under high rate-growth conditions reaches the order of 10^{12} cm^{-2}, being corresponding to the order of 10^{18} cm^{-3} as volume-defect density. As the defect rich surface of intrinsic μc-Si:H forms defective p/i interface when fabricating n-i-p type solar cells, control of surface-defect density during film growth or appropriate treatment after film growth is required for a drastic improvement of photovoltaic performance in the solar cells. In this section, we shows a novel approach to decrease the surface-defect density by means of silicon-thin-layer deposition followed by thermal-annealing procedure just after the μc-Si:H growth at high rate.

Figure 7 shows the μc-Si:H film-thickness dependence of areal defect density obtained by precise ESR measurement. All the samples in this figure were prepared under identical preparation conditions with growth rate of 2.6 nm/sec. Closed symbols represent as-prepared μc-Si:H without any treatment (A) and with silicon thin layer of 50 nm, having low defect density, deposited on μc-Si:H (B). In Fig. 7, open symbols denote samples after thermal annealing at 200 °C for 1 hour in N_2 atmosphere, i. e., (C) shows the areal defect density of samples after thermal annealing of samples (B). As is seen in this figure, areal defect density increases linearly with film thickness. Volume-defect density in the bulk region of μc-Si:H (slope of fitting line for A) is calculated to be in the order of 10^{15} cm^{-3}. In addition, dangling-bond-defect density at the surface (interface) estimated from the intercept at the vertical axis in Fig. 7 is as high as 6.3 x 10^{12} cm^{-2}, being corresponding to 1.58 x 10^{19} cm^{-3} volume-defect density. (B) is for μc-Si:H films with 50 nm-thick silicon thin layer. The slope of (B) in areal defect density vs. thickness plot exhibits no change as compared to that in (A), indicating that the volume-defect density in the bulk region is not influenced by the presence of silicon thin layer. However, the defect density on the surface

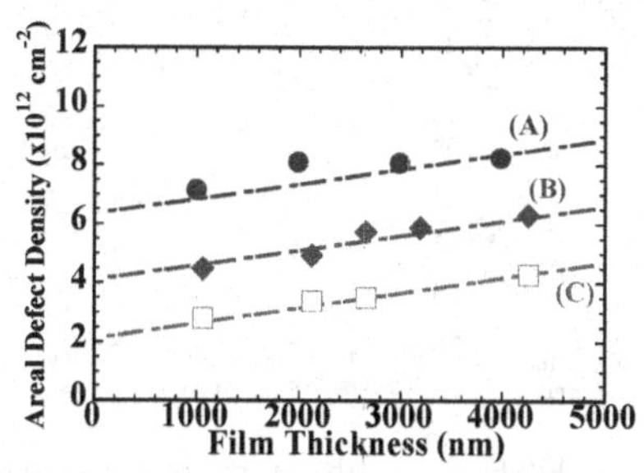

Figure 7. Areal-defect density plotted against film thickness for intrinsic μc-Si:H; as-deposited state (A), just after preparation of 50 nm-thick compress layer (B), and after annealing of sample B (C). The lines are guide for eyes.

(interface) is decreased slightly but strictly by the deposition of silicon thin layer. Surface-defect density in (C), in which (B) was annealed in N_2 atmosphere at 200 °C for 1 hour, is drastically reduced, keeping the same volume-defect density in the bulk region. These results demonstrate that the silicon thin layer deposited over μc-Si:H grown at high rate acts as a defect-removing layer effectively when the post-deposition thermal annealing is performed. Thereby, it is appropriate to name this layer as "compress layer" instead of "cap layer [23,24]" or "buffer layer [25,26]" used conventionally, by taking into account the roles of deposition and annealing procedures for a drastic reduction of surface defects (defects are actually removed by silicon-thin-layer deposition followed by thermal annealing), i. e., this silicon layer behaves as if a hot compress treats the damaged muscle and pain.

Figures 8 (a) and (b) show the V_{OC} and FF of n-i-p μc-Si:H solar cells plotted against thickness of Si-compress layer, respectively. All the preparation conditions of each layer for solar cells were the same and the thickness of i-layer μc-Si:H was fixed at 2.0 μm. As shown in this figure, both V_{OC} and FF are improved by the application of the compress layer, which is thought to be due to an improvement of optoelectronic properties at p/i interface by removing the p/i-interface defects with the pasting of Si-compress layer followed by thermal annealing. On the other hands, Figs. 8 (c) and (d) show the J_{SC} and conversion efficiency of n-i-p μc-Si:H solar cells plotted against the thickness of Si-compress layer, respectively. As shown in this figure, J_{SC} is slightly decreased by applying Si-compress layer, being suggested that absorption loss of incident light occurs in Si-compress layer. In this study, optimum Si-compress layer thickness is determined to be 50 nm from the result of conversion efficiency vs. compress-layer-thickness plot shown in Fig. 8 (d). Detailed discussions for optimum preparation condition of Si-compress layer will be done in the near feature.

Figure 9 (a) shows the photo-J-V characteristics of μc-Si:H single junction solar cells fabricated on Asahi type-U glass substrate with adjusting each layer-growth conditions. Photovoltaic i layer was fabricated by the gradual SiH_4-introduction scheme. Film thickness of i layer and that of compress layer were 2.4 μm and 50 nm, respectively. The film-growth rate of i layer was 2.3 nm/sec. We have demonstrated the conversion efficiency of 9.27% (J_{SC} = 24.1

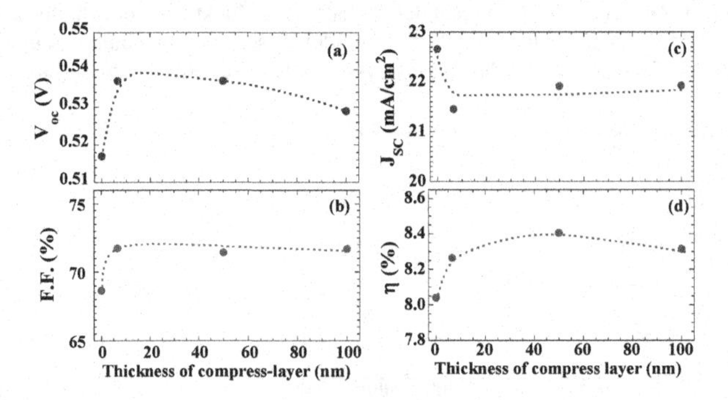

Figure 8. Open circuit voltage; V_{OC} (a), Fill Factor; FF (b), short circuit current density; J_{SC} (c) and conversion efficiency; η of n-i-p single-junction μc-Si:H solar cells with i-layer thickness of 2.0 μm as a function of thickness of compress layer.

mA/cm^2, V_{OC} = 0.54 V, FF = 71.3%). Figure 9 (b) shows the external QE spectra measured under the short circuit condition (solid line) and under reverse-bias-voltage condition of 1 V (dashed line) for μc-Si:H solar cell whose J-V characteristic is shown in Fig. 9 (a). The QE spectrum at short circuit condition shows a peak at around 580 nm and tails into longer wavelength region, giving high J_{SC}. In addition, the QE spectrum under reverse bias-voltage-application condition is almost unchanged as compared to that under short circuit condition.

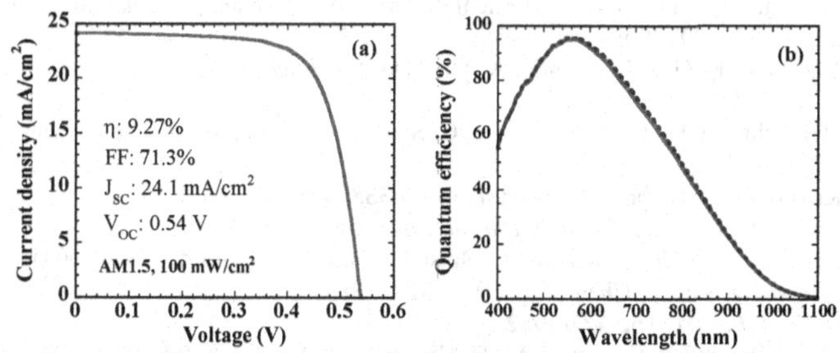

Figure 9. Photo-J-V characteristics (a) and external quantum-efficiency spectrum (b) of n-i-p μc-Si:H solar cells having 2.4 μm-thick-intrinsic layer deposited at high rate of 2.3 nm/sec. In (b), solid line represents spectrum measured under short circuit condition and dashed line spectrum is obtained under reverse bias-voltage-application condition of 1V.

CONCLUSIONS

We have found that dangling-bond defects are uniformly distributed in the bulk region of μc-Si:H films and large number of dangling bonds are located at the surface of μc-Si:H especially when the film is deposited at high growth rate. On the basis of microscopic understandings for the defect-formation process during μc-Si:H-film growth, we have proposed a new starting procedure of film growth for obtaining high quality μc-Si:H films showing low defect density through the control of electron temperature in the plasma. We have also proposed a new method for a drastic reduction of surface-defect density by means of thin-Si-layer deposition on μc-Si:H with thermal annealing. Utilizing these two methods together with several interface-controlling methods, we have demonstrated the fabrication of high efficiency (9.27%) substrate-type (n-i-p) μc-Si:H solar cells whose intrinsic μc-Si:H layer is deposited at high growth rate of 2.3 nm/sec.

ACKNOWLEDGMENTS

This work was supported in part by Global COE Program (Core Research and Engineering of Advanced Materials-Interdisciplinary Education Center for Materials Science) and the New Energy and Industrial Technology Development Organization (NEDO).

REFERENCES

1. E. Vallat-Sauvain, A. Shah, J. Bailat, *Thin Film Solar Cells, Fabrication, Characterization and Applications* (Wiley, Chichester, 2006) Chap. 4, pp. 133-172, and reference therein.
2. A. Matsuda, *J. Non-Cryst. Solids* **338-340,** 1 (2004).
3. T. Matsui, M. Kondo, A. Matsuda, *Jpn. J. Appl. Phys.* **42**, L901 (2003).
4. Y. Sobajima, M. Nishino, T. Fukumori, T. Higuchi, S. Nakano, T. Toyama, H. Okamoto, *Sol. Energy Mater. Sol. Cells* **93**, 980 (2009).
5. M. Fukawa, S. Suzuki, L. H. Guo, M. Kondo, A. Matsuda, *Sol. Energy Mater. Sol. Cells* **66**, (2001) 217.
6. Y. Sobajima, T. Higuchi, J. Chantana, T. Toyama, C. Sada, A. Matsuda, H. Okamoto, *Phys. Status Solidi C* **7**, 521 (2010).
7. C. Niikura, M. Kondo, A. Matsuda, *J. Non-Cryst. Solids* **338–340**, 42 (2004).
8. J. Perrin, O. Leroy, M. C. Bordage, *Contrib. Plasma Phys.*, **36**, 3 (1996).
9. C. Niikura, N. Itagaki, M. Kondo, Y. Kawai, A. Matsuda, *Thin Solid Films* **457**, 84 (2004).
10. M. Tsuda, S. Oikawa, K. Sato, *J. Chem. Phys.* **91**, 6822 (1989).
11. Z. Iqbal, S. Veprek: *J. Phys. C* **15,** 377(1982).
12. R. W. Griffith, F. J. Kampas, P. E. Vanier, M. D. Hirsch, *J. Non-Cryst. Solids* **35–36**, 391 (1980).
13. A. Matsuda, K. Tanaka, *Thin Solid Films* **92**, 171(1982).
14. M. Takai, T. Nishimoto, M. Kondo, A. Matsuda, *Appl. Phys. Lett.* **77**, 2828(2000).
15. M. Takai, T. Nishimoto, M. Kondo, A. Matsuda, *Thin Solid Films* **390**, 83 (2001).
16. J. Perrin, J. F. M. Aarts, *Chem. Phys.* **80**, 351(1983).
17. S. Oikawa, M. Tsuda, J. Yoshida, Y. Jisai, *J. Chem. Phys.* **85**, 2808 (1986).
18. A. Matsuda, M. Takai, T. Nishimoto, M. Kondo, *Sol. Energy Mater. Sol. Cells* **78,** 3 (2003).
19. M. Iwata, M. Tanaka, Y. Sobajima, T. Toyama, C. Sada, A. Matsuda, H. Okamato, *Proceedings of 25th European photovoltaic Solar Energy Conference and Exhibition*, 3AV.1.30, p2967.
20. T. Toyama, T. Kitagawa, Y. Sobajima, H. Okamoto, *Jpn. J. Appl. Phys.* **46**, 5125 (2007).
21. A. Matsuda, *J. Vac. Sci. Technol.* A **16**, 365 (1998).
22. Y. Kawai, H. Ikegami, N. Sato, A. Matsuda, K. Uchino, M. Kuzuka, A. Mizuno, *Industrial Plasma Technology* (Wiley-VCH, Weinhein, 2010) Chap. 18, p. 221.
23. M. A. Green, *Sol. Energy* **74**, 181 (2003).
24. C. Leguijt, P. Lolgen, J. A. Eikelboom, A. W. Weeber, F. M. Schuurmans, W. C. Sinke, P. F. A. Verhoef, *Sol. Energy Mater. Sol. Cells* **40**, 297 (1996).
25. K. S. Lim, M. Konagai, K. Takahashi, *J. Appl. Phys.* **56**, 538 (1984).
26. T. Matsui, A. Matsuda, M. Kondo, *Sol. Energy Mater. Sol. Cells* **90**, 3199 (2006).

Mater. Res. Soc. Symp. Proc. Vol. 1321 © 2011 Materials Research Society
DOI: 10.1557/opl.2011.1289

Monitoring the growth of microcrystalline silicon deposited by plasma-enhanced chemical vapor deposition using in-situ Raman spectroscopy

S. Muthmann*, F. Köhler, M. Hülsbeck, M. Meier, A. Mück, R. Schmitz, W. Appenzeller, R. Carius and A. Gordijn

IEK5-Photovoltaik, Forschungszentrum Jülich, D-52425 Jülich, Germany

ABSTRACT

A novel setup for Raman measurements under small angles of incidence during the parallel plate plasma enhanced chemical vapor deposition of μc-Si:H films is described. The possible influence of disturbances introduced by the setup on growing films is studied. The substrate heating by the probe beam is investigated and reduced as far as possible. It is shown that with optimized experimental parameters the influence of the in-situ measurements on a growing film can be neglected. With optimized settings, in-situ Raman measurements on the intrinsic layer of a microcrystalline silicon solar cell are carried out with a time resolution of about 40 s corresponding to 20 nm of deposited material during each measurement.

INTRODUCTION

Hydrogenated microcrystalline silicon (μc-Si:H) is a material of a great technological relevance. The applications range from thin-film transistors to large-scale devices like solar-cells. It is well known that μc-Si:H is a phase mixture of amorphous and crystalline material. A well established way to deposit thin μc-Si:H films is the plasma enhanced chemical vapor deposition (PECVD). This type of deposition usually leads to an evolution of the complex structure in growth direction [1,2]. A lot of effort has been made to understand and control the growth process of μc-Si:H [3 - 6]. The methods that are available up to now have in common that they are more or less indirect ways to obtain information about material properties. We present results obtained with a new experimental setup that enables Raman measurements on thin silicon films during their deposition with PECVD in parallel plate configuration. With this technique direct information about the composition of a growing layer can be obtained during the growth.

EXPERIMENT

The in-situ Raman setup was included into an existing multi-chamber PECVD system for state-of-the-art thin-film silicon solar cells. Deposition of the intrinsic layers and the Raman measurements were carried out in a chamber containing a showerhead electrode. The doped layers were deposited in separate chambers to avoid cross contamination. Intrinsic layers were deposited at a pressure of 13 mbar with a heater temperature of 200 °C at an excitation frequency of 13.56 MHz which is the industrial standard for large area depositions. The power density was in the range of 0.2 W/cm². In-house produced texture etched zinc oxide covered glass with a size of 10x10 cm² was used as a substrate.

The source gases for the deposition were silane and hydrogen. The parameter used to describe the amount of silane in the source gas composition is the silane concentration SC:

$$SC = \frac{SiH_4}{SiH_4 + H_2} \quad (1)$$

It is well known that the amount of amorphous phase in a μc-Si:H layer can be controlled by adjusting the SC [1]. All the Raman measurements were performed using a 532 nm solid state laser with a cw-output power of 1 W and a beam diameter of 0.7 mm.

RESULTS

To use in-situ Raman spectroscopy as a process control in large area deposition systems an optical feed through has to be integrated into the showerhead electrode. This modification is necessary since the Raman scattered light has to be collected in a small angle to the surface normal. Hence at typical electrode distances of 10 mm it is only possible to monitor central regions of a layer using a pierced electrode surface.

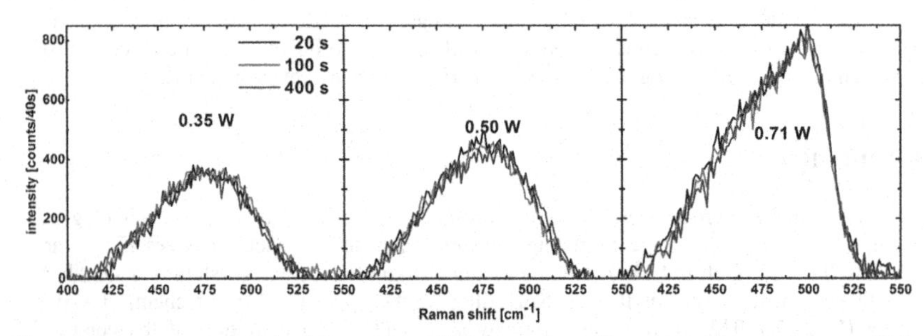

Figure 1: Raman spectra obtained from a 1 μm thick amorphous silicon layer. The spectra were obtained after 20, 100 and 400 s of illumination at different laser powers with a pulse length of 65 μs.

Above a certain diameter a hole in a PECVD electrode induces inhomogeneities in the deposited layers. This effect has been studied in [7].

The experimental configuration used for the in-situ Raman measurements is described in detail in [8]. An electrical shielding by metallic grids is proposed to minimize the effect of the disturbed electrode [7]. Thereby it is possible to strongly reduce the influence of the disturbance in the electrode surface on film growth. This enables to draw conclusions from measurements of the observed fraction of the layer that are valid for the whole coated substrate.

Since a large background signal emitted from the deposition plasma will also contribute to an in-situ Raman spectrum the laser power used to generate the Raman signal also has to be relatively high. This means that the temperature increase (ΔT) caused by the laser can not be neglected. To limit ΔT pulsed illumination is applied. Typical pulse lengths in the range of 100 μs generate high Raman signals at a limited heating of the sample.

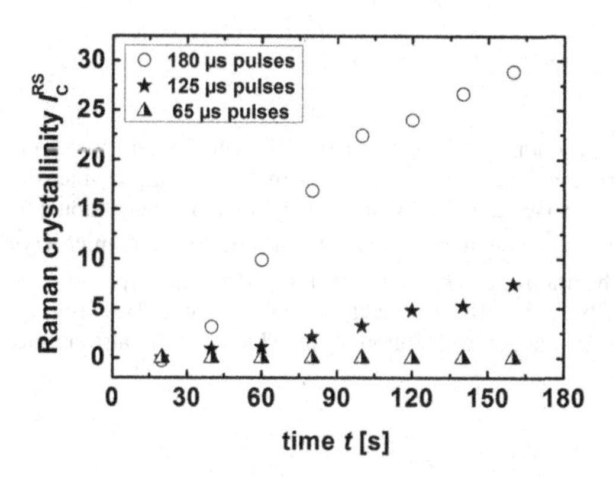

Figure 2: Raman crystallinities obtained after different times of illumination for pulse lengths of 65, 125 and 180 μs at a laser power of 0.5 W

This is important for in-situ measurements since the temperature of a growing thin silicon film is known to influence its properties [9]. To obtain information about ΔT Raman measurements with the in-situ setup were performed on an hydrogenated amorphous silicon (a-Si:H) layer with a thickness of 1 μm. Due to the larger absorption coefficient of a-Si:H for 532 nm radiation compared to μc-Si:H ΔT is expected to be largest for a-Si:H.

Figure 1 shows Raman measurements performed with different laser powers after different times of illumination. Here all the pulses had a length of 65 μs. It can be clearly seen that neither for 0.35 W nor for 0.50 W crystallization of the silicon takes place. For a larger laser power a microcrystalline contribution is clearly visible in the Raman spectra, as indicated by the new contribution at 520 cm^{-1}. This means that the temperature increase caused by the laser was large enough at 0.71 W to introduce a partial crystallization of the layer.

Due to the relatively short pulses it cannot be expected that a steady state temperature will be reached during a single laser pulse. Hence another parameter determining ΔT is the pulse length τ. The influence of different pulse durations at an input power of 0.5 W is illustrated in Figure 2. It can be seen that for pulses longer than 65 μs a crystallization of the layers takes place after some time. This effect is stronger for longer pulses. For a pulse length of 125 μs the crystallization rate is much lower than for the larger pulse length. Thus this pulse length can be estimated to represent the crystallization threshold for a laser power of 0.5 W.

Based on these results in-situ Raman measurements were performed during the deposition of a standard μc-Si:H solar cell. To record spectra that allow a reasonable good evaluation of the Raman crystallinity the applied parameters allow for a temporal resolution of about 40 s. Together with an average deposition rate of about 5 Å/s this means that about 20 nm of silicon were deposited during each measurement. Figure 3 shows the obtained values for the crystalline volume fraction I_C^{RS} which is determined according to equation 2:

$$I_C^{RS} = \frac{I_{520}}{I_{520} + I_{480}} \quad (2)$$

where I_{520} is the integrated intensity of the crystalline contribution of a spectrum. It is obtained by subtracting an amorphous reference spectrum with the integrated intensity I_{480} from the original data. It can be clearly seen in Figure 3 that after an incubation phase with a low crystalline volume fraction I_C^{RS} stabilizes at about 45 %. The very thin microcrystalline p-doped which is present at the beginning contributes very little to the first spectra. Hence they appear completely amorphous. The relatively large scatter of data is due to the noise present in the data. This could be reduced by longer integration time which would also reduce the temporal resolution.

DISCUSSION

It was shown that by choosing the appropriate measurement parameters the temperature increase caused by the Raman laser can be minimized and can be kept well below the crystallization threshold. This means that its influence on the properties of the deposited layer is expected to be very small. Since optimal μc-Si:H solar cells are deposited in a small process window, very close to the phase transition towards amorphous growth, small changes in the process are expected to have a large impact. Due to the fact that a minimal heating by the laser can not be avoided completely small changes of the electronic and optical properties in the area illuminated by the laser can not be excluded. A comparison of in-situ results with depth dependent measurements obtained after the deposition shows that the in-situ measurements of the Raman crystallinity provide results that are also valid for the parts of the layer that are not influenced by the in-situ setup [8].

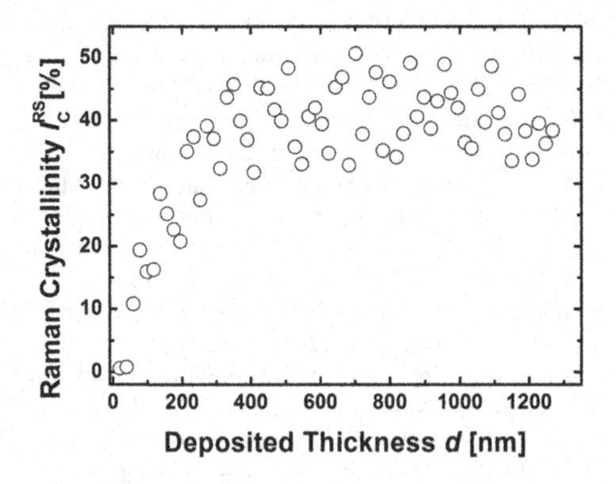

Figure 3: Raman crystallinities obtained during the deposition of a microcrystalline silicon solar cell.

When looking at Figure 3 the incubation layer seems to be rather thick. For conclusions regarding the actual composition of a silicon layer the information depth of the 532 nm laser radiation also has to be considered. According to Vetterl less than 1/e of the Raman signal is contributed by fractions of a layer in a depth of more than $1/(2\alpha)$ beneath the surface [10]. Assuming an absorption coefficient of 4×10^4 cm^{-1} for a wavelength of 532 nm in microcrystalline silicon [11] with a high crystalline volume fraction the recorded Raman spectra mainly contain information from the topmost 125 nm. This information depth is reduced to 56 nm for amorphous silicon with an absorption coefficient of approximately 9×10^4 cm^{-1} and between the two values for moderate crystallinities as in Figure 3.

When considering the fact that roughly 20 nm are deposited during each measurement interval it can be concluded that the I_C^{RS} values shown in Figure 3 are strongly influenced by the underlying layers. This effect is expected to be most pronounced during the initial phase of deposition since there the crystallinity is expected to change strongly with the deposited layer thickness.

CONCLUSIONS

It has been shown that in-situ measurements of the Raman crystallinity of a thin microcrystalline silicon films are possible during the PECVD deposition in parallel plate configuration with a temporal resolution of about 40 s. The influence of the measurement setup and especially of the probe laser has been identified and could be reduced such that in-situ values for crystallinity can be obtained that are representative for the complete area of a coated substrate.

REFERENCES

1. O. Vetterl, F. Finger, R. Carius, P. Hapke, L. Houben, O. Kluth, A. Lambertz, A. Mück, B.Rech, H. Wagner, Sol. Energy Mater. Sol. Cells **62**, 97 (2000)
2. R.W. Collins, A.S. Ferlauto, G.M. Ferreira,C. Chen, J. Koh, R.J. Koval, Y. Lee, J.M. Pearce, C.R. Wronski, Sol. Energy Mater. Sol. Cells **78**, 143 (2003)
3. L. Guo, M. Kondo, A. Matsuda, Jpn. J. Appl. Phys. **37**, L1116 (1998)
4. B. Strahm, A.A. Howling, C. Hollenstein, Plasma Source Sci T **49**, B411 (2007)
5. T. Kilper, M.N. van den Donker, R. Carius, B. Rech, G. Bauer, T. Repmann, Thin Solid Films **516**, 4633 (2008)
6. B. Strahm , A.A. Howling, L. Sansonnens, C. Hollenstein, Plasma Sources Sci. Technol. **16**, 80 (2007)
7. S. Muthmann, M. Meier, R. Schmitz, W. Appenzeller, A. Mück ,A. Gordijn, Surf. Coat. Tech. doi: 10.1016/j.surfcoat.2011.02.037
8. S. Muthmann, F. Köhler, M. Meier, M. Hülsbeck, R. Carius, A. Gordijn, Phys. Status Solidi-R **5**, 144 (2011)
9. A. Matsuda, Thin Solid Films **337**, 1 (1999)
10. O. Vetterl, PhD Thesis RWTH Aachen (2001)
11. R. Carius. F. Finger, U. Backhausen, M. Luysberg, P. Hapke, L. Houben, M. Otte, H. Overhof Material Research Soc. Proc. **467**, 283 (1997)

Mater. Res. Soc. Symp. Proc. Vol. 1321 © 2011 Materials Research Society
DOI: 10.1557/opl.2011.929

Deposition of P-Type Nanocrystalline Silicon Using High Pressure in a VHF-PECVD Single Chamber System

Xiaodan Zhang, Guanghong Wang, Xinxia Zheng, Shengzhi Xu, Changchun Wei, Jian Sun, Xinhua Geng, Shaozhen Xiong and Ying Zhao

Institute of Photo-electronic Thin Film Devices and Technology of Nankai University, Weijin Road 94#, Nankai District, Tianjin 300071, P.R.China

ABSTRACT

In this article, we present a study of boron-doped hydrogenated nanocrystalline silicon (nc-Si: H) films by very high frequency-plasma enhanced chemical vapor deposition (VHF-PECVD) using high deposition pressure. Electrical, structural and optical properties of the films were investigated. Dark conductivity as high as 2.75S/cm of p-type nc-Si: H prepared at 2.5Torr pressure has been achieved at a deposition rate of 1.75Å/s for 25nm thin film. By controlling boron and phosphorus contamination, single junction nc-Si: H solar cells incorporated p-layers prepared under high pressure and low pressure, respectively, were deposited. It has been proven that nanocrystalline silicon solar cells with incorporation of p layer prepared at high pressure has resulted in enhanced open circuit voltage, short circuit current density and subsequently high conversion efficiency. Through the optimization of the bottom solar cell and application of ZnO/Al back reflector, 10.59% initial conversion efficiency of micromorph tandem solar cell (1.027cm^2) with an open circuit voltage of 1.3864V, has been fabricated, where the bottom solar cell using a high pressure p layer was deposited in a single chamber.

INTRODUCTION

Plasma-enhanced chemical vapour deposition (PECVD) of p-i-n type solar cells in a single plasma reactor offers advantages of low cost compared to multi-chamber processes which use separate reactors for fabrication of the p-, i- and n-layers, respectively [1]. In addition, to further reduce costs to industry, it is essential that solar cells be deposited at high rate by high deposition pressure and the same, small electrode distance. In general, nc-Si: H p-layers are prepared by very high frequency-plasma enhanced chemical vapor deposition (VHF-PECVD) at low pressure or RF deposition at low or high pressure, respectively [2-5]. No further information on the fabrication and structural properties of nc-Si: H p-layers prepared by VHF-PECVD deposition at high pressure have been reported so far.

Conditions for a better p-layer suitable for p-i-n type solar cell application are quite stringent. It must have low absorption, high conductivity, and sufficient crystallinity to promote the nucleation and homogeneous growth of the subsequent nanocrystalline intrinsic layer at the p/i interface [6]. In this study, dark conductivity, crystallinity, surface morphology as well as absorption coefficient of p-type nc-Si: H had been investigated varying deposition pressure and thickness as well as the performance of these films in single junction and micromorph solar cells.

EXPERIMENT

The nc-Si: H films were deposited using a capacitively coupled ultrahigh vacuum (UHV) parallel plate single-chamber reactor in a multi-chamber system. The details of this setup were reported in our previous paper [7]. The excitation frequency of the electrical source was 75MHz. Pressure and flow rates were independently controlled by a downstream throttle valve controller and upstream mass flow controllers, respectively. Deposition of solar cells uses silane (SiH_4) and hydrogen (H_2) as source gases, as well as diborane (B_2H_6) and phosphine (PH_3) as doping gases.

The films were deposited on Eagle 2000 glass substrates. ZnO was sputtered on glass substrates and etched by 0.5% HCl, which was used for the fabrication of single-junction nc-Si: H solar cells. The cell structure is glass/ZnO/p-nc-Si: H/i-nc-Si: H /n-a-Si: H/ZnO/Al. For a-Si: H/nc-Si: H tandem solar cell, front TCO is commercial SnO_2 coated glass substrate. The electrical properties of nc-Si: H samples were measured with a co-planar Al electrode configuration. The length of Al electrode was about 1.60cm, and the distance between the two Al electrodes was about 0.045cm. The thickness of the films was measured by a XP-2 surface profiler. The structural properties of nc-Si: H samples were characterized with Raman (LabRAM UV, including 325nm, 488nm and 632.8nm wavelength laser) measurements. Surface roughness was analyzed by atom force microscopy (AFM-DI Nanoscope IV Multimode Scanning Probe Microscope). Surface morphology has also been identified using by field emission scanning electron microscopy (FE-SEM, JEOL JSM-6700) with 10.0kV power at 100,000x magnification. The optical reflection and absorption of the films were measured by UV-3600 UV-VIS-NIR Spectrophotometer from Shimadzu, and from these data the band gap of the films have been estimated. The solar cell performance was characterized with current versus voltage (I-V) measurements under an AM1.5 solar simulator at 25℃.

RESULTS AND DISCUSSION

3.1 Study of p-type nc-Si: H films

The deposition pressure was varied under otherwise constant conditions. Figure 1 shows the variation of the dark conductivity for 60nm p-type nc-Si: H films deposited at different pressures between 1.5Torr and 3Torr. All thin films show conductivity above 1S/cm. With increasing deposition pressure, collision probability of precursors in the plasma increases, which lead to a decrease of electron temperature. The low electron temperature, low ion energy, and high radical flux facilitate fast growth of crystallites. The compact films are formed due to nanocrystalline nuclei coalescence, and lower grain boundary scattering leads to an increase in the conductivity [8]. However, the conductivity decreases with the increase of the pressure above 2Torr. This arises from the lower effective doping

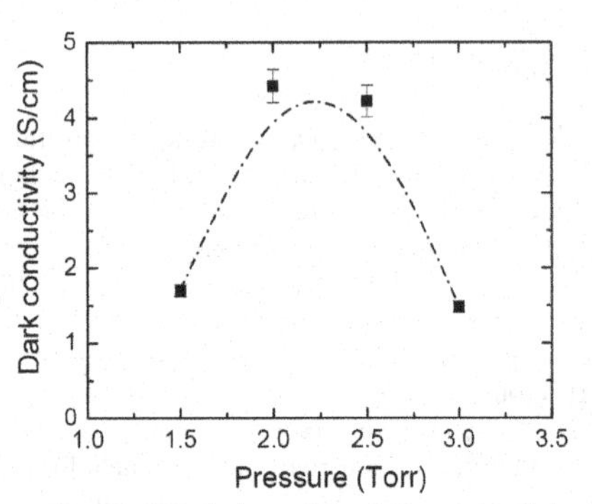

Figure 1 Variation of the dark conductivity for 60nm p-type nc-Si: H thin films deposited under different pressures.

under high pressure, resulting in the formation of almost amorphous silicon, which has been proved from the Raman spectra. The dark conductivity is reduced with the amorphous network [9], so it can be concluded that the growth with crystallinity control depends on the optimum combination of pressure with other deposition conditions.

The above characterizations are made on 60nm thick films, but for window layer of solar cells thin p-layer is necessary with good conductivity and low absorption. A thicker p-layer will absorb more sunlight and introduce less light into the active layer of the cell and thus produce less current. The objective of the following is mainly to make a thin nanocrystalline silicon p-layer with optoelectronic properties suitable for application in nanocrystalline silicon solar cells and micromorph tandem solar cells. Figure 2 shows the variation of dark conductivity with film thickness at 2.5Torr pressure. It is seen that the dark conductivity increases with the increase of film thickness. The initial process of p-layer growth involves an island film, and so carriers cannot get the percolation path. With the increase of film thickness, the onset of coalescence of islands occurs, and the islands evolve into a continuous film, which enhances the carrier mobility and results higher conductivity [10].

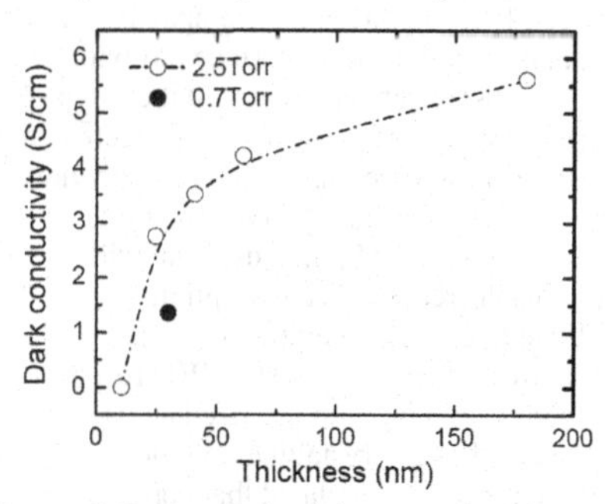

Figure 2 Dark conductivity as a function of film thickness for p-type nc-Si: H films deposited at 2.5Torr pressure.

In practice, the film with only 11nm thickness also showed the initial crystalline growth. The growth mechanism of the nc-Si: H strongly depends on the surface condition of the substrate as well as the deposition condition, and the properties of p-layer are key factors for controlling the subsequent crystalline growth of the i-layer, and ultimately influence the performances of solar cells. The high crystalline volume fraction in the p-layer may partly lead to initial crystalline growth of i-layer and then reduce the thickness of amorphous incubation layer at the p/i interface. In Figure 2, it has been noted that a dark conductivity as high as 2.75S/cm of p-type nc-Si: H film with 1.75Å/s deposition rate prepared at 2.5Torr high pressure has been achieved for 25nm thin film. However, only 1.36 S/cm for p-layer prepared at 0.7Torr with 30nm thickness can be obtained.

To analyze the growth mechanism of films, surface morphologies of 25nm thickness p-type nc-Si: H films deposited at low and high pressures were characterized by SEM. The film prepared at high pressure shows a more compact surface microstructure which also can be seen in Figure 3. The lower grain boundary scattering for samples prepared under high pressure led to an increase in the conductivity.

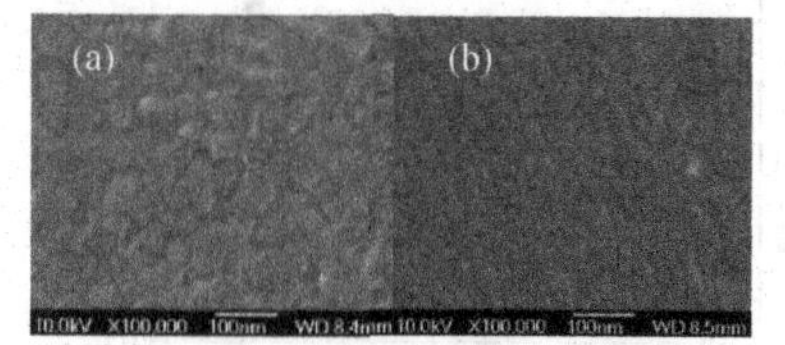

Figure 3 SEM images of p nc-Si: H thin films with 25nm thickness deposited at low pressure 0.7Torr (a) and high pressure 2.5Torr (b).

From the above we can see that crystalline nuclei growth behavior of silicon p-layer can be divided into two categories. High pressure deposition of p-layer is a homogeneous nucleation condition where small nanocrystalline nuclei appear in amorphous matrix, and low pressure deposition results in a heterogeneous nucleation condition resulting in larger. Microstructure compactness is important for fabricating high efficiency nanocrystalline silicon solar cell [11]. Nevertheless, compactness of these thin films cannot be easily measured directly. We can only deduce from surface morphology and roughness qualitatively.

The absorption coefficient of the p-type films as a function of photon energy is shown in Figure 4. The absorption coefficient decreases in the visible range of photon energy for p-layer deposited at high pressure. The above phenomenon was attributed to the high crystallinity in the film. An increase in crystalline volume fraction in this mixed phase material will result in a decrease in the absorption coefficient because absorption of nc-Si: H film is lower than a-Si: H. The XRD results have shown that the film prepared at high pressure with high crystalline volume fraction. It can be calculated that band gap of the films increases compared to that of film deposited at 0.7Torr. The application of the p-type nc-Si: H film in p-i-n solar cell may lead to better open circuit voltage and short-circuit current density.

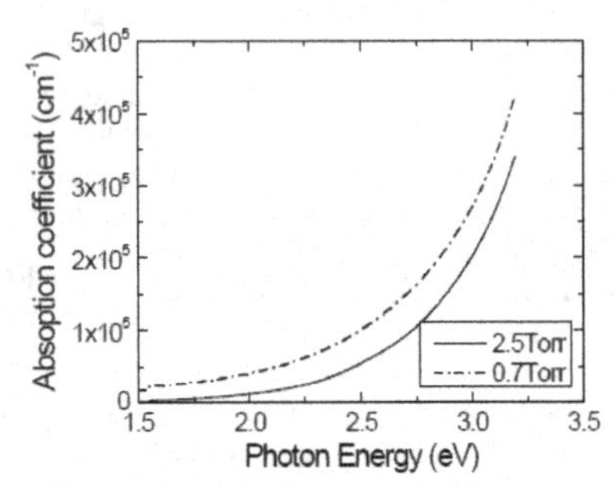

Figure 4 Optical absorption spectra of p-type nc-Si: H thin films deposited under 0.7Torr and 2.5Torr pressure.

Figure 5 shows the dependence of the films hole concentration and hall mobility on the deposition pressure. As is evident that with an increase in deposition pressure from 0.7Torr to 3.0Torr, hole concentration and mobility increase and reach the maximum at 2.0Torr and then decrease. At pressures larger than 2.0Torr, a low crystalline peak intensity indicates that the material becomes more amorphous, and the electrical properties decreased as shown for dark conductivity (in Figure 1), hole concentration, and hall mobility (in Figure 5).

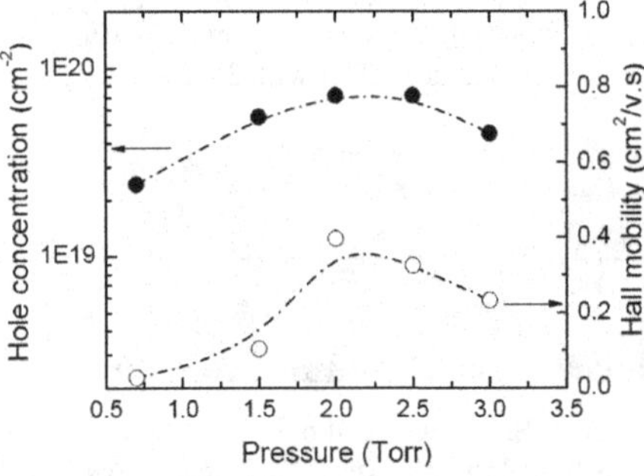

Figure 5 Hall mobility and carrier density of samples varied with deposition pressures.

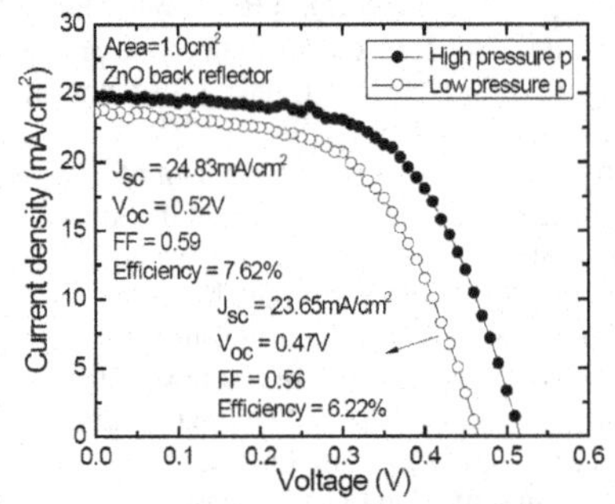

Figure 6 J-V curve of single junction nanocrystalline silicon solar cells incorporated p-layer prepared under low pressure and high pressure in a single chamber system.

3.2 Application in solar cells

The optimization of deposition parameters at completely high pressure deposition techniques resulted in p-type nc-Si: H films with high dark conductivity, high hole concentration, high hall mobility, and lower absorption coefficient while maintaining the nanoocrystallinity. In order to check the advantage for above p-layer, p-layers prepared under low pressure (0.7Torr) and high pressure (2.5Torr) were incorporated into single junction nanocrystalline silicon solar cells which were completely deposited in a single chamber system for p-, i-, and n- layers, respectively. Boron contamination between p- and i- layer, phosphor contamination between n- and p- layer during the deposition process of cell has been effectively controlled through methods which have been researched in our previous papers [12, 13]. Figure 6 shows the J-V curve of nanocrystalline silicon solar cell. It is evident that nanocrystalline silicon solar cell with incorporation of p-layer prepared at high pressure has resulted in increased open circuit voltage, short circuit current density and conversion efficiency.

To check the quality for above materials prepared under high pressure condition, we also deposited micromorph tandem solar cells in the configuration of a-Si pin/nc-Si p-i-n on glass substrate coated with SnO_2 transparent conductive oxide. As we know, another very important advantage of using high conductivity p-nc-Si: H layer at the n-p junction between the top and bottom cells in a multi-junction solar cell is to reduce the contact resistance significantly, which improves the conversion efficiency of the device. Through modifying the microstructure and thickness of nanocrystalline silicon bottom cell deposited in a single chamber system, micromorph tandem solar cell with 10.59% initial conversion efficiency has been fabricated (shown in Figure 7). It is noticed that 1.3864 V open circuit voltage was obtained for micromorph tandem solar cell prepared in a single chamber system. The cell area is $1.027cm^2$. The above performance has been confirmed by NREL.

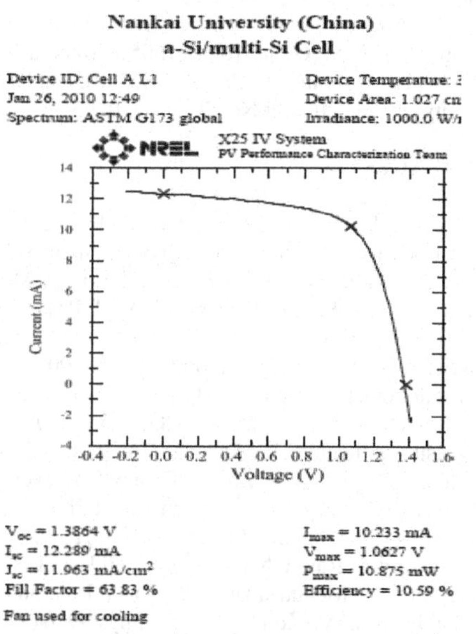

Figure 7 J-V curve of micromorph tandem solar cell which nanocrystalline silicon bottom solar cell prepared in a single chamber.

CONCLUSIONS

P-type nc-Si: H thin films prepared under high pressure demonstrated high conductivity, high hole concentration, high mobility, and low absorption coefficient while maintaining high crystallinity. In addition, surface morphology of thin film also indicates that more compact thin films can be obtained under high deposition pressure. Incorporation of p-layer deposited under high pressure into single junction nanocrystalline silicon solar cell led to an increased open circuit voltage and short circuit current density and then conversion efficiency. Micromorph tandem solar cell with 10.59% initial conversion efficiency has also been fabricated and confirmed by NREL.

ACKNOWLEDGEMENTS

Authors greatly thank Qi Wang and Keith Emery in National renewable energy laboratory (NREL) for I-V measurement of micromorph tandem solar cells. The authors also gratefully acknowledge William Nemeth of National Renewable Energy Laboratory for helpful discussion and proofreading. Project supported by Hi-Tech Research and Development Program of China (Grant No.2007AA05Z436, 2009AA050602), Science and technology support project of Tianjin (Grant No.08ZCKFGX03500), National Basic Research Program of China (Grant No.2011CBA00705, Grant No.2011CBA00706, No.2011CBA00707), National Natural Science Foundation of China (Grant No.60976051), International Cooperation Project between China-Greece Government (2009DFA62580), and Program for New Century Excellent Talents in University of China (Grant No.NCET-08-0295).

REFERENCES

1. U. Kroll, C. Bucher, S. Benagli, I. Schönbächler, J. Meier, A. Shah, J. Ballutaud, A. Howling, C. Hollenstein, A. Büchel, M. Poppeller, Thin Solid Films 451, 525 (2004)
2. Y. Nasuno, M. Kondo, A. Matsuda, Sol. Energy Mater. Sol. Cells 74, 497 (2000)
3. F.Demichelis, C.F. Pirri, E.Tresso, J. Appl. Phys 72(4), 1327 (1992)
4. T. Roschek, nanocrystalline silicon solar cells prepared by 13.56MHz, Ph.D. Thesis, IPV, Berichte des Forschungszentrums Jülich, 2003, pp.41-43.
5. S. A. Filonovich, H.Águas, I. Bernacka-Wojcik, C.Gaspar, M.Vilarigues, L.B.Silva, E.Forunato, R.Martins, Vacuum 83, 1253 (2009)
6. T. Matsui, M. Kondo, A. Matsuda, J. Non-crystal. Solid 338-340, 646 (2004)
7. Y. Zhao, X. D. Zhang, F. Zhu, Y.T. Gao, C.C. Wei, J.M. Xue, H.Z. Ren,. D.K. Zhang, G.F. Hou, J. Sun, X. H. Geng, in: 15th International Photovoltaic Science & Engineering Conference (PVSEC-15), Shanghai, PR China, 2005, pp. 65.
8. K. Adhikary, S. Ray, J. Non-Crystal. Solids 353, 2289 (2007)
9. Debajyoti Das, Madhusudan Jana, Mater. Lett. 58, 980 (2004)
10. T. Fujibayashi, M. Kondo, J. Appl. Phys 99, 043703 (2006)
11. T. Matsui, A. Matsuda, M. Kondo, Sol. Energy Mater. Sol. Cells 90, 3199 (2006)
12. X.D. Zhang, F.H. Sun, G.H. Wang, S.Z. Xu, C.C. Wei, G.F. Hou, J. Sun, S.Z. Xiong, X.H. Geng, and Y. Zhao, Phys. Status Solidi C 7(3-4), 1073(2010)
13. G.H. Wang, X.D. Zhang, S.Z. Xu, C.C. Wei, J. Sun, S.Z. Xiong, X.H. Geng, Y. Zhao, Phys. Status Solidi C 7(3-4), 1116(2010)

Mater. Res. Soc. Symp. Proc. Vol. 1321 © 2011 Materials Research Society
DOI: 10.1557/opl.2011.953

Influence of the Electrode Spacing on the Plasma Characteristics and Hydrogenated Amorphous Silicon Film Properties Grown in the DC Saddle Field PECVD System

Keith R. Leong, Nazir P. Kherani, and Stefan Zukotynski
Department of Electrical and Computer Engineering
University of Toronto, Toronto, Ontario, M5S 3G4, Canada

ABSTRACT

A new plasma deposition system was built with the capability of varying the electrode spacing in the DC Saddle Field plasma enhanced chemical vapor deposition system. An ion mass spectrometer was installed just below the substrate holder to sample the ion species travelling towards the substrate. Silane plasma and amorphous silicon film studies were conducted to shed light on the impinging ion species, ion energy distributions, and film properties with varying electrode spacing. The results indicate that decreasing the distance between the substrate and cathode leads to a reduction in the high energy ion bombardment.

INTRODUCTION

Hydrogenated amorphous silicon (a-Si:H) is a versatile optoelectronic material that is used in devices such as thin film transistors and solar cells. Hydrogenated amorphous silicon has been deposited by a wide range of methods, some of which include radio frequency (rf) Plasma Enhanced Chemical Vapor Deposition (PECVD) [1], Hot Wire CVD [2], microwave plasma [3], DC PECVD [1], and DC Saddle Field PECVD [4]. However, the majority of systems used to deposit low defect density, device quality, a-Si:H are capacitively coupled rf PECVD systems. Further, the optimization of the properties of a-Si:H films has largely proceeded by experimental trial and error.

The inter electrode spacing has been found to be a critical parameter. Weakliem et al. [5] reported that increasing the electrode spacing increases the hydrogen content of the resulting film and the number density of ions consisting of four or five silicon atoms within the plasma. Ross and Jaklik [6] found that plasma polymerization increases with increasing inter electrode spacing. This was thought to be due to higher order radicals (Si_nH_m) [7]. With decreasing electrode spacing the silane dissociation rate increases because of a higher power per silane molecule [8, 9]. This increased dissociation rate was also observed at different pressures [7].

The DC Saddle Field (DCSF) PECVD system utilizes a central semitransparent grid anode held at a positive potential. The anode is sandwiched between two semitransparent grid cathodes. The cathodes can be electrically biased, floating, or grounded. Beyond each cathode lies a grounded substrate holder. Once the plasma is ignited, electrons are forced by the electric field towards the anode. Due to the transparency of the anode, a fraction of the electrons will miss the anode and pass through to the other side. By symmetry this process continues and the electrons tend to oscillate about the anode. Thus, the electron path length is increased and its interaction with the precursor gas leads to effective excitation, ionization and dissociation.

It has been shown that the plasma ignition pressure for the DCSF is lower than that for a planar electrode DC diode system [10]. Previously, the DCSF system has been used to deposit a-Si:H films [4]. Ultrathin a-Si:H films (8 to 40 nm) deposited by the DCSF have also been used to effectively passivate the surfaces of crystalline silicon wafers [11]. The latter films were deposited using a "Tetrode" configuration (see Figure 1) in which there were four electrodes (one anode, one semitransparent cathode, one solid cathode, one substrate holder). In this study we report on a first systematic study of the effect of the grid spacing in the tetrode configuration.

EXPERIMENT

In this experiment a deposition chamber is built as shown in Figure 1. The showerhead, anode, and cathode electrodes are mounted to the chamber by linear motion controlled bellows feedthoughs to vary the electrode spacing. The showerhead itself can be electrically biased with an rf (13.56 MHz) or DC power supply. As well, the gas is supplied from the gas manifold system though the showerhead electrode. The pressure is set by controlling the butterfly valve. A substrate holder can be placed on the heater plate to heat the substrates. The heater plate has a 1 cm hole in the middle that permits the Hiden EQP1000 Ion Mass Spectrometer (IMS) head to sample the plasma. The ion mass spectrometer was differentially pumped with a 70 l/s turbo drag pump.

For the plasma characterization and film growth, the chamber was filled with 160 mTorr of pure silane (SiH_4) flowing at 30 sccm. The heater temperature was kept at 100°C for plasma analysis and 300°C for film growth. The anode current was kept constant at 100 mA. Amorphous silicon films were grown on double side polished high resistivity (> 20 Ω cm) crystalline silicon (c-Si) substrates of thickness greater than 500 μm. The c-Si substrates were cleaned in a piranha solution (H_2SO_4:H_2O_2 = 4:1) for 10 minutes at 80°C, then rinsed in de-ionized (DI) water for 10 minutes in an ultrasonic bath. After the DI water rinse the substrates were blow dried with compressed dry air. a-Si:H films where then characterized by spectroscopic ellipsometry to extract the thickness and optical constants and FTIR to obtain the microstructural parameter (R) and hydrogen content.

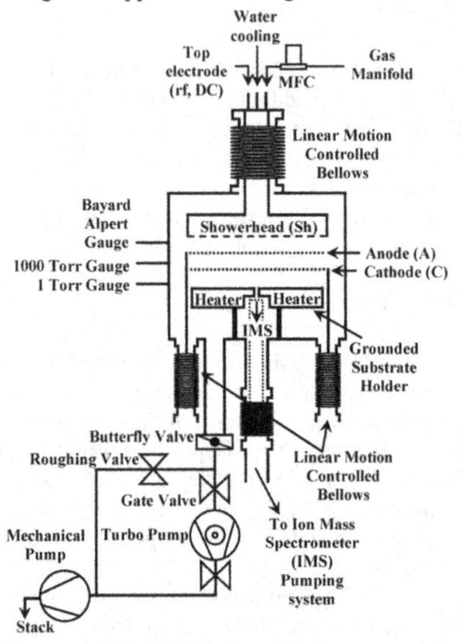

Figure 1: The DCSF deposition system with electronically controlled linear motion feedthroughs for three of the electrodes (anode, cathode, and showerhead). An Ion Mass Spectrometer (IMS) is placed below the heater to sample the plasma.

For the plasma characterization a mass spectrum (with ion filament off) was measured along with an energy scan for different masses of interest. The anode to cathode distance (A-C) was kept constant at 33 mm and the cathode to substrate distance (C-Sub), or anode to showerhead (A-Sh) distance, was varied.

RESULTS AND DISCUSSION

Plasma Characterization

A mass scan (relative intensity as a function of m/z (mass/charge) at zero energy) with the filament off and plasma on is shown in Figure 2. We observe that ions containing one to five silicon atoms are present for all electrode distances used in this study. The SiH_3^+ ion is the dominant species followed by $Si_2H_5^+$. This is in contrast to Weakliem et al. [5], where at larger electrode distances di-silicon containing ions increased above mono-silicon containing ions. This may either be due to the lower pressure (160 mTorr compared to 310 mTorr) or the DCSF system itself.

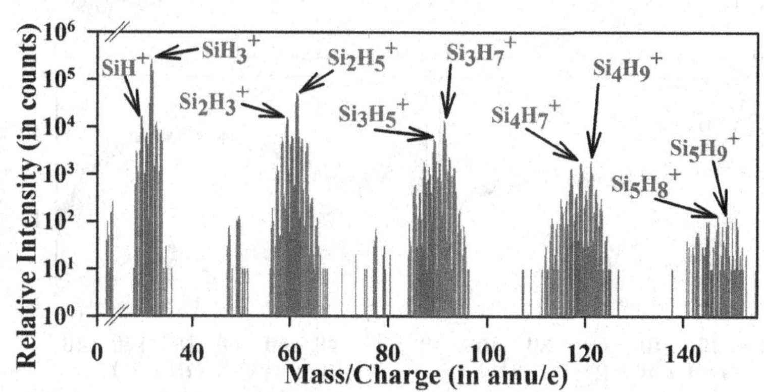

Figure 2: Relative ion intensity at zero energy (with the ion filament off) as a function of m/z for a SiH_4 plasma in the DCSF in the tetrode configuration for a C-Sub (A-Sh) distance of 42 mm (21 mm). All spectra in these experiments showed the presence of higher order silicon containing ions $Si_nH_m^+$ (n=1..5, m=1..13). SiH_3^+ ions are the dominant ions.

An energy scan was performed from 0 eV to 200 eV for the following ions: SiH_5^+, SiH_4^+, SiH_3^+, SiH_2^+, SiH^+, Si^+, $Si_2H_2^+$, $Si_2H_3^+$, $Si_2H_4^+$, $Si_2H_5^+$, $Si_2H_6^+$, $Si_2H_7^+$, $Si_3H_7^+$, $Si_3H_8^+$, and $Si_3H_9^+$. Figure 3 shows the relative intensity as a function of ion energy for SiH_3^+ (top), SiH^+ (lower left) and Si^+ (lower right) ions. For larger cathode to substrate (C-Sub) distances, or smaller anode to showerhead distances (A-Sh), we find a high energy tail (as in SiH_3^+) that slowly decreases for most species. This high energy tail decreases as we move from SiH_m^+ ions to $Si_2H_m^+$ ions, and disappears for $Si_nH_m^+$ ions where $n \geq 3$ (not shown). The SiH^+ ions exhibit an almost constant high energy distribution that extends to almost 150 eV. The Si^+ ion displays a distribution which first peaks around 9 eV, shows a long extended tail that peaks again around 110 eV. Moreover, the relative intensity of the low energy Si+ peak and low energy SiH+ peak are of the same order of magnitude. It is worthy to note that the observed ion energy distributions for Si^+ and SiH^+ ions are similar to those observed in an rf diode discharge operating at an electrode distance greater than 30 mm at 30 mTorr and a power density of 10 mW/cm^2.

Figure 3 also shows that as the cathode to substrate (C-Sub) distance is decreased, the high energy tail decreases. As well, the peak energy also decreases. This behavior was also observed to a similar degree but less pronounced when the cathode is moved further from the

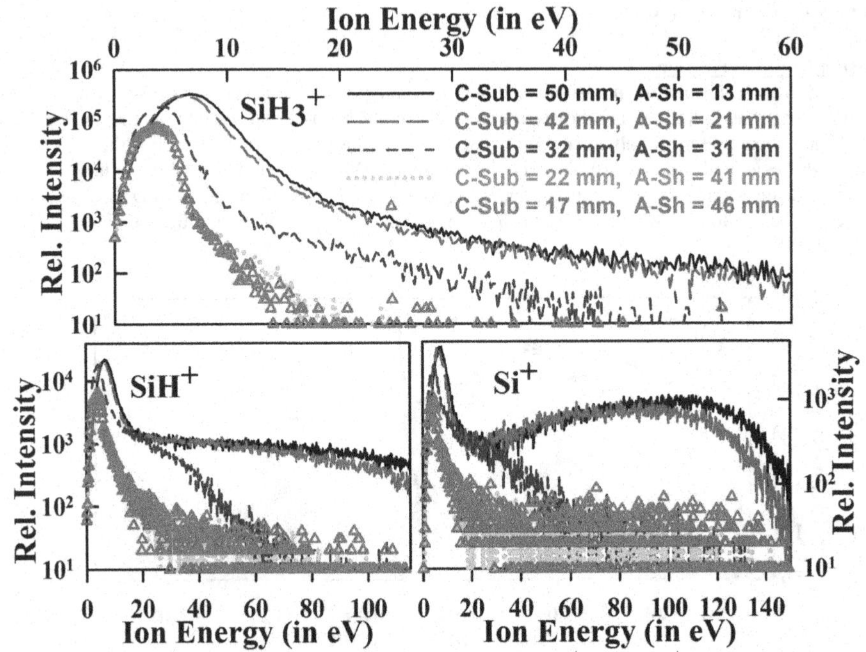

Figure 3: Relative intensity as a function of ion energy for SiH_3^+ (top), SiH^+ (bottom left), and Si^+ (bottom right) ions. As the anode-cathode set moves further from the showerhead (and closer to the substrate holder) the peak energy decreases and the intensity of high energy tail is reduced.

anode (towards the substrate) while keeping all other distances constant (anode to showerhead and anode to substrate). The decreasing of the peak energy and the loss of the high energy tail is attributed to the increasing of the anode to showerhead (A-Sh) distance. Since the showerhead is solid (when compared to a grid cathode), ion bombardment of the showerhead cathode releases secondary electrons that will be accelerated by the sheath in front of the grounded showerhead. These high energy electrons cause ionization and dissociation. Since the pressure is relatively large (160 mTorr) the ions can dissipate more energy in the increased volume between anode and showerhead. At smaller A-Sh distances, ions may pass though the anode (since at high pressure the anode fall is relatively weak [12]), be accelerated by the transparent cathode sheath, pass though the transparent cathode and continue to flow to the substrate holder (hence IMS).

The peak intensity as a function of the cathode to substrate (C-Sub) distance for different ions is shown in Figure 4. We find that the peak value shows a minimum at C-Sub distance of about 20 mm. The $Si_2H_7^+$ peak intensity increases as the C-Sub distance decreases, whereas the $Si_3H_7^+$ peak intensity remains constant. This result will be investigated at a later date.

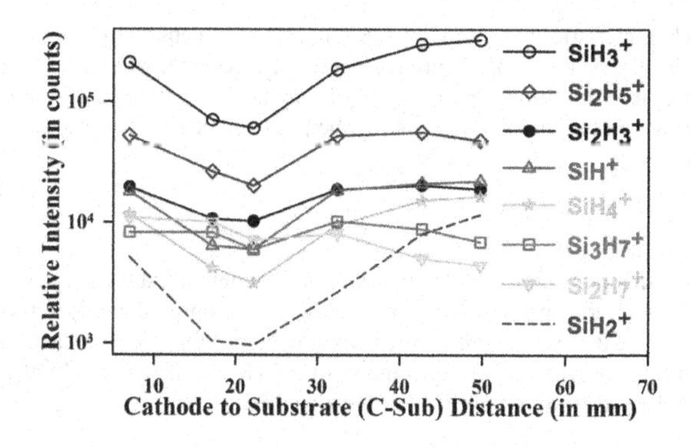

Figure 4: Relative peak intensity as a function of the cathode to substrate (C-Sub) distance for various ion species. There is a minimum at C-Sub distances of around 20 mm. The $Si_2H_7^+$ peak increases as the C-Sub distance decreases, whereas the $Si_3H_7^+$ peak intensity remains constant.

Film Characterization

Table I summarizes the a-Si:H film properties obtained from FTIR and spectroscopic ellipsometry (SE). For the SE measurements the Tauc-Lorentz (TL) model was used to fit the data from 2.0 to 5.0 eV. From the TL fit the refractive index (*n*) and extinction coefficient (*k*) were extracted and the Tauc model was used to find the Tauc gap.

Table I: Summary of a-Si:H films deposited under different electrode distances.

Sample	C-A (mm)	A-Sh (mm)	C-Sub (mm)	Tauc Gap (eV)	Growth Rate (nm/min)	Hydrogen Content [H] %	Microstructural Parameter R $I_{2100}/(I_{2000}+I_{2100})$	Average C parameter from the TL model
1	33	31	32	1.76	0.462	17.8	0.201	2.32
2	33	41	22	1.71	0.400	16.8	0.156	2.29

Two samples were deposited. The microstructural parameter (R) decreased which could be reflected in the slight decrease in the C parameter of the SE fit. A decrease in both of these parameters could point to an increase in the structural ordering of the films. The increased ordering is thought to be due to the decrease in the high energy ion bombardment. The growth rate has decreased which could correlate with the decrease in the relative ion intensity as the C-Sub distance is decreased. The Tauc gap decreased which could be reflected in the slight decrease in the hydrogen content and increase in the structural ordering. The relatively high hydrogen content in both films could be due to the large overall electrode distance (Sub-Sh = 96 mm). Weakliem et al. [5] found that large electrode distances increase higher order Si containing radicals and ions and hence the hydrogen content of the resulting films. Higher order Si containing ions ($Si_nH_m^+$, where n = 2 to 5) were present for all electrode distances investigated in this study.

The heater temperature at which plasma characterization takes place is different from that used during film growth. The heater does provide additional energy to the plasma which could then affect the plasma characteristics (electron temperature and/or ion energy distribution). Takai et al. [13] observed that as the heater temperature increases the electron temperature decreases. Hence, the observed change in the film properties may or may not be correlated with the changes observed in the ion energy and ion intensity. Films will be prepared at 100°C to investigate this effect more closely.

CONCLUSIONS

It has been shown that a minimum exists in the peak ion energy and the peak intensity with varying distance between the anode-cathode and substrate. An optimal anode-cathode to substrate distance of about 22 mm will decrease the peak ion energy for most ions and the high energy tail for SiH^+ and Si^+ ions. Ion mass spectrometry of a SiH_4 plasma formed by the DC Saddle Field shows that the predominant ion is the silyl ion (SiH_3^+).

ACKNOWLEDGMENTS

This work was supported through grants from Natural Science and Engineering Research Council (NSERC) of Canada, Ontario Centres of Excellence, and the Ontario Research Fund – Research Excellence, a tripartite program. The assistance of Drs. T. Kosteski and D. Yeghikyan and the support of Sopra are also acknowledged.

REFERENCES

1. R. Platz, C. Hof, S. Wieder, B. Rech, D. Fischer, A. Shah, A. Payne, and S. Wagner, *Materials Research Society Symposium Proceedings*, San Francisco, 1998, **507**, pp. 565-570.
2. P. Alpuim, V. Chu, and J. P. Conde, *Journal of Applied Physics*, **86**, pp. 3812–3821 (1999).
3. W. J. Soppe, H. J. Muffler, A. C. Biebericher, C. Devilee, A. R. Burgers, A. Poruba, L. Hodakova, and M. Vanecek, *Proceedings of the twentieth European Photovoltaics Solar Energy Conference*, Barcelona, Spain, (June 2005), pp. 1604–1607.
4. R. V. Kruzelecky, S. Zukotynski, C. I. Ukah, F. Gaspari, and J. M. Perz, *Journal of Vacuum Science and Technology A*, **7** (4), pp. 2632–2638 (1989).
5. H. A. Weakliem, R. D. Estes, and P. A. Longeway, *Journal of Vacuum Science and Technology A* **5** (1), pp. 29-36 (1987).
6. R. C. Ross, J. Jaklik Jr., *Journal of Applied Physics*, **55**, pp. 3785-3794 (1984).
7. Y. Maemura, H. Fujiyama, T. Takagi, R. Hayashi, W. Futako, M. Kondo, and A. Matsuda, *Thin Solid Films* **345**, pp. 80-84 (1999).
8. Shin-ichiro Ishihara, M. Kitagawa, T. Hirao, and K. Wasa, *Journal of Applied Physics* **62**, pp. 485-491 (1987).
9. D. Mataras, S. Cavadias, and D. Rapakoulias, *Journal of Vacuum Science and Technology A* **11**, pp. 664-671 (1993).
10. E. Sagnes, J. Szurmak, D. Manage, and S. Zukotynski, *Journal of Vacuum Science and Technology A* **17** (3), pp. 713-720 (1999).
11. B. Bahardoust, A. Chutinan, K. Leong, A. B. Gougam, D. Yeghikyan, T. Kosteski, and N. P. Kherani, S. Zukotynski, *Physica Status Solidi A*, **207**, pp. 539–543 (2010).
12. J. Wong, N. P. Kherani, and S. Zukotynski, *Journal of Applied Physics* **101**, pp. 013308 (2007).
13. M. Takai, T. Nishimoto, M. Kondo, and A. Matsuda, *Applied Physics Letters* **77**, pp. 2828-2830 (2000).

Sensors and Novel Devices

Mater. Res. Soc. Symp. Proc. Vol. 1321 © 2011 Materials Research Society
DOI: 10.1557/opl.2011.937

Development of Si Microliquid Processing using Piezo Actuator

Muneki Akazawa, Shunki Koyanagi and Seiichiro Higashi
Department of Semiconductor Electronics and Integration Science,
Graduate School of Advanced Sciences of Matter, Hiroshima University,
Kagamiyama 1-3-1, Higashi-Hiroshima, 739-8530, Japan

ABSTRACT

The tip of a Si rod was melted by laser diode (LD) irradiation and we succeeded in dropping small Si droplets by vibration of the Si rod using a piezo actuator. We confirmed multiple small Si droplets under a condition of the resonance frequency of 5.8 kHz for the rod length of 6.0 mm. We observed ejection of droplets from a cone edge of molten Si and the minimum width of the solidified Si was ~ 1 μm in diameter. The solidified Si show high crystallinity with the Raman scattering TO phonon band of 515.6 cm^{-1}.

INTRODUCTION

The formation of highly-crystallized Si films at a low temperature has attracted much attention because of their potential advantages in the application to active and contact layers of thin film transistors (TFTs) and solar cells [1]. In that regard, many studies based on plasma enhanced chemical vapor deposition (PECVD) techniques have so far been made to enhance the growth rate and to improve the film quality, especially the crystallinity [2,3]. Recently, we have proposed a new rapid crystallization technique using a molten Si droplet ejected from a boron nitride cylinder with a 100 μmφ nozzle heated at 1850 K [4]. By dropping multiple small droplets like ink-jet printing, this technique has the possibility to form high crystallinity Si films without severe heat damage to the substrate [5,6]. In our previous work, we have found that pseudo-epitaxial growth occurs by dropping Si microliquids on hydrogen(H)-terminated Si wafer surfaces[6]. It is quite interesting that solidified Si droplets are easily removed from H-terminated Si wafer, however, they show (100) and (110) orientations when dropped on H-terminated (100) and (110) surfaces, respectively. Therefore, if we could deposit multiple small droplets on H-terminated wafer, we can fabricate orientation-controlled Si thin films by this method. However, there are two critical issues. One is the Boron contamination from the cylinder wall, and the other one is large droplet size (~ 300 μm).

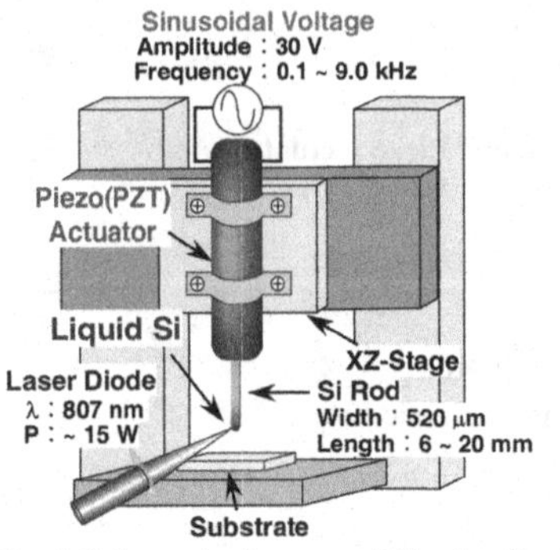

Fig. 1 Schematic diagram of Si microliquid processing equipments using a piezo actuator.

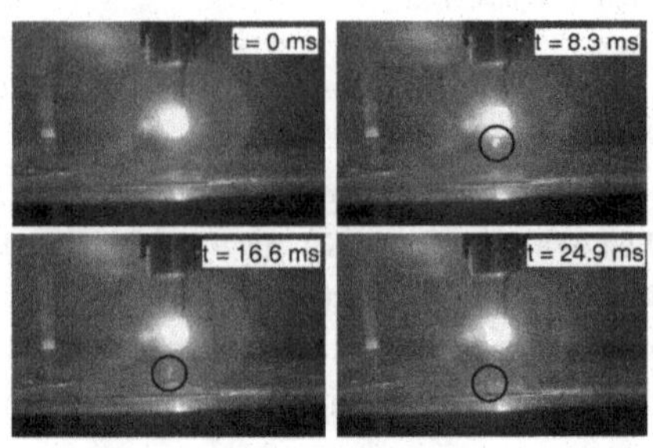

Fig. 2 Snapshots of a Si microliquid ejecting from the tip of the rod.

In this work, the tip of a single crystalline Si rod was melted by laser diode (LD) irradiation and we attempted to drop smaller pure Si liquid by vibration of the Si rod using a piezo actuator.

EXPERIMENT

The experimental setup is shown in Fig. 1. A 525 μm-thick n-type Si(100) wafer was cut into 6 ~ 20 mm long rods with the width of 520 μm. The Si rod was connected to a piezo actuator. The distance between the tip of the Si rod and a quartz substrate was set at 1 mm. A laser diode (LD) light, with a wavelength of 807 nm and a power of ~ 15 W was irradiated to the tip of the Si rod to form molten Si by non-contact method. The piezo actuator was operated by applying sinusoidal voltage at an amplitude of 30 V in the frequency (f) range of 0.1 ~ 9.0 kHz. Ejection of Si droplets from the rod tip was observed by a video camera with a frame rate of 240 per second. The crystallinity of the solidified Si droplets on quartz was investigated by Raman scattering spectra in which a p-polarized 514.5 nm laser light was used for excitation.

RESULTS AND DISCUSSION

We observed small Si droplets ejected from the tip of the Si rod under f of 5.4 kHz as shown in Fig. 2. From the snapshots in Fig. 2, we confirmed a dropping Si microliquid within a time less than 25 ms. The velocity of the Si microliquid was calculated as 30 cm/s in this specific

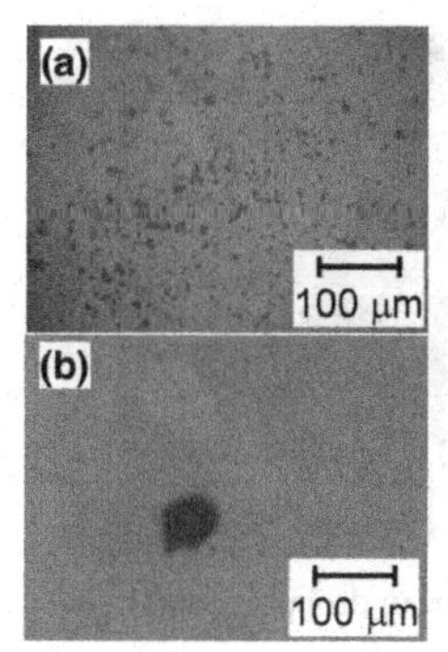

Fig. 3 Photographs of (a) small Si droplets, and (b) a large Si droplet.

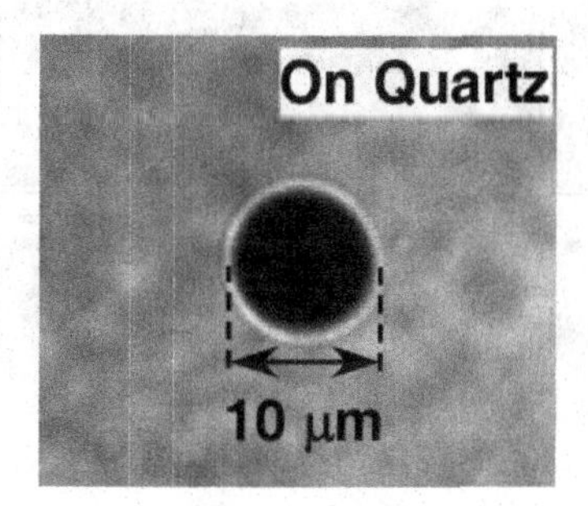

Fig. 4 Photographs of small droplets shown in Fig. 3(a).

case. We scanned the *f* from 0.1 to 9.0 kHz, and under a certain frequency, we confirmed significant number of small Si droplets ejected on a quartz substrate (Fig. 3(a)). In some cases, we observed only a few large Si droplets (0.1 ~ 1.0 mm in width) (Fig. 3(b)). We observed large droplets solidified in asymmetric shape, while small droplets showed a tendency to have circular shape. Fig. 4 shows microscope image of a small Si droplet. The minimum width of the solidified Si was ~ 1 μm. We plotted the number of Si droplets as a function of *f* (Fig. 5). It is clearly seen that the number of droplet show a sharp peak at a certain frequency (5.4 kHz), which suggests the resonance of piezo-oscillator and Si rod (6 mm) gives formation of large number of micro-droplets. Finally, we obtained a huge number (~ 450) of small droplets (1 ~ 50 μm in diameter) on quartz surface. This specific frequency depends on the rod length.

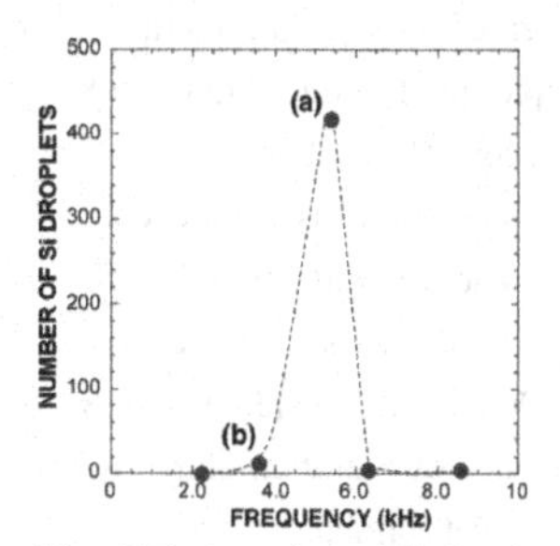

Fig. 5 The number of Si droplets as a function of oscillator frequency.

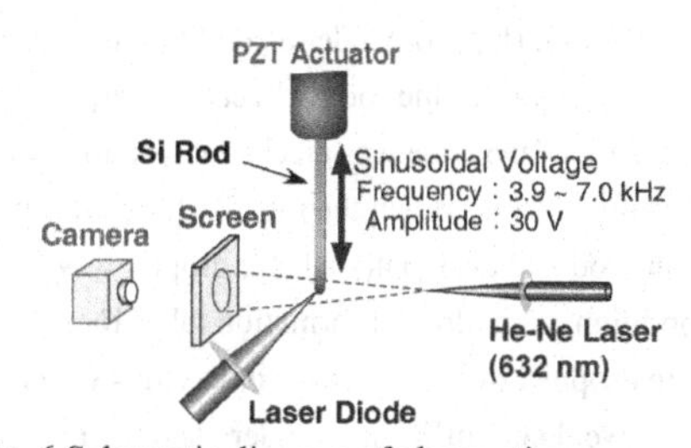

Fig. 6 Schematic diagram of observation on shadow of the tip of the Si rod by video camera.

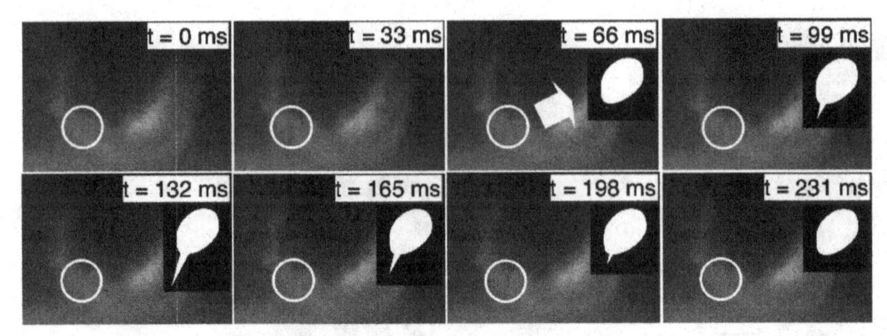

Fig. 7 Selected frames of photographs for shadow of the tip of Si rod during dropping small Si droplet. White drawing shows the sketch of the tip shape indicated by a circle.

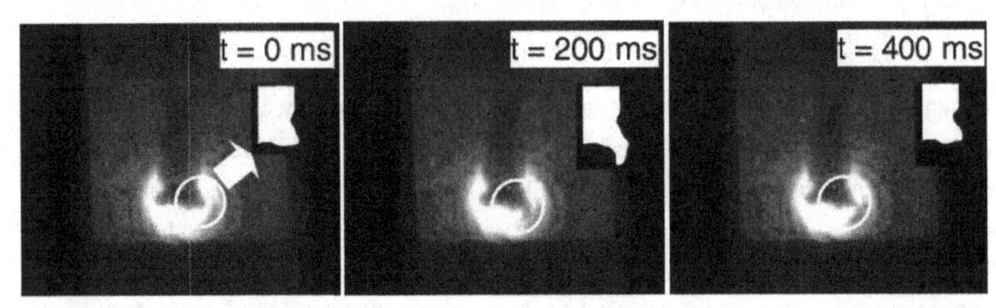

Fig. 8 Selected frames of photographs for shadow of the tip of Si rod during dropping large Si droplet.

In order to better understand the behavior and formation process of droplets, we introduced a shadow observation method. The molten Si at the tip of the rod was irradiated by a He-Ne laser light through a lens, and the magnified shadow on a screen was observed by a video camera during PZT actuation (Fig. 6). Under a condition of dropping multiple small Si microliquid (5.8 kHz, in this case), a part of the rod tip became sharper and formed a cone (Fig.7. 99ms). Eventually, the tip of the cone was ejected and apart from the melt (Fig. 7). This is a fast event which took about 100 ms. When the f is off-resonance, we observed a deformation of molten Si hanging on the rod and eventually a large droplet was formed (Fig. 8). In contrast with the resonant condition, large droplet formation takes longer time of ~ 400 ms. We considered that Si droplets were dropped by sound waves or cavities which were generated by vibration of the Si rod. However, we don't sufficiently understand the mechanism of droplet formation.

In order to evaluate crystallinity of the each droplet, we measured Raman scattering spectra. The solidified Si (1 μm in width) shows high crystallinity with Raman TO phonon peak position of 515.6 cm^{-1} and the full width at half maximum (FWHM) of 8.0 cm^{-1} (single crystalline Si: 520.0 cm^{-1} and 6.3 cm^{-1}, respectively) (Fig. 9). We found decreasing peak wavenumber with decreasing droplet size. This is attributed to either tensile stress due to the difference of thermal expansion between Si and quartz, or slight temperature rise of Si crystals induce by excitation laser.

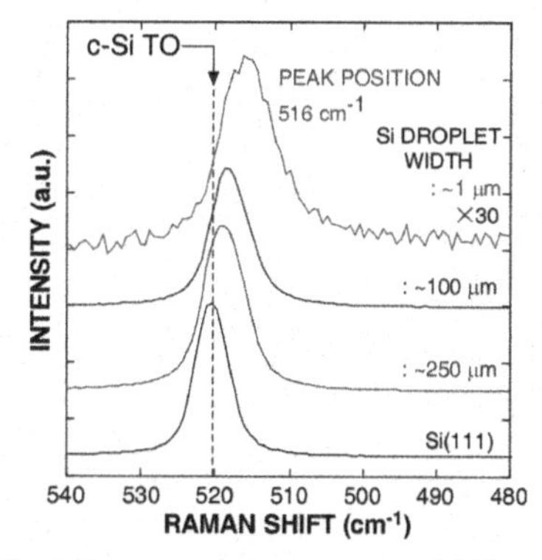

Fig. 9 Raman scattering spectra of different diameter Si droplets.

CONCLUSIONS

We succeeded in a rapid crystallization technique of Si microliquid on quartz substrates using a piezo actuator. We confirmed significant number (~ 450) of small Si droplets dropped at a certain piezo-oscillator frequency, which suggested the resonance of the oscillator and Si rod gave formation of large number of micro-droplets. The minimum width of the solidified Si was ~ 1 μm. In addition, the solidified Si films show high crystallinity by Raman Scattering Analysis.

These results suggest that piezo-driven microliquid formation is a very promising technique for low temperature formation of high-crystallinity Si films.

ACKNOWLEDGEMENTS

The authors would like to thank staff members of the Research Institute for Nanodevice and Bio Systems (RNBS), Hiroshima University for their experimental cooperation. The measurement of Si crystallinity was made using Laser Raman microscope at the

Natural Science Center for Basic Research and Development (N-BARD), Hiroshima University.

REFERENCES

1. M. Kondo et al., J. Non-Cryst. Solids 266–269 (2000) 84.
2. A. Shah et al., Thin Solid Films 403–404 (2002) 179.
3. N. Kosku et al., Appl. Surf. Sci. 244 (2005) 39.
4. N. Koba et al., Proc. Int. TFT Conf. (2009) 263.
5. S. Higashi et al., Abst. ICANS23 (2009) 252.
6. T. Matsumoto et al., Abst. of 19th PVSEC (2009) 295.

Mater. Res. Soc. Symp. Proc. Vol. 1321 © 2011 Materials Research Society
DOI: 10.1557/opl.2011.1330

Thin Film Power Harvesting System for Displays

Arman Ahnood[1], Reza Chaji[2], Arokia Nathan[1]
[1] London Center for Nanotechnology, University College London, UK
[2] IGNIS Innovation Inc., Kitchener, Ontario, Canada

ABSTRACT
An amorphous silicon (a-Si:H) thin film transistor (TFT) circuit designed for charging of an energy storage device using a photovoltaic (PV) array is presented. The TFT circuit is fabricated at plastic compatible temperatures (~150°C) and as such can easily be integrated within a range of platforms including flexible displays. The circuit provides a high degree of output voltage stability over a range of light intensities and device stress.

INTRODUCTION
Consumer requirements for handheld mobile devices include longer battery life, so that mains recharging can be performed less frequently. To this end a key component of any mobile system is a high power and high energy density battery. An alternative approach to extend battery lifetime for handheld devices is to recycle some of their own energy consumption or harvest energy from ambient sources. Displays are one of the most power consuming components of mobile devices. In modern mobile devices, the display occupies a substantial portion of the exposed area. By designing a display such that it also functions as a "solar cell" it is possible to harvest some of the ambient light energy [1-3]. Given the scope for power harvesting, it is important to investigate their implementation and integration. In this work we present a TFT based circuit capable of regulating the output power of a PV array for storage in a battery or supercapacitor. The schematic shown in Figure 1 indicates how the TFT charging circuit fits within the system.

The power conditioning circuit considered here fulfils two roles:

a) Limits the charging voltage to avoid damage to the energy storage device. Furthermore, a narrower voltage range at the energy storage device will allow for a more efficient design of the power circuitry in mobile devices.

b) Provides input impedance to the PV array such that it would operate close to its maximum power point.

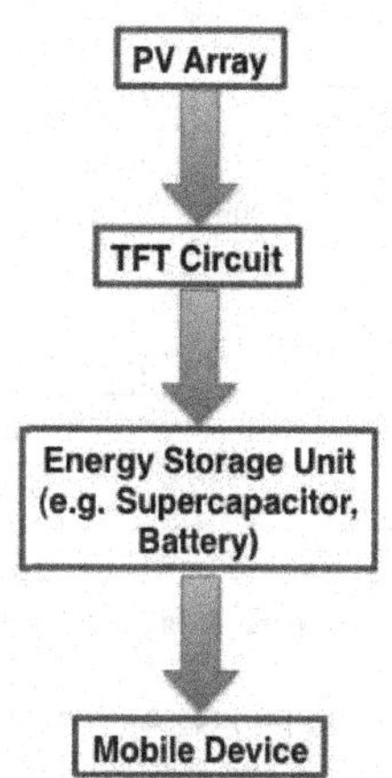

Figure 1. Overview of on-panel power harvesting system for displays.

EXPERIMENTAL
Figure 2 shows the cross sectional diagram of the TFT used in the circuit. The TFTs were fabricated using the etch stop (ES) process[4]. The device structure consists of 50nm amorphous silicon (a-Si:H) channel layer, 300nm silicon nitride (SiNx) gate dielectric layer, and 50nm n-type nano-crystalline silicon (nc-Si:H) contact layers. All of the films were deposited using radio-frequency plasma enhanced chemical vapour deposition (RF-PECVD) at the 150°C in a two chamber system. The a-Si:H was deposited in a dedicated intrinsic chamber using $SiH_4/H_2/He$ with gas flow rates of [2]/[15]/[5] sccm at 0.8 Torr pressure and 21mW/cm^2 RF power density. The SiNx was

deposited at a pressure of 1Torr using a mixture of $SiH_4/H_2/NH_3$ with gas flow ratios of [4]/[40]/[120] sccm at 125mW/cm^2, 200Hz pulsed-RF (50% duty cycle) power. The deposition of n-type nc-Si:H was performed in two stages with an initial seed layer growth using H_2/SiH_4 of 150 followed by deposition nc-Si:H using a lower hydrogen dilution of 100. Here PH_3 was used as the n-type dopant gas source with PH_3/SiH_4 ratios of [1]/[2] and [2]/[2] for the seed and nc-Si:H layers, respectively. Thermally evaporated chromium (Cr) and aluminium (Al) were used as the gate electrode and as the drain and source contacts as well as interconnect electrode in the circuit, respectively.

Figure 3 (a) shows the schematic of the charging circuit used in this work. The circuit's input would be connected to the PV array and the output to the energy storage unit. Transistor T3 provides the shunt path, which acts as an overvoltage protection mechanism for the energy storage unit. Transistor T4 (with T3) sets the output voltage. It also prevents the discharge of the capacitor, via the PV array. Initially the load capacitor is fully discharged (V_{out} = 0). Connecting the PV array (under illumination) results in its charging leading to an increase in V_{out} accompanied with a rise in V_{in}. The increase in V_{in} leads to two outcomes:

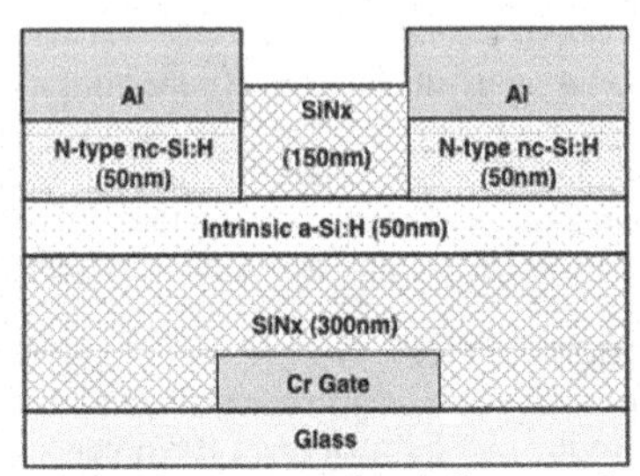

Figure 2. Cross sectional diagram of TFT used in this work.

a) Shift in the PV array's operating point towards higher voltages (and lower currents).

b) Higher $V_{control}$ leading to higher conduction via the shunt TFT (T3).

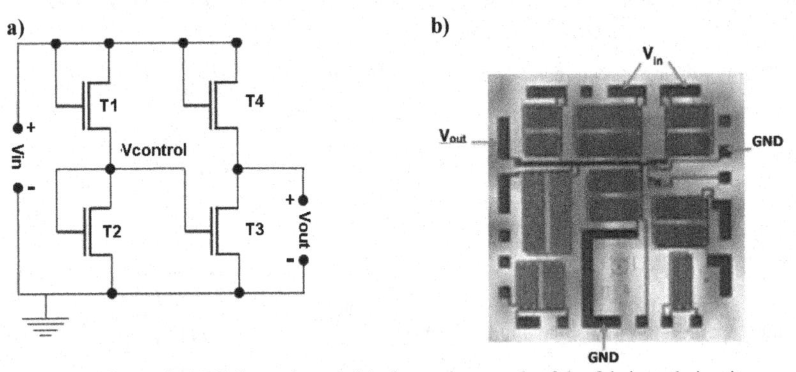

Figure 3. (a) Schematic and **(b)** photomicrograph of the fabricated circuit.

As charging progresses, the circuit would finally reach a point when T3 conducts all of the current, and V_{out} reaches a maximum value. This value of V_{out} can be chosen to be the maximum voltage the energy storage unit can reach without any damage. A correctly designed system would act in two ways:

a) Limit the variation in the PV array's operating point, allowing it to maintain operation close to the maximum power point.

b) Minimise the shunt conduction in T3.

Figure 3(b) shows the photomicrograph of the fabricated circuit, where the positions of various circuit connections are indicated. TFTs with the following W/L dimensions (μm) were used in this circuit: T1:5000/23; T2:40/80; T3:6000/23; and T4:7500/23.

RESULTS AND DISCUSSION

Figure 4 (a) shows the transfer characteristics of a typical TFT from which the key device parameters are extracted. The device has a field effect mobility of 1.1 $cm^2/V.s$, subthreshold voltage slope of 0.3 V/dec, threshold voltage of 1.8 V and ON/OFF ratio of 10^6. The dependence of the field effect mobility on gate voltage is shown in Figure 4 (b). It can be seen that the value of the mobility does not drop in higher gate voltages suggesting negligible influence of contact resistance[5-6].

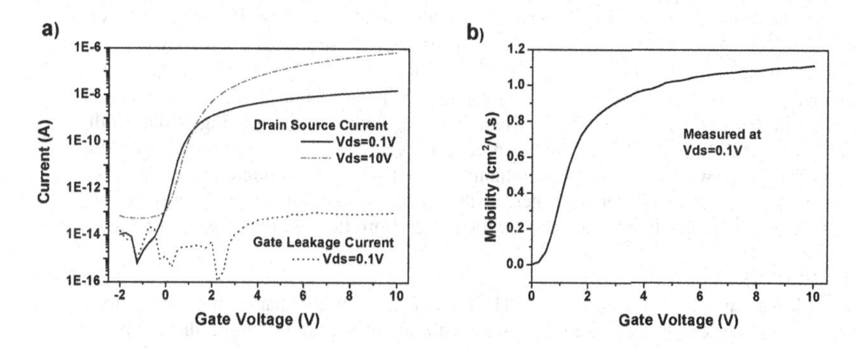

Figure 4. (a) The transfer characteristics of a typical TFT in the linear (V_{ds}=0.1V) and saturation (V_{ds}=10V) regimes. **(b)** The dependence of the extracted field effect mobility on gate voltage in the linear regime.

Figures 5 shows the effect of gate voltage biased stress on the threshold voltage. It can be seen that stressing at 10V leads to a substantial shift in threshold voltage. There was no significant change in the subthreshold slope, suggesting charge trapping in the dielectric as the dominant cause of the instability [7].

The V_{out} stability under various illumination conditions and circuit operation duration was investigated by connecting the circuit's input terminal to a PV emulator, consisting of a programmable current source (Keithley 2400) in parallel with 16 serially connected diodes. A PV emulator was used in order to decouple the effect of PV array's instabilities from the circuit instabilities. Effect of the variation in the light intensity was emulated by varying the short circuit current (I_{sc}) using the Keithley 2400 current source. The circuit stressing was performed over a $2x10^4$ sec time period using a stress current I_{sc} of 1μA and under the fully charged load condition. Figure 6 shows the variation of V_{out} with I_{sc} before and after the circuit stressing was applied. The stressing inevitably causes shift in the threshold voltage of the TFTs in circuit. The two orders of magnitude variation in the I_{sc} corresponds to a similar order of variation in light intensity.

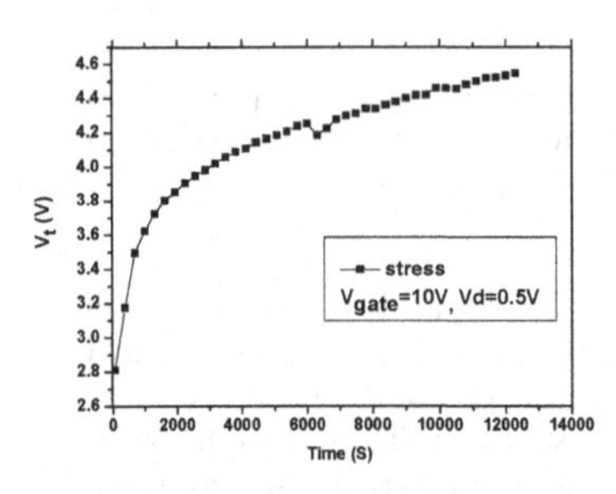

Figure 5. Effect of the gate bias stressing on the a-Si:H TFT on threshold voltage.

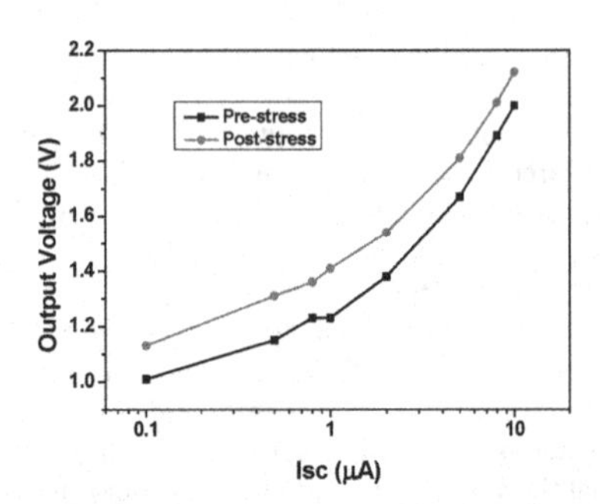

Figure 6. Dependence of V_{out} on short circuit current (I_{sc}) of the PV emulator before and after $2x10^4$ sec stress of $I_{sc} = 1\mu A$.

The relatively good stability of V_{out} in the I_{sc} range of 0.1~1μA is desirable for selection of the size of the PV array elements. The variation in V_{out} with circuit stressing is due to the threshold voltage shift of the TFTs which can be improved by optimising the SiN_x layer and using nc-Si:H TFTs which exhibit higher stability than their a-Si:H counterparts. However, despite this the V_{out} stability demonstrated in this work is sufficient when energy storage devices such as supercapacitors[8] are used, which offer more flexible charging requirements.

CONCLUSIONS

In this work we have demonstrated an a-Si:H TFT circuit for power harvesting applications. The low fabrication temperature used allows its seamless integration within a display panel, making it an attractive solution for display power scavenging. Furthermore the possibility of using low deposition temperature a-Si:H PV array and thin film super-capacitors opens the possibility of a fully integrated energy harvesting system.

ACKNOWLEDGEMENTS

This work was partially funded by the IGNIS Innovation Inc.

REFERENCES

1. P. Hyehyang and C. Byoungdeog, U.S. Patent No. 20080163923 (n.d.).
2. H. Tani, T. Kyoji, and A. Shigeru, U.S. Patent No. 4095217 (n.d.).
3. C.-J. Yang, T.-Y. Cho, C.-L. Lin, and C.-C. Wu, Appl. Phys. Lett. **90**, 173507 (2007).
4. A. Nathan, A. Kumar, K. Sakariya, P. Servati, S. Sambandan, K.S. Karim, D. Striakhilev, IEEE J. Solid-State Circuits, vol. 39 (2004) 1477-1486.
5. M. Shur, M. Hack, and J.G. Shaw, Journal of Applied Physics **66**, 3371-3380 (1989).
6. A. Ahnood, K. Ghaffarzadeh, A. Nathan, P. Servati, F. Li, M.R. Esmaeili-Rad, and A. Sazonov, Appl. Phys. Lett. **93**, 163503-3 (2008).
7. M.J. Powell, C. van Berkel, and J.R. Hughes, Appl. Phys. Lett. **54**, 1323 (1989).
8. P. Barrade, S. Pittet, and A. Rufer, PCIM2000 Power Conversion and Intelligent Motion, (Nurnberg, Germany, 2000).

Mater. Res. Soc. Symp. Proc. Vol. 1321 © 2011 Materials Research Society
DOI: 10.1557/opl.2011.814

Optical bias controlled amplification in tandem Si-C pinpin devices

M. Vieira[1,2,3], M. A Vieira[1,2], P. Louro[1,2], M. Fernandes[1,2], A. Fantoni[1,2], M. Barata[1,2]

[1]Electronics Telecommunication and Computer Dept. ISEL, R. Conselheiro Emídio Navarro, 1949-014 Lisboa, Portugal Tel: +351 21 8317290, Fax: +351 21 8317114, mv@isel.ipl.pt.
[2] CTS-UNINOVA, Quinta da Torre, Monte da Caparica, 2829-516, Caparica, Portugal.
[3] DEE-FCT-UNL, Quinta da Torre, Monte da Caparica, 2829-516, Caparica, Portugal.

ABSTRACT

A monolithic double pi'n/pin a-SiC:H device that combines the demultiplexing operation with the simultaneous photodetection and self amplification of the signal is analyzed under different electrical and optical bias conditions at low and high excitation frequencies. Results show that the transducer is a bias wavelength current-controlled device that make use of changes in the wavelength of the background to control the power delivered to the load. Self optical bias amplification or quenching under uniform irradiation and transient conditions is achieved. The device acts as an optical amplifier whose gain depends on the background wavelength and frequency. An optoelectronic model supported by an electrical simulation explains the operation of the optical system.

INTRODUCTION

There has been much research on semiconductor optical amplifiers [1]. Here, a specific band or frequency need to be filtered from a wider range of mixed signals. Amorphous silicon carbon tandem structures, through an adequate engineering design of the multiple layers' thickness, absorption coefficient and dark conductivities can accomplish this function. Those devices have a nonlinear amplitude-dependent response to each incident light wave. Under controlled wavelength backgrounds the light-to-dark sensitivity in a specific wavelength range can be enhanced and quenched in the others, tuning a specific band.

This paper reports results on the use of a double pi'n/pin a-SiC:H Wavelength Division Multiplex (WDM) heterostructure as an active band-pass filter transfer function dependent on the wavelength of the trigger light and on the electrical and optical applied bias.

DEVICE PREPARATION, CHARACTERIZATION AND OPERATION

Optical amplifiers were produced and optimized for a fine tuning of a specific wavelength. The active device consists of a p-i'(a-SiC:H)-n / p-i(a-Si:H)-n heterostructure with low conductivity doped layers. The configuration of the device is shown in Figure 1. Experimental details on the preparation, characterization and optoelectronic properties of the amorphous silicon carbide films and junctions were described elsewhere [2].
The thicknesses and optical gap of the thin i'- (200 nm; 2.1 eV) and thick i- (1000 nm; 1.8 eV) layers are optimized for light absorption in the

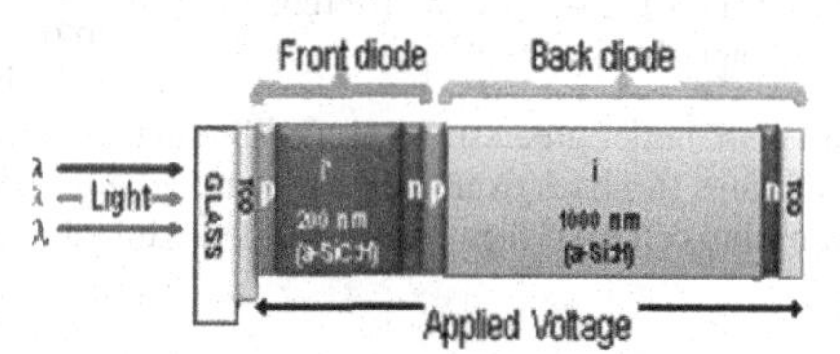

Figure 1 Device configuration.

blue and red ranges, respectively [2]. As a result, both front and back structures act as optical filters confining, respectively, the blue and the red optical carriers. The device operates within the visible range using as optical signals the modulated light (external regulation of frequency and intensity) supplied by a red (R: 626 nm; 51 μW/cm^2) a green (G: 524 nm; 73 μW/cm^2) and a blue (B: 470 nm; 115 μW/cm^2) LED. Additionally, steady state red, green blue and violet illumination (background) was superimposed using similar LEDs. The devices were characterized through data transmission measurements under different carrier frequencies (50 Hz<f<3500 Hz) with (R, G, B) and without applied steady state bias. To analyze the device under information-modulated wave and uniform irradiation, three monochromatic pulsed lights separately (input color channels) or their combination (multiplexed signal) illuminated the device. Steady state optical bias was superimposed separately and the photocurrent generated measured at -8 V.

SELF BIAS AMPLIFICATION UNDER UNIFORM IRRADIATION

In Figure 2, the spectral photocurrent at 250 Hz (a) and 3500 Hz (b) modulated carriers is displayed under red, green blue and violet background irradiations (color symbols) and without it (black symbols). Results show that, whatever the frequency, the violet background amplifies the spectral sensitivity in the visible range while the blue optical bias only enhances the spectral sensitivity in the long wavelength range and quenches it in the short wavelength region. The red and green optical bias effects depend on the frequency. In the low frequency regime (Figure 2a), under red irradiation, the photocurrent is strongly enhanced at short wavelengths and disappears for wavelengths higher than 550 nm. Under green irradiation the spectral sensitivity is strongly reduced. In the high frequency regime (Figure 2b), under red irradiation, the spectral sensitivity is enhanced only in the blue region of the spectrum while under green background, the blue and the red sensitivities are enhanced and the green one decreased. In Figure 3 the spectral gain, defined as the ratio between the spectral photocurrents, under red (α^R), green (α^G) and blue (α^B) steady state illumination and without it (no background) are plotted for the same two frequencies. Results confirm the wavelength controlled spectral sensitivity of the device, under steady state illumination.

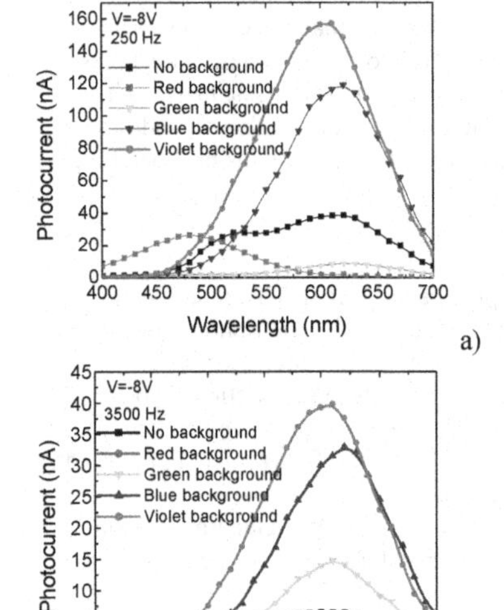

Figure 2 Spectral photocurrent at: a) 250Hz and b) 3500Hz and different wavelength background.

It is known that when an external electrical bias is applied to a double pin structure, its main influence is in the field distribution within the less photo excited sub-cell [3].The front cell, under red irradiation; the back cell, under violet or blue light and both, under green steady state illumination. In comparison with thermodynamic equilibrium conditions (no background), the electrical field under illumination is lowered in the most absorbing cell (self forward bias effect) while the less absorbing reacts by assuming a reverse

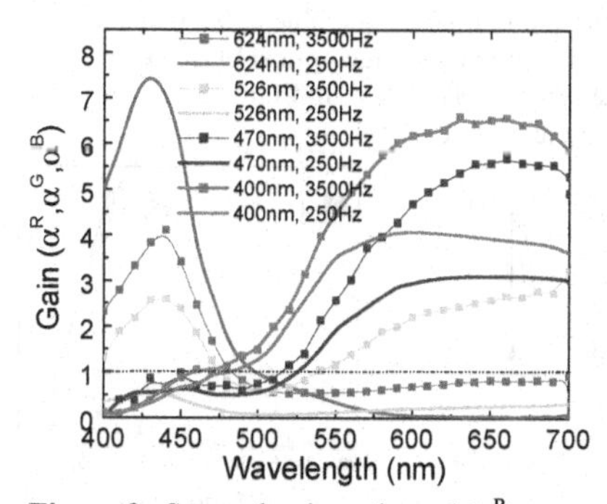

Figure 3 Spectral gain under red (α^R), green (α^G), and blue (α^B), optical bias and without it for different frequencies.

bias configuration (self reverse bias effect). The sensor is a bias wavelength current- controlled device that makes use of changes in the wavelength of the background to control the power delivered to the load, acting as an optical amplifier. Its gain depends on the background illumination wavelength. If the electrical field increases locally (self optical amplification) the collection is enhanced and the gain is higher than one. If the field is reduced (self optical quench) the collection is reduced and the gain becomes lower than one. This optical nonlinearity makes the transducer attractive for optical communications and can be used to distinguish a wavelength, to suppress a color channel or to multiplex or demultiplex an information-modulated wave.

FREQUENCY ANALYSIS

In Figure 4 the spectral gain as a function of the frequency is displayed under red (α^R), green (α^G) and blue (α^B) backgrounds at 624 nm (α_R, red channel), at 526nm (α_G, green channel) and at 470nm (α_B, blue channel). Results show that the background wavelength also controls the frequency dependence of the color channels. Under red and green irradiations two frequency

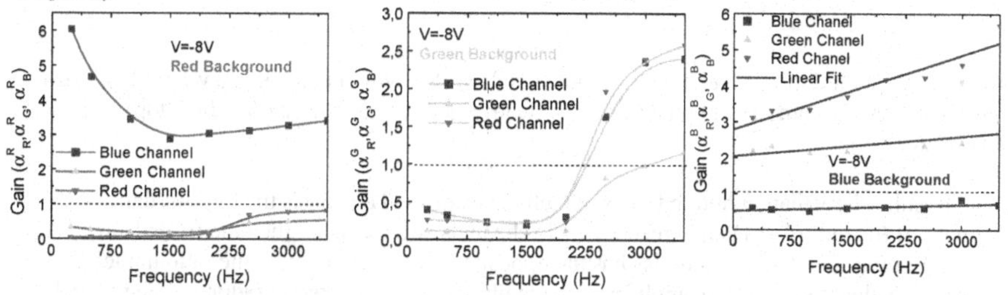

Figure 4 Spectral gain under red (α^R), green (α^G) and blue (α^B) backgrounds as a function of the frequency at 624nm (red channel), at 526nm (green channel) and at 470nm (blue channel).

regimes can be considered. One, for frequencies lower than 2000Hz, were either under red and green backgrounds the green and the red channel gains are very low (<<1). The blue gain is strongly enhanced (>>1) under red background or reduced (<1) under green irradiations. The other regime, for frequencies higher than 2000Hz, the gain increases with the frequency, gradually under red and quickly under green steady state illumination. Under blue, the gain increases slowly with the frequency being higher than one for the red and green channels and lower for the blue one.

TRANSIENT OPTICAL BIAS AMPLIFICATION

Three monochromatic pulsed lights (input channels): red, green and blue at 6000 bps and 250 bps illuminated separately the device. Steady state red, green and blue optical bias was superimposed separately and the photocurrent generated measured at -8 V. In Figure 5 the transient photocurrent is displayed for each monochromatic input channel.

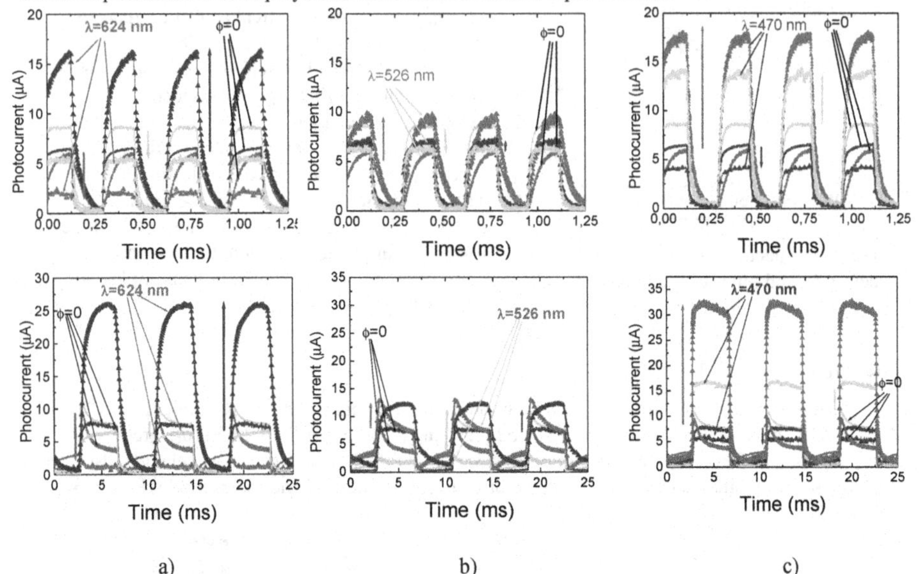

Figure 5 Input red, green and blue transient signals at -8V without (φ=0) and with: a) red (624 nm), b) green (526 nm) and c) blue (470 nm) steady state optical bias at 6000 bps (top) and 250 bps (bottom).

Results show that, at both bit rates, the blue background enhances the light-to-dark sensitivity of the red ($\alpha^B{}_R>1$) and green ($\alpha^B{}_G>1$) channels and quenches the blue ($\alpha^B{}_B<1$) as shown in Figures 3 and 4. The red bias has the opposite behavior; it reduces the ratio in the red/green wavelength range and amplifies it in the blue one. Under green irradiation the red and blue signals are enhanced and the green reduced by α factors that depend strongly on the bit rate used for the transmission (see color arrows trend in the figures).

A chromatic time dependent wavelength combination of the same input pulsed channels but with different bit sequences was used to generate a multiplexed signal in the device. The output photocurrents at -8V with (color lines) and without (dark lines) background is displayed in Figure 6 at 6000 bps (a) and 250 bps (b). The bit sequences are shown at the top of the figure to guide the eyes. Results show that, even under transient input signals, the background wavelength controls the output signal due to the asymmetrical light penetration of the input channels across the device together with the modification on the electrical field profile due to the optical bias. This high optical nonlinearity makes the optimized devices attractive for the amplification or quenching of all optical signals and enables the device to demultiplex an optical encoded data stream. To recover the transmitted information (8 bit per wavelength channel) the output wave

under red irradiation and without it was used. Both multiplexed signals, during a complete cycle (T), were divided into eight bit time slots (Δ=1/bps) corresponding to one bit where the independent optical signals can be ON (1) or OFF (0).

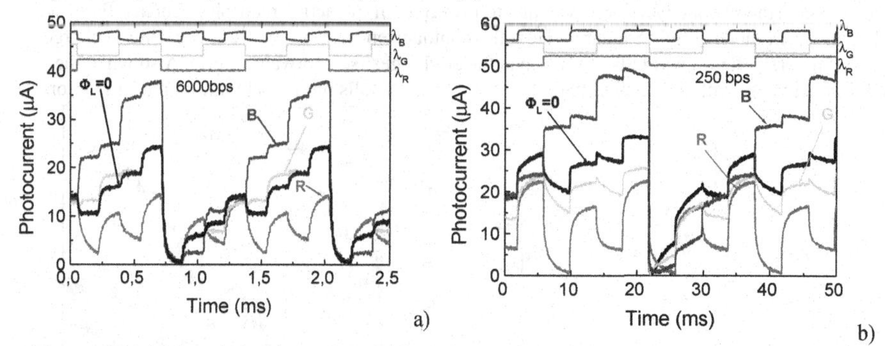

Figure 6 Multiplexed signals at -8 V; without (φ_L=0) and with red (R) green (G) and blue (B) optical bias at 6000 bps (a) and at 250 bps(b).

In Figure 6 all the possible combinations of the three input channels are present, so, the waveform of the output without optical bias is an 8-level encoding (2^3) to which it corresponds 8 different photocurrent thresholds. Taking into account Figures 3-5, under red background the red channel is strongly quenched ($\alpha^R{}_R$<<1) and the blue enhanced ($\alpha^R{}_B$>>1) for both bit rates. So, the output waveform becomes a main 4-level encoding (2^2). Here, the higher level corresponds to both blue and green ON (_11) and the lower to the absence of both (_00). The other two intermediate levels are ascribed as follow, the upper level to the ON state of the blue (_01) channel and the lower to the green channel ON (_10). To decode the red channel, similar 4-levels encoding with and without red background have to be compared, the higher ones have the red channel ON. A demux algorithm was implemented in Matlab that receives as input the measured photocurrent and derives the sequence of bits that originated it. The independent red, green and blue bit sequences were decoded as: (a) B[01010101], G[00110011] and R[1111000] or (b) B[10101010], G[10011001] and R[01111000].

CAPACITIVE MODEL

The device was modeled by a two single-tuned stages circuit with two variable capacitors and interconnected phototransistors through a resistor [4] as displayed in Figure 7. Two optical gate connections ascribed to the different light penetration depths across the front and back phototransistors were considered to allow independent blue (I_B), red (I_R) and green ($I_{Gpi'n}$, I_{Gpin}) channels transmission. The operation is based upon the following principle: the flow of current through the resistor connecting the two front and back transistor bases is proportional to the difference in the voltages across both capacitors (charge storage buckets). In Figure 7 the simulated current without and under red backgrounds is displayed (symbols). The input channels (I_R, I_G, I_B, Figure 5) are also displayed (lines). To simulate the red background, current sources intensities (input channels) were multiplied by the on/off ratio between the input channels with and without red optical bias ($\alpha^R{}_R$, $\alpha^R{}_G$, $\alpha^R{}_B$, Figure 4). The same bit sequence and bit rate of

Figure 6a was used. Good agreement between experimental and simulated data was observed. The eight expected levels, under reversed bias, and their reduction under red irradiation, are clearly seen. Under red background the expected optical amplification of the blue channel and the quenching of the red one were observed due to the effect of the active multiple-feedback filter when the back diode is light triggered. Here the photocurrent rises as the front capacitor is charged up (self reverse effect), and falls as the capacitor discharges (self forward effect). Also when the red channel is on, with red background, the back capacitor falls more than without red irradiation.

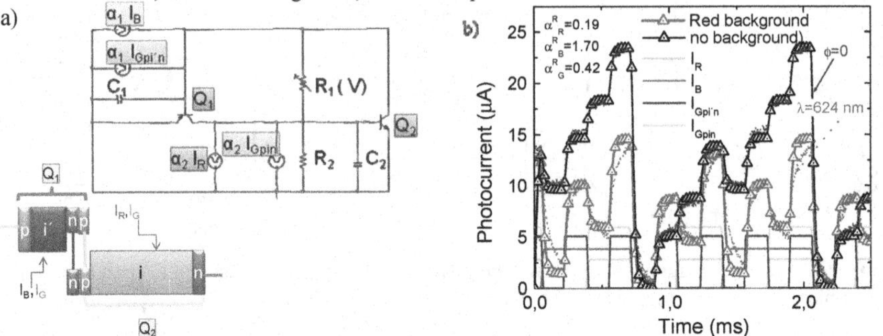

Figure 7 a) Optoelectronic model. b) Multiplexed simulated (symbols), input channels (dashed lines) and experimental (solid lines): under negative dc bias and red background.

CONCLUSIONS

Optical bias controlled amplification in tandem SiC pinpin devices was analyzed under different steady state optical bias conditions and frequencies. A capacitive model was presented to support the analysis. Results show that devices have a nonlinear amplitude-dependent response to each incident light wave. The background wavelength controls the output signal due to the asymmetrical light penetration of the input channels together with the modification on the electrical field profile due to the optical bias. This optical nonlinearity makes them attractive for the amplification or quenching of all optical signals. An algorithm based on optoelectronic logic programming was present to allow prediction over relational multiplexed signals.

ACKNOWLEDGEMENTS

This work was supported by FCT (CTS multi annual funding) through the PIDDAC Program funds and PTDC/EEA-ELC/111854/2009.

REFERENCES

1. M. J. Connelly, Semiconductor Optical Amplifiers. Boston, MA: Springer-Verlag, 2002. ISBN 978-0-7923-7657-6.
2. M. Vieira, M. Fernandes, P. Louro, A. Fantoni, M. A. Vieira, J. Costa, M. Barata, Mat. Res. Soc. Symp. Proc.Volume 1245, A08-02 (2010) .
3. M. Vieira, A. Fantoni, P. Louro, M. Fernandes, R. Schwarz, G. Lavareda, and C. N. Carvalho, Vacuum, Vol. 82, Issue 12, 8 August 2008, pp: 1512-1516.
4. M. A. Vieira, M. Vieira, J. Costa, P. Louro, M. Fernandes, A. Fantoni, in Sensors & Transducers Journal Vol. 9, Special Issue, December 2010, pp.96-120.

Mater. Res. Soc. Symp. Proc. Vol. 1321 © 2011 Materials Research Society
DOI: 10.1557/opl.2011.938

Amorphous Silicon Based Particle Detectors

N. Wyrsch[1], A. Franco[1], Y. Riesen[1], M. Despeisse[1], S. Dunand[1], F. Powolny[2], P. Jarron[2] and C. Ballif[1]
[1]Ecole Polytechnique Fédérale de Lausanne (EPFL), Institute of Microengineering (IMT),
[2]Photovoltaics and thin film electronics laboratory, Breguet 2, 2000 Neuchâtel, Switzerland.
CERN, CERN Meyrin, 1211 Genève 23, Switzerland.

ABSTRACT

Radiation hard monolithic particle sensors can be fabricated by a vertical integration of amorphous silicon particle sensors on top of CMOS readout chip. Two types of such particle sensors are presented here using either thick diodes or microchannel plates. The first type based on amorphous silicon diodes exhibits high spatial resolution due to the short lateral carrier collection. Combination of an amorphous silicon thick diode with microstrip detector geometries permits to achieve micrometer spatial resolution beneficial for high accuracy beam positioning. Microchannel plates based on amorphous silicon were successfully fabricated and multiplication of electrons was observed. This material may solve some of the problems related to conventional microchannel devices. Issues, potential and limits of these detectors are presented and discussed.

INTRODUCTION

Hydrogenated amorphous silicon (a-Si:H) exhibits two advantages for applications as particle sensors: it is one of the most radiation hard semiconductors [1] and can be deposited as thin layers on various types of substrates over large areas. For these reasons this material has attracted much attention for direct and indirect (using a scintillating layer) particle detection [2]. Using a direct detection scheme, various types of particles such as protons, neutron, electrons or X-ray were successfully detected using thick a-Si:H diode [3,4,5]. Due to the low interaction of minimum ionizing particle (MIP) with silicon (due to the low atomic mass of this material) very thick diodes are necessary. Given the high voltage needed to fully deplete the device, such thick diodes are challenging in terms of the fabrication process and of the quality the material.

While a-Si:H has been successfully implemented in indirect particle detectors for X-ray radiography [6], a Si:H detector has never been practically used for direct particle detection in a physics experiment or in a commercial product. Insufficient performance, failure to detect single MIP, metastability of a-Si:H and fabrication issues are some of the reasons for this lack of success. Nevertheless, as demonstrated here, a-Si:H thick diodes can be improved for MIP detection and selected applications could greatly benefit from present a-Si:H based detectors. Vertical integration of a thick a Si:H diode array on top of a readout ASIC (so-called "thin-film on ASIC" or "TFA" technology) (see Fig. 1) enables the fabrication of very radiation hard detector and may help achieving single MIP detection [7]. Using microstrip pixel geometries, very high spatial resolution can be obtained that are comparable or superior to the ones of state-of-the-art c-Si detectors.

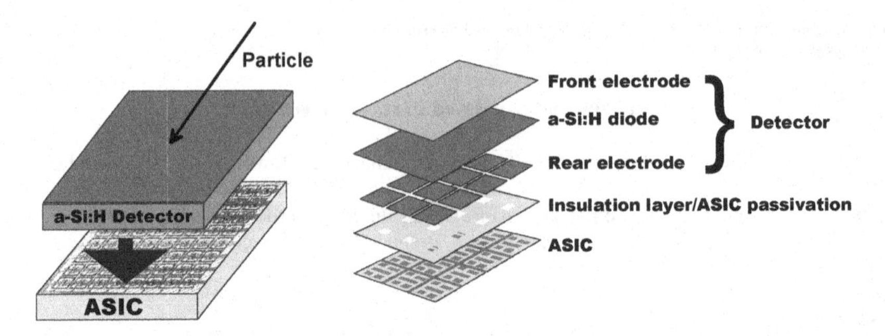

Figure 1. Schematic view of a thin-film on ASIC (TFA) particle detector.

A specific test chip was designed in the past by CERN to study the effect of pixel geometry on the performance of TFA sensors [8]. This chip includes pixels of various sizes and shapes, microstrips of various width and pitch values and two different geometries for the openings in the passivation layer of the chip (see Fig. 2). With this chip (as well as for most TFA devices), the back contacts of the individual pixels are provided by the last metal layer of the ASIC through openings in the ASIC passivation. The latter can be local or global. For the local case, the opening corresponds to the size of the metal pad with a positive or negative overlap (the size of the opening being slightly larger or slightly smaller than the size of the local metal pad). A global opening corresponds to a case where we have a single opening for several metal pads or pixels. For all microstrips present on this test chip, the openings are global. Several studies have demonstrated that the edges from the passivation openings lead to important leakage current at the periphery of the pixels in case of local openings. Global opening is therefore necessary to keep low leakage current values [9,10]. In the same studies, it was also demonstrated that diode without n-layer (i.e. an i-p layer structure directly deposited on the metallic back contact –i.e. metal-i-p configuration) further reduces the leakage. In this paper, the spatial response of microstrips and the carrier collection is analyzed for various strip and diode geometries and diode configurations.

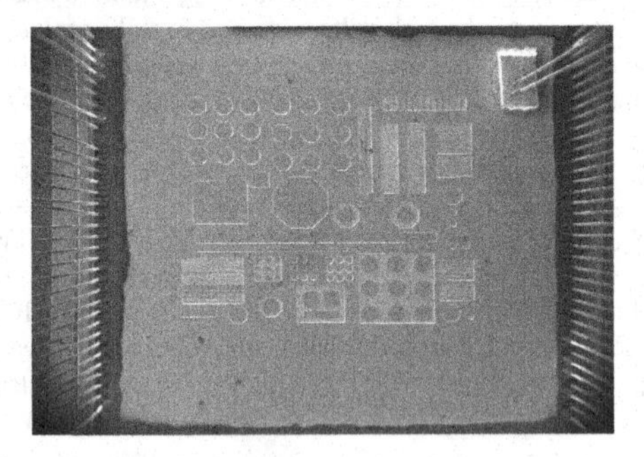

Figure 2. Picture of a 4x4 mm^2 test circuit (aSiHTest) designed for TFA performance analysis coated with a thick a-Si:H n-i-p diode. Various pixel and microstrip structures can be seen.

We recently proposed microchannel plate (MCP) detectors based on a-Si:H as an alternative to particle detection using thick diodes [11,12]. Such detectors consist of thick plates with very narrow micro-channels drilled throughout the device. When a primary electron hits the channel wall secondary electrons are emitted. Since a high electric field is applied between the two faces of the plate, an avalanche mechanism takes place in the microchannels. These avalanches thus lead to multiplication of the primary electrons. MCP are commonly used as image intensifying devices [13]. For MCP, a-Si:H material properties permit to overcome some performance limitations imposed by the lead glass or c-Si material used in commercial MCP fabrication [14,15].a-Si:H technology would also greatly simplify the fabrication of MCPs, would allow a vertical integration of such detectors on the readout electronics, thus broadening the range of applications. The typical structure of an AMCP (a-Si:H MCP) and its operation principle is given in Fig. 3. Issues regarding the fabrication of AMCP and latest characterization using EBIC techniques will be presented and discussed.

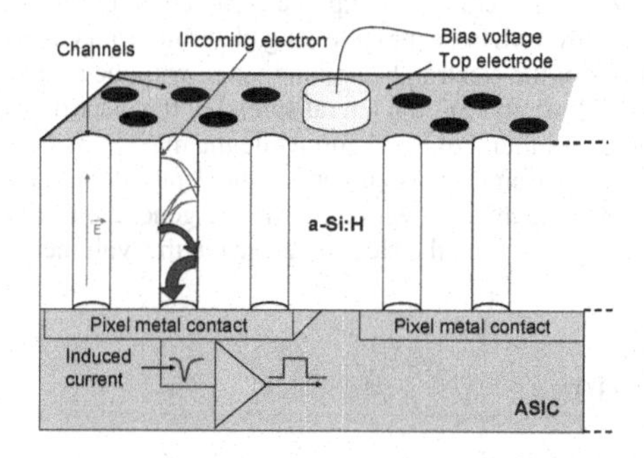

Figure 3. Schematic view of an a-Si:H based MCP (AMCP) structure, vertically integrated on a readout ASIC.

EXPERIMENT

Deposition of thick a-Si:H layers (as needed for the devices discussed here) requires careful optimization in order to achieve high deposition rate with reasonable material quality while maintaining low internal mechanical stress. In this context VHF PECVD (very high frequency plasma enhanced chemical vapor deposition) enables such deposition [16] and was used for the fabrication of all thick a-Si:H based devices discussed in this paper. Depositions were performed at a plasma excitation frequency of 70 MHz and at deposition temperatures between 200 and 235 °C. While almost stress free layer can be deposited at temperatures around 170°C on c-Si [16], higher deposition temperature was needed to keep the film integrity and to avoid delamination during and after the growth for film thicknesses up to 100 μm while obtaining acceptable compressive stress in the material. Typical deposition rate of 2 nm/s was achieved during deposition.

All optimizations of thick diodes were performed on Cr-coated glass substrates with patterned ITO top contacts. Identical process parameters were used for the fabrication of all TFA devices. In this case the Al pads of the CMOS readout chips were used as the diode array back contacts and an ITO layer was used as common top contact. Diodes without n-layer were also used both in test and TFA devices. For metal-i-p (M-i-p) TFA configurations (using the chip metal pads as metal back contacts), a chemical cleaning of the CMOS pads was necessary to obtain reproducible and uniform device performance.

AMCP test devices were fabricated on oxidized 4" Si wafers with an oxide layer thickness of 1-2 μm covered with the Al back contact plane. 80 -100 μm thick a-Si:H intrinsic layer was then deposited followed by a 0.1-0.5 μm thick n-doped a-Si:H layer used as a common top contact of the MCP pixels. A mesa patterning of the a-Si:H layers was then performed by reactive ion etching (RIE) using a Cr hard mask before the deep reactive ion etching (DRIE) of the micro-channels. For some AMCP an additional SiO_2 was deposited between the Al contact layer and the a-Si:H thick layer as an etch stop layer to improve the quality of the fabricated devices. Additional information on the fabrication of the AMCP as well as the measurement configuration can be found in [12].

An EBIC characterization system was set up in a commercial Philips ESEM XL30 (Environmental Scanning Electron Microscope) allowing electron beam energy values ranging from 1 to 30 keV. Simulations of the electron beam interaction with the a-Si:H based device were done using CASINO [17]. At 30 keV, the lateral spread of the electrons (coming from the beam) inside the a-Si:H layer was found to be <1 μm while the maximum stopping power as a function of penetration of the electron beam is attained at the depth of 2 μm and most of the electron energy is transferred to the material within 5 μm. The generation rate of electron-hole pairs (resulting from the energy loss from the electron beam) in this volume is estimated to be larger than 10^{22} $cm^{-3}s^{-1}$.

RESULTS AND DISCUSSION

TFA sensors – thick diode optimization

The intrinsic layer of the diode used for the collection of the electro-hole pairs generated by the ionizing particle should be fully depleted. The externally applied field should be high enough to allow the depletion region to extend through the entire intrinsic layer and to maximize charge collection. The depletion region is the active part of the diode. The voltage needed for full depletion increases as the square root of the thickness and voltage of ≈600 V is necessary to deplete a 30 μm thick diode [18]. Practically one usually observes a supra linear increase of the diode leakage current as the voltage is raised and this effect is more pronounced when the diode thickness is increased. Concentration of the internal field at the p-i interface, Poole-Frenkel effect, and additional current injection from the p-layer are the main reasons for such an increase in the leakage current [7].

In a previous study [18] we showed that the introduction of a buffer layer at the p-i interface is, to some extend, effective at limiting the strong increase of the leakage current with bias voltage. However, leakage is in most cases still too high to get any benefit from diodes with thicknesses larger than 15 μm; the required voltage for full depletion leads to unacceptably high leakage current.

Recently p-doped SiO_x layers have been introduced in solar cells [19,20]. Such layers enable the quenching of local shunts by introducing an anistropic resistive layer between the contacts and the intrinsic layer. This material consists of c-Si filaments embedded into a silicon oxide matrix. It exhibits a much higher in plane resistivity than its transversal resistivity, therefore limiting the spatial spreading of the effect of shunts. Such layers have been introduced in thin (1-2 μm) and thick (10 μm) sensors. As one can observe in Fig. 4, such p-doped SiO_x layers are effective at reducing current leakage in thin diodes. However, for thick ones, the benefit is not yet clear. Further study will be needed to clarify this observation.

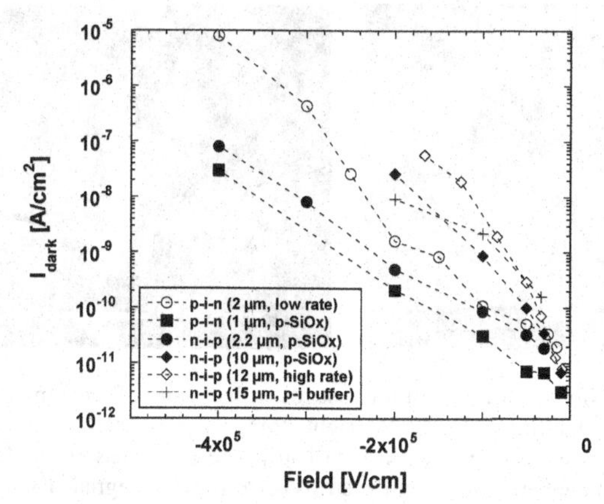

Figure 4. Dark current (or leakage current) of dioded of various thicknesses as a function of electrical field values. Full symbols correspond to diode incorporating a p-doped SiO_x layer.

TFA sensors – microstrip characterization

Preliminary characterizations of the microstrips present in the chip shown in Fig. 2 were already performed using beta particles [8]. It was demonstrated that almost no cross-tall takes place for microstrips only 4 μm apart and that the lateral collection of the charges generated by the ionizing particles is very short. Such microstrip structures could therefore permit very high spatial resolution, unmatched by the state-of-the-art detector technologies. In the case of c-Si based technology, localization of a beam/particle is given by measuring the charge sharing between two adjacent strips or pixels. The transport and collection is provided by diffusion process over relatively large distance of several tens or hundreds of microns. In the case of a-Si:H, collection is ensured by drift process and therefore an electric field must be present. The electric field outside the pixels or microstrip is low and that explains the low lateral collection. The spatial resolution is therefore strongly linked to the density of the pixels or microstrips.

To get more insight about the lateral collection of carriers in a-Si:H based microstrips, various microstrips of the chip presented above were characterized by EBIC on several diode thicknesses and geometries. In Fig. 5, the SEM picture (a) and the corresponding EBIC image (b) are shown for microstrips with a width of 16.5 μm, spacing of 33.5 μm and a-Si:H p-i-n diode of 5 μm. A one dimensional EBIC scan through the microstrips is also plotted on Fig. 6. All microstrips of this structure are connected in parallel except the central one. One can observe that the charge collection between the strips is very limited, as expected. The dark lines in the EBIC pictures are due to previous line scan through the microstrips which locally degraded the material (due to very high generation rate of electron-hole pairs induced by the SEM electron beam) and affected charge collection. This degradation, similar to the effect of light-soaking in a-Si:H, is reversible and the initial state of the diode can be recovered by thermal annealing. The fluctuation of the signal "on the microstrips" (peaks of the signal in Fig. 6) are caused by local non uniformities that can also be observed in detailed EBIC maps or SEM pictures.

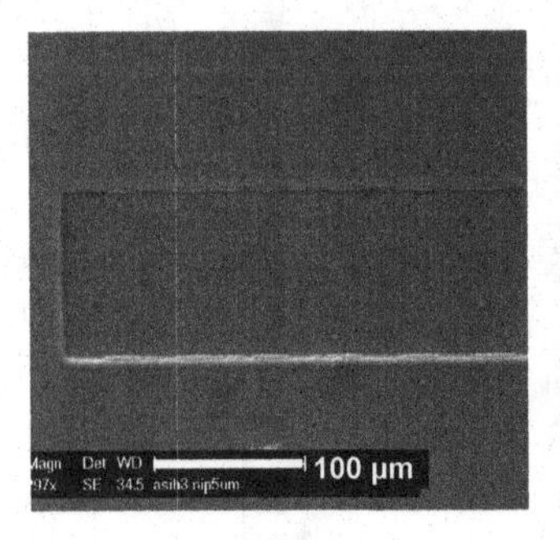

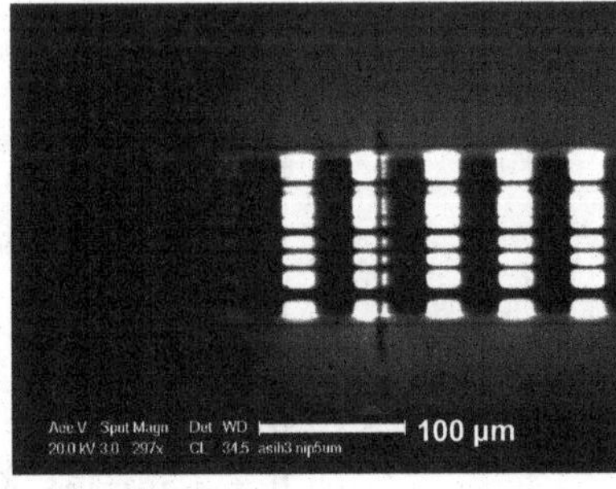

Figure 5. (Left) SEM image of a set of microstrips with a width of 16.5 µm and spacing of 33.5 µm covered with 5 µm a-Si:H n-i-p diode; (right) corresponding EBIC image obtained at applied voltage of 30 V and beam energy of 20 keV. All strips are connected in parallel except the left one. Dark regions in the EBIC image of the strips correspond to degraded area by previous EBIC line scans.

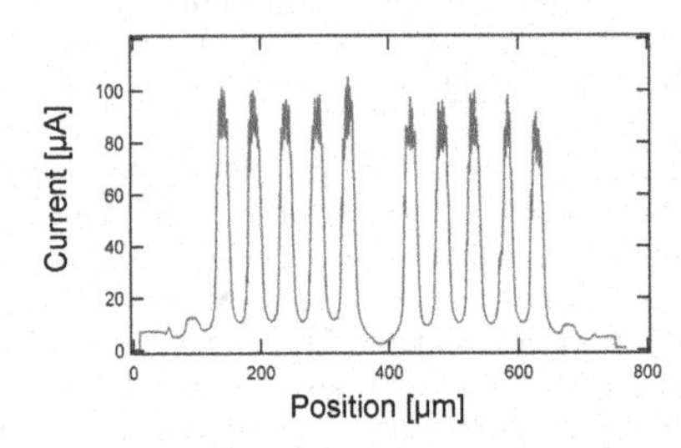

Figure 6. Signal from the EBIC line scans through the microstrip set of Fig. 5.

For detailed analysis, a Lateral Charge Collection (*LCC*) value was defined as

$$LCC[\%] = \frac{(I_{res} - I_{noise})}{(I_{strip} - I_{noise})} \cdot 100 \qquad \text{(Eq. 1)}$$

where I_{strip} is the average signal on a strip, I_{res} the signal measured at mid-point between two microstrips, and I_{noise} the residual signal far from a strip. A signal/noise ratio (*S/N)* value was also defined as

$$\frac{S}{N} = \frac{(I_{strip} - I_{noise})}{I_{noise}} \qquad \text{(Eq. 2)}$$

For the same diode thickness and configuration, EBIC pictures and one dimensional scan for microstrip with a width of 6.6 μm and spacing of 16.4 μm and for microstrips with a with of 1.5 μm width and spacing of 3.5 μm are shown in Fig. 7 and Fig. 8, respectively. *LCC* and *S/N* values measured on various microstrips are given in Table 1. One can see that with narrower strips and smaller spacing the contrast between I_{strip} and I_{res} is reduced due to the fact that some charge is collected by the adjacent microstrips for smaller spacing. Nevertheless, the individual microstrips with 1.5 μm width and only 3.5 μm spacing can be easily resolved, both in the maps and one dimensional scans. This demonstrates that almost micrometric spatial resolution can be achieved with a-Si:H microstrips.

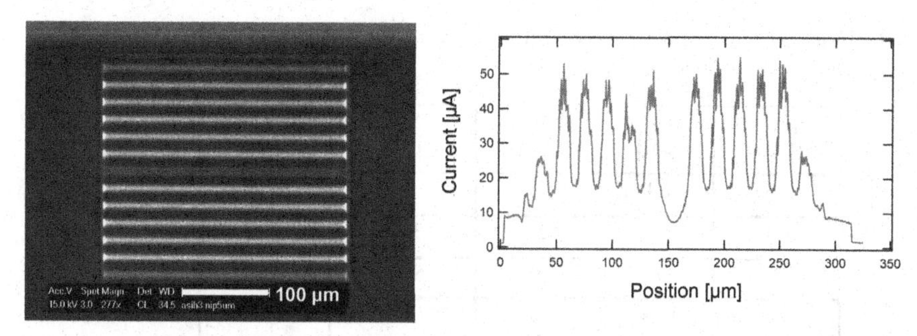

Figure 7. (Left) EBIC image of a set of microstrips with a width of 6.6 μm and spacing of 13.4 μm covered with 5 μm a-Si:H n-i-p diode at applied voltage of 30 V and beam energy of 20 keV and (right) corresponding signal from EBIC line scans through the microstrip set. All strips are connected in parallel except the middle one.

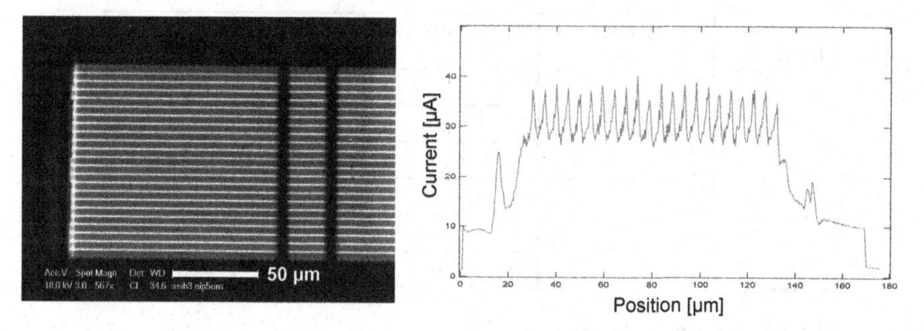

Figure 8. (Left) EBIC image of a set of microstrips with a width of 1.5 μm and spacing of 3.5 μm covered with 5 μm a-Si:H 5 μm n-i-p diode at applied voltage of 30 V and beam energy of 20 keV and (right) corresponding signal from EBIC line scans through the microstrip set. All strips are connected in parallel. Dark regions in the EBIC image of the strips correspond to degraded area by previous EBIC line scans.

Effect of the a-Si:H diode thickness as well and configuration (n-i-p or M-i-p) are summarized in Table 2 for 2 different microstrip geometries (wide and narrow). One can see that thicker diodes (in the present case) increase the *S/N* ratio while keeping the *LCC* almost constant. Replacing the n-i-p configuration with M-i-p helps to improve the *S/N* ratio for large microstrips, but is not effective for narrow ones. Even though the average electric field was kept constant for all measurements, it is expected that the field distribution in the intrinsic layer may be quite different for the three types of diodes as discussed in the diode optimization section above. The applied voltage of 60 V for 10 μm thick devices was probably not sufficient to achieve full depletion and could therefore result in a lower I_{strip} and corresponding lower *S/N*.

Table 1. *S/N* and *LCC* values determined for various microstrip geometries (width and spacing) and a 5 μm a-Si:H n-i-p diode.

Strip width [μm]	Spacing [μm]	Pitch [μm]	*S/N*	*LCC* [%]
0.6	1.4	2	-	-
1.5	3.5	5	28.3 ± 1.4	62.0 ± 0.3
3	7	10	18 ± 1	57.0 ± 0.2
6.6	13.4	20	39.0 ± 1.7	34.0 ± 0.2
16.5	33.5	50	200 ± 30	11.5 ± 0.1

Table 2. *S/N* and *LCC* values determined for two microstrip geometries (width and spacing) and 3 different a-Si:H n-i-p diodes (with different thickness values and diode configuration).

Strip width [μm]	Spacing [μm]	Diode structure	S/N	*LCC* [%]
16.5	33.5	n-i-p 5 μm	200 ± 30	11.5 ± 0.1
		n-i-p 10 μm	1.2 ± 0.3	14.2 ± 0.8
		M-i-p 10 μm	40 ± 1	7.4 ± 0.4
1.5	3.5	n-i-p 5 μm	28.3 ± 1.4	62.0 ± 0.3
		n-i-p 10 μm	3.2 ± 0.4	80 ± 1
		M-i-p 10 μm	2.2 ± 0.5	80.6 ± 1.5

Microchanel plate detectors - fabrication

For AMCP testing purposes, 15x15 mm^2 reticles have been defined with 24 pixels of 0.5x0.5, 1x1 and 2x2 mm^2 (see Fig. 9). Various reticles were designed with micro-channels (over each pixel) with a (nominal) diameter of 1.5 to 5 μm separated by a gap of (nominally) 2.5 to 3.5 μm. In Fig. 10, one can see SEM pictures of the AMCP surface of the corner of one pixel, as well as the cross section of some channel drilled by DRIE. Parallel channel walls can be obtained with high aspect ratios.

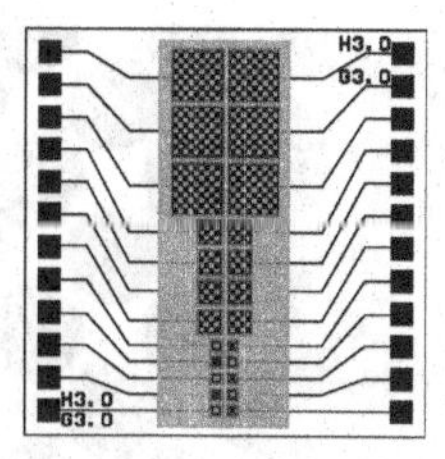

Figure 9. Schematic view of an AMCP test reticle. The back electrodes and bonding pads are in black; the colored/shaded area is covered by the thick a-Si:H layer while the patterned squares correspond to the area where the micro-channels are drilled.

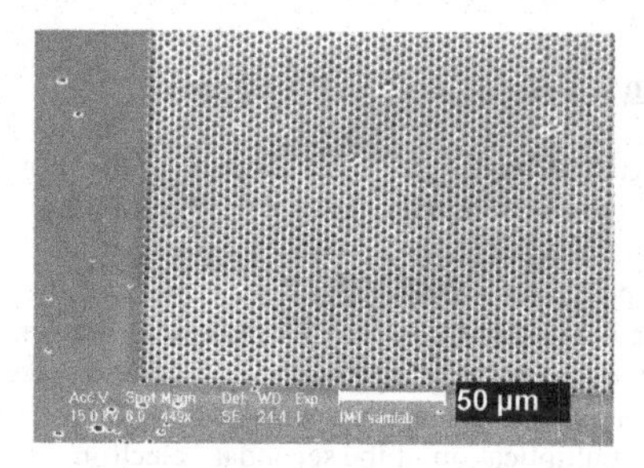

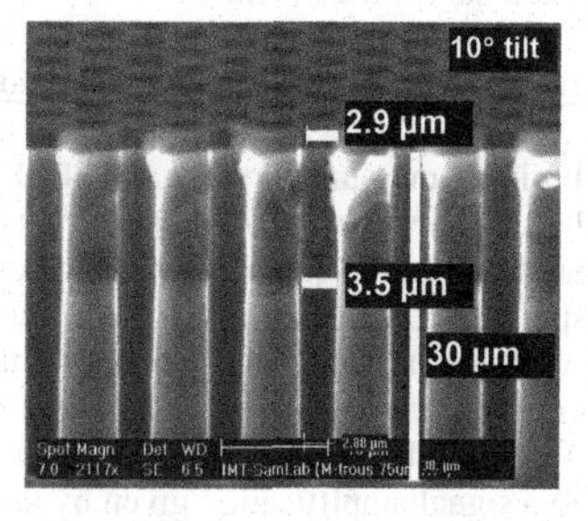

Figure 10. SEM pictures of the surface and cross section of thick a-Si:H layers with holes drilled by DRIE.

Fabrication of AMPC still results in the formation of local defects for some devices (see Fig. 11 and 12). In other devices an erosion of the front surface is sometime observed (Fig. 11 right). Even though both types of defects seem to take place at the end of the DRIE process, there is no direct link between the two as demonstrated by Fig. 12 where the surface is smooth and intact but local defects are present. Additional SiO_x layer between the Al back contact and the intrinsic layer was introduced to improve the fabrication process.

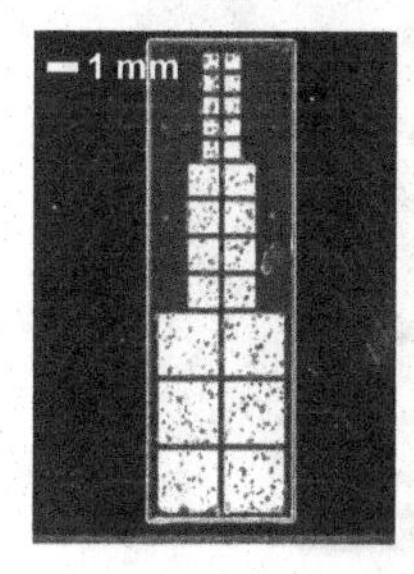

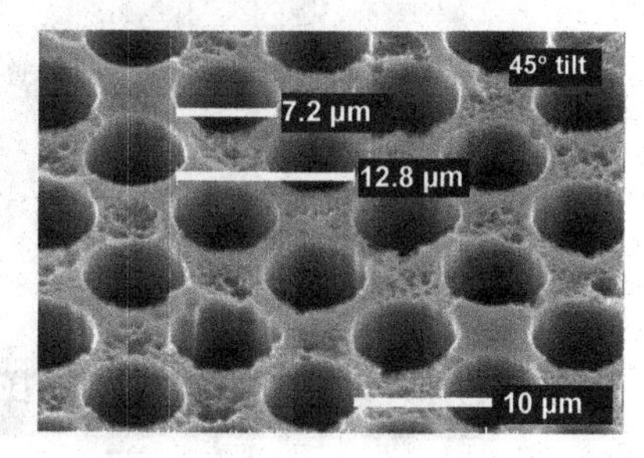

Figure 11. (Left) Picture of an AMCP reticle with many local defects on the pixel areas. (Right) SEM detail view of the pixel area showing the eroded surface.

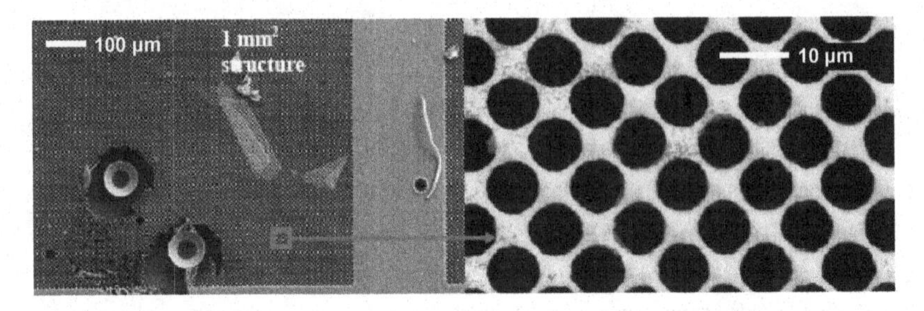

Figure 12. (Left) SEM picture of a pixel with two local defects and (Right) zoom over a defect free non-eroded part of the pixel.

Microchanel plate detectors - characterization

Preliminary measurements of AMCP already demonstrated multiplication of the primary electrons [11,12]. However, all AMCP reticles characterized in the past suffered from the erosion of the surface. resulting in etching of the top contact (a-Si:H n-layer), leading to very non uniform response of the microchannels over the pixel areas [11,12]. For devices without such erosion we can see that the response of each channel (see EBIC image in Fig. 13) is uniform. The local defects are clearly visible but do not contribute to the pixel response. If we zoom in (see Fig. 14 left) we can observe that only the periphery of the channel openings leads to EBIC signal, i.e. to a signal amplification given by the multiplication of the secondary electrons emitted from the impact of the EBIC electron beam with the channels. This result can be easily understood by looking at 4 possible interactions of the incoming electron beam with the microchannels (Fig. 15). Highest signal with uniform response of the entire channel opening is expected when the channel axis is tilted with respect to the incoming beam and this is what is observed experimentally (see Fig. 14 right).

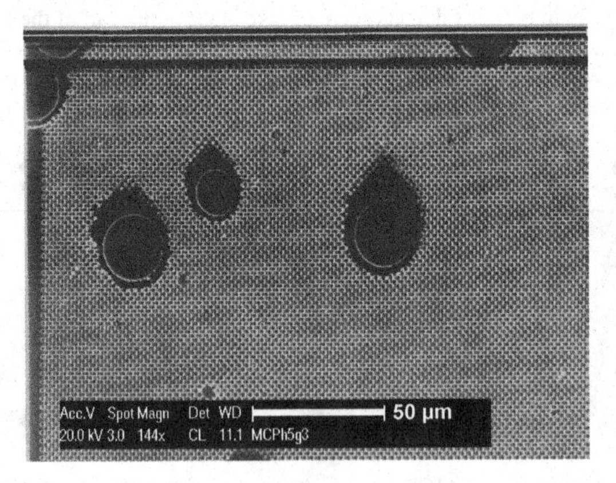

Figure 13. Partial EBIC image of a 1 mm^2 pixel of AMCP with non-eroded top surface.

Figure 14. Detailed EBIC image of the microchannel openings for AMCP perpendicular to the electron beam (no tilt, left) and with 10° tilt (right). Pictures are taken on the same pixel as in Fig. 13.

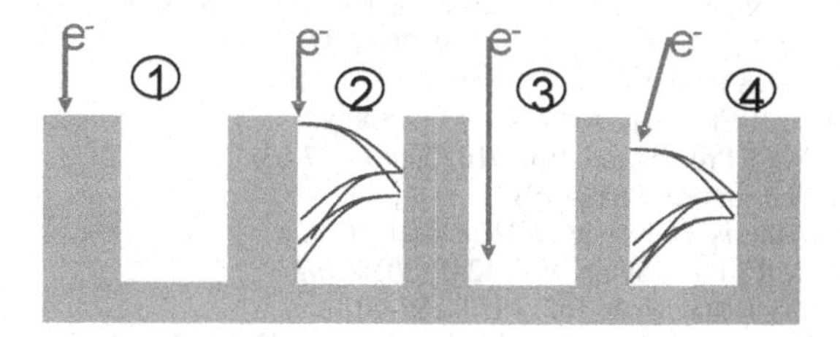

Figure 15. Possible interaction of an incoming electron with one microchannel. Only cases 2 and 4 lead to multiplication of the primary electrons. Such multiplication can only take place when the incoming electron hit the periphery of the channel opening or enter that opening with a certain angle.

CONCLUSIONS

Despite the fact that a-Si:H has not yet been implemented in any application for the direct detection of particles, we believe that this material remains very attractive when a large radiation hardness is required. Pixel or microstrip detectors involving a-Si:H thick diodes can provide unmatched spatial resolution and are suitable for very accurate beam monitoring or positioning. Such vertical integration would also be appropriate for high resolution particle imaging. However, detection of single minimum ionizing particle has not yet been achieved and may limit the potential of a-Si:H in high energy physics experiments. Further device improvement is here necessary.

a-Si:H is also a promising material candidate for MCP. In contrast to current MCP technology (using lead glass or c-Si wafer), a-Si:H has a bulk resistivity that enables fast channel charge replenishment. It allows less cumbersome fabrication process (compared to conventional

MCP) and permits a vertical integration of MCP on the readout electronics. First AMCP were successfully fabricated and multiplication of secondary electrons has been observed as expected. Some processing issues have to be solved in order to optimize the device performance and get better understanding of AMCP potential and limitations.

ACKNOWLEDGMENTS

This work was partially supported by the Swiss National Science Foundation under Contract 200021_126926/1

REFERENCES

1. H. Schade, in "Thin Film Silicon Solar" edited by A. Shah. EPFL Press, 2010
2. V. Perez-Mendez et al., J. of Non-Cryst. Sol. 137&138 (1991) 1291.
3. N. Kishimoto et al., J. Nucl. Mater. 258-263 (1998) 1908.
4. J. R. Srour et al., IEEE Trans. Nucl. Sci. 45 (1998) 2624.
5. V. P. Mendez et al., Nucl. Instrum. Methods Phys. Res. A273 (1988) 127.
6. Kim et al., Int. J. of Prec. Engin. and Manuf. 9 (2008) 86.
7. N. Wyrsch et al., MRS Proc. Symp. Vol. 869, 2005, 3-14.
8. M. Despeisse et al., IEEE Trans. Nucl. Sci. 55 (2008) 805.
9. C. Miazza et al., MRS Proc. Symp. Vol. 910 (2006) A17-03.
10. N. Wyrsch et al., Sensors 8 (2008) 4656.
11. F. Powolny, Ph.D. thesis, University of Neuchâtel, 2009.
12. N. Wyrsch et al., MRS Proc. Symp. Vol. 1245, 2010, 193.
13. J. L. Wiza, Nucl. Instr. and Meth. 162, 1979, 587-601.
14. C. P. Beetz et al., Nucl. Instr. and Meth. In Phys. Res. A 442, 2000, 443-451.
15. Q. Duanmu et al., Proc. of SPIE Vol. 4601, 2001, 284-287.
16 N. Wyrsch et al., MRS Proc. Symp. Vol. 869, 2005, 3-14.
17 http://www.gel.usherbrooke.ca/casino/What.html
18. N. Wyrsch et al., MRS Proc. Symp. Vol. 808 (2004) 441.
19. P.Cuony et al., Appl. Phys. Lett. 97 (2010) 213502.
20. P.Cuony et al., this volume.

Mater. Res. Soc. Symp. Proc. Vol. 1321 © 2011 Materials Research Society
DOI: 10.1557/opl.2011.1230

Amorphous Silicon Photosensors for Detection of Intrinsic Cell Fluorophores

A. Joskowiak[1,2,3], V. Chu[1], D.M.F. Prazeres[2,3], J.P. Conde[1,2]

[1]INESC Microsistemas e Nanotecnologias and IN-Institute of Nanoscience and Nanotechnology, Rua Alves Redol 9, Lisbon, Portugal.
[2]Department of Bioengineering, Instituto Superior Tecnico, Av. Rovisco Pais, Lisbon, Portugal
[3]IBB – Institute for Biotechnology and Bioengineering, Instituto Superior Tecnico, Av. Rovisco Pais, Lisbon, Portugal

ABSTRACT

An amorphous silicon (a-Si:H) photoconductor array with two distinct integrated amorphous silicon carbon alloy (a-SiC:H) high pass filters is used to detect two of the cell intrinsic fluorophores. The cutoff wavelength of the filters is tuned by the carbon content in the film. The fluorophores of interest – reduced nicotinamide adenine dinucleotide (NADH) and flavin adenine dinucleotide (FAD) are indicative of the redox state of the cells. Concentrations down to 1 μM for NADH and 50 μM for FAD were detected.

INTRODUCTION

Monitoring the cells' intrinsic fluorescence has been successfully used in fields ranging from bioprocess monitoring to cancer diagnostics. This characterization focuses on the changes of the fluorescence intensity, and thereby molecular concentration, of aromatic amino acids (tryptophan, phenylalanine and tyrosine), some of the coenzymes (like FAD or NADH) and vitamins (A, B2, B6, D, E) [1].

Due to their characteristics, a-Si:H based devices – photodiodes and photoconductors – are suitable for detection of fluorescence of biological species. These devices have low dark conductivity, high internal quantum efficiency in the visible, and low processing temperatures, which allows use of variety of substrate materials, including glass and polymers [2].

In this paper, microfabricated a-Si:H photoconductors with integrated amorphous silicon carbon (a-SiC:H) alloy filters were used to detect two of the cell intrinsic fluorophores – FAD and NADH. Measurements were made by placing a 5 μL drop of solution on top of the sensor which is protected by a 100 μm PDMS layer. In the future, this sensing system will be coupled with a microfluidic device to obtain a fully functional lab-on-a-chip device for realtime monitoring of the intrinsic fluorescence of cell cultures.

MATERIALS AND METHODS

Photodiode fabrication

Glass is used as a substrate for microfabrication of a 200 x 200 μm thin-film intrinsic amorphous silicon (i-a-Si:H) photoconductor. Al contacts are first deposited by sputtering and then defined by photolitography and wet etching. The distance between the two parallel contacts is 20 μm and the width of the contact line is 200 μm. 5000 Å of i-a-Si:H is deposited on top of the contact lines by radio frequency (rf) plasma enhanced chemical vapour deposition (PECVD). Sensor islands (200 x 200 μm) are defined by photolitography and reactive ion etching (RIE). Amorphous silicon carbon alloy thin-films (a-SiC:H), ~2 μm thick, are tailored for the fluorophores of interest by changing the carbon concentration, and are also deposited by radio frequency plasma enhanced chemical vapour deposition (rf PECVD) (details of the filter characteristics and fabrication to be found in [3]). Access to the contact pads is opened by lift-off. Contact pads are wire bonded to a PCB plate and the wires are protected by epoxy glue. Geometry of the device and a PCB-bonded chip are shown in Figure 1.

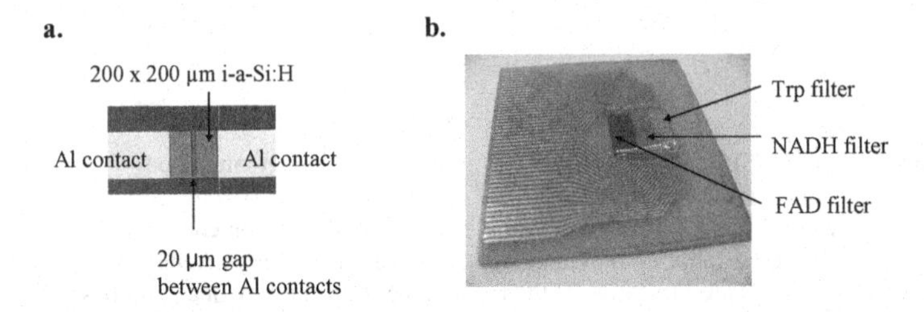

Figure 1. (a) Single a-Si:H photodetector schematic with dimensions of the structure and (b) an integrated three filter chip wirebonded on a PCB.

PDMS sheet fabrication

PDMS was prepared by mixing the base and the curing agent in a 10:1 weight ratio. The prepared mixture was then degassed in a vacuum environment. A 100 μm thick layer of the polymer was achieved by spinning the mixture for 5 s at 500 rpm followed by further 20 s at 1200 rpm. The PDMS sheet was then cured for 2 h at 60°C. Before the measurements the PDMS sheet was oxidized with corona discharge to make its surface hydrophilic.

Fluorophore detection

All single fluorophore solutions were aqueous. Fluorescence calibration curves for different fluorophore concentrations were determined using a commercial spectrofluorimeter (Varian Cary Eclipse).

Integrated detection with the microfabricated silicon photoconductors was made using devices with suitable a-SiC:H filters as detailed above. The light was provided by a tungsten-halogen lamp and the wavelength was selected using a monochromator. Initially the device functionality was tested by measuring the photocurrent at the wavelength of full transparency of a given filter, and at the emission and excitation wavelengths of the fluorophore, respectively – for FAD – at 630 nm, 525 nm and 450 and for NADH – at 530 nm, 460 nm and 377 nm. The measurements were repeated after placing the PDMS sheet on top of the sensors. Subsequently, a

5 μL of sample solution was placed on top of the sensor and the signal was recorded. A double washing with 10 μL of distilled water was always performed between measurements. A schematic representation of the measurement setup is depicted in Figure 2.

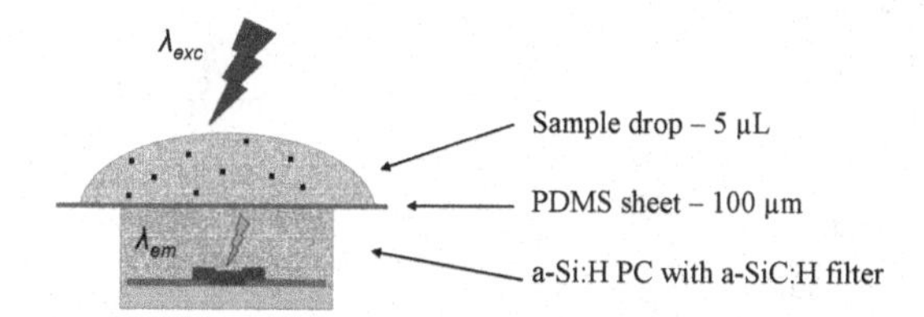

Figure 2. Schematic representation of the measurement setup.

RESULTS AND DISCUSSION

Photoconductor characteristics

Figure 3 shows the sensor response with the integrated filter for FAD at various wavelengths as an example. The photocurrent (J_{ph}) is approximately linearly dependent on the incident photon flux (Φ), independently of the incident light wavelength. The decrease of J_{ph} with decreasing wavelength is due to the filtering properties of the a-SiC:H layer, which reduces the light reaching the amorphous silicon detector. The external quantum efficiency of the detectors is shown in Figure 4. At full filter transparency the *EQE* is ~0.04 for the FAD sensor and 0.1 for the NADH sensor. *EQE* decreases to 2×10^{-5} and 10^{-4} for the devices with the FAD and NADH filters, respectively, at lower wavelengths, when the filter becomes opaque.

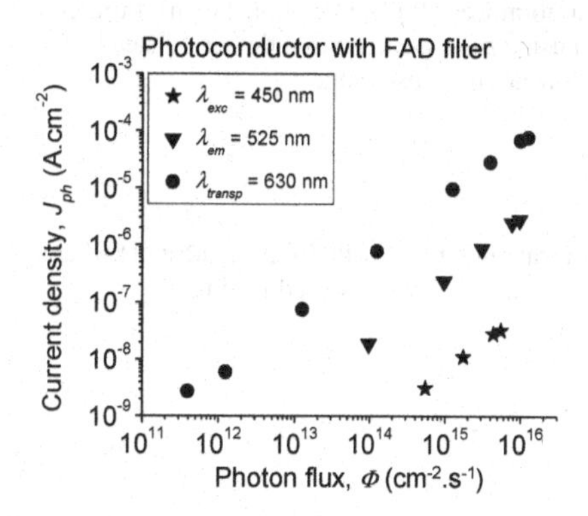

Figure 3. Photocurrent dependance on the photon flux for a miniaturized i-a-Si:H photoconductor with an integrated a-SiC:H fluorescence filter for FAD.

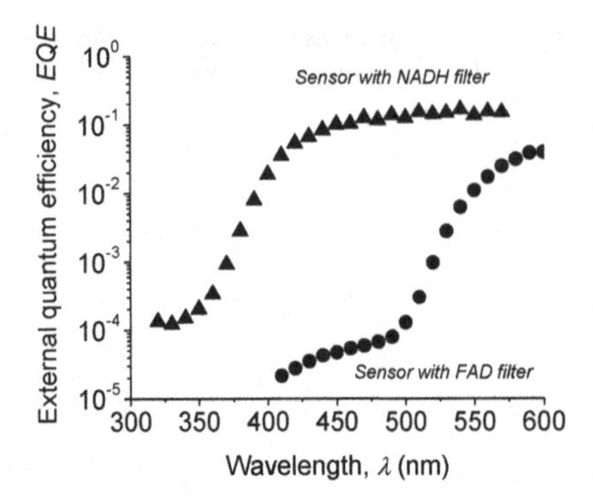

Figure 4. External quantum efficiency of the miniaturized i-a-Si:H photoconductor detectors with integrated NADH (triangles) and FAD (circles) a-SiC:H filters.

Detection of intrinsic fluorophores

Figure 5 shows typical photocurrent density changes during measurements of fluorescence of intrinsic fluorophores NAD and FAD. After placement of the protective PDMS sheet on top of the sensors, an increase in the signal can be observed for all the tested wavelengths. This can be attributed to changes in the optical path of the incident light due to the presence of the polymeric film. Further increase is observed after placing a drop of water or an aqueous solution on top of the polymer, which can be explained by a lens effect of the drop. Both of the fluorophores were successfully detected with dedicated a-SiC:H filter/i-a-Si:H sensor pair (as demonstrated in Figure 5). In the case of NADH the lowest measured concentration was 1 μM, whereas for FAD the lowest measured concentration was 50 μM (data not shown). Further improvement in the detection limit can be expected using *p-i-n* a-Si:H based devices instead of photoconductors due to their lower dark current and wider response range.

CONCLUSIONS

A microfabricated i-a-Si:H photoconductor array integrated with different a-SiC:H filters was demonstrated. The integrated devices were used to detect two of the cell intrinsic fluorophores – NADH and FAD in aqueous solution.

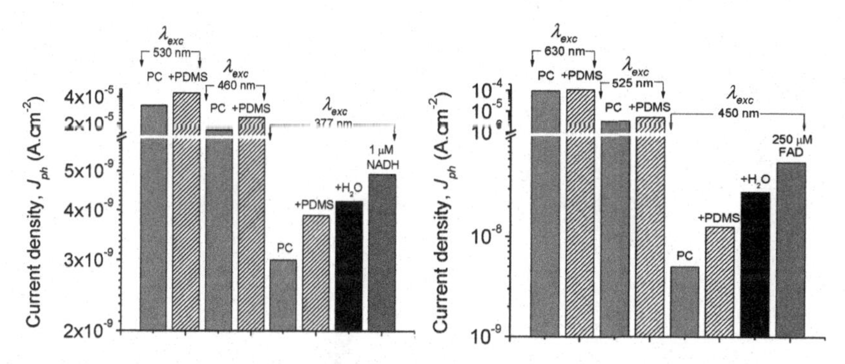

Figure 5. Detection of fluorescence of NADH (left) and FAD (right) in aqueous solution. Response of the simple photoconductor-filter device tandem is represented with grey bars. Response with the addition of the PDMS sheet top is shown using striped bars, and the response with a water drop on the top of the PDMS with black bars. The response of the sensor with a drop of fluorophore solution is shown in green for NADH and in red for FAD (first bar on the right in each of the plots).

ACKNOWLEDGMENTS

The authors would like to thank F. Silva, V. Soares and J. Bernardo for help in cleanroom processing and device packaging. This work was supported by Fundação para a Ciencia e a Tecnologia through research project MICRO-2D FS (PTDC/CTM/104387/2008) and Ph.D. Grant SFRH/BD/39081/2007.

REFERENCES

1. C. Lindemann, S. Marose, H.O. Nielsen, T. Scheper. *Sensor Actuat. B Chem*, **51**, pp. 273-277 (1998).
2. R.A. Street, *Hydrogenated amorphous silicon.* Cambridge University Press (1991).
3. B. Lipovšek, A. Jóskowiak, J. Krč, M. Topič, D.M.F. Prazeres, V. Chu, J.P. Conde. *Sensor Actuat. A Phys,* **163** 1, pp. 96-100 (2010).

Mater. Res. Soc. Symp. Proc. Vol. 1321 © 2011 Materials Research Society
DOI: 10.1557/opl.2011.819

Self optical gain in multilayered silicon-carbon heterostructures: A capacitive active band-pass filter model

M. A. Vieira[1,3], M. Vieira[1,2], P. Louro[1,2], M. Fernandes[1,2], J. Costa[1,2], A. S. Garção[2,3]
[1]Electronics Telecommunication and Computer Dept. ISEL, R. Conselheiro Emídio Navarro, 1949-014 Lisboa, Portugal Tel: +351 21 8317290, Fax: +351 21 8317114, mv@isel.ipl.pt .
[2] CTS-UNINOVA, Quinta da Torre, Monte da Caparica, 2829-516, Caparica, Portugal.
[3] DEE-FCT-UNL, Quinta da Torre, Monte da Caparica, 2829-516, Caparica, Portugal.

ABSTRACT

This paper reports results on the use of a pi'n/pin a-SiC:H heterostructure as an active band-pass filter transfer function whose operation depends on the wavelength of the trigger light, on the applied voltage and on the wavelength of the additional optical bias.

Results show that the device combines the demultiplexing operation with the simultaneous photodetection and self amplification of the signal. Experimental and simulated results show that the output signal has a strong nonlinear dependence on the light absorption profile. The device, modeled by a simple circuit with variable capacitors and interconnected phototransistors through a resistor, is a current-controlled device. It uses a changing capacitance to control the power delivered to the load acting as a state variable filter circuit. It combines the properties of active high-pass and low-pass filter sections into a capacitive active band-pass filter.

INTRODUCTION

There has been much research on semiconductor devices as elements for optical communication when a band or frequency needs to be filtered from a wider range of mixed signals. Amorphous silicon carbon tandem structures, through an adequate engineering design of the multiple layers' thickness, absorption coefficient and dark conductivities can accomplish this function.

In this paper, light-activated multiplexer/demultiplexer silicon-carbon devices are analyzed. Characteristics of tunable wavelength filters based on a-SiC:H multilayered stacked cells are studied both theoretically and experimentally. A capacitive active band-pass filter model supports the experimental data. An algorithm to decode the multiplexed signal is established.

DEVICE CONFIGURATION

The sensor element is a multilayered heterostructure based on a-Si:H and a-SiC:H produced by PE-CVD at 13.56 MHz radio frequency. The configuration of the device, shown in Figure 1, includes two stacked p-i-n structures (p(a-SiC:H)-i'(a-SiC:H)-n(a-SiC:H)-p(a-SiC:H)-i(a-Si:H)-n(a-Si:H)) sandwiched between two transparent contacts. The thicknesses and optical gap of the front i'- (200nm; 2.1 eV) and back i- (1000nm; 1.8eV) layers are optimized for light absorption in the blue and red ranges, respectively [1]. As a

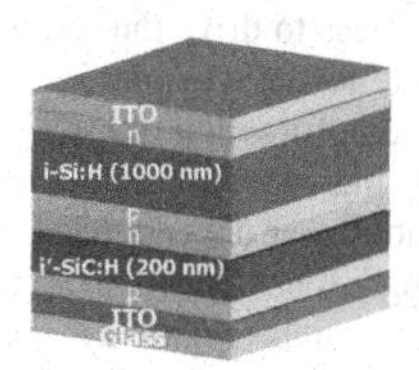

Figure 1 Device configuration.

result, both front and back pin structures act as optical filters confining, respectively, the blue and the red optical carriers to their active areas.

SELF-OPTICAL BIAS AMPLIFICATION

Three monochromatic pulsed lights (input channels): red, green and blue illuminate separately the device. Steady state red, green and blue optical bias was superimposed separately and the photocurrent generated measured at -8V. Results have shown that due to the self bias effect [2] the sensor is a wavelength current-controlled device that makes use of changes in the wavelength of the optical bias to control the power delivered to the load. The red background enhances the light-to-dark sensitivity in the short wavelength range and quenches it in the long wavelength range. The blue bias has an opposite behavior; it amplifies the light to dark sensitivity in the red/green wavelength range and reduces it in the blue one. Under green irradiation the light-to-dark sensitivity is mainly quenched in the green spectral range.

In Figure 2 the input red (R: 624 nm), green (G: 524 nm) and blue (B: 470nm) channels, under negative bias, without (φ=0) and with red steady state optical bias (624nm, φ=73mWcm^{-2}) are displayed for each monochromatic input channel at a bit rate of 6000bps. Since, under red illumination, the electrical field increases locally in the least absorbing cell, the front diode (self optical amplification) the collection due to the carriers generated by the blue channel (absorbed in the front diode) is enhanced. The channel gain, defined as the ratio between the channel signal with and without steady state optical bias becomes higher than one (α_B>>1, see arrows). In the back diode the electric field is reduced (self optical quench) so, the collection of the carriers generated by the red channel is reduced. Here, the gain becomes lower than one (α_R<<1). The green channel is mostly absorbed across the back diode due to its higher thickness, so its gain is also lower than one (α_G<1) but higher than α_R due to the front carrier collection.

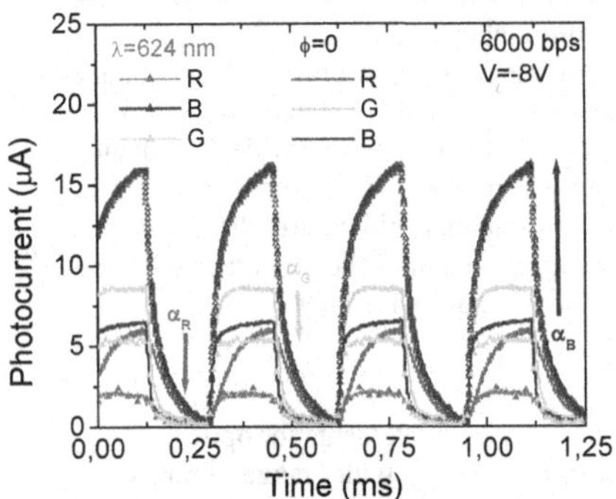

Figure 2 Input red (R) green (G) and blue (B) channels under negative bias, without (φ=0) and with (λ=624 nm) red background.

DATA INPUTS PREDICTION

A chromatic time dependent wavelength combination of the same pulsed input channels but with different bit sequences was used to generate a multiplexed signal. The output waveforms are shown in Figure 3, under red steady state illumination (red line) and without it (dark line). The bit sequences to drive the channels are shown in the top of the figure to guide the eyes. Results show that the device combines the demultiplexing operation with the simultaneous photodetection and self amplification of the signal. When a polychromatic combination of different pulsed channels impinges on the device, under steady state additional optical bias, the output signal has a strong nonlinear dependence on the light absorption profile in a mode that induces a nonlinear wavelength bias dependent gain. This gain depends on the background wavelength that controls the input signals (Figure 2). To recover the transmitted information (6000bps, 8 bit per wavelength channel) the output waveforms, under red irradiation and without it, were used. Both multiplexed signals, during a complete cycle (T), were divided into eight time slots (Δ=133 μs), each one with a bandwidth of 7.5 kHz, corresponding to one bit where the independent optical signals can be ON (1) or OFF (0). In Figure 3 all possible combinations of the three input channels are present. The waveform of the output without optical bias is an 8-level encoding (2^3) to which it corresponds 8 different photocurrent thresholds. Taking into account Figure 2, under red background the red channel is strongly quenched and the blue enhanced. So, the output waveform becomes a main 4-level encoding (2^4). Here, the higher level corresponds to both blue and green ON (_11) and the lower to the absence of both (_00). The other two intermediate levels are ascribed, as follows: the upper level to the ON state of the blue (_01) channel and the lower to the green channel ON (_10). To decode the red channel, similar time slots with red background have to be compared with the ones without it, being higher the ones with the red channel ON. Using this simple algorithm the independent red, green and blue bit sequences were decoded as: B[01010101], G[00110011] and R[00001111].

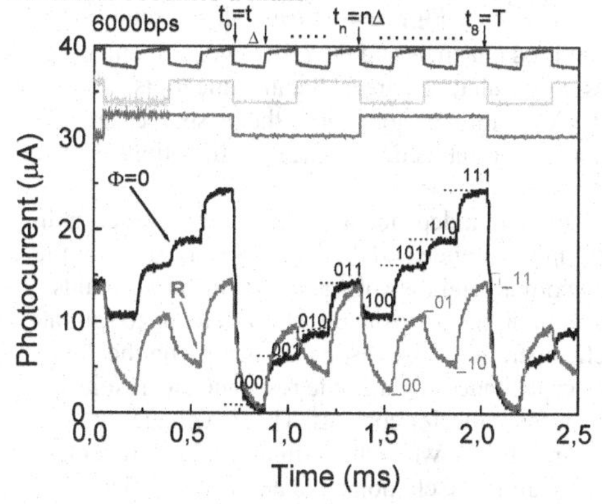

Figure 3 Output waveform signal at -8V; without (φ=0) and with red (R) optical bias. The bit sequences are shown at the top of the figure.

CAPACITIVE MODEL AND VALIDATION

Based on the experimental results and device configuration an optoelectronic model was developed [3, 4] and displayed in Figure 4. It is supported by the complete dynamical large signal Ebers-Moll model. Here, the charge stored in the space-charge layers is modeled by the capacitors C_1 and C_2. R_1 and R_2 model the dynamical resistances of the internal and back junctions under different *dc* bias conditions. The operation is based upon the following principle: the flow of current through the resistor connecting the two transistor bases is proportional to the difference in the voltages across both capacitors (charge storage buckets). The capacitors exhibit time-varying charge/voltage characteristics, being the current across them the instantaneous rate of change of charge. If the device is biased negatively Q_1 and Q_2 are in their reverse active regions. Two optical gate connections ascribed to the different light penetration depths across the front (Q_1) and back (Q_2) phototransistors were considered to allow independent blue, red and green channels transmission. Four square-

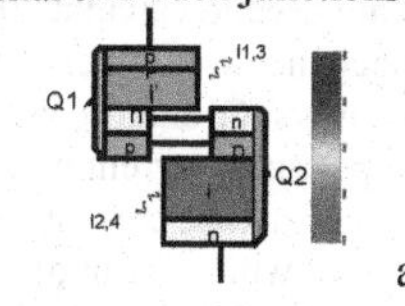

a)

wave current sources with different intensities are used; two of them, α_1 I_1 and $\alpha_2 I_2$, with

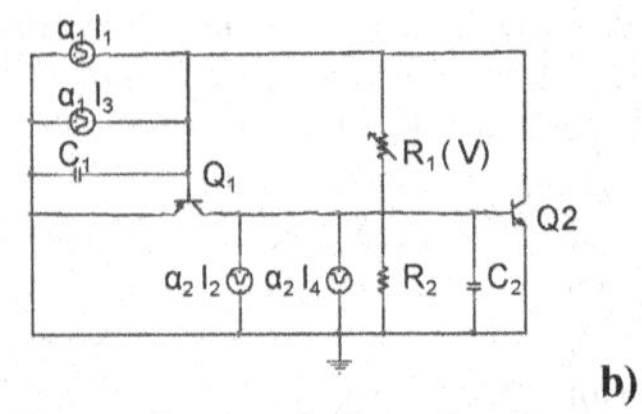

b)

Figure 4 ac equivalent circuit.

different frequencies to simulate the input blue and red channels and the other two, $\alpha_1 I_3$ and $\alpha_2 I_4$, with the same frequency but different intensities, to simulate the green channel due to its asymmetrical absorption across both front and back phototransistors. Once the *ac* sources are connected in the load loop an *ac* current flows through, establishing voltage modifications across the two capacitors.

During the simultaneous transmission of the three independent bit sequences, the set-up in this capacitive circuit loop is constantly changing in magnitude and direction. This means that the voltage across one capacitor builds up until its maximum and the voltage across the other builds up to a minimum. The system collapses and builds up in the opposite direction. It tends to saturate and then leave the saturation because of the cyclic operation. This results in changes on the reactance of both capacitors. The use of separate capacitances on a single resistance R_1 results in a charging current gain proportional to the ratio between collector currents. The *dc* voltage, according to its strength, aids or opposes the *ac* currents. So, when the pi'npin device is reverse-biased, the base-emitter junction of both transistors are inversely polarized and conceived as phototransistors, taking, so, advantage of the amplifier action of adjacent collector junctions which are polarized directly. This results in a current gain proportional to the ratio between both collector currents. Under positive bias the internal junction becomes reverse-biased and no amplification effect is observed. The external current depends not only on the balance between blue, green and red photocurrents but also on the end of each half-cycle of each modulated current. Here, the movement of charge carriers with an increase/decrease in the irradiation, results in a charging or a displacement current similar to the current (i=C dV/dt) that charges the capacitors C_1 and C_2 in opposite ways.

Assuming that the frequency is low enough so that quasi-statics are valid, the incremental capacitances are used to calculate the total charge stored. So, the junction capacitance and diffusion capacitance would behave as if they have been kept in parallel. Because inherently they are leaky capacitors, C_1 and C_2, they need to store higher charge that has also to be supplied. This charge will appear as a current component. The current would be the superposition of two components: one leakage current that is in-phase with the voltage and one displacement-capacitive-current which is out of phase with the voltage.

The Kirchhoff's laws give the state-space realization of the equivalent circuit. The time periodic linearized state equations are given by:

$$\frac{dv_{1,2}}{dt} = A\times v_{1,2}(t)+B\times i_{1,2}(t) \qquad i(t)=C\times v_{1,2}(t)+D\times i_{1,2}(t) \qquad 1$$

where A is the system matrix, and relates how the current state affects the state change $v'_{1,2}(t)$.

B is the control matrix, and determines how the system input affects the state change. C is the output matrix, and determines the relationship between the system state and the system output. D

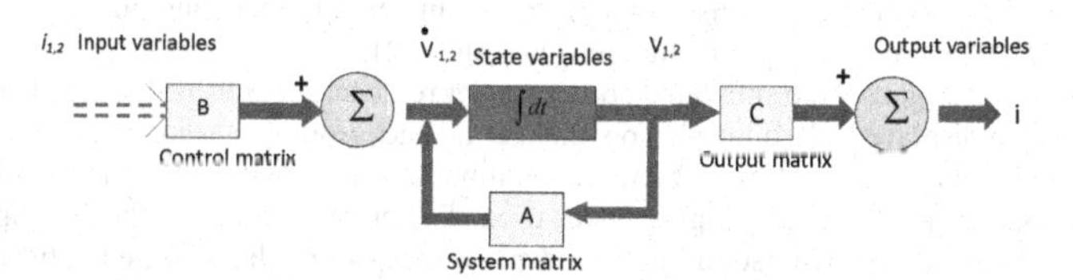

Figure 5 Block diagram of the state model considered as a parallel current bucket connection inflow $[i_1(t) // i_2(t)]$.

is the feed-forward matrix, and allows for the system input affecting the system output directly. Here D matrix is considered the zero matrix. Figure 5 show the block diagram of the state model considered as a parallel current bucket connection inflow $[i_1(t) // i_2(t)]$.Taking into account Figure 4 and assuming as state variables the voltage $v_1(t)$across the capacitor C_1 and the voltage $v_2(t)$ across the capacitor C_2 the current will be given by:

$$i_{c1}(t) = C_1 \frac{dv_1(t)}{dt} \qquad i_{c2}(t) = C_2 \frac{dv_2(t)}{dt} \qquad i(t) = \frac{v_2(t)}{R_2} \qquad 2$$

Under optical bias the control matrix B should take into account the enhancement or quenching of the channels due to the steady state irradiation (Figure 2). It will be affected by the α_1 and α_2 coefficients that determine how the background affects the state change. Under red light background we consider $\alpha_1 > 1$ ($\alpha_B + \alpha_G > 1$) and $\alpha_2 < 1$($\alpha_R + \alpha_G < 1$). The opposite will occur under blue irradiation. Under green background both are balanced. The time periodic linearized state equations are rewritten in the form:

$$\frac{dv_{1,2}}{dt} = \begin{bmatrix} -\frac{1}{R_1C_1} & \frac{1}{R_1C_1} \\ \frac{1}{R_1C_2} & -\frac{1}{R_1C_2} - \frac{1}{R_2C_2} \end{bmatrix} v_{1,2}(t) + \begin{bmatrix} \frac{\alpha_1}{C_1} \\ \frac{\alpha_2}{C_2} \end{bmatrix} i_{1,2}(t) \qquad i(t) = \begin{bmatrix} 0 & \frac{1}{R_2} \end{bmatrix} v_{1,2}(t) \qquad 3$$

In Figure 6 it is displayed the block diagram of the optoelectronic state model for a WDM pi´npin device under different electrical and optical bias conditions. λ_1, λ_2, λ_3 represent the color channels, R_1 and R_2 the dynamic internal and back resistances and α_1 and α_2 are the coefficients for the steady state irradiation. Based on the optoelectronic model,

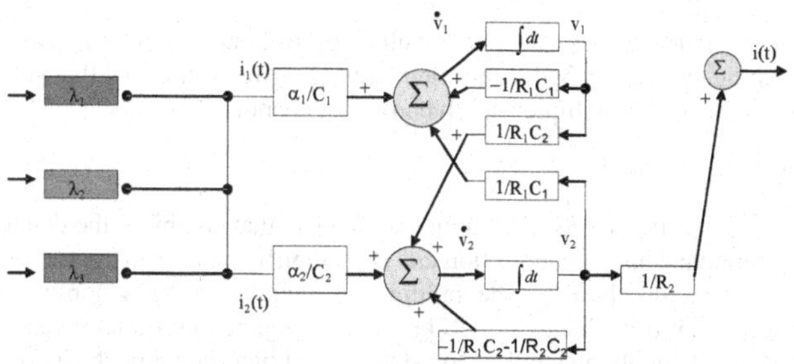

Figure 6 Block diagram of the optoelectronic state model for a WDM pi´npin device

the multiplexed signal was simulated by applying the Kirchhoff's laws for the *ac* equivalent circuit (Figure 4b) and the four order Runge-Kutta method to solve the corresponding state equations. MATLAB was used as a programming environment and the input parameters chosen in compliance with the experimental results (Figure 2).

In Figure 7 the simulated current under red backgrounds is displayed. The current sources are also displayed (dash lines). To simulate the background, current source intensities were multiplied by the on/off ratio between the input channels with and without optical bias (Figure 2). The same bit sequence of Figure 3 was used. To validate the model the experimental multiplexed signals are also shown (solid lines).Under red background the expected optical amplification in the short wavelength range is observed due to the effect of the active multiple-feedback filter when the back diode is light triggered. Under negative bias both transistors are inversely polarized and conceived as phototransistors. This results in a charging current gain proportional to the ratio between both collector currents ($\alpha_2 C_1 / \alpha_1 C_2$). The device behaves like an optoelectronic controlled transmission system that stores, amplifies and transports the minority carriers generated by the current pulses, through the capacitors C_1 and C_2. To validate the model the experimental multiplexed signals are also shown (solid lines). Here, the *dc* voltage control creates a voltage across one or both capacitors which, when superimposed with an *ac* pulse, collectively saturates the circuit. No additional change in the voltage across the capacitors occurs. Depending on its wavelength the optical bias changes the amplitude of the *ac* current sources by a α factor, and so the voltages across one or both capacitors. Blue, red or green irradiations move asymmetrically the bases of Q_1, Q_2 or both toward (away) their emitter voltages, self-forward (reverse) effect, resulting, respectively in lower (higher) values of I_1, I_2, I_3 and I_4 when compared with no optical bias. The circuit can leave the saturation resulting in a wavelength controlled power transfer to the load that allows tuning an input channel or to optically demultiplex a polychromatic channel.

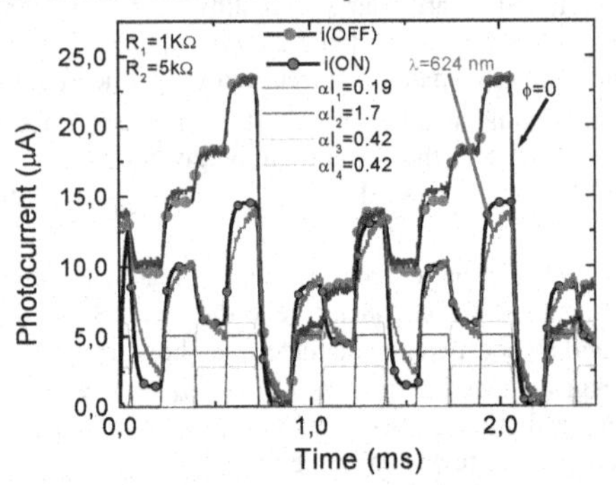

Figure 7 Multiplexed simulated (symbols), current sources (dash lines) and experimental (solid lines) under negative (R1=1kΩ; -8V) *dc* bias and red background.

CONCLUSIONS

A pi'n/pin a-SiC:H multiplexer device that combines the demultiplexing operation with the simultaneous photodetection and self amplification of an optical signal is analyzed.

Results show that the multiplexed signal has a strong nonlinear dependence on the light absorption profile due to the self biasing of the junctions under unbalanced light generation profiles. By applying a red steady state optical bias the input channels were recovered. A capacitive active band-pass filter model supports the experimental data. An algorithm to decode the multiplexed signal was established.

ACKNOWLEDGEMENTS

This work was supported by FCT (CTS multi annual funding) through the PIDDAC Program funds and PTDC/EEA-ELC/111854/2009.

REFERENCES

1. M. Vieira, A. Fantoni, M. Fernandes, P. Louro, G. Lavareda, C. N. Carvalho, Journal of Nanoscience and Nanotechnology, Vol. 9 , Number 7, July 2009 , pp. 4022-4027(6).
2. M. Vieira, P. Louro, M. A. Vieira, J. Costa, M. Fernandes, Procedia Engineering, Volume 5, 2010, Pages 232-235.
3. M. A. Vieira, M. Vieira, M. Fernandes, A. Fantoni, P. Louro, M. Barata, 2009, MRS Proceedings, Volume 1153, A08-03.
4. M. A. Vieira, M. Vieira, J. Costa, P. Louro, M. Fernandes, A. Fantoni, in Sensors & Transducers Journal Vol. 9, Special Issue, December 2010, pp.96-120.

Mater. Res. Soc. Symp. Proc. Vol. 1321 © 2011 Materials Research Society
DOI: 10.1557/opl.2011.951

Optical Demultiplexer Device: Frequency and optical bias analysis

P. Louro[1,2], M. Vieira[1,2,3], M. A. Vieira[1,2], T. Silva[1]
[1] Electronics Telecommunications and Computer Dept, ISEL, Lisbon, Portugal.
[2] CTS-UNINOVA, Lisbon, Portugal.
3 DEE-FCT-UNL, Quinta da Torre, Monte da Caparica, 2829-516, Caparica, Portugal

ABSTRACT

In this paper we present results on the use of a multilayered a-SiC:H heterostructure as a device for wavelength-division demultiplexing of optical signals. This device is useful in optical communications applications that use the wavelength division multiplexing technique to encode multiple signals into the same transmission medium. The device is composed of two stacked p-i-n photodiodes, both optimized for the selective collection of photo generated carriers. Band gap engineering was used to adjust the photogeneration and recombination rates profiles of the intrinsic absorber regions of each photodiode to short and long wavelength absorption and carrier collection in the visible spectrum. The photocurrent signal using different input optical channels was analyzed at reverse and forward bias and under steady state illumination. A demux algorithm based on the voltage controlled sensitivity of the device was proposed and tested. The operation frequency of the device was analyzed under different optical bias conditions. An electrical model of the WDM device is presented and supported by the solution of the respective circuit equations.

INTRODUCTION

Wavelength division multiplexing (WDM) devices are used when different optical signals are encoded in the same optical transmission path, in order to enhance the transmission capacity and the application flexibility of optical communication and sensor systems. The use of WDM technologies not only provides high speed optical communication links, but also offers advantages such as higher data rates, format transparency, and self-routing. Various types of available WDM devices include prisms, interference filters, and diffraction gratings. Currently modern optical networks use Arrayed Waveguide Grating (AWG) as optical wavelength (de)multiplexers [1] that use multiple waveguides to carry the optical signals. In this paper we report the use of a monolithic WDM device based on an a-Si:H/a-SiC:H multilayered semiconductor heterostructure. The device makes use of the fact that the optical absorption of the different wavelengths can be tuned by means of electrical bias changes or optical bias variations. This capability was obtained using adequate engineering design of the multiple layers thickness, absorption coefficient and dark conductivities [2].

DEVICE CONFIGURATION

The device described herein operates from 400 to 700 nm which makes it suitable for operation at visible wavelengths in optical communication applications. The device is a multilayered heterostructure based on a-Si:H and a-SiC:H. The configuration of the device includes two stacked p-i-n structures between two electrical and transparent contacts (Fig. 1). Both front (pin1) and back (pin2) structures act as optical filters confining, respectively, the short and the long optical carriers, while the intermediate wavelengths are absorbed across both [3,4]. The devices were produced by PECVD and optimized for a proper fine tuning of a specific wavelength. The active device consists of a p-i'(a-SiC:H)-n / p-i(a-Si:H)-n heterostructure with low conductivity doped layers. The thicknesses and optical gap of the thin i'- (200nm; 2.1 eV) and thick i- (1000nm; 1.8 eV) layers are optimized for light absorption in the blue and red ranges, respectively.

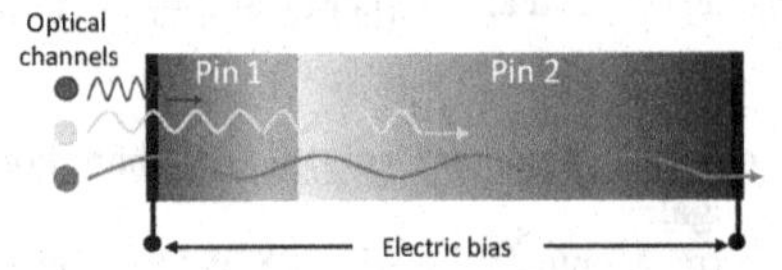

Figure 1 WDM device configuration.

The device was operated within the visible range using as optical signals, to simulate the transmission optical channel, the modulated light (external regulation of frequency and intensity) supplied by a red, a green and a blue LED with wavelengths of 470 nm, 524 nm and 626 nm, respectively.

Figure 2 displays the spectral photocurrent under reverse and forward bias. Results show that in the long wavelengths range (> 600 nm) the spectral response is independent on the applied bias while in the short wavelength the collection strongly increases with the reverse bias. This means that for the used transmission channels both blue and green channels will show a strong dependence on the applied voltage, on opposite to the red channel that is insensitive to the electric bias.

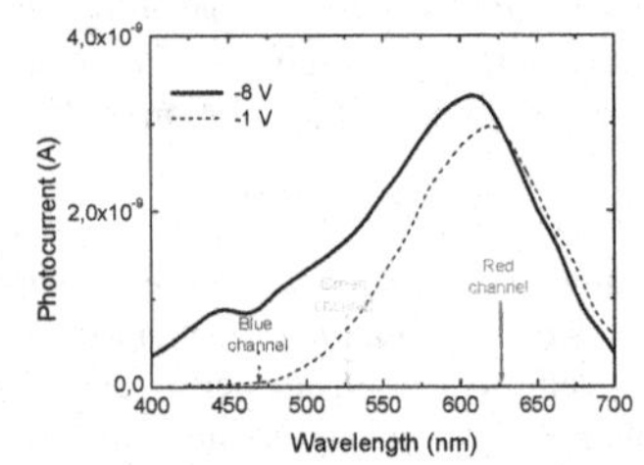

Figure 2 Spectral photocurrent under reverse and forward bias.

Figure 3 displays the single and multiplexed signals under reverse (-8V) and forward (+1V) electrical bias.

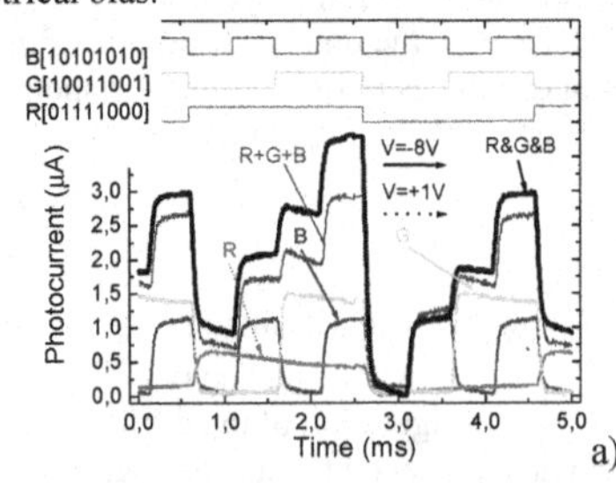

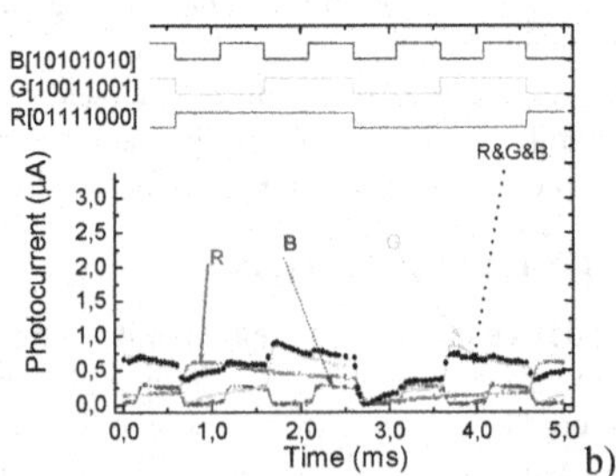

Figure 3 Device spectral photocurrent under a) reverse and b) forward bias.

As expected from Fig. 2, the input red signal remains constant while the blue and the green ones decrease as the voltage changes from negative to positive. The output multiplexed signal, obtained with the combination of the three optical sources, depends on both the applied voltage and on the ON-OFF state of each input optical channel. Under negative bias, there are eight separate levels while under positive bias they were reduced to one half. The highest level appears when all the channels are ON and the lowest if they are OFF. Furthermore, the levels ascribed to the mixture of three or two input channels are higher than the ones due to the presence of only one (R, G, B). Optical nonlinearity was detected; the sum of the input channels (R+B+G) is lower than the correspondent multiplexed signals (R&G&B). This optical amplification, mainly on the ON-ON states, suggests capacitive charging currents due to the time-varying nature of the incident lights. Under positive bias the levels are reduced to one half since and the blue component of the combined spectra falls into the dark level, the red remains constant and the green one decreases.

To recover the transmitted information (8 bit per wavelength channel) the multiplexed signal, during a complete cycle, was divided into eight time slots, each corresponding to one bit where the independent optical signals can be ON (1) or OFF (0). Under positive bias, the device has no sensitivity to the blue channel (Fig. 1-2), so the red and green transmitted information are identified. The highest level corresponds to both channels ON (R&G: R=1, G=1), and the lowest to the OFF-OFF stage (R=0; G=0). The two levels in-between are related with the presence of only one channel ON, the red (R=1, G=0) or the green (R=0, G=1). To distinguish between these two situations and to decode the blue channel, the correspondent sub-levels, under reverse bias, have to be analyzed. The highest increase at -8V corresponds to the blue channel ON (B=1), the lowest to the ON stage of the red channel (R=1) and the intermediate one to the ON stage of the green (G=1). Using this simple key algorithm the independent red, green and blue bit sequences can be decoded as: R[01111000], G[10011001] and B[10101010], as shown on the top of Fig. 2, which are in agreement with the signals acquired for the independent channels.

INFLUENCE OF THE STEADY STATE OPTICAL BIAS

Figure 4 shows the time dependent photocurrent signal measured under reverse (-8V, symbols) and forward (+1V, straight lines) voltage for the different input optical signals. The figure also shows the effect of applying green (λ_L) steady state optical bias in addition to the input optical signals. The optical signals were obtained by wave square modulation of the LED driving currents. The optical power intensity of the red, green and blue channels was adjusted to 51, 90, 150 μW/cm^2, respectively. The steady state optical bias was generated using a green LED driven at a constant intensity (150 μW/cm^2).

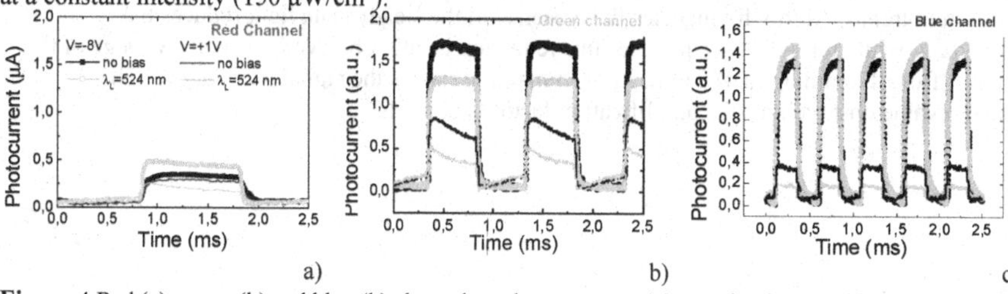

a) b) c)

Figure 4 Red (a), green (b) and blue (b) channels under reverse and forward voltages without and with (λ_L) green steady state bias.

Results show that the green optical bias affects mainly the output of the green channel, which is significantly reduced. The signals of the red and blue channels show negligible changes. The behavior of the device under steady state optical bias can be explained taking in consideration the variation of the internal electric field distribution. When an optical bias is applied its effect is to enhance the field distribution within the less photo excited sub-cell: the back under blue irradiation and the front under red steady bias. Therefore, the reinforcement of the electric field under blue irradiation and negative bias increases the collection.

FREQUENCY ANALYSIS

The study of the frequency influence on the device performance was analyzed through the spectral response of the device without and with steady state optical light. Results are displayed in Fig. 5. Data from Fig. 5 show that that without background light the curves measured under different frequencies exhibit the same trend with two peaks located at 500 nm and 600 nm.

The signal is reduced with the increase of the frequency. Under green background the spectral response shows two different regimes depending on the operation frequency. In the low frequency range the signal is similar to the trends obtained under red steady state light, while at higher frequencies it follows the behavior obtained without background light.

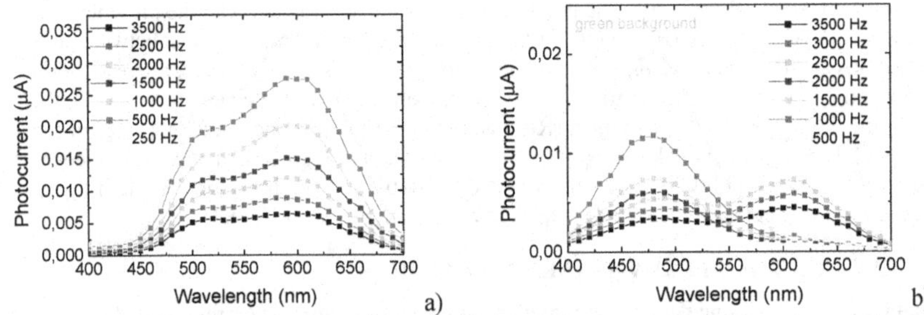

Figure 5 Photocurrent variation with the wavelength for different frequencies at -8 V obtained: a) without and b) with green background.

In Fig. 6 it is displayed the ratio of the signal measured under steady light and without it at different frequencies.

Results show that for the green background light the ratio between the photocurrents shows two different trends. At low frequencies the presence of the steady state light reduces the photocurrent, while for higher frequencies this reduction is only observed in a narrow range of the spectrum (from 470 nm up to 480 nm). In the remaining ranges the signal is enhanced. The maximum value observed for the amplification factor is around 3.

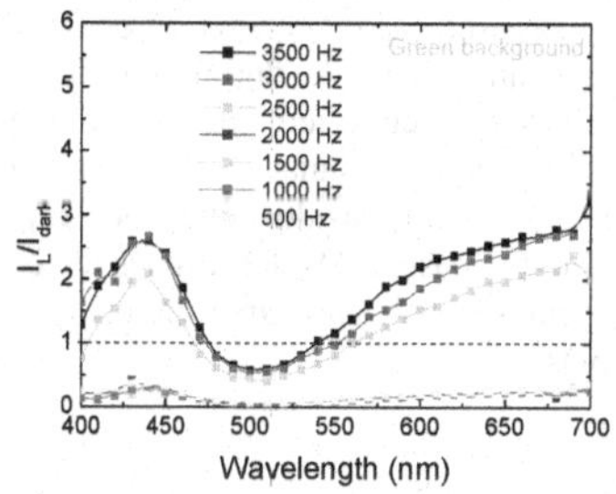

Figure 6 Ratio between the photocurrents under green steady state illumination and without it (dark) at -8 V under different frequencies.

ELECTRICAL SIMULATION

The silicon-carbon pi'npin device can be considered as a monolithic double pin photodiode structure with two red and blue optical connections for light triggering. Based on the experimental results and device configuration an electrical model was developed [5]. Operation is explained in terms of the compound connected phototransistor equivalent model (Fig. 7 a).

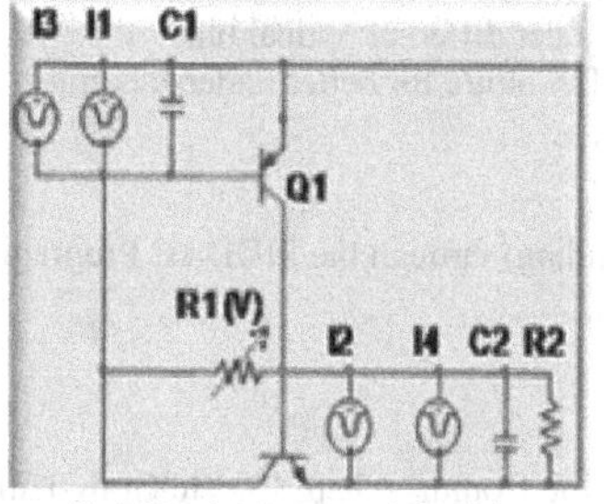

a)

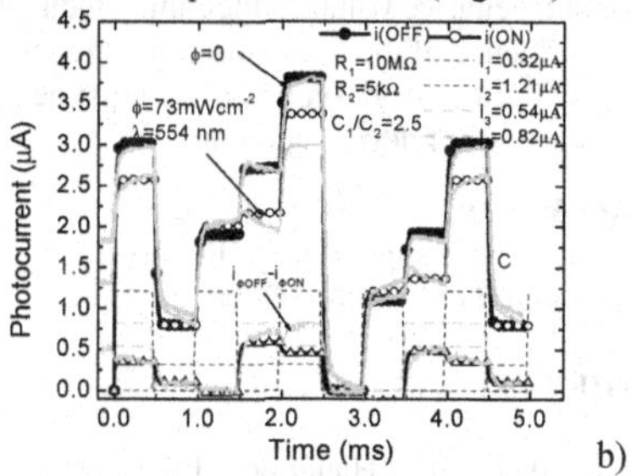

b)

Figure 7 a) Equivalent model of the compound connected phototransistor; b) Multiplexed simulated (symbols), current sources (dash lines) and experimental (solid lines) under forward (R_1=10MΩ; +1V) and reverse dc bias (R_1=1kΩ; -8V) with and without green light.

To simulate the green background, current sources intensities were multiplied by the on/off ratio between the input channels with and without optical bias (Figure 6). The same bit sequence of Figure 3 was used in both figures. To validate the model the experimental multiplexed signals are also shown (solid lines).Good agreement between experimental and simulated data was observed. The eight expected levels, under reversed bias, and their reduction under green irradiation are clearly seen. When the pi'npin device is reverse-biased, the base emitter junction of both transistors are inversely polarized and conceived as phototransistors, taking, so, advantage of the amplifier action of neighboring collector junctions which are polarized directly. This results in a charging current gain proportional to the ratio between both collector currents (C_1/C_2). Under positive bias the internal junction becomes always reverse-biased. If not triggered ON it is nonconducting, when turned ON by light it conducts like a photodiode, for one polarity of current. Green irradiation moves asymmetrically the Q_1 and Q_2 bases toward their emitter voltages, resulting in lower values of I_3 and I_4 when compared without optical bias (Figure 7). I_1

and I_2 slightly increase due to the increased carrier generation on the less absorbing phototransistors. Under negative bias and during the duration of the red and blue pulses (I_1 or I_2 ON), as without optical bias, the internal junction remains forward biased and the transferred charge between C_1 and C_2 reaches the output terminal as a capacitive charging current.

During the green pulse (I_3 and I_4 ON) only residual charges are transferred between C_1 and C_2. So, only the charges generated in the base of Q_2 (I_4) reaches the output terminal as can be confirmed by the good fitting between simulate and experimental differences of both multiplexed signals without and with optical bias.

CONCLUSIONS

A double pi'n/pin a-SiC:H heterostructure with two optical gate connections for light triggering in different spectral regions was presented. Multiple monochromatic pulsed communication channels, in the visible range, were transmitted together, each one with a specific bit sequence. The combined optical signal was analyzed by reading out, under positive and negative voltages and optical green bias, the generated photocurrent across the device. Results show that the output multiplexed signal has a strong nonlinear dependence on the light absorption profile, i.e. on the incident light wavelength, bit rate, intensity and optical bias due to the self biasing of the junctions under unbalanced light generation profiles. By switching between positive and negative voltages the input channels can be recovered or removed.

The influence of the operation frequency was analyzed under different optical bias conditions. Further work on this topic must be accomplished in future for better understanding and optimization of device operation.

ACKNOWLEDGEMENTS

This work was supported by FCT (CTS multi annual funding) through the PIDDAC Program funds and POCTI/FIS/70843/2006.

REFERENCES

1 M.Bas, Fiber Optics Handbook, Fiber, Dev.and Syst. for Opt. Comm., Chap, 13, Mc Graw-Hill, 2002.

2 M. Vieira, M. Fernandes, P. Louro, A. Fantoni, Y. Vygranenko, G. Lavareda, C. Nunes de Carvalho, Mat. Res. Soc. Symp. Proc., Vol. 862 (2005) A13.4.

3 P. Louro, M. Vieira, M.A. Vieira, M. Fernandes, A. Fantoni, C. Francisco, M. Barata, Physica E: Low-dimensional Systems and Nanostructures, 41 (2009) 1082-1085.

4 P. Louro, M. Vieira, M. Fernandes, J. Costa, M. A. Vieira, J. Caeiro, N. Neves, M. Barata, Phys. Status Solidi C 7, No. 3–4, 1188– 1191 (2010).

5 M. A. Vieira, M. Vieira, M. Fernandes, A. Fantoni, P. Louro, and M. Barata, Amorphous and Polycrystalline Thin-Film Silicon Science and Technology 2009, MRS Proceedings Vo. 1153, A08-0.

Mater. Res. Soc. Symp. Proc. Vol. 1321 © 2011 Materials Research Society
DOI: 10.1557/opl.2011.952

Thin-film Photodiode with an a-Si:H/nc-Si:H Absorption Bilayer

Y. Vygranenko[1, 2*], M. Vieira[1,2], A. Sazonov[3]

[1]Electronics, Telecommunications and Computer Engineering, ISEL, 1949-014 Lisbon, Portugal
[2]CTS-UNINOVA, 2829-516 Caparica, Portugal
[3]Electrical and Computer Engineering, University of Waterloo, Waterloo, N2L 3G1, Canada

ABSTRACT

We report on the fabrication and characterization of n^+-n-i-δ_i-p thin-film photodiodes with an active region comprising a hydrogenated nanocrystalline silicon (nc-Si:H) *n*-layer and a hydrogenated amorphous silicon (a-Si:H) *i*-layer. The combination of wide- and narrow-gap absorption layers enables the spectral response extending from the near-ultraviolet (NUV) to the near-infrared (NIR) region. Moreover, in the low-bias range, when only the *i*-layer is depleted, the leakage current is significantly lower than that in the conventional nc-Si:H n^+-n-p^+ photodiode deposited under the same deposition conditions. Device with the 900nm/400nm thick *n*-*i*-layers exhibits a reverse dark current density of 3 nA/cm^2 at -1V. In the high-bias range, when the depletion region expands within the *n*-layer, the magnitude of the leakage current depends on electronic properties of nc-Si:H. The density of shallow and deep states, and diffusion length of holes in the *n*-layer have been estimated from the capacitance-voltage characteristics and from the bias dependence of the long-wavelength response, respectively. To improve the quantum efficiency in the NIR-region, we have also implemented a Cr / ZnO:Al back reflector. The observed long-wavelength spectral response is about twice as high as that for a reference photodiode without ZnO:Al layer. Results demonstrate the feasibility of the photodiode for low-level light detection in the NUV-to-NIR spectral range.

INTRODUCTION

Hydrogenated amorphous silicon (a-Si:H) photodiodes are conventionally used in flat-panel digital X-ray image sensors due to large-area capability and high sensitivity in the visible spectral range [1]. A similar technology based on hydrogenated nanocrystalline silicon (nc-Si:H) can be developed for imaging applications. The optical bandgap of highly crystalline nc-Si:H films is close to the bandgap of crystalline silicon thus allowing to extend the spectral response in the near-infrared (NIR). However, the absorption coefficient of nc-Si:H is low in the NIR-range, i.e., the development of thin-film photodiodes requires a light management to increase the effective optical path within the absorber [2]. Another technological issue is associated with the complex microstructure of this heterogeneous material, which determines the electronic properties and effects the device performance. In particular, defects at grain boundaries, voids, and other structural defects are the source of excessive leakage current limiting the lowest detectable light level.

In this paper, we report on the fabrication and characterization of nc-Si:H-based photodiodes of different configurations. The discussed technological developments are aimed to minimize the leakage current and to enhance the external quantum efficiency (EQE) in the NIR-region. The electronic properties of nc-Si:H are also evaluated analyzing their impact on the device performance.

EXPERIMENT

Figure 1 shows the cross-sectional diagrams of fabricated photodiodes. Here, n^+- and n-layers denote doped and undoped nc-Si:H, respectively. The n-type behavior of undoped nc-Si:H films is a known technological issue ascribed to oxygen-related donors [3]. Sample #1 is a fully nanocrystalline n^+-n-p^+ photodiode. Sample #2 is an n^+-n-δ_i-p heterojunction photodiode with an 8 nm thick a-Si:H buffer (δ_i) inserted between the nc-Si:H n-layer and the a-SiC:H p-layer to passivate interface defects. The similar interface design has been used for HIT-type solar cells [4]. In samples #3 and #4, a nc-Si:H/a-Si:H absorption bilayer is combined with an a-SiC:H p-layer. In this n^+-n-i-δ_i-p structure, a thin (~4 nm) undoped a-SiC:H δ_i-layer is implemented to reduce the reverse dark current and recombination losses at the a-SiC:H/a-Si:H p-i interface [5]. To enhance the backside reflection of infrared radiation, a 130 nm thick ZnO:Al layer is implemented in sample #4.

The photodiodes were fabricated by the following deposition sequence. First, Cr or Cr/ZnO:Al films were sputtered on the Corning glass substrate, followed by the deposition of semiconductor stack. Then, a 65 nm thick ZnO:Al film was sputtered and patterned to form the top electrodes with an area ranging from 2×2 to 5×5 mm^2.

The semiconductor layers were deposited at 150°C using a multichamber, 13.56 MHz PECVD system, manufactured by MVSystems Inc. The distance between the RF electrodes and the electrode area were 2.1 cm and 232 cm^2, respectively. Trimethylboron, ($B(CH_3)_3$) (TMB), and phosphine (PH_3) diluted in hydrogen to a concentration of 1%, were used as the doping gases. Doped nc-Si:H was deposited using a 1:100:0.01 mixture of SiH_4 / H_2 / doping gas. The process pressure and RF power were 900 mTorr and 2 W, respectively. Undoped nc-Si:H was deposited in SiH_4+H_2 plasma at a hydrogen dilution ratio of 98%, a pressure of 4 Torr, and an RF power of 25 W. Details on tailoring of the p-i heterojunction interface along with deposition conditions of a-Si:H, a-SiC:H, and ZnO:Al films have been reported elsewhere [6].

The current-voltage measurements were performed at room temperature using a Keithley 4200-SCS semiconductor characterization system. The capacitance-voltage (C-V) characteristics of selected devices were measured at 10 kHz frequency using an Agilent 4284A LCR-meter. The spectral response measurements were performed with a PC-controlled setup based on an Oriel 77 200 grating monochromator, a Stanford Research System SR540 light chopper, and an SR530 DSP lock-in amplifier. The system was calibrated in the spectral range of 300–1100 nm using a Newport 818-UV detector.

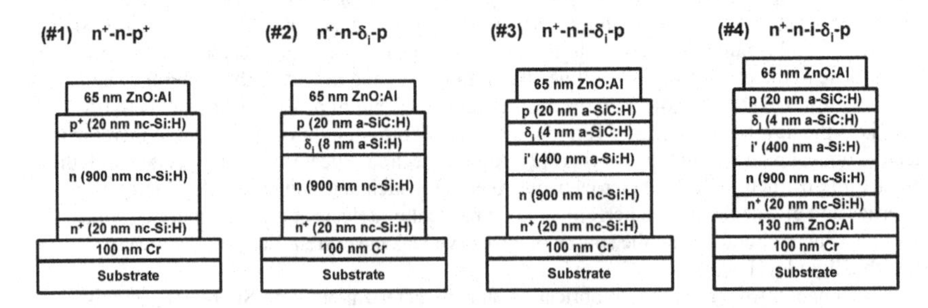

Figure 1. Cross-sectional diagrams of the fabricated photodiodes.

RESULTS AND DISCUSSION

I-V and C-V characteristics

Figure 2 shows the typical quasi-static current-voltage characteristics of the fabricated photodiodes. Samples #1 and #2 with nc-Si:H absorbers show an exponential dependence of the forward current over five orders of magnitude in the bias range from 0.1 to 0.45 V. At higher biases, the TCO series resistance and the space-charge limited current effect are the factors defining the current-voltage dependence. Sample #3 exhibits the exponential growth of forward current in a narrower range from 0.2 to 0.4 V, and, at higher biases, the electron transport through the *n-i* heterojunction is likely a limiting factor. The diode ideality factor (n) and saturation current density (J_0) values, determined through a fitting procedure, are shown in Table 1. The table also includes a reverse dark current density at –1 V and an equivalent shunt resistance (R_0) at the bias ±50 mV.

Table I. Comparison of device parameters deduced from current-voltage characteristics.

Sample		n	J_0, A/cm^2	R_0, Ω·cm^2	J(-1V), A/cm^2
#1	n^+-n-p^+	1.5	$2.2 \cdot 10^{-8}$	$1.18 \cdot 10^{6}$	$5 \cdot 10^{-7}$
#2	n^+-n-δ_i-p	1.12	$1.1 \cdot 10^{-9}$	$1.2 \cdot 10^{7}$	$1.1 \cdot 10^{-8}$
#3	n^+-n-i-δ_i-p	2.04	$3.4 \cdot 10^{-10}$	$1.1 \cdot 10^{8}$	$3.3 \cdot 10^{-9}$

The nc-Si:H n^+-n-p^+ photodiode shows the highest level of the leakage yielding a reverse dark current density of 5 μA/cm^2 at –5 V. The n^+-n-δ_i-p photodiode with the a-SiC:H *p*-layer shows two orders of magnitude lower leakage indicating excellent properties of the n-δ_i-p interface. The performance of the n^+-n-i-δ_i-p photodiode is the best in terms of J_0, R_0, and the reverse dark current. In this device, the depletion region expands with increasing reverse bias from the *i-p* region towards the *n-i* heterojunction, then, into the *n*-layer thus causing a steep increase of the dark current in the bias range from 0 to –2 V. At higher reverse biases, the leakage current is predominantly due to thermal generation of electron-hole pairs in the *n*-layer.

Figure 3 shows the $1/C^2$ versus V plot of the n^+-n-i-δ_i-p photodiode. Two linear regions of

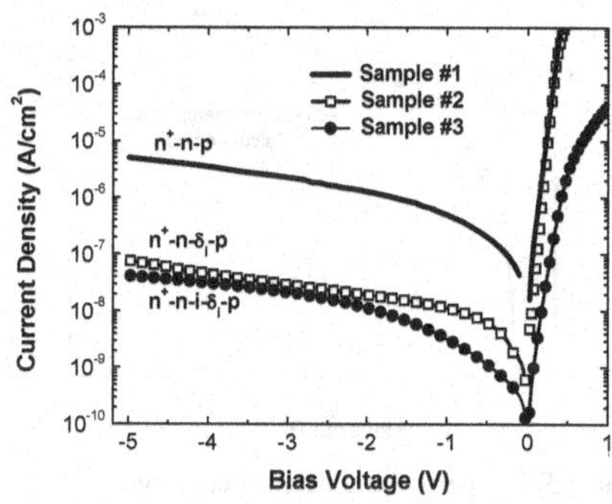

Figure 2. Current-voltage characteristics of the fabricated photodiodes.

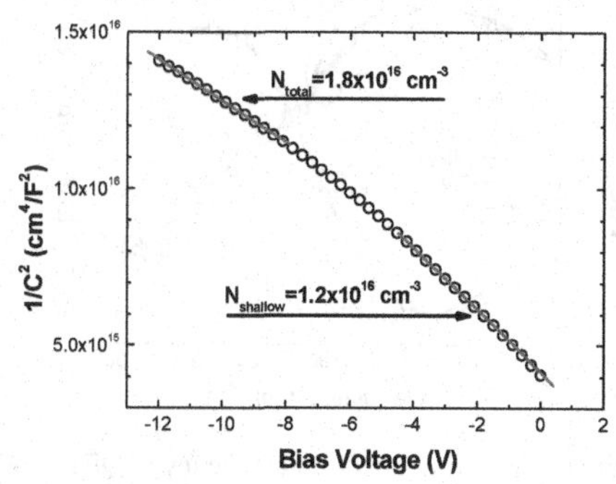

Figure 3. Capacitance-voltage characteristics of the n^+-n-i-δ_i-p photodiode.

the curve can be distinguished. Similar C-V characteristics of the nc-Si:H solar cells have been reported elsewhere [7]. The slope of the curve in the low-bias range gives the density of shallow states ($\sim 1.2\cdot10^{16}$ cm^{-3}), whereas the slope at large reverse biases yields an estimate of both deep and shallow states ($\sim 1.8\cdot10^{16}$ cm^{-3}).

Measurement of diffusion length of holes in nc-Si:H

The diffusion length of minority carries in *p-n* or Schottky-barrier photodiodes can be estimated analyzing the bias dependence of long-wavelength quantum efficiency [8]. This technique was used here to measure the diffusion length of holes in nc-Si:H.

Figure 4 shows the comparison of external quantum efficiency spectra measured at –2 V and zero bias. To reveal the difference in the spectra, EQE(–2 V) / EQE(0 V) ratio curve is also included in the plot. Under the reverse bias conditions, the EQE enhancement is significant in the long-wavelength region, which can be ascribed to the expansion of the space-charge region. Note that at $\lambda > 850$ nm, the EQE ratio is weakly dependent on the wavelength because of the quasi-uniform charge generation in the nc-Si:H *n*-layer. Considering negligibly low carrier recombination at the n^+-n interface, the photocurrent generated within the *n*-layer is [9]

$$J = q\Phi[1-(1-z)\exp(-\alpha W)],$$

$$z = \frac{1}{1-(\alpha L_p)^{-2}}\left[1-\frac{\frac{1}{\alpha L_p}\sinh\left(\frac{y}{L_p}\right)+\exp(-\alpha y)}{\cosh\left(\frac{y}{L_p}\right)}\right],$$

$$y = d_n - W, \qquad (1)$$

where Φ is the photon flux density passing through the a-Si:H/nc-Si:H interface, α the absorption coefficient of nc-Si:H at a wavelength of the incident monochromatic light, L_p the hole diffusion length, d_n the thickness of the *n*-layer, and W the depletion width at given bias V.

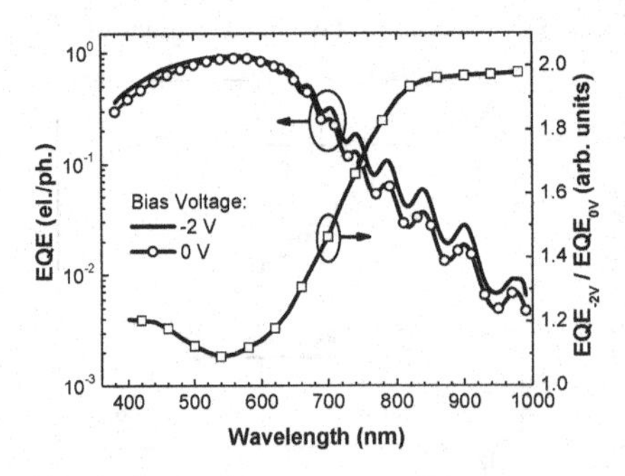

Figure 4. External quantum efficiency (EQE) spectra of the n^+-n-i-δ_i-p photodiode at –2 V and zero bias. The EQE ratio is also shown.

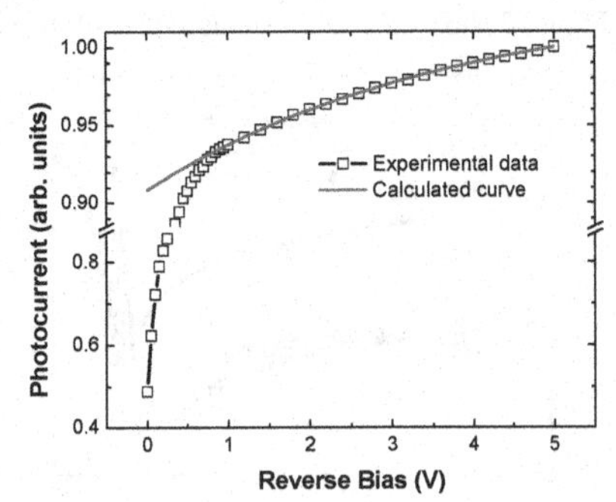

Figure 5. Measured and calculated photocurrent as a function of the reverse bias at a wavelength of 850 nm.

The depletion width can be determined from the measured capacitance-voltage characteristics according to

$$W(V) = \frac{\varepsilon\varepsilon_0 A}{C(V)} - d_i, \tag{2}$$

where ε is the static dielectric constant of the semiconductor material, ε_0 the permittivity of free space, A the area of the junction, and d_i the i-layer thickness.

Figure 5 shows the measured and calculated photocurrent of the n^+-n-i-δ_i-p photodiode as a function of the reverse bias at a wavelength of 850 nm. Excellent fitting of experimental points by the simulated curve was achieved at L_p = 1 μm. A curve mismatch in the low-bias range is because of recombination losses (up to 50%) at the nc-Si:H / a-Si:H heterojunction interface and within the partially depleted i-layer.

Spectral-response characteristics

Figure 6 shows the EQE spectra of the photodiodes at a reverse bias of 2 V. Peak values of the curves are 0.89 at λ = 510 nm, 0.85 at λ = 520 nm, and 0.93 at λ = 560 nm for samples #1, #2, and #3, respectively. The n^+-n-p^+ photodiode exhibits a sensitivity enhancement in the blue range due to low absorption losses in the nc-Si:H p^+-layer. The spectral-response of the n^+-n-i-δ_i-p photodiode is enhanced in the green-red range due to better absorption efficiency of the a-Si:H i-layer.

The long-wavelength sensitivity of nc-Si:H-based photodiodes can be improved by increasing the absorber layer thickness up to a double value of the hole diffusion length. Another approach is to increase the effective optical path within the device by proper light trapping schemes. In sample #4 the Cr / ZnO:Al layers serve as a back reflector (see Fig. 1). EQE spectra of the n^+-n-i-δ_i-p photodiodes with and without back reflector are shown in Figure 7. At λ > 850 nm, the amplitude of interference fringes is larger by a factor of 2 due to the back reflector.

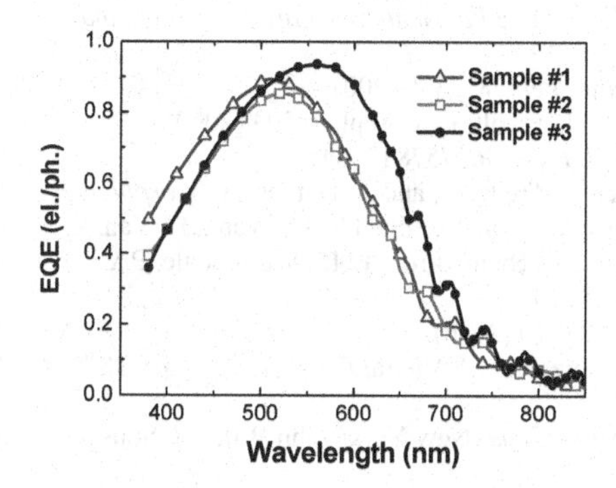

Figure 6. EQE spectra of the fabricated photodiodes at a reverse bias of 2 V.

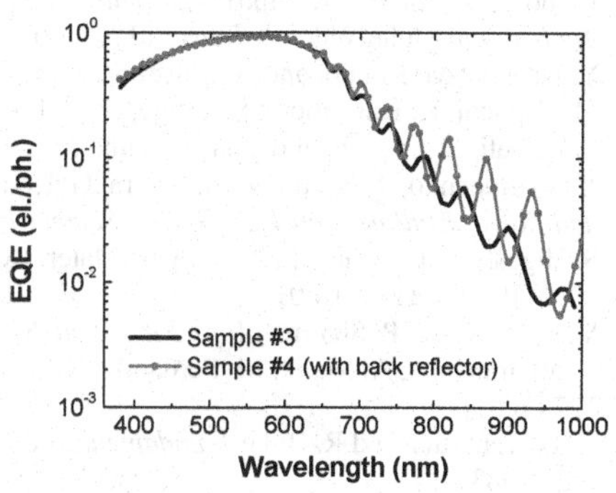

Figure 7. Comparison of EQE spectra of the n^+-n-i-δ_i-p photodiodes with and without back reflector.

CONCLUSIONS

Applying PECVD nc-Si:H grown in the high-pressure mode as a light absorbing material, the photodiodes of different configurations have been fabricated and characterized. The n^+-n-δ_i-p heterojunction photodiode with a-SiC:H p-layer showed significantly lower leakage than that observed in the nc-Si:H n^+-n-p^+ homojunction. The n^+-n-i-δ_i-p photodiode with nc-Si:H/a-Si:H absorption bilayer exhibited the best performance yielding a reverse dark current density of 3 nA/cm^2 at −1 V. The quality of undoped nc-Si:H was estimated analyzing C-V characteristics and bias dependence of long-wavelength quantum efficiency. The density of shallow (~ $1.2 \cdot 10^{16}$ cm^{-3}) and deep (~ $6 \cdot 10^{15}$ cm^{-3}) donor states along with a diffusion length of holes of 1 μm were determined. The charge collection within the n-layer was shown to be diffusion controlled due to high donor density in nc-Si:H. The sensitivity enhancement in the NIR-region was achieved incorporating Cr/ZnO:Al back reflector. Results demonstrate the feasibility of the n^+-n-i-δ_i-p photodiode for low-level light detection in the NUV-to-NIR spectral range.

ACKNOWLEDGMENTS

The authors are grateful to the Portuguese Foundation of Science and Technology through fellowship BPD20264/2004 for financial support of this research, and to the Giga-to-Nanoelectronics Centre at the University of Waterloo for providing the necessary equipment and technical help to carry out this work.

REFERENCES

1. R. A. Street, Ed., *Technology and Applications of Amorphous Silicon* (Berlin: Springer-Verlag, 2000).
2. J. Poortmans and V. Arkhipov, Ed., *Thin Film Solar Cells Fabrication, Characterization and Applications* (John Wiley & Sons Ltd., 2006).
3. Y. Nasuno, M. Kondo, and A. Matsuda, *Appl. Phys. Lett.* **78**, 2330 (2001).
4. M. Taguchi, K. Kawamoto, S. Tsuge, et al., Prog. Photovolt: Res. Appl. 8, 503 (2000).
5. P. Servati, Y. Vygranenko, and A. Nathan, *J. Appl. Phys.* **96**, 7578 (2004).
6. Y. Vygranenko, A. Sazonov, M. Vieira, G. Heiler, T. Tredwell, and A. Nathan in *Amorphous and Polycrystalline Thin-Film Silicon Science and Technology*, edited by Q. Wang, B. Yan, S. Higashi, C.C. Tsai, and A. Flewitt (Mater. Res. Soc. Symp. Proc. **1245**, Warrendale, PA, 2010) Paper 1245-A18-01.
7. V. L. Dalal and P. Sharma, *Appl. Phys. Lett* .**86**, 103510 (2006).
8. Y. Vygranenko, A. Malik, M. Fernandes, R. Schwarz, and M. Vieira, *Phys. Stat. Sol. (a)* **185**, 137 (2001).
9. A. Fahrenbruch and R. Bube, *Fundamentals of Solar Cells* (New York: John Wiley & Sons Ltd., 1983).

AUTHOR INDEX

SUBJECT INDEX

Printed in the United States
by Baker & Taylor Publisher Services

Printed in the United States
by Baker & Taylor Publisher Services